Lecture Notes in

Founding Editors

Gerhard Goos
Juris Hartmanis

Editorial Board Members

Elisa Bertino, *Purdue University, West Lafayette, IN, USA*
Wen Gao, *Peking University, Beijing, China*
Bernhard Steffen, *TU Dortmund University, Dortmund, Germany*
Moti Yung, *Columbia University, New York, NY, USA*

The series Lecture Notes in Computer Science (LNCS), including its subseries Lecture Notes in Artificial Intelligence (LNAI) and Lecture Notes in Bioinformatics (LNBI), has established itself as a medium for the publication of new developments in computer science and information technology research, teaching, and education.

LNCS enjoys close cooperation with the computer science R & D community, the series counts many renowned academics among its volume editors and paper authors, and collaborates with prestigious societies. Its mission is to serve this international community by providing an invaluable service, mainly focused on the publication of conference and workshop proceedings and postproceedings. LNCS commenced publication in 1973.

Maciej Paszynski · Amanda S. Barnard ·
Yongjie Jessica Zhang
Editors

Computational Science – ICCS 2025 Workshops

25th International Conference
Singapore, Singapore, July 7–9, 2025
Proceedings, Part II

Springer

Editors
Maciej Paszynski ⓘ
AGH University of Krakow
Krakow, Poland

Amanda S. Barnard ⓘ
Australian National University
Canberra, ACT, Australia

Yongjie Jessica Zhang ⓘ
Carnegie Mellon University
Pittsburgh, PA, USA

ISSN 0302-9743　　　　　　ISSN 1611-3349 (electronic)
Lecture Notes in Computer Science
ISBN 978-3-031-97556-1　　ISBN 978-3-031-97557-8 (eBook)
https://doi.org/10.1007/978-3-031-97557-8

© The Editor(s) (if applicable) and The Author(s), under exclusive license to Springer Nature Switzerland AG 2025

This work is subject to copyright. All rights are solely and exclusively licensed by the Publisher, whether the whole or part of the material is concerned, specifically the rights of translation, reprinting, reuse of illustrations, recitation, broadcasting, reproduction on microfilms or in any other physical way, and transmission or information storage and retrieval, electronic adaptation, computer software, or by similar or dissimilar methodology now known or hereafter developed.
The use of general descriptive names, registered names, trademarks, service marks, etc. in this publication does not imply, even in the absence of a specific statement, that such names are exempt from the relevant protective laws and regulations and therefore free for general use.
The publisher, the authors and the editors are safe to assume that the advice and information in this book are believed to be true and accurate at the date of publication. Neither the publisher nor the authors or the editors give a warranty, expressed or implied, with respect to the material contained herein or for any errors or omissions that may have been made. The publisher remains neutral with regard to jurisdictional claims in published maps and institutional affiliations.

This Springer imprint is published by the registered company Springer Nature Switzerland AG
The registered company address is: Gewerbestrasse 11, 6330 Cham, Switzerland

If disposing of this product, please recycle the paper.

Preface

Welcome to the Workshops on Computational Science, which were co-organized with the 25th International Conference on Computational Science (ICCS - https://www.iccs-meeting.org/iccs2025/), held on July 7–9, 2025 at the Nanyang Technological University (NTU), Singapore.

This 25th edition in Singapore marked our return to a fully in-person event. Although the challenges of our present times are manifold, we have always tried our best to keep the ICCS community as dynamic, creative, and productive as possible. We are proud to present the proceedings you are reading as a result.

ICCS 2025 was jointly organized by Nanyang Technological University, the A*STAR Institute of High Performance Computing, the University of Amsterdam, and the University of Tennessee.

Considered one of the most developed countries in the world, the island country of Singapore is a major aviation, financial, and maritime shipping hub in Asia. Singapore is multilingual, multiethnic, and multicultural, and as such a very popular, safe tourism destination.

NTU Singapore is a public university ranked among the world's best, with 35,000 students, and home to the world-renowned autonomous National Institute of Education and S. Rajaratnam School of International Studies. In addition to many research institutes and centers at the university, college, and school levels, NTU also hosts two National Research Foundation (NRF) and Ministry of Education (MOE) Research Centers of Excellence, namely the Singapore Center for Environmental Life Sciences Engineering (SCELSE) and the Institute for Digital Molecular Analytics & Science (IDMxS), and 11 Corporate Labs in partnership with various industries. ICCS 2025 took place on the One-north campus.

The Institute of High Performance Computing (IHPC) is a national research institute under the Agency for Science, Technology and Research (A*STAR), dedicated to advancing science and technology through computational modeling, simulation, AI, and high-performance computing. With a multidisciplinary team of scientists and engineers, IHPC drives innovation across sectors such as advanced manufacturing, microelectronics, sustainability, maritime, and biomedical sciences. It leads Singapore's national efforts in hybrid quantum-classical computing and digital twin platforms, and partners extensively with industry and government agencies to translate deep tech into real-world impact.

The Workshops on Computational Science are a set of thematic workshops organized by experts in a particular area of Computational Science. These workshops are specifically intended to provide a forum for the discussion of novel and more focused topics in the field of Computational Science among an international group of researchers, and to strengthen the application of Computational Science in particular disciplines.

We are proud to note that this 25th edition of ICCS, with 23 workshops (the Workshops on Computational Science), one co-located event (the Asian Network of Complexity Scientists Workshop), and over 300 participants, kept to the tradition and high standards of previous editions.

The theme for 2025, "**Making Complex Systems tractable through Computational Science**", highlighted the role of Computational Science in tackling the complex problems of today and tomorrow.

ICCS is well known for its lineup of keynote speakers. The keynotes for 2025 were:

- **Johan Bollen**, Indiana University Bloomington, USA
- **Jack Dongarra**, University of Tennessee, USA
- **Mile Gu**, Nanyang Technological University, Singapore
- **Erika Fille Legara**, Center for AI Research | Asian Institute of Management, Philippines
- **Yong-Wei Zhang**, Institute of High Performance Computing, A*STAR, Singapore

This year, the Workshops on Computational Science registered 322 submissions, of which 137 were accepted as full papers, and 32 as short papers. There were on average 2.7 single-blind reviews per submission.

We would like to thank all committee members from the main track and workshops for their contribution to ensuring a high standard for the accepted papers. We would also like to thank *Springer, Elsevier,* and *Intellegibilis* for their support. Finally, we appreciate all the local organizing committee members for their hard work in preparing this conference.

We hope you enjoyed the conference and the beautiful country of Singapore.

July 2025

Maciej Paszynski
Amanda S. Barnard
Yongjie Jessica Zhang

Organization

Program Committee Chair – Workshops on Computational Science

Maciej Paszynski — AGH University of Krakow, Poland

Program Committee – Workshops on Computational Science

Amanda S. Barnard — Australian National University, Australia
Yongjie Jessica Zhang — Carnegie Mellon University, USA

Program Committee Chairs – ICCS Conference

Peter M. A. Sloot — University of Amsterdam, The Netherlands
Jack J. Dongarra — University of Tennessee, USA
Michael H. Lees — University of Amsterdam, The Netherlands
David Abramson — University of Queensland, Australia
Wentong Cai — Nanyang Technological University, Singapore
Cheong Siew Ann — Nanyang Technological University, Singapore
Su Yi — Institute for High Performance Computing, A*Star, Singapore

Local Program Committee at NTU Singapore

Ee Hou Yong — Nanyang Technological University, Singapore
Kang Hao — Nanyang Technological University, Singapore

Publicity Chairs

Leonardo Franco — University of Málaga, Spain
Muhamad Azfar Ramli — Institute for High Performance Computing, A*Star, Singapore

Impact Chair

Valeria Krzhizhanovskaya — University of Amsterdam, The Netherlands

Outreach Chair

Alfons Hoekstra — University of Amsterdam, The Netherlands

Workshop Chairs

Advances in High-Performance Computational Earth Sciences: Numerical Methods, Frameworks & Applications – IHPCES

Takashi Shimokawabe	University of Tokyo, Japan
Kohei Fujita	University of Tokyo, Japan
Dominik Bartuschat	FAU Erlangen-Nürnberg, Germany

Artificial Intelligence Approaches for Network Analysis – AIxNA

Marianna Milano	University Magna Graecia of Catanzaro, Italy
Pietro Hiram Guzzi	University Magna Graecia of Catanzaro, Italy
Giuseppe Agapito	University Magna Graecia of Catanzaro, Italy
Pietro Cinaglia	University Magna Graecia of Catanzaro, Italy

Artificial Intelligence and High-Performance Computing for Advanced Simulations – AIHPC4AS

Maciej Paszynski	AGH University of Krakow, Poland
Maciej Woźniak	AGH University of Krakow, Poland
Victor Calo	Curtin University, Australia
David Pardo	Basque Center for Applied Mathematics, Spain
Quanling Deng	Australian National University, Australia

Biomedical and Bioinformatics Challenges for Computer Science – BBC

Mario Cannataro	University Magna Graecia of Catanzaro, Italy
Giuseppe Agapito	University Magna Graecia of Catanzaro, Italy
Mauro Castelli	NOVA IMS, Universidade Nova de Lisboa, Portugal
Riccardo Dondi	Università degli Studi di Bergamo, Italy
Rodrigo Weber dos Santos	Universidade Federal de Juiz de Fora, Brazil
Italo Zoppis	University of Milano-Bicocca, Italy

Computational Diplomacy and Policy – CoDiP

Michael Lees	University of Amsterdam, Netherlands
Roland Bouffanais	University of Geneva, Switzerland
Brian Castellani	University of Durham, UK

Computational Health – CompHealth

Sergey Kovalchuk	Huawei, Russia
Georgiy Bobashev	RTI International, USA
Anastasia Angelopoulou	University of Westminster, UK

Computational Modeling and Artificial Intelligence for Social Systems – CMAISS

Tanzhe Tang	University of Amsterdam, Netherlands
Jaeyoung Kwak	Nanyang Technological University, Singapore
Takashi Kamihigashi	Kobe University, Japan
Thomas Feliciani	Politecnico di Milano, Italy

Computational Optimization, Modelling and Simulation – COMS

Xin-She Yang	Middlesex University London, UK
Slawomir Koziel	Reykjavik University, Iceland
Leifur Leifsson	Purdue University, USA

Computational Science and AI for Addressing Complex and Dynamic Societal Challenges Equitably – CASCADE

Ilkay Altintas	University of California, San Diego, USA
Manish Parashar	University of Utah, USA
David Abramson	Monash University, Australia
Melissa Floca	UC San Diego, USA

Computer Graphics, Image Processing and Artificial Intelligence – CGIPAI

Andres Iglesias Prieto	Universidad de Cantabria, Spain
Farhan Mohamed	Universiti Teknologi Malaysia, Malaysia
Chan Vei Siang	Universiti Teknologi Malaysia, Malaysia
Lihua You	Bournemouth University, UK
Akemi Galvez-Tomida	University of Cantabria, Spain

Computing and Data Science for Materials Discovery and Design – CDMDD

Ulf Schiller	University of Delaware, USA
Derek Groen	Brunel University London, UK
Xiao Xue	University College London, UK

Large Language Models and Intelligent Decision-Making within the Digital Economy – LLM-IDDE

Wei Li	National University of Singapore, Singapore
Luyao Zhu	Nanyang Technological University, Singapore
Yi Qu	Chinese Academy of Sciences, China
Muyang Li	University of Chinese Academy of Sciences, China
Yunlong Mi	University of Chinese Academy of Sciences, China

Machine Learning and Data Assimilation for Dynamical Systems – MLDADS

Rossella Arcucci	Imperial College London, UK
Sibo Cheng	ENPC, France

Multi-Criteria Decision-Making: Methods, Applications, and Innovations – MCDM

Wojciech Sałabun	National Institute of Telecommunications, Poland
Jarosław Wątróbski	University of Szczecin, Poland

(Credible) Multiscale Modelling and Simulation – MMS

Derek Groen	Brunel University London, UK
Diana Suleimenova	Brunel University London, UK
Bartosz Bosak	PSNC, Poland

Numerical Algorithms and Computer Arithmetic for Computational Science – NACA

Paweł Gepner	Warsaw University of Technology, Poland
Ewa Deelman	USC Information Sciences Institute, USA
Hatem Ltaief	King Abdullah University of Science and Technology, Saudi Arabia

Quantum Computing – QCW

Katarzyna Rycerz	AGH University of Technology, Poland
Marian Bubak	AGH University of Krakow, Poland and University of Amsterdam, Netherlands

Quantum-Enhanced Agents and Recurrent Computation – QuARC

Mile Gu	Nanyang Technological University, Singapore

Retrieval-Augmented Generation – RAGW

Aleksander Smywiński-Pohl	AGH University of Krakow, Poland
Magdalena Król	AGH University of Krakow, Poland

Simulations of Flow and Transport: Modeling, Algorithms and Computation – SOFTMAC

Shuyu Sun	King Abdullah University of Science and Technology, Saudi Arabia
Jingfa Li	Yangtze University, China
James Liu	Colorado State University, USA

Smart Systems: Bringing Together Computer Vision, Sensor Networks and Artificial Intelligence – SmartSys

Pedro J. S. Cardoso	University of the Algarve, Portugal
Roberto Lam	University of the Algarve, Portugal
Jânio Monteiro	University of the Algarve, Portugal
João M. F. Rodrigues	University of the Algarve, Portugal
Jaime A. Martins	University of the Algarve, Portugal

Solving Problems with Uncertainty – SPU

Vassil Alexandrov	Hartree Centre STFC, UK
Aneta Karaivanova	IPP-BAS, Bulgaria

Teaching Computational Science – WTCS

Evguenia Alexandrova	STFC - Hartree Centre, UK
Tseden Taddese	Hartree Centre - STFC, UK

Reviewers

Zeeshan Abbas	Sungkyunkwan University, South Korea
David Abramson	Monash University, Australia
Tesfamariam Mulugeta Abuhay	University of Gondar, Ethiopia
Giuseppe Agapito	Università Magna Graecia Catanzaro, Italy
Adriano Agnello	STFC Hartree Centre, UK
Elisabete Alberdi	University of the Basque Country UPV/EHU, Spain
Vassil Alexandrov	Hartree Centre STFC, UK
Evguenia Alexandrova	STFC - Hartree Centre, UK
Rayner Alfred	Universiti Malaysia Sabah, Malaysia
Shaukat Ali	Simula Research Laboratory and Oslo Metropolitan University, Norway
Ilkay Altintas	University of California, San Diego, USA
Julen Alvarez-Aramberri	University of the Basque Country (UPV/EHU), Spain
Domingos Alves	University of São Paulo, Brazil
Sergey Alyaev	NORCE, Norway
Anastasia Anagnostou	Brunel University, UK
Anastasia Angelopoulou	University of Westminster, UK
Hideo Aochi	BRGM, France
Rossella Arcucci	Imperial College of London, UK
Paula Bajdor	Częstochowa University of Technology, Poland
Krzysztof Banaś	AGH University of Krakow, Poland
Luca Barillaro	Magna Graecia University of Catanzaro, Italy
Dominik Bartuschat	FAU Erlangen-Nürnberg, Germany

Pouria Behnoudfar	Curtin University, Australia
Jörn Behrens	University of Hamburg, Germany
Gebrail Bekdas	Istanbul University, Turkey
Mehmet Ali Belen	Iskenderun Technical University, Turkey
Sana Ben Abdallah Ben Lamine	University of Manouba, Tunisia
Stefano Beretta	San Raffaele Telethon Institute for Gene Therapy, Italy
Benjamin Berkels	RWTH Aachen University, Germany
Gabriele Bertoli	University of Florence, Italy
John Betts	Monash University, Australia
Kishor Bharti	National University of Singapore, Singapore
Asniyani Nur Haidar Binti Abdullah	Universiti Teknikal Malaysia Melaka, Malaysia
Piotr Biskupski	IBM, Poland
Georgiy Bobashev	RTI International, USA
Klavdiya Bochenina	ITMO University, Russia
Carlos Bordons	University of Seville, Spain
Kamil Bortko	West Pomeranian University of Technology in Szczecin, Poland
Bartosz Bosak	PSNC, Poland
Lorella Bottino	University Magna Graecia Catanzaro, Italy
Roland Bouffanais	University of Geneva, Switzerland
Marian Bubak	AGH Krakow, Poland and University of Amsterdam, Netherlands
Keith Butler	University College London, UK
Aleksander Byrski	AGH University of Krakow, Poland
Cristiano Cabrita	Universidade do Algarve, Portugal
Xing Cai	Simula Research Laboratory, Norway
Carlos T. Calafate	Universitat Politècnica de València, Spain
Victor Calo	Curtin University, Australia
Almudena Campuzano	University of Amsterdam, Netherlands
Mario Cannataro	University Magna Graecia of Catanzaro, Italy
Karol Capała	AGH University of Krakow, Poland
Pedro J. S. Cardoso	Universidade do Algarve, Portugal
Brian Castellani	University of Durham, UK
Mauro Castelli	NOVA IMS, Universidade Nova de Lisboa, Portugal
Nicholas Chancellor	Durham University, UK
Prasenjit Chatterjee	MCKV Institute of Engineering, India
Boyang Chen	Imperial College London, UK
Sibo Cheng	ENPC, France
Su-Fong Chien	MIMOS Berhad, Malaysia

Marta Chinnici	ENEA - Italian National Agency for New Technologies, Energy and Sustainable Economic Development, Italy
Witold Chmielarz	University of Warsaw, Poland
Bastien Chopard	University of Geneva, Switzerland
Maciej Ciesielski	University of Massachusetts, USA
Pietro Cinaglia	University Magna Graecia of Catanzaro, Italy
Noélia Correia	Universidade do Algarve, Portugal
Adriano Cortes	University of Rio de Janeiro, Brazil
Ana Cortes	Universitat Autònoma de Barcelona, Spain
Enrique Costa-Montenegro	Universidad de Vigo, Spain
David Coster	Max Planck Institute for Plasma Physics, Germany
Carlos Cotta	Universidad de Málaga, Spain
Matteo Croci	Basque Center for Applied Mathematics, Spain
Daan Crommelin	CWI Amsterdam, Netherlands
Attila Csikasz-Nagy	King's College London, UK/Pázmány Péter Catholic University, Hungary
António Cunha	UTAD, Portugal
Pawel Czarnul	Gdańsk University of Technology, Poland
Pasqua D'Ambra	IAC-CNR, Italy
Dong Dai	University of Delaware, USA
Lisandro Dalcin	KAUST, Saudi Arabia
Haluk Damgacioglu	Medical University of South Carolina, USA
Bhaskar Dasgupta	University of Illinois Chicago, USA
Subhasis Dasgupta	University of California, San Diego, USA
Ewa Deelman	USC Information Sciences Institute, USA
Quanling Deng	Australian National University, Australia
Muhammet Deveci	National Defense University, Turkey
Jean Dezert	Onera, France
Jacek Długopolski	AGH University of Science and Technology, Poland
Riccardo Dondi	Università degli Studi di Bergamo, Italy
Rafal Drezewski	AGH University of Krakow, Poland
Hans du Buf	University of the Algarve, Portugal
Witold Dzwinel	AGH University of Science and Technology, Poland
Rob E. Loke	Amsterdam University of Applied Sciences, Netherlands
Wouter Edeling	Vrije Universiteit Amsterdam, Netherlands
Nahid Emad	Paris-Saclay University, France
Christian Engelmann	ORNL, USA
Aniello Esposito	Hewlett Packard Enterprise, Switzerland

Fedra Rosita Falvo	University Magna Graecia of Catanzaro, Italy
Thomas Feliciani	Politecnico di Milano, Italy
Jinyuan Feng	Research Center on Fictitious Economy & Data Science, China
Anna Fernandez	Basque Center for Applied Mathematics, Spain
Nicola Ferrier	Argonne National Laboratory, USA
Melissa Floca	University of California San Diego, USA
Marcin Fojcik	Western Norway University of Applied Sciences, Norway
Piotr Frąckiewicz	Pomeranian University, Poland
Alberto Freitas	University of Porto, Portugal
Kohei Fujita	University of Tokyo, Japan
Takeshi Fukaya	Hokkaido University, Japan
Wlodzimierz Funika	AGH University of Krakow, Poland
Ernst Fusch	Siemens Labs, Siemens AG, Germany
Teresa Galvão	University of Porto, Portugal
Akemi Galvez-Tomida	University of Cantabria, Spain
Luis Garcia-Castillo	Universidad Carlos III de Madrid, Spain
Bartłomiej Gardas	Jagiellonian University, Poland
Piotr Gawron	Nicolaus Copernicus Astronomical Centre, Polish Academy of Sciences, Poland
Bernhard Geiger	Know-Center GmbH, Austria
Paweł Gepner	Warsaw University of Technology, Poland
Maziar Ghorbani	Brunel University London, UK
Alexandrino Gonçalves	Polytechnic University of Leiria, Portugal
Simon Goodchild	STFC, UK
Yuriy Gorbachev	Soft-Impact LLC, Russia
Pawel Gorecki	University of Warsaw, Poland
Derek Groen	Brunel University London, UK
Mile Gu	Nanyang Technological University, Singapore
Joel Guerreiro	University of the Algarve, Portugal
Manish Gupta	Harish-Chandra Research Institute, India
Piotr Gurgul	Snapchat, Switzerland
Oscar Gustafsson	Linköping University, Sweden
Pietro Hiram Guzzi	University Magna Graecia of Catanzaro, Italy
Zulfiqar Habib	COMSATS University Islamabad, Lahore Campus, Pakistan
Laura Harbach	Brunel University London, UK
Mohamed Hassan	Virginia Tech, USA
Alexander Heinecke	Intel Parallel Computing Lab, USA
Teiko Heinosaari	University of Jyväskylä, Finland

Marcin Hernes	Wrocław University of Economics and Business, Poland
Maximilian Höb	Leibniz-Rechenzentrum der Bayerischen Akademie der Wissenschaften, Germany
Xiao Hou	Beihang University, China
Huda Ibeid	Intel Corporation, USA
Andres Iglesias Prieto	Universidad de Cantabria, Spain
Alireza Jahani	Brunel University London, UK
Jaroslaw Jankowski	West Pomeranian University of Technology, Poland
Peter Janku	Tomas Bata University in Zlín, Czechia
Jiří Jaroš	Brno University of Technology, Czechia
Hang-Hyun Jo	Catholic University of Korea, South Korea
Takashi Kamihigashi	Kobe University, Japan
John Kang	San Diego State University, USA
Aneta Karaivanova	IPP-BAS, Bulgaria
Haruo Kobayashi	Gunma University, Japan
Łukasz Kobyliński	Institute of Computer Science, Polish Academy of Sciences, Poland
Marcel Koch	KIT, Germany
Bartłomiej Kocot	AMD, USA
Ivana Kolingerova	University of West Bohemia, Czechia
Joanna Kołodziejczyk	National Institute of Telecommunications, Poland
Georgy Kopanitsa	Tomsk Polytechnic University, Russia
Sergey Kovalchuk	Huawei, Russia
Slawomir Koziel	Reykjavik University, Iceland
Marek Kozłowski	OPI PIB, Poland
Ronald Kriemann	Max Planck Institute for Mathematics in the Sciences, Germany
Magdalena Król	AGH University of Krakow, Poland
Valeria Krzhizhanovskaya	University of Amsterdam, Netherlands
Marek Kubalcik	Tomas Bata University in Zlín, Czechia
Sebastian Kuckuk	Friedrich-Alexander-Universität Erlangen-Nürnberg, Germany
Krzysztof Kurowski	Poznań Supercomputing and Networking Center, Poland
Halim Kusumaatmaja	Durham University, UK
Marcin Kuta	AGH University of Science and Technology, Poland
Jaeyoung Kwak	Nanyang Technological University, Singapore
Roberto Lam	ISE - Universidade do Algarve, Portugal
Marek Lampart	VŠB-Technical University of Ostrava, Czechia

Ilaria Lazzaro	Università degli studi Magna Graecia di Catanzaro, Italy
Paola Lecca	Free University of Bozen-Bolzano, Italy
Michael Lees	University of Amsterdam, Netherlands
Leifur Leifsson	Purdue University, USA
Kenneth Leiter	Army Research Laboratory, USA
Yu Leng	Los Alamos National Laboratory, USA
Paulina Lewandowska	IT4Innovations National Supercomputing Center, Czechia
Muyang Li	University of Chinese Academy of Sciences, China
Wei Li	National University of Singapore, Singapore
Jingfa Li	Yangtze University, China
Tomer Libal	University of Luxembourg, Luxembourg
Che Liu	Imperial College London, UK
Zhao Liu	National Supercomputing Center in Wuxi, China
James Liu	Colorado State University, USA
Marcelo Lobosco	Federal University of Juiz de Fora, Brazil
Jay Lofstead	Sandia National Laboratories, USA
Chu Kiong Loo	University of Malaya, Malaysia
Marcin Łoś	AGH University of Krakow, Poland
Hatem Ltaief	King Abdullah University of Science and Technology, Saudi Arabia
Stefan Luding	University of Twente, Netherlands
Piotr Luszczek	University of Tennessee Knoxville, USA
Pedro M. M. Guerreiro	Universidade do Algarve, Portugal
Raghu Machiraju	Ohio State University, USA
Peyman Mahouti	Yildiz Technical University, Turkey
Krzysztof Małecki	West Pomeranian University of Technology, Poland
Alexander Malyshev	UiB, Norway
Anirban Mandal	Renaissance Computing Institute, USA
Livia Marcellino	University of Naples Parthenope, Italy
Tomas Margalef	Universitat Autònoma de Barcelona, Spain
Osni Marques	Lawrence Berkeley National Laboratory, USA
Ignacio Martinez-Moyano	Argonne National Laboratory, USA
Maria Chiara Martinis	Università Magna Graecia di Catanzaro, Italy
Jaime A. Martins	University of the Algarve, Portugal
Michele Martone	Max-Planck-Institut für Plasmaphysik, Germany
Pawel Matuszyk	Baker Hughes Inc., USA
Valerie Maxville	Curtin University, Australia
Francesca Mazzia	Università di Bari Aldo Moro, Italy

Wagner Meira Jr.	Universidade Federal de Minas Gerais, Brazil
Roderick Melnik	Wilfrid Laurier University, Canada
Ivan Merelli	Institute for Biomedical Technologies - National Research Council, Italy
Yunlong Mi	University of Chinese Academy of Sciences, China
Jakub Mielczarek	Jagiellonian University, Poland
Marianna Milano	Università Magna Græcia di Catanzaro, Italy
Jaroslaw Miszczak	Institute of Theoretical and Applied Informatics, Polish Academy of Sciences, Poland
Farhan Mohamed	Universiti Teknologi Malaysia, Malaysia
Mohd Khalid Mokhtar	Universiti Teknikal Malaysia Melaka, Malaysia
Fernando Monteiro	Polytechnic Institute of Bragança, Portugal
Jânio Monteiro	University of the Algarve, Portugal
Andrew Moore	University of California Santa Cruz, USA
Anabela Moreira Bernardino	Polytechnic Institute of Leiria, Portugal
Eugénia Moreira Bernardino	Polytechnic Institute of Leiria, Portugal
Leonid Moroz	Warsaw Technology University, Poland
Dariusz Mrozek	Silesian University of Technology, Poland
Peter Mueller	IBM Zurich Research Laboratory, Switzerland
Judit Munoz-Matute	University of the Basque Country, Spain
Ikmal Faiq Albakri Mustafa Albakri	Universiti Teknikal Malaysia Melaka, Malaysia
Hiromichi Nagao	University of Tokyo, Japan
Fallah Najjar	Al-Furat Al-Awsat Technical University, Iraq
Kengo Nakajima	University of Tokyo, Japan
Philipp Neumann	Helmut-Schmidt-Universität, Germany
Sinan Melih Nigdeli	Istanbul University, Turkey
Anita Nikolich	University of Illinois Urbana-Champaign, USA
Hitoshi Nishizawa	Sony Research Labs., Japan
Joseph O'Connor	University of Edinburgh, UK
Lidia Ogiela	AGH University of Krakow, Poland
Ángel Javier Omella	University of the Basque Country (UPV/EHU), Spain
Kenji Ono	Kyushu University, Japan
Hiroyuki Ootomo	NVIDIA, Japan
Eneko Osaba	TECNALIA Research & Innovation, Spain
Joanna Paliszkiewicz	Warsaw University of Life Sciences, Poland
Dragan Pamučar	University of Belgrade, Serbia
George Papadimitriou	Apple, USA
Nikela Papadopoulou	University of Glasgow, UK
Manish Parashar	University of Utah, USA

David Pardo	University of the Basque Country and IKERBASQUE, Spain
Dário Passos	CENTRA-IST, Portugal
Zbigniew Pastuszak	Maria Curie-Skłodowska University, Poland
Anna Paszynska	Jagiellonian University, Poland
Maciej Paszynski	AGH University of Krakow, Poland
Abani Patra	Tufts University, USA
Lukasz Pawela	Institute of Theoretical and Applied Informatics, Polish Academy of Sciences, Poland
Sara Perez Carabaza	Universidad de Cantabria, Spain
Piotr Pęzik	University of Łódź, Poland
Frank Phillipson	TNO, Netherlands
Anna Pietrenko-Dabrowska	Gdańsk University of Technology, Poland
Armando Pinho	University of Aveiro, Portugal
Yuri Pirola	Università degli Studi di Milano-Bicocca, Italy
Ollie Pitts	Imperial College London, UK
Beth Plale	Indiana University, USA
Robert Platt	Imperial College London, UK
Paweł Pławiak	Cracow University of Technology, Poland
Michal Pluhacek	AGH University of Krakow, Poland
Paweł Poczekajło	Koszalin University of Technology, Poland
Valeria Popello	University Magna Graecia of Catanzaro, Italy
Cristina Portales	Universidad de Valencia, Spain
Simon Portegies Zwart	Leiden University, Netherlands
Anna Procopio	Università Magna Graecia di Catanzaro, Italy
Małgorzata Przybyła-Kasperek	Uniwersytet Śląski w Katowicach, Poland
Ubaid Ali Qadri	Science and Technology Facilities Council, UK
Yipeng Qin	Cardiff University, UK
Yi Qu	Chinese Academy of Sciences, China
Andrianirina Rakotoharisoa	Imperial College London, UK
Raul Ramirez	Tecnológico de Monterrey, Mexico
Célia Ramos	University of the Algarve, Portugal
Georgia Ray	Imperial College London, UK
Robin Richardson	Netherlands eScience Center, Netherlands
Heike Riel	IBM Research - Zurich, Switzerland
João M. F. Rodrigues	Universidade do Algarve, Portugal
Daniel Rodriguez	University of Alcalá, Spain
Marcin Rogowski	Saudi Aramco, Saudi Arabia
Sergio Rojas	Monash University, Australia
Albert Romkes	South Dakota School of Mines and Technology, USA
Tomasz Rybotycki	IBS PAN, CAMK PAN, AGH, Poland

Katarzyna Rycerz	AGH University of Technology, Poland
Emre Sahin	UKRI STFC Hartree Centre, UK
Wojciech Sałabun	National Institute of Telecommunications, Poland
Özlem Salehi	Özyeğin University, Turkey
Ayşin Sancı	altinay, Turkey
Allah Bux Sargano	COMSATS University Islamabad, Lahore Campus, Pakistan
Jaromir Savelka	Carnegie Mellon University, USA
Robert Schaefer	AGH University of Krakow, Poland
Ulf Schiller	University of Delaware, USA
Tapan Senapati	Southwest University, China
Paulina Sepúlveda-Salas	Pontificia Universidad Católica de Valparaíso, Chile
Marzia Settino	Universita Magna Graecia Catanzaro, Italy
Mostafa Shahriari	Basque Center for Applied Mathematics, Spain
Pengyuan Shao	Imperial College London, UK
Takashi Shimokawabe	University of Tokyo, Japan
Bhargav Sriram Siddani	Lawrence Berkeley National Laboratory, USA
Marcin Sieniek	Google Brain, USA
Haozhen Situ	South China Agricultural University, China
Leszek Siwik	AGH University of Krakow, Poland
Grażyna Ślusarczyk	Jagiellonian University, Poland
Maciej Smołka	AGH University of Krakow, Poland
Michalis Smyrnakis	STFC, UK
Aleksander Smywiński-Pohl	AGH University of Krakow, Poland
Isabel Sofia Brito	Instituto Politécnico de Beja, Portugal
Chengxi Song	University of Chinese Academy of Sciences, China
Robert Staszewski	University College Dublin, Ireland
Željko Stević	University of East Sarajevo, Bosnia and Herzegovina
Magdalena Stobinska	University of Gdansk and Institute of Physics, Polish Academy of Sciences, Poland
Barbara Strug	Jagiellonian University, Poland
Karol Struniawski	Warsaw University of Life Sciences - SGGW, Poland
Diana Suleimenova	Brunel University London, UK
Shuyu Sun	King Abdullah University of Science and Technology, Saudi Arabia
Martin Swain	Aberystwyth University, UK
Tseden Taddese	Hartree Centre - STFC, UK
Claude Tadonki	Mines ParisTech/CRI - Centre de Recherche en Informatique, France

Chi Wee Tan	Tunku Abdul Rahman University of Management and Technology, Malaysia
Tanzhe Tang	University of Amsterdam, Netherlands
Andrew Tangarra	Centre for Quantum Technologies, Singapore
Osamu Tatebe	University of Tsukuba, Japan
Michela Taufer	University of Tennessee Knoxville, USA
Jamie Taylor	CUNEF Universidad, Spain
Andrei Tchernikh	CICESE Research Center, Mexico
Marco ten Eikelder	TU Darmstadt, Germany
Kasim Terzic	University of St Andrews, UK
Jannis Teunissen	Centrum Wiskunde & Informatica, Netherlands, KU Leuven, Belgium
Jayne Thompson	National University of Singapore, Singapore
Sue Thorne	UKRI Science and Technology Facilities Council, UK
Vinod Tipparaju	ByteDance, USA
Pawel Topa	AGH University of Science and Technology, Poland
Ola Torudbakken	Meta, Norway
Paolo Trunfio	University of Calabria, Italy
Eirik Valseth	South Dakota School of Mines and Technology, USA
Aleksandra Vatian	ITMO University, Russia
Chan Vei Siang	Universiti Teknologi Malaysia, Malaysia
Milana Vuckovic	European Centre for Medium-Range Weather Forecasts, UK
Jianwu Wang	University of Maryland, Baltimore County, USA
Shaoni Wang	University of Groningen, Netherlands
Peng Wang	NVIDIA, China
Ximing Wang	Nanyang Technological University, Singapore
Jarosław Wąs	AGH University of Krakow, Poland
Jarosław Wątróbski	University of Szczecin, Poland
Rodrigo Weber dos Santos	Universidade Federal de Juiz de Fora, Brazil
Mei Wen	National University of Defense Technology, China
Wendy Winnard	UKRI STFC, UK
Konrad Wojtasik	Politechnika Wrocławska, Poland
Maciej Wołoszyn	AGH University of Science and Technology, Poland
Maciej Woźniak	AGH University of Krakow, Poland
Michał Wroński	NASK National Research Institute, Poland
Dunhui Xiao	Tongji University, China
Huilin Xing	University of Queensland, Australia

Xiao Xue	University College London, UK
Jiayu Xue	UCAS, China
Yani Xue	Brunel University London, UK
Xin-She Yang	Middlesex University, UK
Lihua You	Bournemouth University, UK
Sebastian Zając	SGH Warsaw School of Economics, Poland
Małgorzata Zajęcka	AGH University of Krakow, Poland
Gabor Závodszky	University of Amsterdam, Netherlands
Justyna Zawalska	ACC Cyfronet AGH, Poland
Wei Zhang	Huazhong University of Science and Technology, China
Tianchi Zhao	University of Chinese Academy of Sciences, China
Lei Zheng	University of Chinese Academy of Sciences, China
Luyao Zhu	Nanyang Technological University, Singapore
Kewei Zhu	University College London, UK
Beata Zielosko	University of Silesia in Katowice, Poland
Ewa Ziemba	University of Economics in Katowice, Poland
Italo Zoppis	University of Milano-Bicocca, Italy
Chiara Zucco	Università degli Studi "Magna Graecia" di Catanzaro, Italy
Pavel Zun	ITMO University, Russia
Karol Życzkowski	Jagiellonian University, Poland

Contents – Part II

Computational Health

Simulation of Blood Flow in the Left Ventricle Considering Purkinje Fibers 3
  Misaki Iwai, Yusuke Nishiya, Masashi Yamakawa, Ayato Takii, Yusei Kobayashi, Takahiro Ikeda, Shinichi Asao, and Seiichi Takeuchi

ViT-SE_Res: A Hybrid Vision Transformer and ResNet50V2 with Squeeze-And-Excitation Block for Cervical Cell Classification 11
  Betelhem Zewdu Wubineh, Andrzej Rusiecki, and Krzysztof Halawa

MedCT: A Clinical Terminology Graph for Generative AI Applications in Healthcare 24
  Ye Chen, Dongdong Huang, Yuqiang Shen, Haoyun Xu, Lin Sheng, Qingli Zhou, and Kai Wang

A Fractional Computation Based Deep Learning Framework for Silicosis Detection 39
  Rinki Sharma and Priyanka Harjule

Combining XAI and Graph Cuts for Skin-Lesion Segmentation 54
  Zyad Husni, Uwe Jaekel, and Babette Dellen

Accelerating Two-Dimensional k-Wave Ultrasound Simulations Through Pruned FFT: A Treatment Planning Optimisation 67
  Ondrej Olsak, David Bayer, and Jiri Jaros

A Computational Framework for Modelling Biomechanical Tumour Dynamics and Tissue Interactions: A Proof-of-Concept in Pleural Mesothelioma 83
  Sacha Gijsbers, Valeria Krzhizhanovskaya, Stefano Trebeschi, and Vivek M. Sheraton

Towards Sensitivity Analysis: 3D Venous Modelling in the Lower Limb 98
  Magdalena Otta, Karol Zając, Maciej Malawski, Ian Halliday, Chung Lim, Janice Tsui, and Andrew Narracott

Cross-Scale Modeling of Healthcare Norms and Patient Features Dynamics with Interpretable Machine Learning 113
  Chao Li, Dutao Zhang, Fei Ren, and Sergey Kovalchuk

Automatic Detection and Segmentation of Coronary Artery Stenosis in Coronary Angiography Images 128
Dmitrii Evtyukhov, Georgy Kopanitsa, Oleg Metsker, Aleksandr Mogilevskii, Alexey Yakovlev, and Sergey Kovalchuk

Explainable Artificial Intelligence for Doctors Decision Support in Diagnosing Spinal Pathologies 141
Aleksandra Vatian, Alexey Zubanenko, Pavel Ulyanov, Alexander Golubev, Artem Beresnev, and Natalia Gusarova

Predicting Disease Transmission Rates for Hybrid Modeling of Epidemic Outbreaks: Statistical and Machine Learning Approaches 149
Maria Koshkareva, Elizabetty Guseva, Alyona Sharova, and Vasiliy Leonenko

Lightweight Heterogeneous SEIR Models for Epidemic Surveillance in Russian Cities: Turning Synthetic Populations Into Equations 164
Andrey Korzin and Vasiliy Leonenko

Is Health Systems Sustainability Measurable? - Operationalizing SDG Targets Using SSP-TOPSIS Approach 176
Aleksandra Bączkiewicz, Jarosław Wątróbski, and Iga Rudawska

Computational Modeling and Artificial Intelligence for Social Systems

Automatic Detection and Identification of Causal Relationships in Polish Legal Texts 193
Łukasz Kurant

A Parameter-Free Model for the Online Spread of Far-Right Messages: Combining Agent-Based Models with Large-Language Models 208
Stephen Zhong, Nathalie Japkowicz, Frédéric Amblard, and Philippe J. Giabbanelli

Accelerated Approximation of Bellman Equation Solutions: Agent Policy Optimization With a Feedforward Neural Network 224
Victoria M. Garibay

Simulation-Based Inference in Agent-Based Models Using Spatio-Temporal Summary Statistics 239
Eric Dignum, Harshita Choudhary, and Mike Lees

Emergent Communication in Merging Artificial Agent Populations 255
Piotr M. Kosela

MAVS: An Ensemble-Based Multi-agent Framework for Fake News
Detection 269
   Dhruv Tyagi, Anurag Singh, and Hocine Cherifi

Evolutionary Game Selection Leads to Emergent Inequality 284
   Isaak Mengesha and Debraj Roy

Computational Optimization, Modelling and Simulation

Enhancing Gaussian Mixture Model Fitting via Equiprobable Binning
and Adaptive Differential Evolution 301
   Wojciech Achtelik and Maciej Smołka

Asymptotics in Curve Estimation by Modified Cubic Spline
and Exponential Parameterization 316
   Ryszard Kozera, Lyle Noakes, and Magdalena Wilkołazka

Physics Informed Neural Networks for Non-stationary Material Science
Problems 332
   *Paweł Maczuga, Tomasz Służalec, Łukasz Sztangret, Danuta Szeliga,
Marcin Łoś, and Maciej Paszyński*

Adaptive Global Modeling Using Neural Networks with Deep Ensembles
and Space-Filling Sequences 347
   *Pavankumar Koratikere, Leifur Leifsson, Slawomir Koziel,
and Anna Pietrenko-Dabrowska*

Automated Antenna Design Using Computational Intelligence
and Numerical Optimization 362
   Slawomir Koziel, Anna Pietrenko-Dabrowska, and Leifur Leifsson

Near-Optimal Mixed Partial Replications Versus Uniform Replication 377
   Ralf Vamosi

Hybrid Subgradient and Simulated Annealing Method for Hemivariational
Inequalities 392
   Piotr Bartman-Szwarc, Adil M. Bagirov, and Anna Ochal

Reduced-Order Modeling of Compressible Flows Using Supervised
Dimensionality Reduction 406
   *Abhijnan Dikshit, Leifur Leifsson, Slawomir Koziel,
and Anna Pietrenko-Dabrowska*

Exact and Approximate Methods for Solving the Edge-Strength Problem 421
Eduardo Rodriguez-Tello, Eric Monfroy, and Claudia Vasconcellos-Gaete

Author Index 433

Computational Health

Simulation of Blood Flow in the Left Ventricle Considering Purkinje Fibers

Misaki Iwai[1]($\boxtimes$), Yusuke Nishiya[1], Masashi Yamakawa[1], Ayato Takii[2], Yusei Kobayashi[1], Takahiro Ikeda[1], Shinichi Asao[3], and Seiichi Takeuchi[3]

[1] Kyoto Institute of Technology, Matsugasaki, Sakyo-Ku, Kyoto 606-8585, Japan
m4623004@kit.edu.ac.jp
[2] Kobe University, 1-1, Rokkodai-Tyo, Nada-Ku, Kobe 657-8501, Hyogo, Japan
[3] College of Industrial Technology, 1-27-1, Amagasaki 661-0047, Hyogo, Japan

Abstract. Heart disease is the leading cause of death worldwide. To determine the factors contributing to the development of cardiac disease, computational fluid dynamics (CFD) and *in vivo* data, such as MRI of blood flow, are being compared and validated to better understand the hemodynamics of the heart in detail. The cardiac conduction system, which transmits electrical signals and controls the heart's beating, is also being studied. However, no studies have examined the relationship between the cardiac conduction system and left ventricular hemodynamics. In this study, we focused on the Purkinje fibers located in the left ventricle within the cardiac conduction system. The simulation results of left ventricular models with and without Purkinje fibers were compared. The results showed differences in blood flow within the left ventricle. Thus, changes in contraction caused by the Purkinje fibers affect the hemodynamics of the left ventricle and can contribute to the development of heart disease.

Keywords: CFD · Purkinje fiber · cardiac conduction system (CCS) · left ventricular

1 Introduction

In the human body, blood is continuously circulated through blood vessels by the beating of the heart, playing a crucial role in sustaining life. As a result, heart diseases are often life-threatening. According to the WHO [1], ischemic heart disease was the leading cause of death worldwide for 18 consecutive years up to 2019. Even in 2021, when COVID-19 caused global disruption and became the second leading cause of death, ischemic heart disease remained the top cause.

A relationship between heart disease and blood flow has been suggested, and detailed understanding of cardiac hemodynamics has been utilized to explore the mechanisms of disease development. *In silico* studies, such as numerical simulations, are non-invasive and allow for intuitive flow visualization. In recent years, advanced imaging technologies such as *in vivo* CT and MRI have enabled the acquisition of detailed data on the dynamic shape and motion of the heart. These data have been incorporated into simulations to perform calculations under physiologically realistic conditions. Many CFD studies [2–4] have focused on the left ventricle, which is particularly prone to cardiac diseases.

The cardiac conduction system (CCS), including the Purkinje fibers, generates and transmits electrical signals in cardiac muscle. There are two types of cardiac muscle: specialized cardiac muscle that transmits electrical impulses, and intrinsic cardiac muscle that contracts in response. Electrical signals generated by the sinus node are transmitted through the His bundle to the Purkinje fibers, which spread across the left ventricle [5]. Since the CCS governs the heart's rhythm, it plays a critical role in cardiac motion and should be considered in motion analysis. However, the relationship between CCS activity and heart function is difficult to investigate experimentally, and has thus been explored through simulations [6, 7]. Previous studies have mainly focused on the resulting mechanical motion, rather than how the electrical stimulus propagation affects blood flow. Conventional CFD simulations often assume synchronous contraction across the left ventricular wall; however, actual contraction timing is dictated by the conduction sequence. Previous studies have mainly focused on the resulting mechanical motion, rather than how the propagation of electrical stimuli affects hemodynamics. Conventional CFD simulations often assume that both systole and diastole begin simultaneously across the entire left ventricular wall; however, in reality, their timing is governed by the conduction sequence. This study aims to clarify whether incorporating the CCS into blood flow simulations improves the reproducibility of left ventricular hemodynamics. To this end, simulations were conducted using two models—one with and one without the Purkinje network.

The unstructured moving grid finite volume method [8, 9] is applied for calculations using the fractional step method, with the LU-SGS method [10] employed as the iterative method for the first step and the Bi-CGSTAB method [11] for the second step, as both are implicit solution methods. To accommodate complex movements such as valve opening and closing, ventricular systole and diastole, and torsion, a grid movement method that simulates the effect of a torsion spring is employed. Due to the increased computational cost, we aim to accelerate computations by utilizing OpenMP for parallel processing [12]. By considering whether Purkinje fibers contribute to the accuracy of blood flow reproduction, we hope to gain a more detailed understanding of hemodynamics and investigate heart diseases.

2 Numerical Approach

2.1 Governing Equation

Blood viscosity remains nearly constant regardless of hematocrit when the shear rate exceeds $100 \, \text{s}^{-1}$ (corresponding to a Reynolds number of several hundred). In the region from the left ventricle to the aortic arch, the Reynolds number exceeds 1000, making the assumption of Newtonian fluid behavior valid. Therefore, in this study, blood is treated as an incompressible Newtonian fluid, and the three-dimensional incompressible Navier-Stokes and continuity Eqs. (1)-(3) are used as the governing equations.

Here, q is the conserved quantity vector, and E, F, G are the inviscid flux vectors in the x, y, z directions, E_v, F_v, G_v are the viscous flux vectors in the same directions. In addition, u, v, w are velocity components, p is pressure, and Re is the Reynolds number. The subscripts x, y, z in the E_v, F_v, G_v components represent derivatives in each direction. Note that the above equations are expressed in dimensionless form.

$$\frac{\partial q}{\partial t} + \frac{\partial E}{\partial x} + \frac{\partial F}{\partial y} + \frac{\partial G}{\partial z} = \frac{1}{\text{Re}} \left(\frac{\partial E_v}{\partial x} + \frac{\partial F_v}{\partial y} + \frac{\partial G_v}{\partial z} \right) \quad (1)$$

$$q = \begin{bmatrix} u \\ v \\ w \end{bmatrix}, E = \begin{bmatrix} u^2 + p \\ uv \\ uw \end{bmatrix}, F = \begin{bmatrix} vu \\ v^2 + p \\ vw \end{bmatrix}, G = \begin{bmatrix} wu \\ wv \\ w^2 + p \end{bmatrix},$$

$$E_v = \begin{bmatrix} u_x \\ v_x \\ w_x \end{bmatrix}, F_v = \begin{bmatrix} u_y \\ v_y \\ w_y \end{bmatrix}, G_v = \begin{bmatrix} u_z \\ v_z \\ w_z \end{bmatrix}, \quad (2)$$

$$\frac{\partial u}{\partial x} + \frac{\partial v}{\partial y} + \frac{\partial w}{\partial z} = 0. \quad (3)$$

2.2 The Unstructured Moving Grid Finite Volume Method

Pulsatile flow in the left ventricle presents a moving boundary problem, driven by time-varying wall geometry and requiring dynamic mesh deformation. To address this, the unstructured moving grid finite volume method [8, 9] was employed. This method is well-suited for such problems, as it conserves physical quantities using space-time control volumes and strictly satisfies the geometric conservation law.

3 Computational Model

3.1 Left Ventricular Pulsation Conditions

At a heart rate of 60 beats per minute, systole and diastole occupy approximately 0.49 and 0.51 s, respectively. During systole, efficient ejection is achieved primarily through coordinated wall motion, including wall thickening—which causes inward displacement of the endocardial surface—and torsional motion along the long axis. Of these mechanisms, the inward displacement due to wall thickening is indirectly reflected in the volume change, which follows the data presented in [13]. As for the torsional motion, according to [14], assuming normal physiological twisting, the torsional angle is defined as positive in the counterclockwise direction when viewed from the apex, with a maximum of $+3°$ at the base and $-7.8°$ at the apex.

3.2 Conditions for Conduction of Stimulation of the Left Ventricle

Stimulus conduction is calculated on a grid of the left ventricular wall with three patterns: (1) Purkinje to Purkinje (P–P), (2) Purkinje to intrinsic cardiac muscle (P–I), and (3) intrinsic cardiac muscle to intrinsic cardiac muscle (I–I). As shown in Fig. 1, calculations are performed for each pattern individually. The conduction condition is determined based on the distance from the grid point where the stimulus originates, with radii R1 for (1) and R2 for (2) and (3). According to the literature [15, 16], the ratio of conduction velocities is set to R1:R2 = 10:1. In this study, the conduction velocity of the intrinsic cardiac muscle is set to 0.1 cm/step, and that of the Purkinje fibers to 1.0 cm/step.

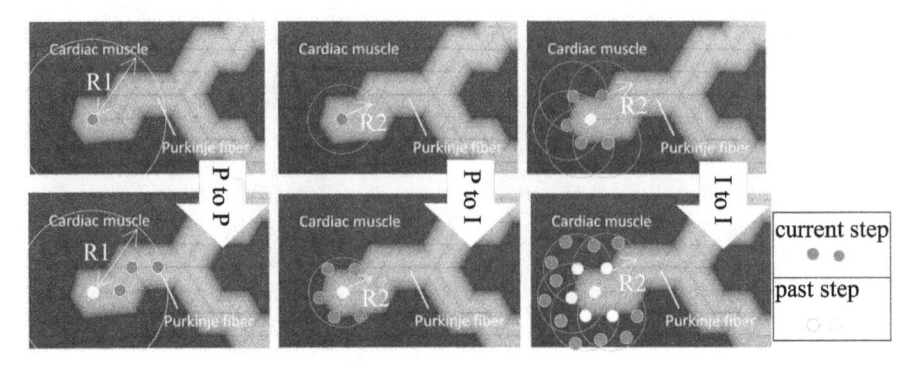

Fig. 1. Three patterns of stimulation conduction

3.3 Left Ventricle Model

The computational grid consists of tetrahedral elements in the interior and prismatic grids near the wall. The left ventricular model contains 1,145,432 elements (755,042 tetrahedral and 390,390 prismatic), and was generated using MEGG3D [17, 18]. At maximum cavity volume, the vessel diameter at the mitral and aortic valves is 3.0 cm, and the distance from the cardiac base to the apex is 7.8 cm. The cross-section is elliptical, with an axial diameter ratio of 1.0:0.8.

According to theories of Purkinje fiber distribution, they first branch from the His bundle into the left anterior and posterior fascicles, which then fan out across the left ventricular wall. Two main branching patterns have been proposed: in one, fibers wrap around the left ventricular apex; in the other, they extend over the anterior and posterior walls after a slight downward turn of the fibers [7]. Based on these patterns, the grid points on the left ventricular wall were classified into specialized and intrinsic cardiac muscle types, as shown in Fig. 2, and modeled accordingly. While the terminal branches of Purkinje fibers form a fine mesh-like network in reality, significant individual variation exists, so a simplified representation is used in this study.

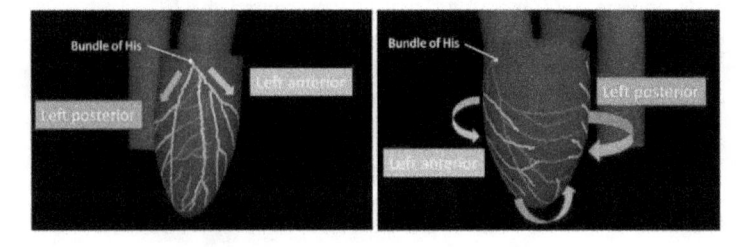

Fig. 2. Purkinje fiber distribution in this model

3.4 Calculation Conditions

Table 1 lists the characteristic values for this simulation.

Initial conditions are: pressure $p = 0.0$ and velocity $u = v = w = 0.0$ for all elements. The time increment is $\Delta t = 0.0005$, and the simulation begins in diastole and runs for three left ventricular beats.

Next, the boundary conditions are described. The left ventricular wall is assumed to be a no-slip wall during both systole and diastole.

- During diastole, the mitral valve is fully open and the aortic valve fully closed.
- During systole, the mitral valve is fully closed and the aortic valve fully open.
- At the open valve, velocity is applied according to a nonlinear function approximated from the graph shown in [20], and the pressure is fixed at $p = 0.0$.
- At wall surfaces including closed valves, the velocity matches that of the moving wall, and the pressure is given by Eq. (4), accounting for acceleration of the wall motion α.
- The opening and closing of the mitral and aortic valves are assumed to occur instantaneously at the onset of diastole and systole.

$$\frac{\partial p}{\partial x} = -\alpha. \qquad (4)$$

Table 1. Characteristic values

characteristic length ($\overline{L_0}$)	0.03 [m]	the vessel diameter at the aortic valve
characteristic kinematic viscosity coefficient ($\overline{v_0}$)	4.43×10^{-6} [m²/s]	the value for blood
characteristic velocity ($\overline{U_0}$)	0.30 [m/s]	the average outflow velocity from the left ventricle during systole

4 Simulation Results

4.1 Validity of the Purkinje Model Results

Hereafter, the conventional model assumes uniform contraction of the entire left ventricle without accounting for electrical conduction, whereas the Purkinje model incorporates conduction dynamics into the contraction process. Figure 3 shows the stimulus transmission behavior. The yellow regions indicate areas where the electrical stimulus has been transmitted. The stimulus rapidly spreads throughout the left ventricle via the Purkinje fibers. The presence of these fibers enables rapid propagation to the apex. This apex-first activation is known to contribute to a squeezing motion, as suggested in literature [15, 19].

Figure 4 shows streamlines during the third diastolic phase in both models. In both models, two asymmetric vortices are generated. The central vortex is larger, consistent with blood flow measurements by Kilner et al. [20]. The dynamics and energy of these vortices have been analyzed in previous studies [21], supporting the validity of the Purkinje model for left ventricular flow simulation .

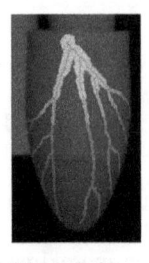

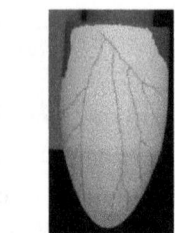

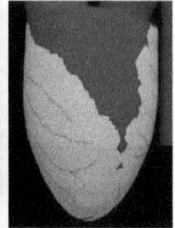

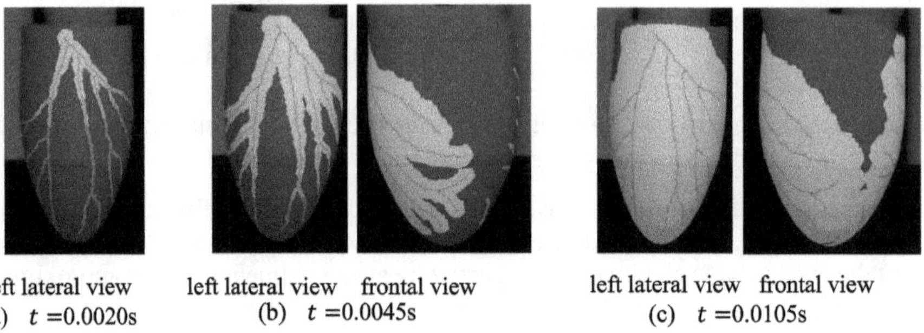

left lateral view   left lateral view   frontal view   left lateral view   frontal view
(a) $t = 0.0020$s   (b) $t = 0.0045$s   (c) $t = 0.0105$s

Fig. 3. The stimulus transmission behavior

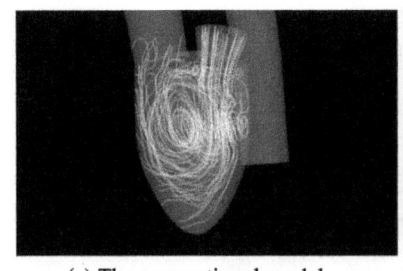

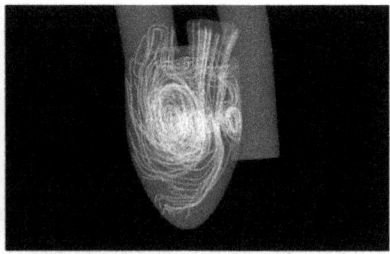

(a) The conventional model   (b) The Purkinje model

Fig. 4. The streamlines in both models

4.2 Differences Between the Purkinje Model and the Conventional Model

Iso-surfaces of the Q-values are commonly used to extract vortex structures. Figure 5 shows the differences in blood flow between the two models during the late systolic phase. The comparison reveals that slight differences in the motion of the left ventricular wall lead to minor variations in the flow field. These differences are particularly noticeable in the central region of the left ventricle. This is likely because the vortex described in Sect. 4.1 disturbs the surrounding flow, making the small differences in wall motion manifest as flow differences in the central area.

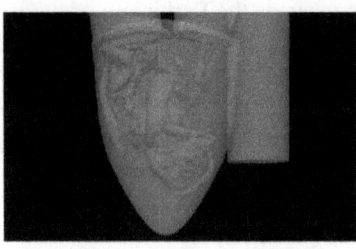

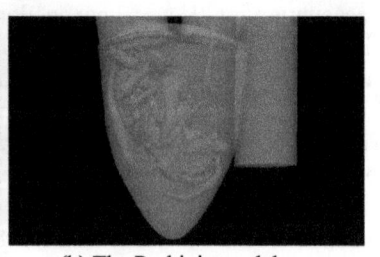

(a) The conventional model   (b) The Purkinje model

Fig. 5. Iso-surface visualization of the Q-values in the late systolic phase

5 Conclusions

This study is the first to incorporate Purkinje fibers into blood flow simulations. The objective was to investigate how the timing of left ventricular contraction, altered by stimulus conduction via Purkinje fibers, affects intraventricular blood flow behavior. By comparing simulation results with and without stimulus conduction, the following conclusions were drawn:

- The results of stimulus conduction using a Purkinje fiber model in the left ventricular wall were consistent with clinical data, confirming the validity of the Purkinje fiber distribution and conduction mechanism.
- Visualization of streamlines revealed the formation of asymmetric vortices in the central region and near the wall of the left ventricle. The characteristics of these vortices matched measurement results and previous simulations, supporting the validity of the fluid dynamics calculations in the Purkinje fiber–considering model.
- Comparison of Q-value iso-surfaces showed that differences in left ventricular motion slightly influenced the flow field.

These results demonstrate that incorporating stimulus conduction via Purkinje fibers leads to different outcomes compared to the conventional model. This approach is essential for accurately reproducing left ventricular blood flow and understanding the causes of cardiac diseases. Future work will extend the simulation domain to the aorta.

References

1. WHO Homepage. https://www.who.int/news-room/fact-sheets/detail/the-top-10-causes-of-death. Accessed 10 Jan 2025
2. Dedè, L., Menghini, F., Quarteroni, A.: Computational fluid dynamics of blood flow in an idealized left human heart. Int. J. Numer. Method Biomed. Eng. **37**(11), e3287 (2021). https://doi.org/10.1002/cnm.3287
3. Xu, F., Kenjereš, S.: Numerical simulations of flow patterns in the human left ventricle model with a novel dynamic mesh morphing approach based on radial basis function. Comput. Biol. Med. **130**, 104184 (2021). https://doi.org/10.1016/j.compbiomed.2020.104184
4. Schenkel, T., Malve, M., Reik, M., Markl, M., Jung, B., Oertel, H.: MRI-based CFD analysis of flow in a human left ventricle: methodology and application to a healthy heart. Ann. Biomed. Eng. **37**, 503–515 (2009). https://doi.org/10.1007/s10439-008-9627-4
5. Cavero, I., Holzgrefe, H.: Remembering the canonical discoverers of the core components of the mammalian cardiac conduction system: Keith and Flack, Aschoff and Tawara, His, and Purkinje. Adv. Physiol. Educ. **46**(4), 549–579 (2022). https://doi.org/10.1152/advan.00072.2022
6. Strocchi, M., Gillette, K., Neic, A., et al.: Effect of scar and His-Purkinje and myocardium conduction on response to conduction system pacing. J. Cardiovasc. Electrophysiol. **34**(4), 984–993 (2023). https://doi.org/10.1111/jce.15847
7. Norouzi, S., Goudarzi, T.: Effects of fiber orientation and the anisotropic behavior of the cardiac tissue on the simulated electrocardiogram. Acta Mech. **233**(10), 3881–3892 (2022). https://doi.org/10.1007/s00707-022-03286-4
8. Yamakawa, M., Takekawa, D., Matsuno, K., Asao, S.: Numerical simulation for a flow around body ejection using an axisymmetric unstructured moving grid method. Comput. Therm. Sci. Int. J. **4**(3), 217–223 (2012). https://doi.org/10.1615/ComputThermalScien.2012004715

9. Asao, S., Matsuno, K., Yamakawa, M.: Simulations of a falling sphere with concentration in an infinite long pipe using a new moving mesh system. Appl. Therm. Eng. **72**(1), 29–33 (2014). https://doi.org/10.1016/j.applthermaleng.2014.06.059
10. Yoon, S., Jameson, A.: Lower-upper symmetric-Gauss-Seidel method for the Euler and Navier-Stokes equations. AIAA J. **26**(9), 1025–1026 (1988). https://doi.org/10.2514/3.10007
11. Van der Vorst, H.A.: Bi-CGSTAB: a fast and smoothly converging variant of Bi-CG for the solution of nonsymmetric linear systems. SIAM J. Sci. Stat. Comput. **13**(2), 631–644 (1992). https://doi.org/10.1137/0913035
12. Yamakawa, M., Kita, Y., Matsuno, K.: Domain decomposition method for unstructured meshes in an OpenMP computing environment. Comput. Fluids **45**(1), 168–171 (2011). https://doi.org/10.1016/j.compfluid.2011.02.008
13. Liang, F., Liu, H.: A closed-loop lumped parameter computational model for human cardiovascular system. JSME Int. J. Ser. C **48**(4), 484–493 (2005). https://doi.org/10.1299/jsmec.48.484
14. Murata, K., Akagawa, E., Tanaka, T.: 2D tissue tracking technique demonstrates attenuated apical endocardial rotation in patients with dilated cardiomyopathy. MEDIX **46**, 8–11 (2007)
15. Durrer, D., Van Dam, R.T., Freud, G.E., Janse, M.J., Meijler, F.L., Arzbaecher, R.C.: Total excitation of the isolated human heart. Circulation **41**(6), 899–912 (1970). https://doi.org/10.1161/01.CIR.41.6.899
16. Elbrønd, V.S., Thomsen, M.B., Isaksen, J.L., et al.: Intramural Purkinje fibers facilitate rapid ventricular activation in the equine heart. Acta Physiol. **237**(3), e13925 (2023). https://doi.org/10.1111/apha.13925
17. Ito, Y., Nakahashi, K.: Surface triangulation for polygonal models based on CAD data. Int. J. Numer. Methods Fluids **39**(1), 75–96 (2002). https://doi.org/10.1002/fld.281
18. Ito, Y.: Challenges in unstructured mesh generation for practical and efficient computational fluid dynamics simulations. Comput. Fluids **85**, 47–52 (2013). https://doi.org/10.1016/j.compfluid.2012.09.031
19. Fantoni, C., Kawabata, M., Massaro, R., et al.: Right and left ventricular activation sequence in patients with heart failure and right bundle branch block: a detailed analysis using three-dimensional non-fluoroscopic electroanatomic mapping system. J. Cardiovasc. Electrophysiol. **16**(2), 112–119 (2005). https://doi.org/10.1046/j.1540-8167.2005.40777.x
20. Kilner, P.J., Yang, G.Z., Wilkes, A.J., Mohiaddin, R.H., Firmin, D.N., Yacoub, M.H.: Asymmetric redirection of flow through the heart. Nature **404**(6779), 759–761 (2000). https://doi.org/10.1038/35008075
21. Kaur, H., Assadi, H., Alabed, S., et al.: Left ventricular blood flow kinetic energy assessment by 4D flow cardiovascular magnetic resonance: a systematic review of the clinical relevance. J. Cardiovasc. Dev. Dis. **7**(3), 37 (2020). https://doi.org/10.3390/jcdd7030037

ViT-SE_Res: A Hybrid Vision Transformer and ResNet50V2 with Squeeze-And-Excitation Block for Cervical Cell Classification

Betelhem Zewdu Wubineh[✉], Andrzej Rusiecki, and Krzysztof Halawa

Faculty of Information and Communication Technology, Wroclaw University of Science and Technology, Wroclaw, Poland
{betelhem.wubineh,andrzej.rusiecki,krzysztof.halawa}@pwr.edu.pl

Abstract. Cervical cancer is the leading cause of mortality among women, and early detection through effective screening is crucial for a better prognosis and treatment. Traditional manual methods, such as the Papanicolaou (Pap) test, are often time-consuming and ineffective. To overcome these limitations, this study proposes a novel hybrid model, Vision Transformer with Squeeze-and-Excitation blocks incorporated into the ResNet50V2 (ViT-SE_Res), for cervical cell classification. The model integrates the local feature extraction capabilities of ResNet50V2 with the global context modeling strength of ViT, effectively capturing both local and global features for improved accuracy. The model was evaluated on two datasets, Pomeranian and SIPaKMeD, achieving an accuracy of 98.80% and 98.51%, respectively, with SIPaKMeD being a binary classification task. The proposed ViT-SE_Res model offers a robust and efficient tool for cervical cancer screening, providing reliable detection of abnormalities to support early intervention.

Keywords: Cervical Cancer · Classification · ResNet50V2 · Vision Transformer · Squeeze-and-Excitation

1 Introduction

Cervical cancer is one of the most widespread diseases among women, and its control is, therefore, a major concern worldwide [1–3]. The prognosis and treatment of cervical cancer have improved significantly with early detection and diagnosis [4]. Papanicolaou (Pap) test is a widely used screening method to detect cervical cancer and precancerous lesions [5–8]. However, manual screening of Pap smears is a tedious, inefficient, and expensive process [9, 10]. Furthermore, there is a subjective judgment of the decision made by different experts, which may be prone to errors [11, 12]. Lastly, traditional techniques are very labor-intensive and have a very long lead time, which makes them undesirable. These limitations led to the intent to investigate automatic screening methods.

Deep learning (DL) has emerged as a promising approach for automated cervical cell classification, offering the potential to learn complex patterns in cervical cytopathology images without manual feature engineering [13, 14]. Deep learning methods have performed very well in the field of medical imaging [15, 16]. Convolutional neural networks (CNNs) have shown considerable success in the classification of cervical cytopathology images [15]. Nevertheless, classical CNNs may lack the ability to model global aspects of cervical cell extraction [1], while Vision Transformers (ViTs), which have shown great potential in computer vision, may fall short in adequately extracting cervical cell morphology information due to the compact size and unique structure of cervical cytopathology images [1]. As such, it is necessary to investigate techniques that could effectively integrate the advantages of CNN and ViT to improve the performance of cervical cell classification [3].

Several studies have explored deep learning-based methods for cervical cell classification. For example, Hemalatha et al. [13] introduced CervixFuzzyFusion for cervical cancer cell image classification. This study uses a feature fusion method that combines features from DenseNet201 and ViT using shifted patch tokenization and locality self-attention models. Yang et al. [1] proposed a pyramid convolutional mixer (PCMixer) to classify cervical cells. The mixer integrates a pyramidal morphology module and a nuclear spatial mixing block to extract information on cervical cytopathology effectively. Furthermore, Anand and Bachhal [2] used the VGG16 architecture, referred to as Cervical-Net, for the classification of cervical cells. Maurya et al. [14] presented an ensemble approach of combining the Vision Transformer network (ViT) and CNN for cervical cell classification. All these studies utilized the publicly available SIPaKMeD datasets.

Although previous studies have explored architectures such as DenseNet201, VGG16, and ensemble methods involving CNNs and Vision Transformers, the potential of combining the ResNet family with ViT for cervical cell classification remains unexplored. This study addresses this gap by leveraging the residual learning of ResNet50V2 with Squeeze-and-Excitation (SE) block and the ViT attention mechanisms to classify cervical cells. ResNet50V2 has shown promising performance in cervical cell classification [17]. The SE block enables the network to focus more on the important features, thereby improving the performance [18]. Additionally, Vision Transformer has demonstrated potential in image classification tasks [5]. By combining the local feature extraction capabilities of ResNet50V2 and the global feature extraction capabilities of ViT [19], this study aims to achieve more accurate and robust classification results. The key contribution of this study is as follows:

- Incorporating SE blocks into the residual block of ResNet50V2 to focus on the important feature and improve performance.
- Proposing a novel hybrid model of SE-ResNet50V2 and Vision Transformer (ViT-SE_Res) for cervical cell classification.

This study aims to develop an effective hybrid deep-learning model for classifying cervical cells to help diagnose cervical cancer. The rest of the paper is organized as follows. Section 2 describes the materials and methods, Sect. 3 discusses the results and findings, and Sect. 4 presents the concluding remarks, followed by the references.

2 Materials and Methods

In this section, we discuss the dataset and its preprocessing technique, the proposed method, and the evaluation metrics.

2.1 Dataset and Preprocessing Technique

The datasets used in this study are from the Pomeranian Medical University in Szczecin, Poland. This dataset contains a total of 419 cervical images, divided into 268 training images, 67 validation images, and 84 testing images. The dataset includes three categories: high squamous intra-epithelial lesion (HSIL), low squamous intra-epithelial lesion (LSIL), and normal squamous intra-epithelial lesion (NSIL) [20]. Most of the images have a resolution of 1130 × 1130 pixels. The second dataset is SIPaKMeD [21], which contains a total of 4049 single-cell images. These are divided into 2,591 training images, 648 validation images, and 810 testing images. This dataset comprises five categories: dyskeratotic, koilocytotic, metaplastic, parabasal, and supra-intermediate. Table 1 provides the details of the datasets.

Table 1. Details of the dataset used for the study

Dataset	Category	No. of image	Category	Training	Validation	Testing
Pomeranian	HSIL	124		268	67	84
	LSIL	61				
	NSIL	234				
SIPaKMeD	Dyskeratotic	713	Abnormal	2591	648	810
	Koilocytotic	725				
	Metaplastic	693				
	Parabasal	687	Normal			
	Superficial-Intermediate	731				

During data preprocessing, we resized the images to a uniform size of 224 × 224 pixels. Then, we normalized the images by scaling the pixel values to the range of [0, 1]. To further optimize input for the Vision Transformer (ViT-B16), the resized images were divided into non-overlapping patches of 16 × 16 pixels, ensuring compatibility with the model architecture. Sample images from both the Pomeranian and the SIPaKMeD dataset are shown in Fig. 1.

As shown in Fig. 1, the first row represents samples from the Pomeranian, while the second row corresponds to the SIPaKMeD dataset. In the Pomeranian dataset, HSIL refers to more serious abnormalities characterized by moderate to severe dysplasia, which carry a higher risk of progressing to cervical cancer if left untreated. LSIL denotes non-cancerous cervical abnormalities where cells exhibit mild dysplasia or slight

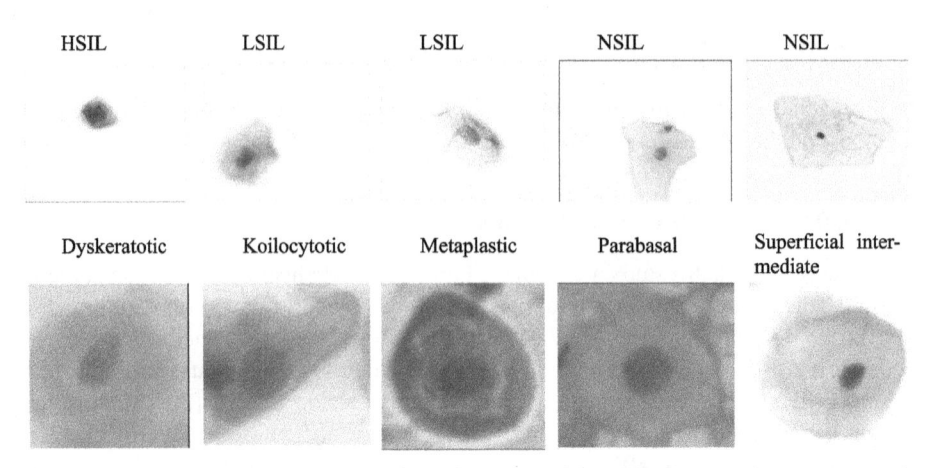

Fig. 1. The first row represents samples from the Pomeranian, while the second row corresponds to the SIPaKMeD dataset.

irregularities. NSIL represents normal cervical cells without precancerous changes. In the SIPaKMeD dataset, dyskeratotic cells are defined by abnormal keratinization and typically appear in dense clusters with orangeophilic cytoplasm and vesicular nuclei, often indicating HPV infection. Koilocytotic cells are characterized by large perinuclear halos, hyperchromatic and irregular nuclei and are pathognomonic for HPV infection, frequently appearing as binucleated or multinucleated cells. Metaplastic cells resemble parabasal cells but exhibit more uniform size and shape, well-defined round cytoplasm, and darker staining; they are often associated with high-grade lesions (HSIL). Parabasal cells are small, immature squamous cells with cyanophilic cytoplasm and large vesicular nuclei, and they can be difficult to distinguish from metaplastic cells. Finally, superficial-intermediate cells are the most abundant type, with large polygonal cytoplasm and small pycnotic or vesicular nuclei, which may display morphological changes in the presence of severe lesions.

2.2 Proposed Method

In this paper, we propose a hybrid cervical cell classification model using ResNet50V2 with SE block and Vision transformer (ViT) architectures. ResNet50V2 is a pre-trained CNN on ImageNet [22], used to extract rich hierarchical spatial features, leveraging its deeper residual structure [15] to improve performance. In our approach, Squeeze-and-Excitation blocks are added following each ResNet block to improve the model's capability to adjust channel-wise feature responses. This allows the network to dynamically focus on more relevant features [23]. It applies global average pooling to squeeze spatial dimensions, followed by two fully connected layers to excite and weight each channel, enhancing the network's focus on important features. On the other hand, ViT, a transformer-based model, captures global contextual relationships and long-range dependencies [19].

Both models are adapted by excluding their classification heads and freezing their pre-trained weights to retain their feature extraction capabilities and prevent overfitting.

The input images are processed through these two branches independently, with features of ResNet50V2 aggregated using global average pooling and SE-enhanced features, and features of ViT flattened into a one-dimensional representation. The outputs from both branches are fused via feature concatenation, creating a unified feature vector. This fused representation is passed through a fully connected classification head, which includes dense layers with ReLU activations, dropout for regularization, and a final softmax layer for class predictions.

By leveraging the complementary strengths of convolutional networks with SE-enhanced residual learning and transformers, this hybrid model can effectively capture both local and global features, providing a robust and efficient framework for cervical cell classification. The overall framework of the proposed model is illustrated in Fig. 2. In our case, we used ViT-B16, which divides the input image of size 224 × 224 into patches of 16 × 16 pixels. This patch size allows the model to capture local information to process the ViT model effectively.

The ViT-B16 architecture used in our model consists of 12 transformer encoder blocks, each comprising a multi-head self-attention mechanism with 12 attention heads and a feed-forward network (FFN) with a hidden size of 3072 [24]. The embedding dimension for each input patch is 768. Each block includes Layer Normalization and residual connections, followed by a GELU activation in the MLP layers. Position embeddings are added to the input patch embeddings to retain spatial information. A dropout rate of 0.1 is applied throughout the transformer blocks to mitigate overfitting.

2.3 Evaluation Metrics

To qualitatively evaluate the proposed model's performance, we used the accuracy, precision, recall, and F1 scores. These metrics provide a comprehensive assessment, with accuracy reflecting overall performance and precision, recall, and F1 score providing information on the model's effectiveness in an imbalanced class [19].

$$Accuracy = \frac{TP + TN}{TP + FP + TN + FN} \tag{1}$$

$$Precision = \frac{TP}{TP + FP} \tag{2}$$

$$Recall = \frac{TP}{TP + FN} \tag{3}$$

$$F1 - Score = \frac{2 * (Precision * Recall)}{(Precision + Recall)} \tag{4}$$

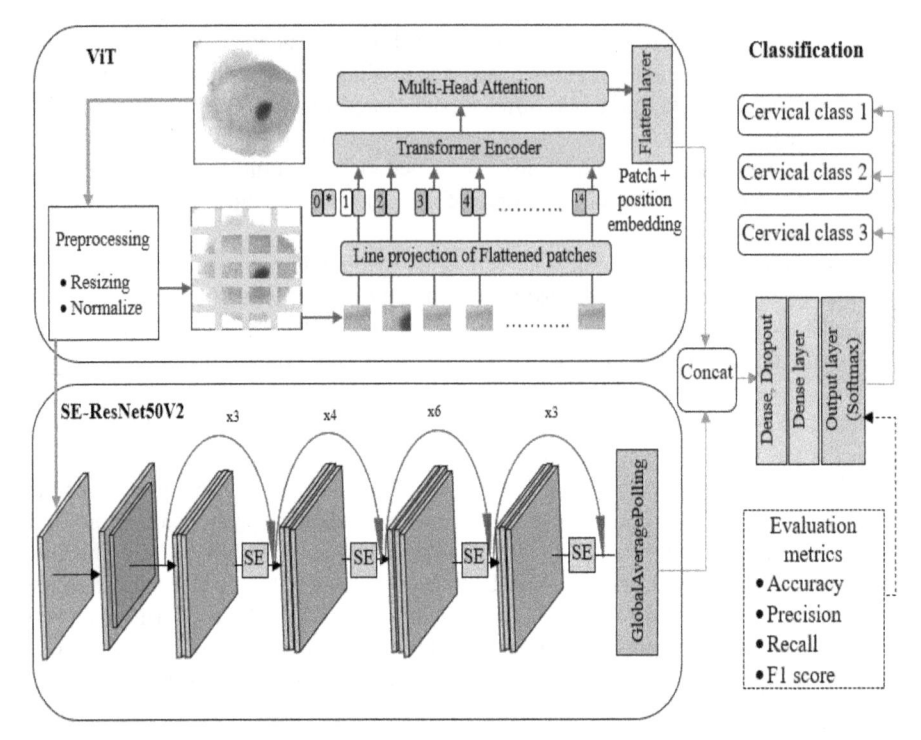

Fig. 2. The ViT-SE_Res proposed method for cervical cell classification

3 Results and Discussion

In this study, to evaluate the proposed model, we used the Pomeranian and SIPaKMeD datasets. The dataset comprising 419 and 4049 images, respectively, were divided into training, validation, and testing sets, as detailed in Table 1. The experiment was conducted using Python programming with the TensorFlow and Keras frameworks. The tuning procedure for the study is described as follows: we used the Adam optimizer with the ReLU activation for the hidden layers and Softmax for the final classification layer, with a learning rate of 0.002. A batch size of 18 was used to process the data before updating the model's weights. The model was trained for 100 epochs, allowing it to pass through the entire training dataset, with early stopping to prevent overfitting. The results of the study using the Pomeranian and the SIPaKMeD dataset are illustrated in Table 2.

As shown in Table 2, the proposed method achieves 98.80% accuracy on the Pomeranian dataset, indicating that the combination of ResNet50V2 with SE and ViT is highly effective in classifying cervical cells. The high precision, recall, and F1 score further demonstrate the robustness of the model for cervical cancer diagnosis. On the SIPaKMeD dataset, the proposed method also performs well, achieving 94.69% and 98.51% accuracy for multiclass and binary classification, respectively. These results highlight the versatility and strong performance of the model across both types of classification tasks. In particular, the model is effective at identifying different types of cervical cells, showing its potential for precise abnormality detection. The high scores in binary classification

Table 2. Results of the study using the Pomeranian and SIPaKMeD dataset

Dataset	Method	Classification type	Accuracy	Precision	Recall	F1 score
Pomeranian	ViT: SE-ResNet50V2	Multiclass	98.80%	98.88%	98.80%	98.81%
SIPaKMeD		Multiclass	94.69%	94.72%	94.69%	94.66%
		Binary	98.51%	98.53%	98.51%	98.52%

also underscore its ability to distinguish between normal and abnormal cells, which is crucial in clinical settings for accurate diagnosis.

The observed difference in performance between multiclass and binary classification on the SIPaKMeD dataset can be attributed to the inherent complexity of the multiclass task. In multiclass classification, the model must distinguish between several visually similar cervical cell types, which introduces a higher level of difficulty and potential for misclassification, especially between classes with subtle morphological differences. In contrast, binary classification simplifies the task by reducing it to identifying whether a sample is normal or abnormal, making it less prone to inter-class confusion. Additionally, class imbalance and overlapping features among certain cell types in the multiclass scenario may further contribute to the lower performance. The confusion matrix of the proposed model is shown in Fig. 3.

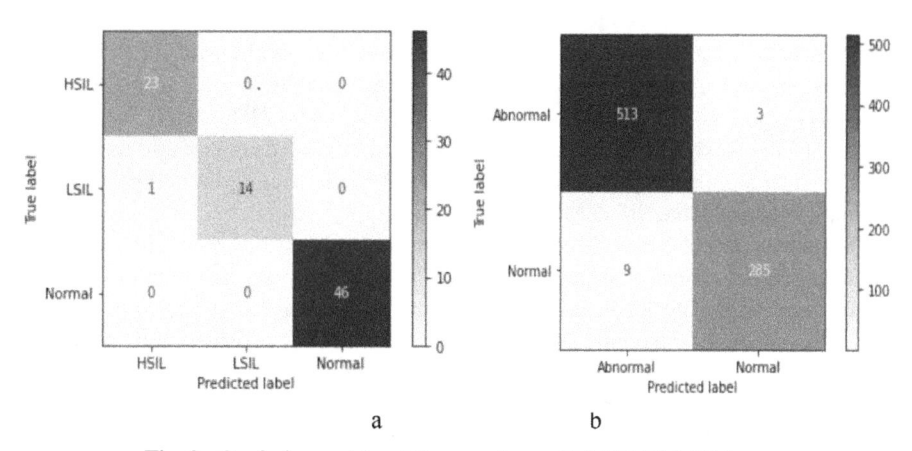

Fig. 3. Confusion matrix: a) Pomeranian and b) SIPaKMeD binary

In the confusion matrix, only one image was incorrectly classified in the Pomeranian dataset, and 12 images were in the binary classification of the SIPaKMeD dataset. This shows how the proposed method effectively classified cervical cells in different datasets. The comparison of our study with the previous work is shown in Table 3.

Table 3. Comparison of our study with previous work

Ref	Method	Dataset	Accuracy	Precision	Recall	F1 score
[1]	PCMixer	SIPaKMeD	96.21	95.70	95.60	95.30
[2]	VGG16	SIPaKMeD	84.33%	–	–	–
[13]	CervixFuzzyFusion	SIPaKMeD	93.36%	–	–	–
[14]	ViT-CNN Max vote	SIPaKMeD	97.6%	99.54	97.65	98.58
[14]	ViT-CNN Stacking	SIPaKMeD	94.11%	–	–	–
[25]	SA-ResNet50V2	SIPaKMeD	92%	92%	92%	92%
Ours	ViT-Res	SIPaKMeD multi	94.69%	94.72%	94.69%	94.66%
Ours	ViT-Res	SIPaKMeD binary	98.51%	98.53%	98.51%	98.52%
Ours	Vit-Res	Pomeranian	**98.80%**	98.88%	98.80%	98.81%

The results in Table 3 show a comparison of the previous works using the SIPaKMeD multiclass dataset. Among these, ViT-CNN demonstrates the highest accuracy, which is 97.6%, outperforming other models such as PCMixer at 96.21%, CervixFuzzyFusion at 93.36%, and ResNet50V2 at 92%. In contrast, traditional deep learning models like VGG16 show significantly lower accuracy at 84.33%, indicating limited performance in complex datasets like SIPaKMeD.

Our proposed ViT-SE_Res hybrid model, however, is distinct from [14], as it integrates the feature extraction capabilities of ResNet50V2 with the attention mechanisms of ViT, providing both local and global feature extraction. On the other hand, [14] ViT-CNN ensembles the predictions of CNN and MobileNet via Max voting, achieving an accuracy of 97.63%, while using the stacking ensemble method results in an accuracy of 94.11%. By concatenating features from ResNet50V2 with SE and ViT, our study performs better than the stacking method but slightly lower than the Max voting ensemble.

This combination improves accuracy, achieving 98.80% on the Pomeranian and 98.51% on the SIPaKMeD binary datasets, and 94.69% accuracy on the SIPaKMeD multiclass dataset, surpassing the model in [13], which achieved an accuracy of 93.36%. Additionally, [25] uses a residual deep convolutional generative adversarial network (Res_DCGAN) for augmentation and adds a self-attention layer at the top of ResNet50V2, achieving an accuracy of 92% which is lower than ours. As a result, our study suggests that the integration of ResNet and ViTs offers significant advantages over standalone approaches, delivering high accuracy and balanced performance metrics across various datasets and configurations.

Moreover, it is important to visualize which parts of the images (features) are crucial for the classification of cervical cancer. This can be achieved through explainable AI (XAI) techniques, which highlight the specific regions of the image that contribute to the model's classification decision. Gradient-weighted class activation mapping (Grad-CAM) is a model-specific XAI method designed for use with DL models that incorporate CNN, where the spatial relationships in the data remain intact following its passes through the convolutional layers [26]. Figure 4 shows the original images alongside their Grad-CAM visualizations, highlighting the area that contributes the most to the model's classification decision.

In Fig. 4, the first and the second column shows images from the Pomeranian dataset and corresponding Grad-CAM, while the third and fourth column displays images from the SIPaKMeD dataset with corresponding Grad-CAM. This visualization helps identify the area of the images that is important for the classification model. For the Pomeranian dataset, the center of the cell background near the cell is significant, whereas in the SIPaKMeD dataset, the center of the cell plays a key role.

In this study, Grad-CAM was employed to visualize and interpret the regions of cervical cell images that contributed most to the model's predictions. Since Grad-CAM is designed specifically for convolutional neural networks, the visualizations were generated using the feature maps from the ResNet50V2 branch of our hybrid model. The Vision Transformer (ViT) branch does not retain the spatial hierarchies required for traditional Grad-CAM, as it operates on flattened patch embeddings and lacks convolutional layers. Therefore, Grad-CAM was not applied to the ViT component. While the transformer branch plays a critical role in the overall classification performance through feature fusion, its interpretability requires alternative methods, such as attention map visualization, Attention Rollout, or Transformer Attribution techniques. Incorporating these transformer-specific explainability approaches will be considered in future work to provide a more comprehensive understanding of the hybrid model's decision-making process.

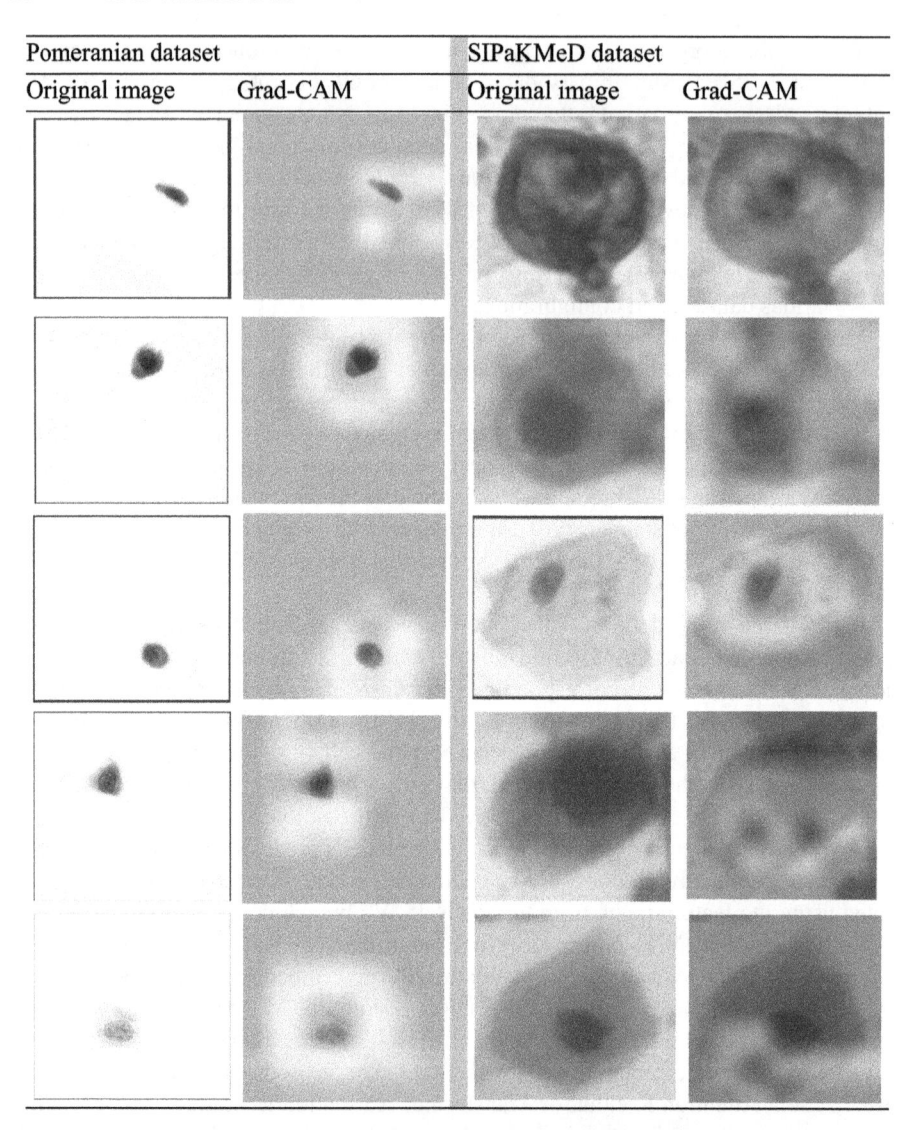

Pomeranian dataset		SIPaKMeD dataset	
Original image	Grad-CAM	Original image	Grad-CAM

Fig. 4. Visualization of images using Grad-CAM

4 Conclusions

This study successfully demonstrates the efficacy of a new hybrid approach that combines SE with ResNet50V2 and Vision Transformer (ViT) for cervical cell classification. By leveraging both the local feature extraction capabilities of ResNet50V2 and the global attention mechanisms of ViT, the proposed ViT-Res model effectively handles the challenges of cervical cell classification and achieves remarkable accuracy and performance across different datasets. The findings highlight the advantage of integrating ResNet50V2 and ViT architectures, which collectively enable improved feature extraction and classification. Furthermore, Grad-CAM was used to identify the important features of the images for decision-making in the classification model. Further studies could explore fine-tuning the model and expanding its application to larger and more diverse datasets, ultimately contributing to the automation and enhancement of cervical cancer screening processes. Additionally, validating the models with the independent dataset would help ensure the generalizability of the model.

References

1. Yang, T., Hu, H., Li, X., Qing, M., Chen, L., Huang, Q.: A pyramid convolutional mixer for cervical pap-smear image classification tasks. Biomed. Signal Process. Control **99**, 106789 (2025). https://doi.org/10.1016/j.bspc.2024.106789
2. Anand, V., Bachhal, P.: Cervical net: an effective convolution neural network for five-class classification of cervical cells. In: Proc. - 2nd IEEE Int. Conf. Device Intell. Comput. Commun. Technol. DICCT 2024, pp. 51–55 (2024). https://doi.org/10.1109/DICCT61038.2024.10532902
3. Fang, M., Fu, M., Liao, B., Lei, X., Wu, F.X.: Deep integrated fusion of local and global features for cervical cell classification. Comput. Biol. Med. **171**, 108153 (2024). https://doi.org/10.1016/j.compbiomed.2024.108153
4. Wubineh, B.Z., Rusiecki, A., Halawa, K.: Cervical cell segmentation and classification using U-Net and Hybrid VGG19-AlexNet architecture. In: 2024 Int. Conf. Inf. Commun. Technol. Dev. Africa, ICT4DA 2024, no. November, pp. 37–42 (2024). https://doi.org/10.1109/ICT4DA62874.2024.10777181
5. Zhao, C., Shuai, R., Ma, L., Liu, W., Wu, M.: Improving cervical cancer classification with imbalanced datasets combining taming transformers with T2T-ViT **81**(17) (2022). https://doi.org/10.1007/s11042-022-12670-0
6. Wubineh, B.Z., Rusiecki, A., Halawa, K.: Segmentation and classification techniques for pap smear images in detecting cervical cancer: a systematic review. IEEE Access **12**(August), 118195–118213 (2024). https://doi.org/10.1109/ACCESS.2024.3447887
7. Wubineh, B.Z., Rusiecki, A., Halawa, K.: Segmentation of cytology images to detect cervical cancer using deep learning techniques. In: Franco, L., de Mulatier, C., Paszynski, M., Krzhizhanovskaya, V.V., Dongarra, J.J., Sloot, P.M.A. (eds.) Computational Science – ICCS 2024. ICCS 2024. Lecture Notes in Computer Science, vol. 14835, pp. 270–278. Springer, Cham (2024). https://doi.org/10.1007/978-3-031-63772-8_25
8. Hussain, E., Mahanta, L.B., Das, C.R., Choudhury, M., Chowdhury, M.: A shape context fully convolutional neural network for segmentation and classification of cervical nuclei in Pap smear images. Artif. Intell. Med. **107**, 101897 (2020). https://doi.org/10.1016/j.artmed.2020.101897
9. Kurnianingsih, et al.: Segmentation and classification of cervical cells using deep learning. IEEE Access **7**, 116925–116941 (2019). https://doi.org/10.1109/ACCESS.2019.2936017
10. Desiani, A., Erwin, M., Suprihatin, B., Yahdin, S., Putri, A.I., Husein, F.R.: Bi-path architecture of CNN segmentation and classification method for cervical cancer disorders based on pap-smear images. IAENG Int. J. Comput. Sci. **48**(3), 1–9 (2021)
11. Madathil, S., Dhouib, M, Lelong, Q., Bourassine, A., Monsonego, J.: A multimodal deep learning model for cervical pre-cancers and cancers prediction: development and internal validation study. Comput. Biol. Med. **186** (2025). https://doi.org/10.1016/j.compbiomed.2025.109710

12. Zewdu Wubineh, B., Jeleń, L., Rusiecki, A.: DCGAN-based cytology image augmentation for cervical cancer cell classification. IEEE Trans. Med. Imaging **50**, 1003–1011 (2020). https://doi.org/10.1016/j.procs.2025.02.206
13. Hemalatha, K., Vetriselvi, V., Aruna Gladys, A.: CervixFuzzyFusion for cervical cancer cell image classification. Biomed. Signal Process. Control **85**, 104920 (2023). https://doi.org/10.1016/j.bspc.2023.104920
14. Maurya, R., Nath Pandey, N., Kishore Dutta, M.: VisionCervix: Papanicolaou cervical smears classification using novel CNN-Vision ensemble approach. Biomed. Signal Process. Control **79**, 104156 (2023). https://doi.org/10.1016/j.bspc.2022.104156
15. Talukder, M.A., Layek, M.A., Kazi, M., Uddin, M.A., Aryal, S.: Empowering COVID-19 detection: optimizing performance through fine-tuned EfficientNet deep learning architecture. Comput. Biol. Med. **168**(August), 2024 (2023). https://doi.org/10.1016/j.compbiomed.2023.107789
16. Wubineh, B.Z., Rusiecki, A., Halawa, K.: Data augmentation techniques to detect cervical cancer using deep learning: a systematic review. In: Zamojski, W., Mazurkiewicz, J., Sugier, J., Walkowiak, T., Kacprzyk, J. (eds.) System Dependability - Theory and Applications. DepCoS-RELCOMEX 2024. Lecture Notes in Networks and Systems, vol. 1026, pp. 325–336. Springer, Cham (2024). https://doi.org/10.1007/978-3-031-61857-4_32
17. Wong, L., Ccopa, A., Diaz, E., Valcarcel, S., Mauricio, D., Villoslada, V.: Deep learning and transfer learning methods to effectively diagnose cervical cancer from liquid-based cytology pap smear images. Int. J. online Biomed. Eng. **19**(4), 77–93 (2023). https://doi.org/10.3991/ijoe.v19i04.37437
18. Sani, U., Suryani, E., Widiarto, W., Salamah, U.: Residual network with squeeze-and-excitation block for white blood cell classification in acute myeloid leukemia. In: Proceedings of International Conference Informatics Comput. Science, pp. 239–244 (2024). https://doi.org/10.1109/ICICoS62600.2024.10636919
19. Jiang, X., Wang, S., Zhang, Y.: Vision transformer promotes cancer diagnosis: a comprehensive review. Expert Syst. Appl. **252**, 124113 (2024). https://doi.org/10.1016/j.eswa.2024.124113
20. Jeleń, Ł, Stankiewicz-Antosz, I., Chosia, M., Jeleń, M.: Optimizing cervical cancer diagnosis with feature selection and deep learning. Appl. Sci. **15**(3), 1–21 (2025). https://doi.org/10.3390/app15031458
21. Plissiti, M.E., Dimitrakopoulos, P., Sfikas, G., Nikou, C., Krikoni, O., Charchanti, A.: Sipakmed: a new dataset for feature and image based classification of normal and pathological cervical cells in pap smear images. In: Proc. - Int. Conf. Image Process. ICIP, pp. 3144–3148 (2018). https://doi.org/10.1109/ICIP.2018.8451588
22. Future Machine Learning, The Significance of F1 Score in Evaluating Algorithms, Future Machine Learning
23. Shahadat, N., Nguyen, A., Lama, R.: Squeeze and hypercomplex networks on leaf disease detection. In: Pattern Recognition, ICPR 2024. Lecture Notes in Computer Science (2025)
24. Dosovitskiy, A., et al.: An Image Is Worth 16X16 Words: transformers for image recognition at scale. In: ICLR 2021 - 9th Int. Conf. Learn. Represent. (2021)

25. Wubineh, B.Z., Rusiecki, A., Halawa, K.: Classification of cervical cells from the Pap smear image using the RES_DCGAN data augmentation and ResNet50V2 with self-attention architecture. Neural Comput. Appl. **0123456789** (2024). https://doi.org/10.1007/s00521-024-10404-x
26. van Zyl, C., Ye, X., Naidoo, R.: Harnessing explainable artificial intelligence for feature selection in time series energy forecasting: a comparative analysis of Grad-CAM and SHAP. Appl. Energy. **353**, 122079 (2024). https://doi.org/10.1016/j.apenergy.2023.122079

MedCT: A Clinical Terminology Graph for Generative AI Applications in Healthcare

Ye Chen[1,3], Dongdong Huang[1,2], Yuqiang Shen[1], Haoyun Xu[3], Lin Sheng[3], Qingli Zhou[1], and Kai Wang[1,2(✉)]

[1] Department of Respiratory and Critical Care Medicine, The Fourth Affiliated Hospital of School of Medicine, Zhejiang University, Hangzhou, China
{8016009,yuqiangs,zhouql,kaiw}@zju.edu.cn
[2] Zhejiang Key Laboratory of Precision Diagnosis and Treatment for Lung Cancer, Hangzhou, China
[3] Tiger Research, Shanghai, China
{yechen,haoyun.xu,lin.sheng}@tigerbot.com

Abstract. While recent advances in large language models (LLMs) offer great promise in healthcare, the safety-critical nature of the domain requires a thoughtful strategy to mitigate risks of hallucinations and potential harms. We propose a graph of domain knowledge and empirically validated its effectiveness in graph-augmented LLM generation. This presents a promising approach to safe, factual, and in turn more precise LLM applications. We first introduce the world's first clinical terminology for the Chinese healthcare community, namely MedCT, accompanied by a clinical foundation model MedBERT and an entity linking model MedLink. The MedCT system enables standardized representation of clinical data, successively stimulating the development of new medicines, treatment pathways, and better patient outcomes. Moreover, the MedCT knowledge graph provides a principled mechanism to minimize the hallucination problem of LLMs, therefore achieving significant levels of accuracy and safety in LLM-based clinical applications. Our experiments show that the MedCT system achieves state-of-the-art (SOTA) performance in semantic matching and entity linking tasks, not only for Chinese but also for English. We also conducted a longitudinal field experiment by applying MedCT and LLMs in a representative spectrum of clinical tasks, including electronic health record (EHR) autogeneration and medical document search. Our study shows a multitude of values of MedCT for clinical workflows and patient outcomes, especially in the new genre of clinical LLM applications. We present our approach in sufficient engineering detail, such that implementing a clinical terminology for other non-English societies should be readily reproducible. To encourage further research on LLM-based healthcare digitalization and promote the wellbeing of humankind, we are releasing our terminology, models, algorithms, and real-world clinical datasets for development.

Keywords: Large Language Model · LLM · Application · Healthcare · Clinical Terminology · Knowledge Graph

1 Introduction

Standard clinical terminologies, e.g., SNOMED CT, LOINC, ICD, can enable a multitude of values for global healthcare systems. For individual patients and clinicians, terminology or ontology coded electronic health records (EHR) greatly boost the consistency and interoperability of clinical data, and in turn increase the opportunities for real-time decision support for care delivering, retrospective reporting and analytics for research, precision medicine, and management [25].

While clinical terminology, ontology, or knowledge graph has been widely perceived as pivotal for healthcare practice and research, the daunting cost of building and optimization has hindered their wide adoption or effective use. SNOMED CT is considered to be the most comprehensive and widely adopted clinical terminology in the world [4]. It was established in 1965, has undergone more than twenty years development with over one hundred million dollar investment [26]. The remarkable advances of large language models (LLMs) [5] have inspired us to explore rapidly building high-quality and reliable terminologies for the healthcare domain. In particular, we address the problem of developing a clinical terminology for the Chinese clinical domain, namely MedCT. Previous work on coding Chinese clinical terms [9, 13] mainly focused on better semantic matching to existing terminology predominantly in English, but none has developed a clinical terminology truly grounded on underserved languages.

As LLMs have been increasingly applied and deployed to real-world healthcare and clinical settings [29], hallucinations or fabricated information, remains one of the prominent challenges. However, the safety-critical nature of the healthcare domain requires a more deterministic approach to restraining hallucination, therefore motivating us to explore the other side of the AI world. In particular, we augment LLM generation with a model of truth, the MedCT knowledge graph. We further deployed the MedCT terminology to a representative spectrum of real world clinical applications. To summarize, we believe that we have made the following contributions to the global healthcare system in the AI era, and the rest of the paper is logically structured likewise.

1. MedCT: the world's first open Chinese clinical terminology at the scale comparable to SNOMED CT.
2. A suite of models and algorithms for readily adoption of the above terminology, namely, MedBERT, a pretrained foundation model, and MedLink, a fine-tuned entity linking model.
3. A holistic approach with implementation details for rapid and cost-efficient development of clinical terminology for other underserved languages.
4. A wide and representative spectrum of real-world clinical applications utilizing the MedCT system, to demonstrate its value propositions and provide a reference framework of truth-augmented LLM applications in healthcare.
5. Finding and observations from the field with regards to the status quo of applying LLMs in real-world clinical setting, e.g., large or small models, LLM or classical NLP techniques, general or domain-specialized models.

2 Methodology

We bootstrap our development from SNOMED CT, that is considered to be the most comprehensive and widely adopted clinical terminology, therefore inheriting decades of its achievements. We first pretrained a LLM, namely Tigerbot-3 [10], continually from Llama-3.1 [1] to strengthen biomedical base knowledge (with medical training data in Table 2) and multilingual coverage (especially Chinese for our applications). We then applied LLM to contextualize and translate the SNOMED concepts into Chinese, thus forming our initial MedCT terminology. Further, we collaborated with a tertiary care hospital for truth-grounding the terminology, through annotating real-world EHRs with MedCT while revising the terminology for correction and localization.

At the core of the models and algorithms to utilize MedCT is a clinical foundation model, called MedBERT. We pretrained MedBERT from scratch using a thoughtfully curated clinical dataset, and yielded SOTA performance in semantic understanding. Next, with the MedCT annotated clinical data, we trained MedLink, fine-tuned models for clinical terminology named entity recognition (NER) and linking (NEL). After we deployed MedCT in the field, the learning process is iteratively reinforced, for both the MedCT terminology and entity linking models. Our work is inspired by the SNOMED CT entity linking challenge [16], and our method largely follows the model-based winning solution SNOBERT [22]. We managed to push the boundary further in model performance and multilingual coverage, by leveraging LLMs and well-curated

Table 1. Clinical Note Snippets with Concept Annotations

Notes	Concept ID (hier.), name & syn.
… 右肾上腺[1] 结节，建议进一步检查。	29392005 (body) Right adrenal gland 右肾上腺
… 支气管扩张试验：吸入沙丁胺醇400ug…	415299008 (procedure) Reversibility trial by bronchodilator 支气管扩张剂可逆性试验
…附见：两侧胸腔少量积液 伴邻近肺不张。	425802001 (finding) Bilateral pleural effusion 双侧胸腔积液
…神志清，精神一般，胃纳差，睡眠差…	64379006 (finding) Decrease in appetite 没胃口，食欲下降

[1] We color-coded concept hierarchy as: green-body, blue-procedure, and red-finding

real-world clinical data. Table 1 illustrates some examples of clinical notes with annotated MedCT concepts.

2.1 MedBERT: A Clinical Foundation Model

Of the central importance for most clinical NLP tasks is a foundation model to encode broad and basic semantics in the domain. Previous work shows that for domains with a copious amount of unlabeled texts, pretraining language models from scratch yielded substantial gains over continual pretraining from general-domain models [17]. Biomedicine is one of such high-resourced domains. Specifically, we pretrain a BERT model, namely MedBERT, from scratch using a biomedical dataset curated with the following design considerations, with statistics and sources outlined in Table 2.

1. A large corpora of biomedical literature and publications with comprehensive and timely coverage of the domain, e.g., the PubMed Central (PMC) repository [2].
2. Data from the field and directly relevant to downstream tasks, e.g., clinical guidelines [7,8,12,27,34] and real-world clinical notes MIMIC-IV [21].
3. Clinical terminologies and their contexts, e.g., SNOMED CT and MedCT terms and descriptions.
4. Multilingual coverage, i.e., English and Chinese.

Table 2. MedBERT Training Data

Source	Dataset (lang)	Examples	Disk size
Publications	PMC abstracts (en) [2]	24,732,786	26G
	PMC full-texts (en) [2]	3,775,772	109G
	PMC patients (en) [35]	167,034	444M
	PubMedQA contexts (en) [20]	211,269	280M
	Open medical books (en) [33]	13,000	11G
	Chinese literature (zh) [23]	27,704	14G
	Trad. Chinese medicine books (zh) [18]	17	13M
Guidelines	Clinical guidelines (en) [7,8]	11,184	527M
	Clinical guidelines (zh) [6,14]	4,364	643M
Clinical notes	MIMIC-IV v2.2 clinical notes (en) [21]	2,653,148	5.8G
	Chinese EHR and clinical notes (zh)	3,109,181	904M
Terminology	SNOMED and MedCT (en) [25]	723,552	23M
Total	—	**35,429,011**	**168G**

We compared the prediction accuracy of the fill-mask task between our MedBERT and other SOTA biomedical and general-domain models, as the results exhibited in Table 3. First, we verified that domain-specific training has advantages, as biomedical models outperform general-domain BERT models by about twenty percentage points. Second, multilingual expansion is critical. Although the evaluation dataset only has less than 10% Chinese data, the multilingual and Chinese BERT surpass the English-only models by a large margin. Furthermore, the scale and quality of the training data tends to yield better model performance, as seen that BiomedBERT trained with PMC full text wins those with PubMed abstracts only.

Table 3. MedBERT Evaluation

Domain	Model	Accuracy
Biomedical	BiomedBERT-base-fulltext [17]	0.5633
	BiomedBERT-large-abstract	0.5100
	BiomedBERT-base-abstract	0.4209
	SciBERT [3]	0.5819
	MedBERT	**0.8344**
General	BERT-base-multilingual [15]	0.5333
	BERT-base-Chinese	0.5582
	BERT-large	0.3199
	BERT-base	0.3440

2.2 MedLink: Clinical Entity Recognition and Linking

We implemented a two-stage approach to recognizing clinical entities from free-text notes and linking the entities to the built MedCT concepts, as follows.

1. First stage: A NER segmentation task to detect spans of texts as clinical entity mentions.
2. Second stage: A NEL ranking task to predict the MedCT concepts for the recognized entities from the first stage.

For the first stage NER task, we fine-tuned a token classification model from the MedBERT foundation model, as described in Sect. 2.1. We classify each token into four classes: {finding, procedure, body, none}, using the BIO format [28], therefore a token tagging task with seven labels: {O, B-find, I-find, B-proc, I-proc, B-body, I-body}. At the second stage NEL task, we need to link segmented entity mentions to concepts in the MedCT ontology. This is a semantic matching task, which we therefore simply formulate it as a ranking problem in the embedding space. Specifically, we chose SapBERT [24] for English embedding, and its cross-lingual extension SapBERT-all-lang for multilingual and Chinese tasks. We measure the performance of trained models with character-level concept-averaged intersection-over-union (IoU). Table 4 exhibits the experimental results. Our MedLink model achieves SOTA performance in both English and Chinese clinical NER and NEL tasks. We conjecture that a stronger multilingual foundation model MedBERT and copious annotated real-world clinical training data largely contribute into the gain.

Table 4. MedLink Evaluation

Type	Base model	English NEL (IoU on MIMIC)	Chinese NEL (IoU on MedCT)
Biomed	BiomedBERT-base-fulltext [17]	0.4797	0.0091
	BiomedBERT-large-abstract	0.4952	0.0005
	BiomedBERT-base-abstract	0.4976	0.0003
	SciBERT [3]	0.4993	0.0026
	MedBERT	**0.5065**	**0.3012**
General	BERT-base-multilingual [15]	0.4717	0.1006
	BERT-base-Chinese	0.4508	0.1516
	BERT-large	0.4868	0.0007
	BERT-base	0.4774	0.0002

3 Experiments and Applications

3.1 Large or Small Models

In this experiment, we compared the two methodologies in the medical NER and NEL applications: small specialized models versus large general models (LLMs). The experimental results are shown in Table 5. Let us first consider the most generalized LLM approach that does not rely on any specialized data. Under this setting, GPT-4o only yields 0.11 IoU on English data and 0.17 IoU on Chinese data, substantially inferior to our MedLink small model approach, that is 0.51 IoU for English and 0.30 IoU for Chinese. Although the LLM results are visually plausible, its numerical measurement of performance is suboptimal. LLM approach with GPT-4o incurs considerably more inference time than small model approach with MedCT, more than 10 times for English and 30 times for Chinese tests. Moreover, for this medical NER task, open-source model (Llama-3) performs comparably as close model (GPT-4o).

3.2 Retrospective Retrieval of Health Records

In this experiment, we wish to validate and measure the value of the clinical terminology MedCT in the application of health record retrieval. A majority of retrospective retrieval of EHRs involves finding cases with similar or related

Table 5. MedLink vs. LLM approach

Model	English NEL (51 MIMIC notes) IoU	Time	Chinese NEL (1860 MedCT notes) IoU	Time
MedLink	**0.5065**	1 m 40 s	**0.30117**	4 m 15 s
GPT-4o[1]	0.1146	13 m 46 s	0.1739	116 m 46 s
Llama-3.1-70B[2]	0.1116	102 m 58 s	0.1689	661 m 26 s

[1,2] Both GPT-4o and Llama-3.1-70B model downloading were executed as of this writing in January, 2025.

diseases, in reference past evidences in testing, diagnosis, treatment and outcome. Therefore, from the `MedCT-clinical-notes` dataset we curated in-house, we took a corpus of discharge summaries for this retrieval experiment. The corpus contains 13,863 examples or discharge notes, entered from all departments during the first quarter of year 2024 in a tertiary care hospital. The data was organized into relevant textual fields including demographics, admission, treatment pathway, discharge summary and instruction. We interviewed a panel of 12 senior physicians to collected a set of 20 queries representative of real-world clinical practice and research. The clinical query set was chosen with non-trivial complexity such that a straightforward keyword match conceivably cannot yield satisfactory results. One example is "post-stroke with pneumonia".

We implemented two retrieval strategies, the classic sparse or dense retrieval and the MedCT-augmented retrieval. For the MedCT-augmented retrieval approach, we first offline tagged each document with MedCT concepts, and then indexed the list of concept ids along with texts per document. These annotated concepts should capture almost all relevant clinical information in the health records. At online retrieval time, we annotated full text queries with MedCT concepts, and then ranked documents with a hybrid strategy based on both text-based sparse or dense retrieval and strict concept id matching. For evaluation, we asked the same panel clinicians to annotate relevant examples from a random sample of 2K discharge notes, for each of the 20 queries. This ground truth allows us to measure precision, recall and the balanced F_1 score as the performance metrics for our retrieval task, as reported in Table 6. Our experiments show that retrieval augmented with MedCT graph substantially outperforms modern text-based search. In particular, MedCT boosts the search recall by a 15% lift over sparse retrieval.

3.3 Health Records Auto-Generation by LLMs

Next we consider the task of health records generation. Nearly half of physician's time is devoted to digital paperwork, rather than direct patient care [30]. The statistics is even worse in regions and countries with shortfall of health workforce. In our field study at a tertiary care hospital in China, reportedly near 90% of residency doctors' time is absorbed in writing clinical notes and

Table 6. EHR retrieval augmented with MedCT

Retrieval method	Precision[1]	Recall[2]	F_1-score[3]
Sparse	0.5294	0.5015	0.5151
Dense	0.0706	0.0995	0.0826
Hybird	0.3882	0.2527	0.3061
MedCT-augmented	**0.6235**	**0.5745**	**0.5980**

[1,2,3] All metrics are measured at top 10 retrieved results, representative of a typical search scenario.

medical records. Among various health records, discharge summary is arguably the most important document a hospitalist writes. The discharge summary is a semi-structured narrative document for communicating clinical information about patients. However, nowadays hospitalists have little time to write good quality discharge summaries, along with often delay to deliver to downstream physicians, causing disruption in the continuity of care and risking poor patient outcomes.

Therefore it is appealing to apply LLMs to generate health record drafts for physicians, to review, mildly edit and submit. This is a text summarization task with moderate complexity yet high practical significance [11]. A good model, likely with domain or task-specific training, would both speed up clinical documentation and improve the quality of health records. In our deployment to a tertiary care hospital in Zhejiang, China, hospitalists reported about 40% reduction in time spent in writing discharge summaries with the help of LLM generation, while observing improvement in both quality and information density.

However, general-purpose LLMs, if used as-is in a vanilla fashion, typically cannot meet the safety requirements of the medical domain [31]. In our controlled experiments, for instance, we observed that vanilla LLMs hallucinated medical misinformation such as made-up procedures (e.g., Laparoscopy for radical pulmonary surgery) and medications (e.g., Cefuroxime) in discharge summary generation. This represents a significant risk of applying LLMs in an ignorant way in mission-critical domains like healthcare. Many of these hallucinations were trivial to identified by qualified physicians, which also symbolizes the large gap between human intelligence and LLMs, especially in domain knowledge.

In order to address the hallucination problem intrinsic to LLMs, we guide the LLM generation with a knowledge graph as source of truth. We believe that our approach brings together the strengths of both worlds, LLMs and specialized small models; and moreover presents a systematic and measurable way to minimize hallucination. As illustrated in the detailed prompts in Fig. 1, the MedCT-guided generation instructs the LLM to attend to major clinical concepts such as "chief complaint" and "physical examination" and therefore should capture key clinical information more comprehensively and accurately.

To evaluate the generation results, we recruited a panel of nine hospitalists to review summary generations from the above two methods, along with the human summary by doctors, in a blind fashion, and then cast Likert-scale to each testing example. Evaluating text summarization models is nontrivial, especially with automatic metrics. In general domains, such as TL;DR Reddits and CNN/DM news article summarization, previous works have used ROUGE or reward models to predict human preference [32]. But these metrics are only rough proxies to real human perceived summary quality, and should not apply indiscriminately to different domains or even different tasks. For example, in book and news article summarization, coherence is often used to measure how easy the summary is to read on its own. But for the task of health record summarization in the clinical domain, with the time pressure and norm use of medical abbreviation and terminology, conciseness and clarity weigh more than coherence. After three sessions of panel discussions, we developed a set of metrics for our health record summarization task. The metrics cover both general language quality and clinical significance, from perspectives of accuracy, completeness, clarity, relevance, conciseness, and clinical depth. Moreover, our evaluation metrics, along with the annotated preference dataset, shall be instrumental to develop automatic metrics for text summarization in the clinical domain.

\# Zero-shot prompt for vanilla LLM approach:
{input context: hospital course, discharge diagnosis}

The above is a detailed hospital course and discharge diagnosis from a medical record. Please summarize it into an accurate and concise medical summary. The summary should include the reason for admission, basis for diagnosis, main treatment measures and their effects, changes in condition, and status at discharge.

Medical summary:

\# Zero-shot prompt for MedCT-guided LLM approach:
{input context: hospital course, discharge diagnosis}

The above is a detailed hospital course and discharge diagnosis from a medical record. Please summarize it into an accurate and concise medical summary. The summary should include the reason for admission, basis for diagnosis, main treatment measures and their effects, changes in condition, and status at discharge.

Also, the above medical records include the following entities, please include these medical entities in the medical course summary.
{entity mentions from the input}

Medical summary:

Fig. 1. Prompts for discharge summary auto-generation

Our evaluation dataset contains 91 examples of discharge notes, with detailed hospital courses and discharge diagnosis as raw input (denoted as raw), along with discharge summaries written by human hospitalists (denoted as human). We then infer the underlying LLM with two prompting approaches as in Fig. 1 (denoted as LLM and MedCT, respectively). The average length in character is 3,545 for the input clinical notes, 542 for the human summary, 274 for the vanilla LLM, and 317 for the MedCT method. LLMs tend to condense more than human, while the MedCT-augmented method conveys richer information than simple LLM prompting. This is as expected, since we instruct the LLM to preserve information regarding clinical entities. For a rapid computerizable evaluation, we compute cosine similarities, in the embedding space projected by MedBERT, between raw input, human and machine generations. With similar compression rates, cosine similarity is a reasonable proxy to how well a summary captures the original text's main points. As shown in Table 7, our MedCT-augmented generation achieved best cosine scores, notably even higher than human summaries. Furthermore once again, for tasks that require deep domain knowledge such as clinical notes summarization, proprietary and open-source models empirically yielded comparable performance.

Table 7. EHR summarization cosine similarity

cosine	Raw			Human	
	Human	LLM	MedCT	LLM	MedCT
GPT-4o	0.8940	0.8984	**0.9288**	0.9242	**0.9257**
Llama-3.1-70B	0.8940	0.8897	**0.9066**	**0.9163**	0.9150

While the programmable cosine similarity gives a rapid proxy to summarization quality, especially for model iteration and comparison, the gold standard is still human review. We distributed the 91 testing examples to nine hospitalists with tenure from ten years or above. For each example, we gave original input clinical notes, and three summary generations, from human doctors, simple LLM, and LLM augmented with MedCT graph (denoted as human, LLM and MedCT, respectively). More importantly, the review was conducted in a blind manner. The three generations were randomly shuffled per example (only organizers held the true orders), and hence reviewers did not know which model or human generated the summary to be scored. The physician reviewers were instructed to rate summaries using 5-point Likert-scale by real-world clinical standards in reference to the above six-dimension metrics. One of the entries is factually human summary by real doctor anyway. The results are shown in Table 8.

Overall, our graph-guided LLM approach achieves highest human ratings, winning five out of six review dimensions. Notably, the gains from the perspectives of "clinical depth" and "relevance" are particularly substantial, over both LLM and human generations. By additionally prompting LLM with MedCT-recognized clinical entities, the underlying models exhibited precise attentions

Table 8. EHR summarization human review scores

	Human	LLM	MedCT
Accuracy	4.42	4.41	**4.42**
Completeness	3.99	4.19	**4.20**
Clarity	4.44	**4.45**	4.40
Relevance	4.56	4.77	**4.84**
Conciseness	4.23	4.27	**4.33**
Cinical depth	3.95	4.07	**4.19**
Overall	25.58	26.15	**26.36**

to clinical concepts in the free-text notes. As illustrated in the snippets in Fig. 2, the MedCT-augmented approach captured the finding of "pulmonary infection" in hospital course summary, while simple LLM and human missed this concept. From the input notes of this patient, "pulmonary infection" is one of the complications through most of her or his hospital course. Both machine generations yielded comparable or better summarization with respect to factual human summaries. The win-or-tie rate of LLM approach is 0.59 over human, and even higher augmented by MedCT with a good odd of 0.68 being at least as good as human.

Input clinical notes:
Chief complaint: The patient's temperature was the highest yesterday at 38řC,...
Diagnosis: 1. Open craniocerebral injury,...3. Contusion of both lungs, bilateral pleural effusion with incomplete expansion of both lower lungs, and pulmonary infection ...

Human summary:
After admission, he was given mannitol to lower intracranial pressure,... insulin combined with metformin to lower blood sugar, and parenteral nutrition support.

LLM summary:
In terms of changes in condition, although the patient suffered from various complications, after careful treatment by the doctor, the patient gradually regained consciousness and .... Finally, the patient was discharged from the hospital on January 28, 2024, and planned to continue rehabilitation training after discharge.

LLM+MedCT summary:
After treatment, the patient's condition gradually stabilized, but there were still some problems, such as repeated fever, low level of consciousness, and lung infection . Finally, the patient was discharged from the hospital on January 28, 2024, and planned to undergo rehabilitation training after discharge.

Fig. 2. A snippet from clinical notes summarization

This finding confirms with previous computerized evaluation of cosine similarity, suggesting that LLMs with domain specialization indeed can help clinical workflow tasks such as health record auto-generation.

4 Conclusions

Healthcare arguably remains one of the most prosperous domains beneficial from the rapid development of artificial intelligence (AI) in general and large language models (LLMs) in particular.

> "We're going to have a family doctor who's seen a hundred million patients and they're going to be a much better family doctor." [19]
>
> *Geoffrey Hinton*

However, even the state-of-the-art general foundation models, e.g., GPT-4o and Llama-3.1, merely scratch the surface encoding deep domain knowledge from high-resourced yet largely private domains such as healthcare. Moreover, the probabilistic root of LLMs, hence the tendency of hallucination, hinders their wide practical adoption in privacy and safety critical tasks. In the context of leveraging LLMs' extraordinary capabilities of semantic understanding, generativeness and interactiveness, while ensuring their safety, unbiasedness, and honesty in real-world applications, we developed and released MedCT. To the best of our knowledge, MedCT is the world's first clinical terminology built for and grounded from non-English community, specifically Chinese. We presented our comprehensive approach to building the clinical terminology knowledge graph, truth grounding and optimizing from real clinical data, training models for named entity recognition and linking to the graph. Our approach leverages LLMs as an integral part of development tools, along with abundant real-world clinical data and annotations by experienced physicians. Consequently, our MedCT models achieve new state-of-the-art in medical NER and NEL tasks (w.r.t. BiomedBERT and SciBERT etc.), especially in a rapid and cost-efficient manner (w.r.t. SNOMED CT etc.).

The values of our MedCT clinical terminology are even more pronounced in complementing LLM applications in clinical setting. A knowledge graph such as MedCT not only injects clinical domain commonsense into foundation models, but also as a source of truth gouges model generations to be more safe, truthful and reliable. We deployed MedCT in a wide variety of clinical applications, including clinical information retrieval and document summarization. Our findings from human blind reviews are inspiring, in that MedCT-augmented LLMs can achieve human-like or even better results in various tasks in clinical workflow and research tasks. Meanwhile, we also found that general-purpose LLM is no silver bullet, especially for domains with knowledge depth. As our approach stands, practical yet mission-critical applications of LLM still require domain specialization, for example in conjunction with classical yet surgical machine learning techniques.

We believe that we are at the dawn of unleashing the values of AI and LLMs for great humanity. In the hope of facilitating further development in the healthcare domain, we open-source release the MedCT suite of models and datasets[1], which include:

1. The MedCT bilingual (English and Chinese) clinical terminology dictionary, with 223K medical concepts.
2. The MedCT named entity recognition (NER) model: MedLink.
3. A biomedical foundation model: MedBERT, that achieves state-of-the-art performance in a variety of downstream tasks, e.g., clinical NER/NEL, search, and summarization.
4. Our MedCT-clinical-notes dataset, including:
    - For the NER and NEL tasks, 7.4K real-world clinical notes in Chinese, and 61K entity mention annotations per MedCT graph.
    - For the search task, 20 clinical queries, and 2K discharge notes with relevance annotations.
    - For the clinical note summarization task, 91 raw discharge notes and summaries by human, LLM and MedCT-augmented generations, along with preference Likert-scale annotated by human physicians.

Acknowledgements. This work was supported by the Natural Science Foundation of China Key Program (U23A20467) and the Zhejiang Provincial Science and Technology Program (Grant No. 2024C03270) under the Pioneer and Leading Goose Research and Development Program.

References

1. AI@Meta: Llama 3.1 model card (2024). https://huggingface.co/meta-llama/Llama-3.1-70B-Instruct
2. Beck, J.: Report from the field: Pubmed central, an xml-based archive of life sciences journal articles. In: International Symposium on XML for the Long Haul: Issues in the Long-term Preservation of XML (2010)
3. Beltagy, I., Lo, K., Cohan, A.: SciBERT: a pretrained language model for scientific text. arXiv preprint arXiv:1903.10676 (2019)
4. Benson, T.: Principles of Health Interoperability HL7 and SNOMED. Springer, London (2010). https://doi.org/10.1007/978-1-84882-803-2
5. Brown, T.B., et al.: Language models are few-shot learners. arXiv preprint arXiv:2005.14165v4 (2020)

---

[1] For a detailed description and ongoing releases, please see MedCT repository: Github: https://github.com/TigerResearch/MedCT; https://huggingface.co/collections/TigerResearch/medct-6744641d6f19b9d70a56f848.

6. Cai, Q., Chai, R., Chen, G.: Clinical practice guidelines for intratympanic drug delivery. Chin. J. Otorhinolaryngol. Skull Base Surg. **30**(1) (2024)
7. Cancer Care Ontario: Cancer care ontario (CCO) guidelines & advice (2024). https://www.cancercareontario.ca/en/guidelines-advice
8. Centers for Disease Control and Prevention: Centers for disease control and prevention: guidelines and recommendations (2024). https://www.guidelinecentral.com/guidelines/CDC/
9. Chen, Y., Hu, D., Li, M., Duan, H., Lu, X.: Automatic SNOMED CT coding of Chinese clinical terms via attention-based semantic matching. Int. J. Med. Inform. **159** (2022)
10. Chen, Y.: Tigerbot 3 model card (2024). https://huggingface.co/TigerResearch/tigerbot-70b-chat-v6
11. Chen, Y., Couto, I., Cai, W., Fu, C., Dorneles, B.: SoftTiger: a clinical foundation model for healthcare workflows. In: AAAI 2024 Spring Symposium on Clinical Foundation Models (2024). https://arxiv.org/pdf/2403.00868.pdf
12. Cochrane Library: Cochrane database of systematic reviews (2024). https://www.cochranelibrary.com/about/about-cochrane-reviews
13. Dai, R., Zhang, X., Li, F., Li, C.: Research on normalization of Chinese clinical terms based on keyword extraction and data augmentation technology. In: Proceedings of the 2023 4th International Symposium on Artificial Intelligence for Medicine Science, ISAIMS 2023, pp. 1291–1298. ACM (2023)
14. Dai, Y., Li, G.: Oncology pharmacy clinic standards (trial). Chin. Pharm. J. **56**(9) (2021)
15. Devlin, J., Chang, M.W., Lee, K., Toutanova, K.: Bert: pre-training of deep bidirectional transformers for language understanding. arXiv preprint arXiv:1810.04805 (2018)
16. Driven Data Inc.: SNOMED CT entity linking challenge (2024). https://www.drivendata.org/competitions/258/competition-snomed-ct/
17. Gu, Y., et al.: Domain-specific language model pretraining for biomedical natural language processing (2020). https://doi.org/10.1145/3458754, https://arxiv.org/pdf/2007.15779.pdf
18. Guo, Z.: Explanation of Huangdi Neijing (1988)
19. Hinton, G.: Large language models in medicine. They understand and have empathy (2023). https://erictopol.substack.com/p/geoffrey-hinton-large-language-models
20. Jin, Q., Dhingra, B., Liu, Z., Cohen, W., Lu, X.: PubMedQA: a dataset for biomedical research question answering. In: Proceedings of the 2019 Conference on Empirical Methods in Natural Language Processing and the 9th International Joint Conference on Natural Language Processing (EMNLP-IJCNLP) (2019)
21. Johnson, A.E.W., et al.: MIMIC-IV, a freely accessible electronic health record dataset. Sci. Data **10**(1) (2023)
22. Kulyabin, M., Sokolov, G., Galaida, A., Maier, A., Arias-Vergara, T.: SNOBERT: a benchmark for clinical notes entity linking in the SNOMED CT clinical terminology. arXiv preprint arXiv:2405.16115 (2024)
23. Library, Z.U.: Chinese medical journal full text database (2024). http://www.yiigle.com/
24. Liu, F., Shareghi, E., Meng, Z., Basaldella, M., Collier, N.: Self-alignment pretraining for biomedical entity representations. In: Proceedings of the 2021 Conference of the North American Chapter of the Association for Computational Linguistics: Human Language Technologies, pp. 4228–4238 (2021). https://arxiv.org/pdf/2010.11784.pdf

25. Liu, K., Hogan, W.R., Crowley, R.S.: Natural language processing methods and systems for biomedical ontology learning. J. Biomed. Inform. **44**(1), 163–179 (2011). https://doi.org/10.1016/j.jbi.2010.07.006
26. National Library of Medicine: United States National Library of Medicine. SNOMED Clinical Terms® To Be Added To UMLS® Metathesaurus® (2003). http://www.nlm.nih.gov/research/umls/Snomed/snomed_announcement.html
27. NICE: National Institute for Health and Care Excellence: NICE guidelines (2024). https://www.nice.org.uk/about/what-we-do/our-programmes/nice-guidance/nice-guidelines
28. Ramshaw, L., Marcus, M.: Text chunking using transformation-based learning. In: Third Workshop on Very Large Corpora (1995). https://aclanthology.org/W95-0107
29. Singhal, K., et al.: Large language models encode clinical knowledge. Nature **620**(7972), 172–180 (2023). https://doi.org/10.1038/s41586-023-06291-2
30. Sinsky, C., et al.: Allocation of physician time in ambulatory practice: a time and motion study in 4 specialties. Ann. Intern. Med. (2016)
31. Stanceski, K., et al.: The quality and safety of using generative AI to produce patient-centred discharge instructions. NPJ Digital Med. **7**(1) (2024). https://doi.org/10.1038/s41746-024-01336-w
32. Stiennon, N., et al.: Learning to summarize from human feedback. arXiv preprint arXiv:2009.01325 (2020)
33. Walther, D.S.: Applied Kinesiology Synopsis, 2nd edn. (1988)
34. World Health Organization: WHO guidelines (2024). https://www.who.int/publications/who-guidelines
35. Zhao, Z., Jin, Q., Chen, F., Peng, T., Yu, S.: PMC-patients: a large-scale dataset of patient summaries and relations for benchmarking retrieval-based clinical decision support systems. Sci Data **10**, 909 (2023). https://doi.org/10.1038/s41597-023-02814-8

A Fractional Computation Based Deep Learning Framework for Silicosis Detection

Rinki Sharma[✉][iD] and Priyanka Harjule[iD]

Malaviya National Institute of Technology, Jaipur, India
rinkysharma0357@gmail.com , priyanka.maths@mnit.ac.in

Abstract. This study presents a new fractional computational approach applied to a new dataset, silicosis. This is a scalable and flexible approach for training neural networks using fractional computation, which conveniently use the conformable fractional derivative. During the training process, the method includes an independent variable, α, which provides additional degree to the framework. Fractional variants of the sigmoid and relu activation functions are explored and compared to conventional activation functions. This method builds on earlier approaches by employing the conformable fractional derivative. The fractional activation functions notably converge to the actions of their standard version when $\alpha = 1$, guaranteeing a smooth integration with conventional neural network models. The study also tackles the problem of managing both positive and negative inputs, which is a crucial prerequisite for the derivative but has been mainly disregarded in earlier studies, underscoring the originality of the current work. The experimental framework incorporates both feedforward neural network and convolutional neural network using fractional activation functions. The findings indicate that the suggested framework performs better and is more accurate for particular values of α. The efficiency of the suggested computational approach is demonstrated by showing that fractional activation on Convolutional Neural Network when paired with transfer learning, performs better for silicosis chest X-ray classification than conventional transfer learning models.

Keywords: Convolutional Neural Network · Activation · Fractional computation · Silicosis Data

1 Introduction

Silicosis is an irreversible and potentially fatal lung disease, which is, entirely preventable. It results from exposure to respirable crystalline silica. Many workers worldwide are at risk of contracting this illness across a variety of industries. Pneumoconiosis refers to a diverse group of occupational interstitial lung diseases resulting from the prolonged inhalation and accumulation of mineral dust in the lungs. Early detection of this is crucial for detect the disease in its prior stages, enabling timely interventions to improve outcomes for affected workers while also facilitating the evaluation of shortcomings in workplace safety protocols [18]. This ailment is very common among Indian workers, particularly in Rajasthan, and early finding is the only way to treat it but there are very few studies for computer aided detection of this [16]. A similar occupational disease, Pneumoconiosis, was also spread in China. A detection model for this condition has been developed using a deep convolutional diagnostic approach [8]. In state of the art the transfer learning models are very helpful for classification of lungs X- rays [20]. Although state-of-the-art research has explored models for silicosis detection, the available datasets often suffer from imbalances and may lack proper authorization [19]. In contrast, this study utilizes a properly labeled, balanced, and authorized dataset to develop a custom model specifically for silicosis medical X- ray images classification. The proposed approach achieves an accuracy of 0.8276, demonstrating superior performance compared to existing methods in the literature.

Activation functions are pivotal in determining the stability of neural networks and significantly impact their performance in modeling and interpreting physical phenomena. They assess the relevance of neuron inputs and decide whether a neuron should be activated through mathematical computations [10]. The study of fractional activation functions for neural networks is a rapidly growing area of research. By extending common functions like sine, cosine, and the logistic function based on the Mittag-Leffler function concept, Ivanov presented a novel method for creating fractional activation functions [12]. The impact of these fractional activation functions on neural network learning and prediction accuracy was assessed through experiments. Optimal parameter settings which occcurs by random search $\alpha \approx 0$ and $\beta = 1$ improved accuracy, occasionally achieving 100% in over epochs.

Recently, the fractional approach in optimization and training of neural networks gained more highlight [9,11,21]. Fractional Adaptive Linear Units, a new generalization of adaptive activation functions, are introduced and the method expands on earlier effective activation function research [26]. Common activation functions can be categorized into families using fractional calculus, which enables the creation of sets using a fractional derivative with a new parameter α denoting the derivative order [27]. The fractional adaption in step and multiquadratic functions make it possible to select one and generate the other by calculating its fractional derivative. Expanding on this idea, a thorough explanation of the three primary activation function families in this way is given [6]. A new definition of the fractional derivative is proposed and considered a natural extension of the conventional derivative and satisfies the arithmetic properties of the classical derivative and also aligns with the well-known fractional derivatives of polynomials [14].

An improved conformable fractional derivative is presented and a sort of historical memory parameter is added to the formulation of this improved conformable fractional derivative, which is also local by definition [7]. The contribution to physics and its physical interpretations validate the conformable fractional derivative [1,22,24].

A family of fractional activation functions using improved fractional derivative and its impact with different $\alpha \in (0,1]$ is represented on some datasets [15]. By using the ideal conformable fractional derivative to create fractional activation functions, this work expands on the earlier approach. This method's ability to guarantee that the fractional activation function acts similarly to the conventional activation function when $\alpha = 1$. This approach helps to compare $\alpha \in (0,1]$ in a single framework. In the experiments it will also make sure that it can handle positive and negative inputs, which is required by the derivative's definition and hasn't been covered in earlier research, underscoring the novelty of this work.

The Highlights of the proposed work are given below:

1. A custom transfer learning model for classification of silicosis dataset has been given.
2. Analysis of effect of fractional values on the computation of learning paradigm of Neural Network.
3. The fractional activation definition is enhanced from other studies as it generalizes with standard one.

The structure of the article is as follows: Section 2 provides mathematical preliminaries about conformable fractional derivative and their fundamental properties. Fractional activation functions viz. fractional sigmoid and fractional relu function with the effect of variable $\alpha \in (0,1]$ is presented in Sect. 3. The experimental work with these proposed functions on wine and silicosis datasets is shown in Sect. 4. Finally, the conclusion is given in Sect. 5.

2 Conformable Fractional Derivative

In 2014, Khalil et al. [14] introduced the conformable fractional derivative, a type of local fractional derivative that retains many properties of classical derivatives. This formulation not only exhibits greater similarity to classical derivatives but also preserves the same arithmetic operations and other properties like Rolle's theorem and Mean value theorem providing a more consistent extension to fractional calculus.

Definition 1: The "conformable fractional derivative" of order α of function Given a function $f : [0, \infty) \to \mathbf{R}$. is defined by

$$D^\alpha(f)(t) = \lim_{\epsilon \to 0} \frac{f(t + \epsilon t^{1-\alpha}) - f(t)}{\epsilon} \quad (1)$$

for all $t > 0$, $\alpha \in (0, 1]$. If f is α-differentiable in some $(0, a)$, $a > 0$, and

$$\lim_{t \to 0^+} f^\alpha(t) \text{ exists, then define } f^\alpha(0) = \lim_{t \to 0^+} f^\alpha(t).$$

here $f^\alpha(t) = D^\alpha f(t)$

Here are theorems related to conformable fractional derivatives, which provide a foundation for the generalization and formal establishment of this derivative concept [28].

Theorem 1 Let $\alpha \in (0,1]$ and f, g be α-differentiable at a point $t > 0$. Then

1. $D^\alpha(af + bg) = aD^\alpha(f) + bD^\alpha(g)$, for all $a, b \in \mathbb{R}$.
2. $D^\alpha(t^p) = pt^{p-\alpha}$ for all $p \in \mathbb{R}$.
3. $D^\alpha(\lambda) = 0$, for all constant functions $f(t) = \lambda$.
4. $D^\alpha(fg) = f(t)D^\alpha(g) + g(t)D^\alpha(f)$.
5. $D^\alpha\left(\frac{f}{g}\right) = \frac{g(t)D^\alpha(f) - f(t)D^\alpha(g)}{g^2}$.
6. If, in addition, f is differentiable, then $D^\alpha(f)(t) = t^{1-\alpha}\frac{df(t)}{dt}$.

Theorem 2 If a function $f : [0, \infty) \to \mathbb{R}$ is α-differentiable at $t_0 > 0$, where $\alpha \in (0, 1]$, then f is continuous at t_0.

3 Methodological Description for Fractional Activation

Fractional activation functions generalize conventional activation functions by introducing fractional exponents in the exponential term. This section presents the mathematical formulation of these functions and analyze their behavior across various fractional $\alpha \in (0, 1]$. To integrate conformable fractional derivatives into the activation functions of neural networks, a fractional generalization of the exponential term is employed which offers an additional degree of freedom that enhances the adaptability of the activation function during network training.

3.1 Fractional Sigmoid Function

The standard sigmoid function is given by $\sigma(x) = \frac{1}{1+e^{-x}}$. The fractional effect is introduced by incorporating the fractional exponent term with $D^\alpha e^{-x}$ in place of e^{-x}. The fractional exponent term increases generalization, adaptability, and memory efficiency in the activation function.

Therefore, the fractional sigmoid function is defined as

$$F^\alpha_{sig}(x) = \frac{1}{1 + D^\alpha e^{-x}} = \frac{1}{1 - e^{-x} x^{1-\alpha}} \tag{2}$$

To guarantee the generalization with $\alpha = 1$ and, to avoid the undefined input for layers the fractional sigmoid activation is modified as in Eq. (3):

$$F^\alpha_{sig}(x) = \frac{1}{1 + e^{-x}|x|^{1-\alpha}} \tag{3}$$

Now, it is well-defined for all x and for $\alpha \in (0, 1]$. And for $\alpha = 1$ it converges to the standard sigmoid function.

Derivative of $F^\alpha_{sig}(x)$ to be used in backpropogation is defined in Eq. (4):

$$D^\alpha F^\alpha_{sig}(x) = F^\alpha_{sig}(x)(1 - F^\alpha_{sig}(x)) \tag{4}$$

The above derivative is well-defined and retains the properties of the original sigmoid derivative, this approach eliminates the complexity introduced by $|x|^{1-\alpha}$ in

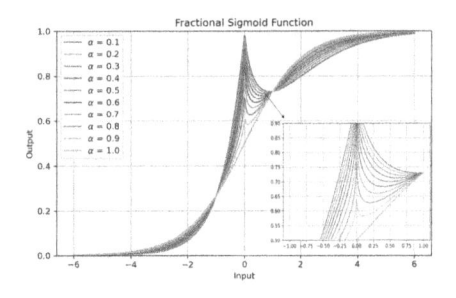

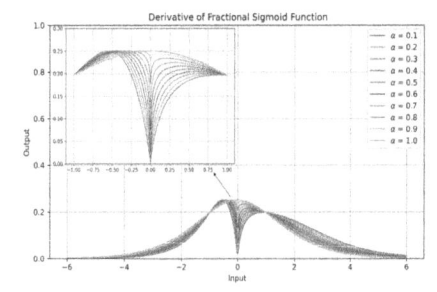

Fig. 1. Fractional sigmoid function and its derivative geometrical shape for $\alpha \in (0, 1]$

the derivative computation. Additionally, this derivative enhances stability during backpropagation and provides a gradient that facilitates learning while leveraging the benefits of fractional calculus.

The geometrical representation of $F_{sig}^{\alpha}(x)$ and its derivative is shown in Fig. 1. For each value of α the curve is different and the curvature of the curve varies for different α values. In the conventional case, as the sigmoid function approaches its extreme values (0 or 1), its gradient vanishes. This problem is mitigated by introducing the fractional term in sigmoid function. By doing so, because of different fractional values of α function reaches to saturation slowly as compared to the standard sigmoid function and as a result the gradient does not vanish easily [5]. As seen in Fig. 1, the higher values of α near 1 provide a smoother transition as compared to its lower values, which have sharper slopes around 0, this accelerates the training even when neurons receive smaller inputs. Steeper gradients accelerate learning by enabling larger updates to the parameters during backpropagation, reducing the number of iterations required for convergence. However, excessively steep gradients can lead to instability or divergence, necessitating careful selection of activation functions and learning rates [25]. The fractional sigmoid allows fine-tuning of gradient behavior, making it beneficial for deep networks where standard sigmoid suffers from saturation using parameter α. The $F_{sig}^{\alpha}(x)$ allows the variable order alpha to have more adaptability as α near 0 ($0.1 - 0.3$) can accelerate the training due to its steep gradients and whenever the training is unstable getting large updates ($\alpha = 0.8 - 1.0$) can smoothen the gradients while other α values (0.4–0.7) provides moderate gradients generally.

3.2 Fractional Relu Activation

The relu activation is very popular in the training of neural networks because of its simplicity and faster computation. It provides nonlinearity in the model despite having the simplest form among the activations. Since relu does not have exponential terms, the derivative of relu in backpropagation will be calculated as a fractional conformable derivative to introduce the fractional effect in the training of the network. The standard relu activation function has a constant gradient for positive inputs, which limits its adaptability in learning [4]. As seen in Fig. 2, the fractional derivative of relu, provides the dynamic gradients rather than flattening, allowing it to adapt smoothly to both large and small input values. This enhances gradient flow, improving the network's ability to learn more effectively across different scales of input. The different values of α will provide an additional degree of adaptability and a better generalization to the neural network. Fractional relu function is symbolically defined in Eq. (5), which is similar to the standard one for all alpha values. The effect of fractional parameter α is observed in its derivative given in Eq. (6). The geometrical shape of this fractional relu and fractional relu derivative is provided in Fig. 2.

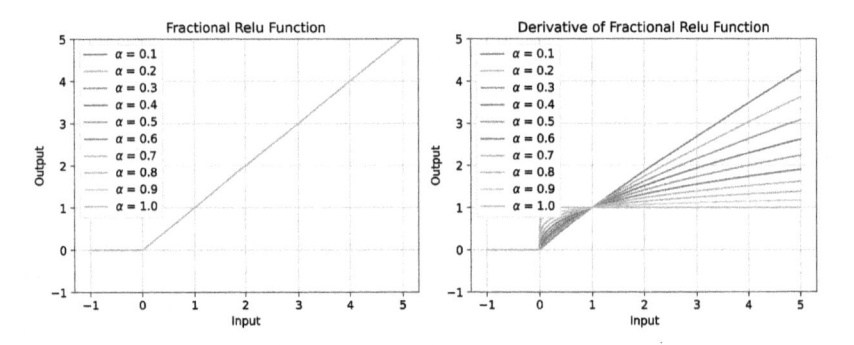

Fig. 2. Fractional relu and its conformable fractional derivative and its geometrical behaviour for $\alpha \in (0, 1]$

$$F_{relu}^{\alpha}(x) = \begin{cases} x, x \geq 0 \\ 0, x < 0 \end{cases} \tag{5}$$

And the conformable fractional derivative of fractional relu is given as:

$$D^{\alpha} F_{relu}^{\alpha} = \begin{cases} x^{1-\alpha}, x \geq 0 \\ 0, x < 0 \end{cases} \tag{6}$$

Figure 2 illustrate that derivative of fractional relu has different properties for different α values. The small α values provide higher slopes for larger inputs. The large α values near to 1 has flattening curve and slower growing derivative. This illustrates that, in learning the fine tuning of α can counter the problem of exploding and vanishing gradient in deep networks, as the term $x^{1-\alpha}$ in Eq. (6) provides a significant output for both smaller and larger inputs. Thereby making it a self adaptive activation function. The derivative of the activation has impact on the gradient flow during training yielding better training results for different α values in experiments.

4 Results and Discussion

The proposed fractional activation functions are integrated in feedforfard neural network and convolutional neural network for experimental analysis. Two datasets have been used, one is a simple wine quality classification dataset and the second is the curated real dataset of X-ray images of silicosis disease. The results were compared through training and testing accuracies achieved for different α values. The details of the data set are given in Table 1.

Table 1. Details of Datasets

Dataset	Type	Classes	Samples
Wine	Numeric	3	178
Silicosis	Image	2	421

4.1 Experiment with Wine Dataset

The wine dataset is a widely used dataset in machine learning, particularly for classification. It contains 178 samples with 13 features, representing the chemical properties of wines prevailing from three distinct cultivars. The data set is categorized into three classes, each representing a specific type of wine. Here, the classification of the wine dataset has been investigated using a neural network incorporating fractional sigmoid and fractional relu activation functions. The neural network architecture consists of two hidden layers, each containing 32 neurons, and remains consistent across training with both fractional activation functions. The learning rate, determined through random search, is set to 0.002, with 16 batch size, 200 epochs were taken for training, Adam optimizer was used for weight updation and loss minimization.

4.1.1 Wine Classification Using Fractional Sigmoid
In this experiment, fractional sigmoid $F_{sig}^{\alpha}(x)$ was applied as activation to train the network. The results of achieved train and test accuracy are given in Fig. 3. Figure 3 illustrates that for $\alpha = 0.3$ and $\alpha = 0.6$, the training accuracy attains highest value. However, $\alpha = 0.3$ also yields the highest testing accuracy among all alpha. Both training and testing accuracies reach 1.0, which can be genuine because dataset is very small and simple and the proposed model can learn the exact patterns of the data instead of general patterns. The dataset exhibits a simple structure and is almost linearly separable, making the results realistic and not an unusual occurrence. The reason for $\alpha = 0.3$ can be justified by the behaviour of fractional sigmoid which shows that, for small $\alpha < 1$ values, the training is faster because of sharper gradients. $\alpha = 0.1, 0.2$ exhibit steep gradients also, but they may lead to unstable gradients in an early phase. In contrast, $\alpha = 0.3$ provides a balance, ensuring both stable training and sufficiently steep gradients.

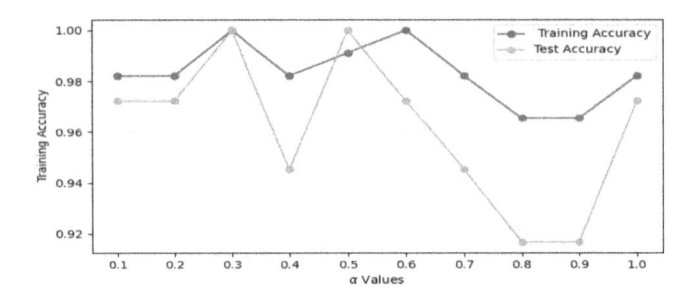

Fig. 3. Training and testing accuracies achieved for different α values with fractional sigmoid on wine dataset

For, $\alpha = 0.5$ results indicates some unusual conduct, with testing accuracy higher than training accuracy. However, this can be regarded as insignificant by considering the test dataset's small size and simple patterns [2]. Notably, $\alpha = 0.3$ and $\alpha = 0.6$ outperform $\alpha = 1$, which shows the standard training approach with the same parameters. This observation shows that the fractional effect on activation enhances generalization on unseen data as compared to the standard activation.

4.1.2 Wine Classification Using Fractional Relu In this experiment, the Fractional relu F^{α}_{relu} activation function and its conformable fractional derivative is used for classification of the wine dataset. All network parameters remain the same as the previous experiment with fractional sigmoid. The results for α values with achieved training and testing accuracies provided in Fig. 4. For $\alpha = 0.2$, the training and testing accuracies both reach 1.0, which is the best result among all α.

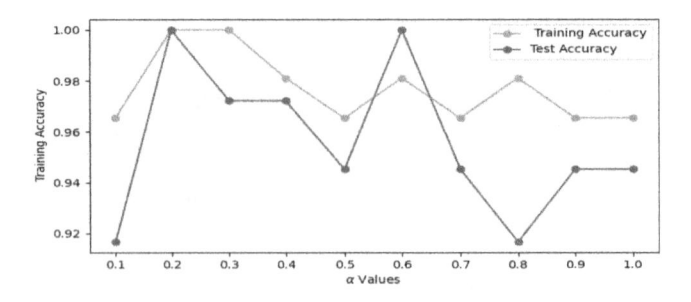

Fig. 4. Training and testing accuracies achieved for different α values with fractional relu on wine dataset

For $\alpha = 0.3$, the training accuracy remains 1.0, while the test accuracy is 0.9722, which is closer to the training accuracy. For $\alpha = 0.6$, the testing accuracy surpasses the training accuracy, which can be attributed to the small test size [2].

The training process has been highly effective, ensuring that the test data aligns well with the network without any errors. Here also, the best result is attained by $\alpha = 0.2$ other than $\alpha = 1$ therefore, it can be concluded that fractional activation behaves better than the standard activation function. The comparison of results with recent state of the arts is given in Table 2.

Table 2. Comparison of wine data accuracy with recent studies

State of The Art	Accuracy	Method
[3](2024)	0.97	Generic Algorithm
[23](2023)	1.00	Deep Neural Network
[13](2024)	0.99	Fractional order Differential evolution
[17](2024)	0.98	BP Neural Network
Proposed	1.00	Fractional Activation

4.2 Experiment on Silicosis Dataset

Silicosis identification in X-ray pictures is mostly dependent on radiologists' skill, which frequently causes delays in diagnosis. Computer-aided identification methods based on machine learning are being developed as a solution to this problem. However, the development of highly accurate deep-learning models for silicosis detection remains challenging due to the limited availability of large databases. This study proposes a novel approach with a new dataset for silicosis detection using transfer learning techniques applied to available X-ray radiographs. The Dataset was curated at Sawai Man Singh hospital Jaipur by a team of radiologists under the project no.1000114110 funded by the Government of Rajasthan. The Dataset division has been given below in Table 3. The dataset maintains a good balance among classes, effectively preventing class dominance. The test data remains unseen by the model throughout training. Due to limited data availability, no separate validation set was used. This study leverages transfer learning model VGG 19 from our previous studies [19] with additional pooling, flattening, two fractional dense layers. The sample images of silicosis dataset are shown in Fig. 5.

The Architecture of the proposed model is given in Fig. 6. This study proposes a custom model specifically for the regional chest X-ray dataset for silicosis classification using fractional sigmoid and fractional relu. The input function for the output layer can be represented by Eq. (7)

$$Z(X) = F^\alpha(F^\alpha(GAP(V(X)) + b_i) + b'_i) \tag{7}$$

Here $V(X)$ is the output of VGG 19 and GAP is global average pooling effect, b_i, b'_i are bias of dense layers and F^α is fractional activation for nonlinearity.

Table 3. Silicosis Dataset Distribution

Silicosis Dataset	Label	Count
Train Data	Silicosis	195
	Normal	197
Test Data	Silicosis	14
	Normal	15
Total -		421

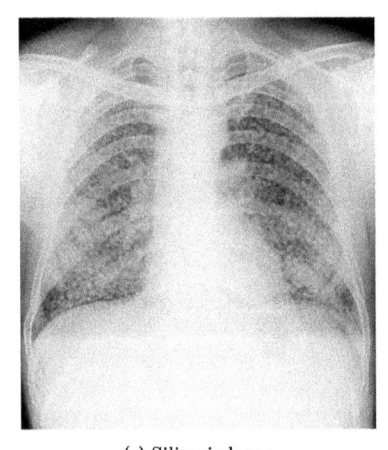

(a) Silicosis lungs

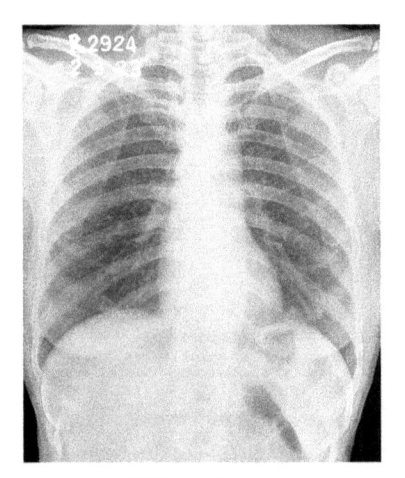

(b) Normal lungs

Fig. 5. Silicosis and normal lungs image of Dataset

4.2.1 Silicosis Classification Using Fractional Sigmoid In this part the fractional sigmoid has been used in dense layers of the model to enhance the classification. The results have been shown in Fig. 7. Graph shows achieved training and testing accuracy for every $\alpha \in (0,1]$. Training has been done for 50 epochs. The learning rate has been selected by random search to be 0.01, batch size is 16, dropout = 0.2 and Adam optimizer is used for weight update. Hyperparameter values were determined through a random search strategy applied to the model. As seen in Fig. 7, $\alpha = 0.3, 0.4, 0.5$ shows best performance with $\alpha = 0.3$ giving the highest training accuracy so this can be considered the best α among others for this experiment, and the best test accuracy is achieved at $\alpha = 0.3$ which is 0.8276.

It is observed that as α is approaching the value 1, performance of the model is decreasing. To show the model generalization performance on unseen test data the receiver operating characteristic (ROC) curve is given in Fig. 9a in which the area under the curve (AUC) is 0.83. The AUC of 0.83 indicates that the model performs well, assigning a higher probability to a positive sample than a negative sample in 83 out of 100 cases. Also it is better than the random guessing which

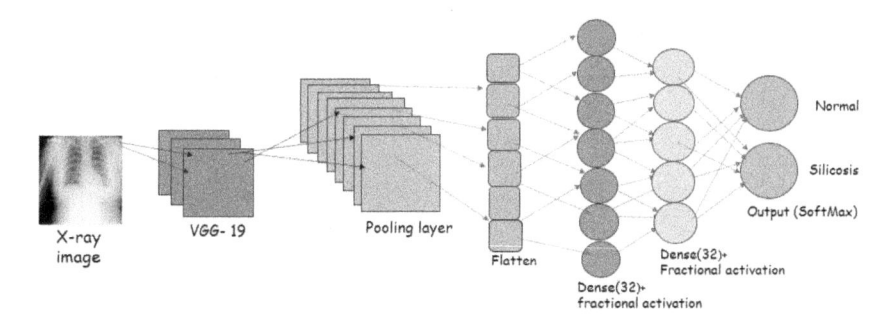

Fig. 6. Architecture of model for silicosis classification

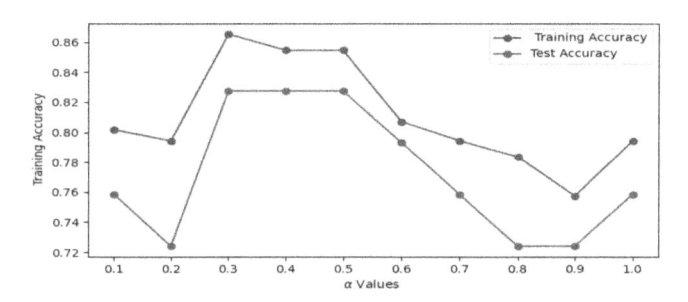

Fig. 7. Training and testing accuracies achieved for different α values with fractional sigmoid on silicosis dataset

has 50% AUC. The model is showing good performance as it achieves high true positive rate by maintaining a low false positive rate.

4.2.2 Silicosis Classification Using Fractional Relu In this section, the fractional relu activation is involved in the custom model for silicosis identification. The model parameters remain consistent as the previous experiments i.e. learning rate of 0.01, batch size of 16, dropout of 0.2, and epochs set to 50. Figure 8 illustrates the training and testing accuracy achieved with fractional relu among different $\alpha \in (0, 1]$.

The findings indicate that the best test accuracy of 0.7931 is attained with $\alpha = 0.3$ and 0.4, that is less than that achieved using the fractional sigmoid activation function. This outcome can be predicted because fractional sigmoid introduces fractional effects in both the function and its derivative, which provides fractional adaptability in forward pass as well as backward pass whereas the fractional relu has fractional value only in the derivative which has effect only in the backward pass. Additionally, fractional relu activation also has the issue of dying neurons for negative inputs like standard relu, which could explain why the performance is slightly lower compared to the fractional sigmoid for this silicosis dataset.

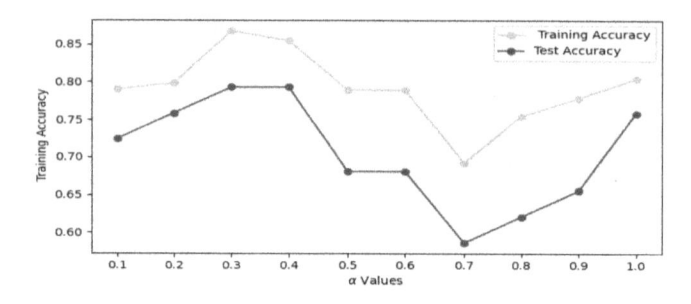

Fig. 8. Training and testing accuracy achieved for different α values with fractional relu on silicosis Dataset

(a) Roc curve using fractional sigmoid

(b) Roc curve using fractional Relu

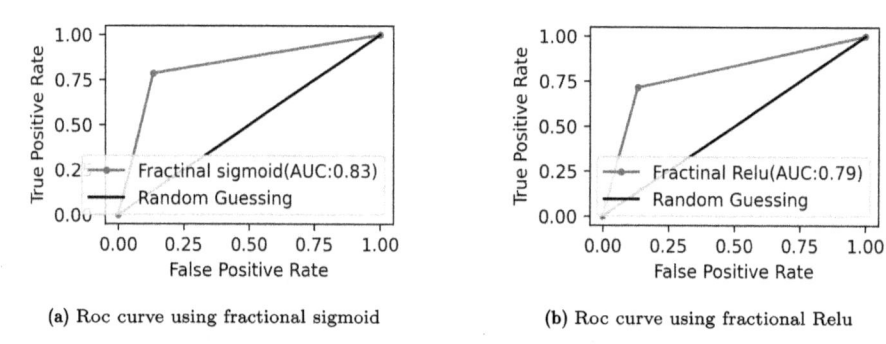

Fig. 9. Roc curves for silicosis classification using fractional activations

The receiver operating characteristic (ROC) curve has been shown in Fig. 9b with area under the curve value equals 0.79. This indicated that the proposed model is performing better than the random guessing. The proposed model with fractional relu has classified the data by providing a high probability to a positive sample than a negative sample 79 cases out of 100. The performance of both activations with different performance metric is given in Table 4.

Table 4. Evaluation of performance metrics for F^α_{relu} and F^α_{Sig} activation functions

Activation	Accuracy	precision	sensitivity	F1-score
F^α_{relu}	0.7931	0.80	0.79	0.79
F^α_{Sig}	0.8276	0.84	0.83	0.83

5 Conclusion

This study introduces a generalized fractional variant of two widely used activation functions, sigmoid and relu. The fractional activation function adds another degree of freedom α to improve generalization capabilities. A promising approach is proposed to objectively improve training performance by providing the capacity to fine-tune α. Using the suitable α, the issue of vanishing and exploding gradients can be mitigated, leading to a more flexible training process. The fractional sigmoid is less prone to the saturation problem, preventing vanishing gradients, while the fractional relu avoids flattening gradients and effectively adapts to inputs due to the varying α values. Additionally, this study experiments on a newly approved silicosis dataset to evaluate the effectiveness of the proposed approach. The empirical results for the silicosis and wine dataset using this approach demonstrate superior performance compared to the state of the art. For the silicosis dataset specifically, the fractional sigmoid outperforms the fractional relu. This may be due to the absence of the fractional α in the forward pass, as well as the inherent issue of dying neurons in relu, which can affect the training process for this data. Future research will focus on developing optimization algorithms for α and further improving silicosis classification, which could surely benefit patients in local communities. Also, a dynamic model that adapts the α as needed for slow or fast training can be made using different α values during training.

Data Availibility. The code and data will be provided as per the individual's request.

Disclosure of Interests. The authors have no competing interests to declare that are relevant to the content of this article.

References

1. Alharbi, F.M., Baleanu, D., Ebaid, A.: Physical properties of the projectile motion using the conformable derivative. Chin. J. Phys. **58**, 18–28 (2019)
2. Aliferis, C., Simon, G.: Overfitting, underfitting and general model overconfidence and under-performance pitfalls and best practices in machine learning and AI. In: Artificial Intelligence and Machine Learning in Health Care and Medical Sciences: Best Practices and Pitfalls, pp. 477–524 (2024)
3. Chai, J., Bi, M., Teng, X., Yang, G., Hu, M.: A mixed mutation strategy genetic algorithm for the effective training and design of optical neural networks. Opt. Fiber Technol. **82**, 103600 (2024)
4. Ding, B., Qian, H., Zhou, J.: Activation functions and their characteristics in deep neural networks. In: 2018 Chinese Control and Decision Conference (CCDC), pp. 1836–1841. IEEE (2018)
5. Dubey, S.R., Singh, S.K., Chaudhuri, B.B.: Activation functions in deep learning: a comprehensive survey and benchmark. Neurocomputing **503**, 92–108 (2022)
6. Esquivel, J.Z., Cruz Vargas, J.A., Lopez-Meyer, P.: Fractional adaptation of activation functions in neural networks. In: 2020 25th International Conference on Pattern Recognition (ICPR), pp. 7544–7550 (2021). https://doi.org/10.1109/ICPR48806.2021.9413338

7. Gao, F., Chi, C.: Improvement on conformable fractional derivative and its applications in fractional differential equations. J. Funct. Spaces **2020**(1), 5852414 (2020)
8. Hao, C., et al.: Balanced convolutional neural networks for pneumoconiosis detection. Int. J. Environ. Res. Public Health **18**(17), 9091 (2021)
9. Harjule, P., Sharma, R., Kumar, R.: Fractional-order gradient approach for optimizing neural networks: a theoretical and empirical analysis. Chaos Solitons Fractals **192**, 116009 (2025)
10. Hayou, S., Doucet, A., Rousseau, J.: On the impact of the activation function on deep neural networks training. In: International Conference on Machine Learning, pp. 2672–2680. PMLR (2019)
11. Herrera-Alcántara, O., Arellano-Balderas, S.: Adaptive morphing activation function for neural networks. Fractal Fractional **8**(8), 444 (2024)
12. Ivanov, A.: Fractional activation functions in feedforward artificial neural networks. In: 2018 20th International Symposium on Electrical Apparatus and Technologies (SIELA), pp. 1–4. IEEE (2018)
13. Jin, T., Su, K., Gao, J., Xia, H., Dai, G., Gao, S.: Fractional-order differential evolution for training dendritic neuron model. SSRN (2024)
14. Khalil, R., Al Horani, M., Yousef, A., Sababheh, M.: A new definition of fractional derivative. J. Comput. Appl. Math. **264**, 65–70 (2014)
15. Kumar, M., Mehta, U., Cirrincione, G.: Enhancing neural network classification using fractional-order activation functions. AI Open **5**, 10–22 (2024)
16. Li, T., Yang, X., Xu, H., Liu, H.: Early identification, accurate diagnosis, and treatment of silicosis. Can. Respir. J. **2022**(1), 3769134 (2022)
17. Liu, W., Liu, M., Yan, C., Qi, M., Zhang, L.: Corporate fraud detection based on improved BP neural network. Comput. Inf. **43**(3), 611–632 (2024)
18. Mushtaq, F., et al.: Artificial intelligence for computer aided detection of pneumoconiosis: a succinct review since 1974. Eng. Appl. Artif. Intell. **133**, 108516 (2024)
19. Sharma, G.K., Harjule, P., Agarwal, B., Kumar, R.: Silicosis detection using extended transfer learning model. In: International Conference on Recent Trends in Image Processing and Pattern Recognition, pp. 111–126. Springer (2023)
20. Sharma, G.K., Harjule, P., Sadhwani, T., Agarwal, B., Kumar, R.: Sequential transfer learning models with additional layers for pneumonia diagnosis. In: 2023 International Conference on Computer, Electronics & Electrical Engineering & their Applications (IC2E3), pp. 1–6. IEEE (2023)
21. Sharma, R., Harjule, P.: Modified gradient descent approach involving caputo fractional derivative with metaheuristic optimizer. J. Vibr. Test. Syst. Dyn. **8**(04), 443–453 (2024)
22. Stojiljkovic, V., et al.: A new conformable fractional derivative and applications. Selecciones Matemáticas **9**(02), 370–380 (2022)
23. Wang, D., Wang, T.: Federated ensemble algorithm based on deep neural network. In: International Conference on Soft Computing in Data Science, pp. 76–91. Springer (2023)
24. Ye, Y., Fan, H., Li, Y., Liu, X., Zhang, H.: Conformable bilinear neural network method: a novel method for time-fractional nonlinear partial differential equations in the sense of conformable derivative. Nonlinear Dyn. **113**(7), 7185–7200 (2024)
25. Yu, X.H., Chen, G.A., Cheng, S.X.: Dynamic learning rate optimization of the backpropagation algorithm. IEEE Trans. Neural Netw. **6**(3), 669–677 (1995)
26. Zamora, J., Rhodes, A.D., Nachman, L.: Fractional adaptive linear units. In: Proceedings of the AAAI Conference on Artificial Intelligence, vol. 36, pp. 8988–8996 (2022)

27. Zamora Esquivel, J., Cruz Vargas, A., Camacho Perez, R., Lopez Meyer, P., Cordourier, H., Tickoo, O.: Adaptive activation functions using fractional calculus. In: Proceedings of the IEEE/CVF International Conference on Computer Vision Workshops (2019)
28. Zhao, D., Luo, M.: General conformable fractional derivative and its physical interpretation. Calcolo **54**(3), 903–917 (2017). https://doi.org/10.1007/s10092-017-0213-8

Combining XAI and Graph Cuts for Skin-Lesion Segmentation

Zyad Husni[(✉)], Uwe Jaekel, and Babette Dellen

University of Applied Sciences Koblenz, Faculty of Mathematics, Informatics, Technology, 53424 Remagen, Germany
{zhusni,jaekel,dellen}@hs-koblenz.de

Abstract. Deep neural networks and supervised machine learning for medical image segmentation, including dermatology [10], require large pixel-wise annotated datasets for training, which can be difficult to obtain. Image classification, on the other hand, only requires a label for each image, which is often automatically provided with a medical diagnosis, but does not provide segmentation maps. However, in image-classification tasks, Explainable-AI (XAI) algorithms provide a means of identifying pixels in the original image that are part of the object or relevant structure. We propose to exploit this information for segmenting the images by building a network graph from XAI explanations and using the graph-cut algorithm for segmentation. Our approach is evaluated using the HAM10k [26] dataset, demonstrating its ability to segment skin lesions in dermatoscopic images without requiring pixel-annotated data for training. This makes our approach a cost-effective alternative in scenarios where annotated images are not available.

Keywords: Image Segmentation · Supervised Machine Learning · Explainable AI · Graph-Cut · SHAP · Grad-CAM · GMM · ResNet152 · CBAM

1 Introduction

Medical image segmentation is a crucial task in the diagnosis and treatment of diseases, especially in dermatology, where accurate delineation of skin lesions is essential for early detection of conditions such as melanoma [4,13]. Supervised machine learning has become an important and frequent paradigm in dermatology [10]. One of the most widely used methods is the U-Net [21], which shows exceptional performance in medical image segmentation. However, it relies heavily on large, annotated datasets, which are costly and time-consuming to produce.

Clustering-based approaches have been proposed to address this challenge [11,23], but a drawback is their dependence on pixel intensity, which often requires careful tuning of parameters. Other approaches rely on user-defined inputs that provide markers for initialization and are therefore not automatic [25]. More recent unsupervised deep learning techniques aim to learn feature representations directly from the data but can be computationally intensive and require large datasets [29].

Skin-lesion segmentation has also been recently formulated as an anomaly detection problem [8]. This has the advantage that no annotated data is required, because the network is trained on images showing only healthy skin. However, this approach does not provide information about the type of anomaly that has been detected, e.g., different stages of a disease or classes.

Our approach addresses these challenges by proposing a fully automated segmentation framework that integrates Explainable-AI techniques [3], specifically SHAP [15,24] and Grad-CAM [17,22], with Gaussian Mixture Models [14,19] (GMM) and Graph-Cut [5,6]. Images are annotated with a single label, representing different classes, and then used to train a classification network. XAI-driven feature maps provide pixel-wise information on the importance of the pixel for the correct classification of the image. These feature maps are used to initialize a graph-cuts method for image segmentation [6]. The XAI-driven feature maps generated by SHAP and Grad-CAM guide the segmentation process, while the GMM model the intensity distributions of the foreground and background. Graph-Cut further refines the segmentation by minimizing the energy function, ensuring that the object boundaries are accurately delineated. This approach avoids tedious pixel-wise annotations of images and therefore provides an automatic and scalable solution for medical image-segmentation tasks where labeled data are scarce or unavailable.

The primary objectives of this study are (i) the development and training of a medical image-segmentation framework combining XAI-driven feature maps from SHAP and Grad-CAM with GMM and graph cuts, (ii) the evaluation of the proposed approach on the HAM10k [26] dataset as well as comparison of the results with segmentation results obtained by U-Net [21], and (iii) the investigation of the impact of CBAM [27] integrations on segmentation quality and model transparency.

2 Methods

The proposed image-segmentation framework consists of the following parts:

(i) The core of the method is a ResNet152 model for image classification, trained with a subset of the HAM10K dataset from the ISIC_2018 challenge [2]. During training, the network is provided only with the class label for each image, not ground-truth pixel-wise annotations. Consequently, the network outputs only a single label prediction for each image.
(ii) SHAP and Grad-CAM explanations, obtained for the ResNet15 model of step (i), are used to create binary masks for each image. Grad-CAM explanations are further improved by integrating CBAM into ResNet152.
(iii) The binary masks from step (ii) are merged and used to initialize the graph-cuts method. The resulting graph-cuts segmentations are the XAI-driven segmentations produced by our method.

A schematic overview of the method is provided in Fig. 1. In the following, we provide further details of the different techniques employed in our framework.

2.1 Datasets and Data Augmentation

The HAM10K dataset from the ISIC_2018 challenge [2] contains dermatoscopic images of skin lesions, categorized into seven classes: melanoma (MEL), nevus (NV), actinic keratosis(AKIEC), basal cell carcinoma (BCC), benign ker-atosis-like lesions (BKL), vascular lesions (VASC), and dermatofibroma (DF). A key characteristic of this dataset is its class imbalance. For example, the most prevalent class, NV, includes over 58 times more images than the least common, DF. The original dataset contained 10,015 training images (each with a corresponding ground-truth segmentation mask), 194 validation images (without ground-truth segmentations), and 1,503 test images (without ground-truth segmentations). Our goal is to generate U-Net and XAI-driven segmentations and evaluate their accuracy using the Dice [30] and Jaccard [20] scores. These scores compare the generated segmentations to the ground-truth segmentations. Since the test and validation set lacked ground-truth segmentations, we restricted our analysis to the original training set of 10,015 images, which was then divided into 3 parts, defining dataset A, consisting of 8,012 images for testing, 202 images for validation during training, and 1,803 images for testing.

To enhance the dataset and improve model performance, image augmentation techniques were applied to the training and validation dataset using the PyTorch deep learning library [18]. For augmentation, the Albumentations library [1,9] was chosen due to its extensive range of augmentation options and efficient implementation. To address class imbalance in the dataset, a weighted augmentation strategy was employed. Using Albumentations, images from underrepresented classes were specifically targeted for augmentation. This strategy was based on inverse class frequency weighting or adaptive data augmentation, which ensured that minority classes received more augmentation than dominant classes. The goal was to balance the distribution of classes in the training data and improve the generalizability of the model. Not all images were augmented; the focus was primarily on the minority classes. Data augmentation resulted in 29,213 training images and 4,190 validation images. The number of testing images remained unchanged, defining dataset B.

2.2 Model Training and Feature Extraction

We use a ResNet152 [12] network as the backbone for feature extraction. ResNet [12] introduces the concept of skip connections, improving training stability and convergence.

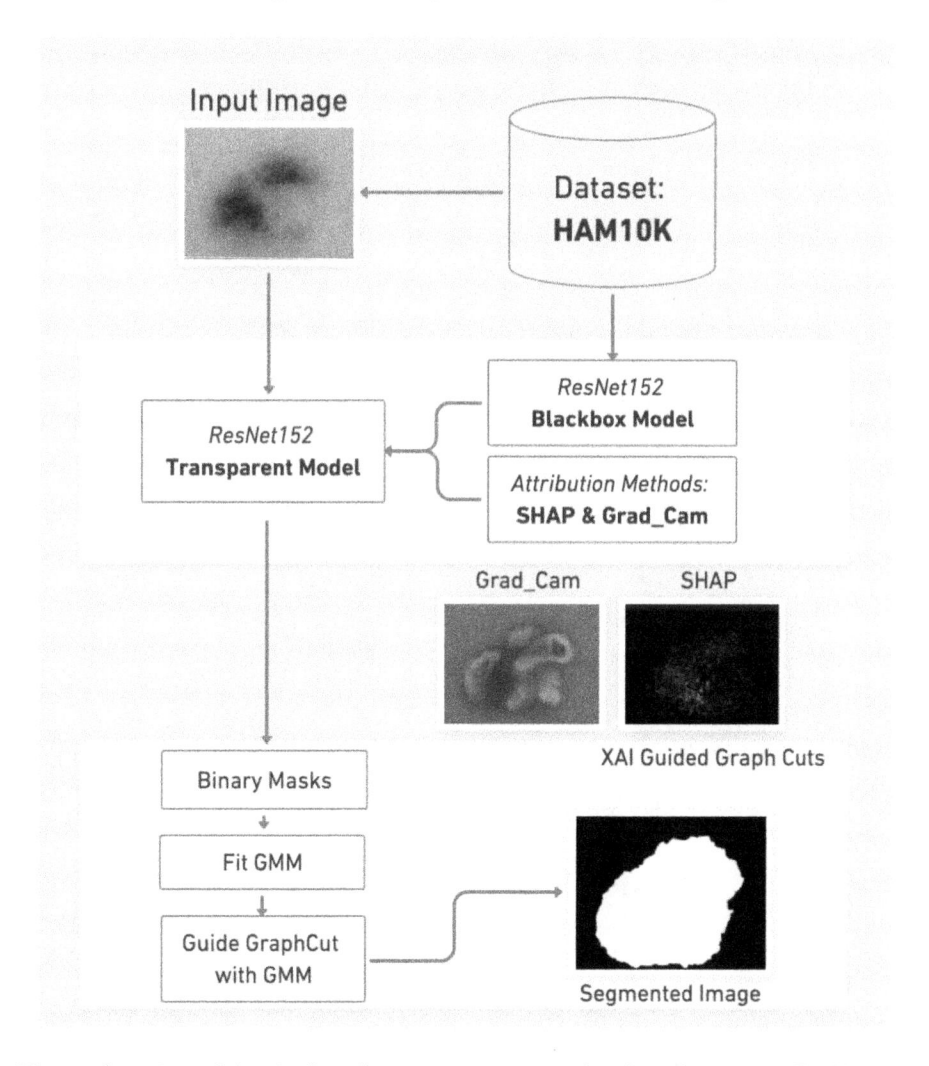

Fig. 1. Overview of the pipeline: Images are augmented and used to train a ResNet-152 as a black-box classifier. The transparent model processes each input to produce Grad-CAM and SHAP attribution maps, which provide an initial binary mask. A Gaussian Mixture Model is then fitted to both foreground and background regions and used to guide a graph-cut algorithm for image segmentation.

The model was trained on dataset A and on the augmented dataset B. Both trainings were conducted on one of the HPC compute nodes of the University of Applied Sciences Koblenz, equipped with 2 AMD EPYC 7713 CPUs (64 cores, 128 threads each), 1 TB RAM, and 4 NVIDIA A100 GPUs (80 GB each) connected via NVLink. The training configuration included a learning rate of 0.1 with a learning rate scheduler, momentum set to 0.8, and a weight decay of

0.001. An early stopping strategy was employed to save the best model during training, preventing overfitting and ensuring better generalization.

Table 1. Classification results of the ResNet152 model, trained with dataset B, for the test set.

Class	Precision	Recall	F1-Score	Support
Class 0 (MEL)	0.71	0.57	0.63	158
Class 1 (NV)	0.93	0.97	0.95	1248
Class 2 (AKEIC)	0.72	0.85	0.78	81
Class 3 (BCC)	0.81	0.57	0.67	68
Class 4 (BKL)	0.83	0.73	0.78	201
Class 5 (DF)	0.78	0.74	0.76	19
Class 6 (VASC)	0.82	0.96	0.89	28
Accuracy			0.88	1803
Macro avg	0.80	0.77	0.78	1803
Weighted avg	0.88	0.88	0.88	1803
Test Loss			0.5447	

2.3 Initial Segmentation-Mask Generation Using XAI

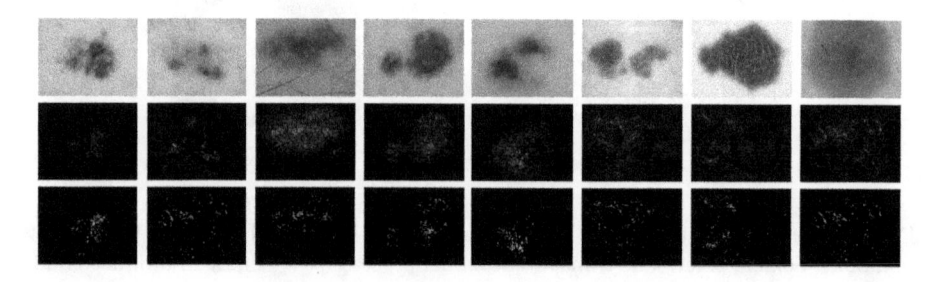

Fig. 2. The first row displays the original images from the HAM10K dataset, followed by the raw SHAP attributions extracted from the transparent model in the second row. The third row shows the corresponding binary masks generated from the SHAP values.

The learned feature maps from the ResNet152 model serve as input for the XAI methods, SHAP and Grad-CAM. SHAP values are used to generate the importance of pixel-level features, highlighting the most influential regions for image classification [15,24]. Grad-CAM utilizes the feature maps generated by the final convolutional layers of the neural network to create coarse localization

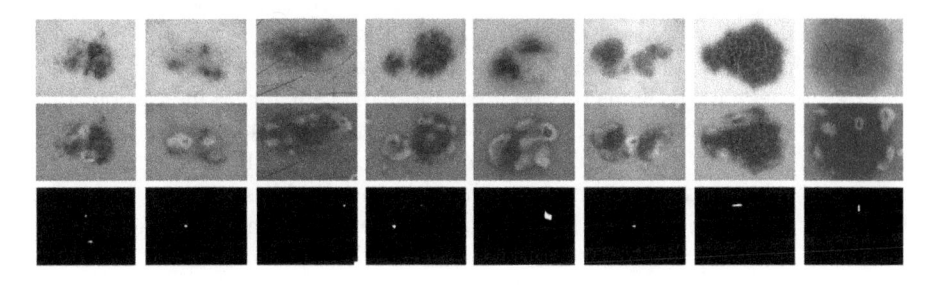

Fig. 3. The first row presents the original images from the HAM10K dataset, followed by the raw Grad-CAM attributions extracted from the transparent model in the second row. The third row displays the corresponding binary masks generated from the Grad-CAM explanations.

maps that highlight areas of interest (see Fig. 3, second row). These maps indicate regions that contributed the most to the predictions, providing valuable insights into where the model is focusing its attention [17]. The outputs from SHAP and Grad-CAM are combined to create an initial binary mask, leveraging SHAP's pixel-level importance and Grad-CAM's spatial localization. This complementary approach addresses the limitations of each method individually, with Grad-CAM providing broader context and SHAP capturing finer details. A binary bitwise AND operation using the OpenCV [7] library is applied to merge the masks, ensuring that only pixels significant to both methods contribute to the final combined mask, improving segmentation accuracy.

2.4 CBAM

To further enhance the feature extraction capabilities of the ResNet152 backbone, we integrated the Convolutional Block Attention Module (CBAM) [27] into the architecture to improve Grad-CAM explanations. CBAM is a lightweight and effective attention mechanism that improves a network's ability to focus on the most informative features within an image. It operates by applying attention mechanisms sequentially along the channel and spatial dimensions of the feature maps, refining the extracted features at each stage. The *channel attention module* identifies the importance of each channel by aggregating spatial information using global average pooling and max pooling. The resulting descriptors are passed through a shared multi-layer perceptron (MLP) that includes a bottleneck layer, where the dimensionality is reduced by a factor known as the reduction ratio. This reduction controls the trade-off between model complexity and the ability to capture fine-grained details. The outputs of the MLP are combined and activated using a sigmoid function to produce the channel attention map, which emphasizes the most relevant channels in the feature map. The *spatial attention module* focuses on the most critical regions in the image. It applies pooling operations along the channel axis to generate spatial descriptors, which are concatenated and passed through a convolutional layer to produce a spatial attention map. This map highlights the important regions in the feature map, guiding the network to focus on the areas most relevant for segmentation. Integrating CBAM into ResNet152 enhances the network's ability to focus on critical areas, such as lesion boundaries, in dermatoscopic images. To accommodate the complexity of the dataset, the reduction ratio in CBAM is set to 4 instead of the default 16, and the kernel size for spatial attention is increased to 5 [16].

2.5 Segmenting with Graph-Cut

The binary masks obtained from SHAP and Grad-CAM are merged and then used to initialize Gaussian Mixture Models (GMM) for foreground and background modeling. GMMs are fitted to the pixel distributions of the image to compute foreground and background probabilities, where the GMM models the intensity and color distribution of the foreground and background regions in the combined mask. This is determined by extracting the coordinates of the black (background) and white (foreground) pixels from the mask and mapping these pixels back onto the original image to identify their respective regions. The resulting foreground and background probability distributions allow estimating the likelihood of each pixel belonging to the foreground or background. The image is treated as a graph where each pixel corresponds to a node. The source and the sink of the graph (terminal nodes) are provided by the foreground and background probabilities computed earlier via the binary XAI mask [6]. Edge weights are calculated from intensity, texture, and gradient similarities between neighboring pixels. These weights influence the energy-minimization process and ensure accurate segmentation boundaries. The Graph-Cut algorithm then minimizes the energy function that defines the relationship of foreground and background regions. This iterative minimization process refines the segmentation mask, effectively separating the object of interest from the background [6].

3 Results

To obtain XAI-guided image segmentation, the ResNet152 model is first trained using the augmented dataset B. The combined outputs of SHAP and Grad-CAM explanations are used to initialize the graph-cut method for image segmentation. We report intermediate results for the different parts of our framework as well as the final XAI-guided segmentation performance.

3.1 ResNet152 Model for Image Classification

The ResNet152 model, trained with dataset A, achieved an accuracy of 88% on the test set. To evaluate model performance, metrics such as precision, recall, and F1-score were used. The minority classes DF and BCC underperformed compared to the other classes. Therefore, we trained with augmented dataset B instead, which improved the results. The overall results are summarized in Table 1. This model provides the core of the proposed framework.

Table 2. Comparison of Dice and Jaccard metrics between U-Net and the proposed approach (XAI segmentation) for the test set.

Method	Dice Score	Jaccard Index
U-Net	0.92	0.85
XAI segmentation	0.84	0.75

3.2 Binary-Mask Generation with SHAP and Grad-CAM

SHAP and Grad-CAM explanations are generated for each image of the test set using the previously trained ResNet152 model. In Figs. 2 and 3, the computed SHAP and Grad-CAM explanations are shown in the second row, respectively. In the third row, the respective binary masks generated from the explanations are presented. The binary masks obtained from Grad-CAM are very sparse, making them less suitable for binary-mask generation. By integrating CBAM [27] into the ResNet152 model, the Grad-CAM explanations and the resulting binary mask could be improved (see Fig. 4, last two rows). These adjustments allow the model to capture subtle features more effectively, improving the overall segmentation performance.

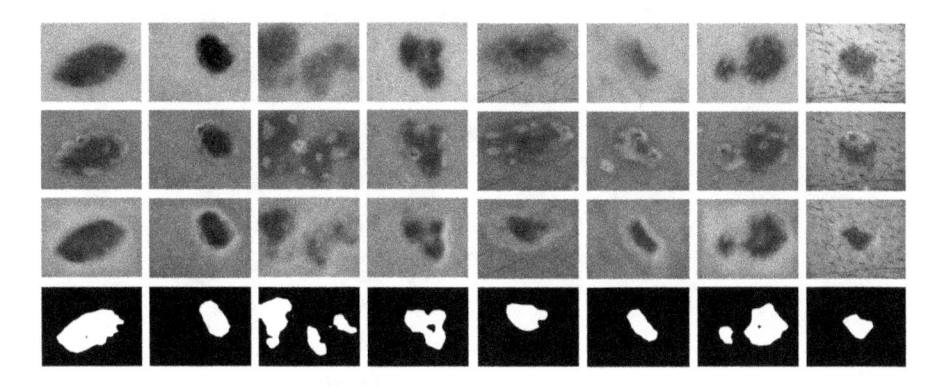

Fig. 4. Improvements of Grad-CAM explanations via integrating CBAM into ResNet152: Original images (first row), Grad-CAM explanations for ResNet152 without CBAM (second row), Grad-CAM explanations for ResNet152 with CBAM (third row), and binary masks generated Grad-CAM explanations for ResNet152 with CBAM (last row).

3.3 XAI-Guided Image Segmentation

Finally, the binary masks obtained in the previous step are merged and used to initialize the graph-cuts method. The resulting binary segmentations are evaluated using standard metrics: Dice Coefficient [30] and Jaccard Index [20]. These

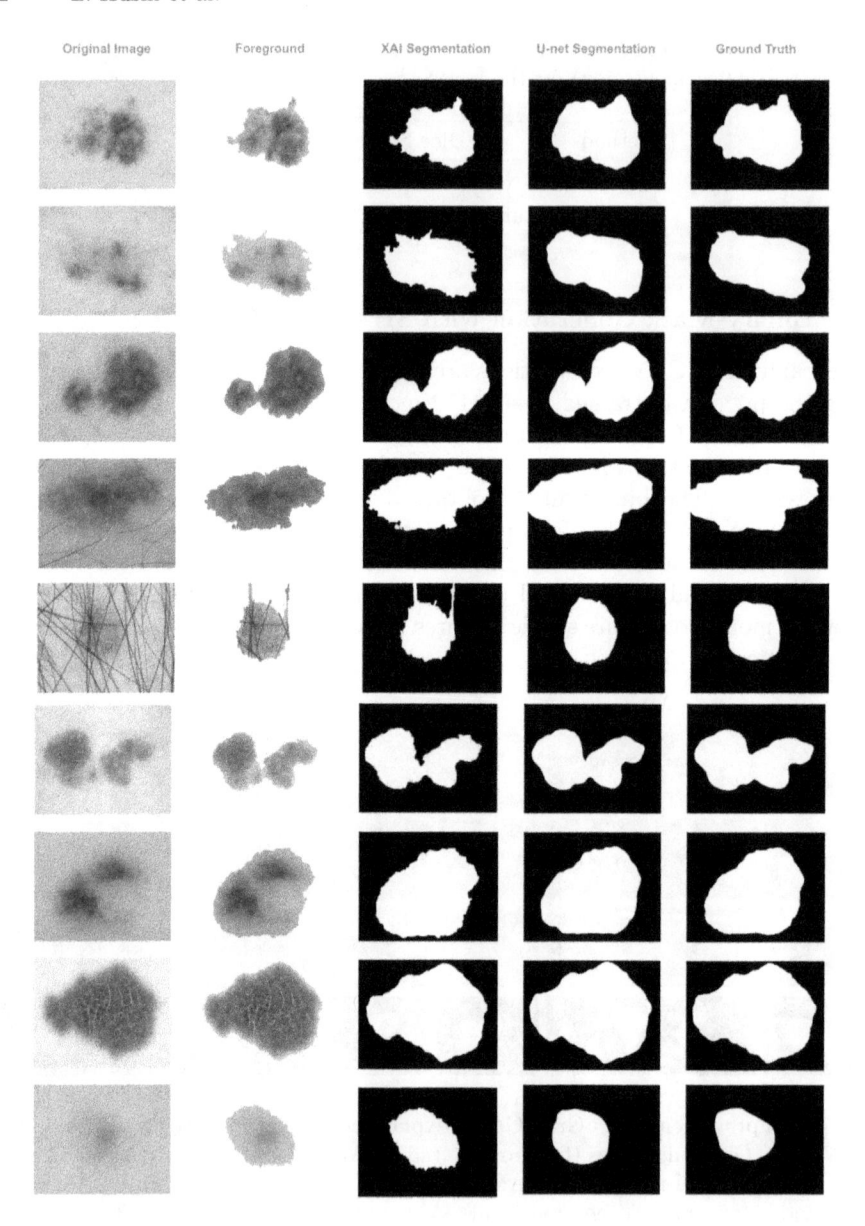

Fig. 5. Example results of our method (XAI-segmentation), U-net and the ground-truth.

metrics quantify the overlap of predicted and ground-truth segmentations provided by experts. Our segmentation framework achieved a Dice score of 0.84 and a Jaccard index of 0.75 (see Table 2). In Fig. 5, segmentation results are shown for images of the test set. These results show that our framework is able to classify and segment skin lesions without using pixel-wise annotations during training.

Fig. 6. Limitations of the method: Segmentation fails when the attribution maps differ significantly from one another. This discrepancy leads to a segmented output that does not adequately cover the region of interest.

The performance of the proposed method is compared with U-Net, a widely adopted model for medical image segmentation. The U-Net architecture uses ResNet152 as the backbone for its encoder-decoder framework, with the implementation sourced from GitHub [28]. Both networks were trained with the augmented dataset B. However, to train the U-Net, the image class labels and the ground-truth segmentations had to be provided during training, while our method is trained only with image class labels. The U-Net achieved a Dice score of 0.92 and a Jaccard index of 0.85 (see Table 2). The score of our method is lower than the one of U-Net, but still in a comparable range. However, a comparison in terms of the Dice and Jaccard score alone is difficult for the following reason: Ground-truth pixel-wise annotations, which are used to train the U-Net and to evaluate segmentation performance of both approaches on the test set, potentially contain a bias. Since the U-Net is trained with this data, it can learn this bias. Therefore, the boundaries of the segments are closer to the precise form of the ground-truth annotations, raising the Dice and the Jaccard scores. When studying the results for individual images in Fig. 5, our method shows, at least visually, competitive results to U-Net.

4 Discussion and Conclusion

This study presents an unsupervised medical image-segmentation approach that leverages XAI techniques, such as SHAP and Grad-CAM, to guide the Gaussian Mixture Model (GMM) initialization for Graph-Cut segmentation. The model was first trained on an augmented dataset, achieving a high accuracy of 88%. XAI explanations were then generated and converted into binary masks to identify key foreground and background pixels. These masks were used to fit GMMs, which provided the initial segmentation. The process was iteratively repeated for a specified number of iterations n or until a convergence threshold was met, measuring the ratio of changed pixels to the total number of pixels. The final output was further refined using morphological operations. For challenging cases, such as lesions that closely resemble healthy skin, improvements were introduced, including the integration of CBAM into the training architecture and adjustments to the number of GMM components for foreground and background modeling. Despite the class imbalance in the HAM10k dataset, the proposed method showed a performance comparable to U-Net. It further eliminates the need for pixel-wise expert annotations, which is an important advantage in scenarios where annotated data are scarce or unavailable.

The approach has limitations due to several parameters that influence its performance, including those related to model training, threshold selection for values generated by XAI methods such as SHAP and Grad-CAM, and parameters of GMM components for foreground and background modeling. These limitations become especially apparent for lesions that are difficult to distinguish from healthy skin. In such cases, the raw attribution maps generated by SHAP and Grad-CAM can differ significantly due to the distinct underlying computation methods of each technique. Consequently, merging these masks can result in suboptimal binary masks, which negatively impact the subsequent GMM segmentation performance (see Fig. 6).

Future research will focus on improving the binary mask generation process, optimizing GMM fitting by testing multiple configurations with the goal of selecting the best combination of components for foreground and background modeling. Furthermore, multiple models could be trained independently for specific lesion classes to obtain a more detailed understanding of each class. By isolating the segmentation process for particular lesion classes, such as malignant versus benign cases, models could capture the unique characteristics of each category in more detail, potentially enhancing overall performance.

Acknowledgement. This research has received funding from the Ministry of Science and Health of Rhineland-Palatinate, Germany, and the Debeka Krankenversicherungsverein a.G. through the Forschungskolleg Data2Health.

References

1. Albumentations Team: Defining a simple augmentation pipeline for image augmentation (2024). https://albumentations.ai/docs/examples/example/
2. ISIC Archive: ISIC challenge archive (nd). https://challenge.isic-archive.com/. Accessed 13 Dec 2024
3. Arrieta, A.B., et al.: Explainable artificial intelligence (XAI): concepts, taxonomies, opportunities and challenges toward responsible AI. Inf. Fus. **58**, 82–115 (2020). https://doi.org/10.1016/j.inffus.2019.12.012
4. Behera, N., Singh, A.P., Rout, J.K., Balabantaray, B.K.: Melanoma skin cancer detection using deep learning-based lesion segmentation. Int. J. Inf. Technol. **16**(6), 3729–3744 (2024)
5. Boykov, Y., Funka-Lea, G.: Graph cuts and efficient nd image segmentation. Int. J. Comput. Vis. **70**(2), 109–131 (2006)
6. Boykov, Y., Jolly, M.P.: Interactive graph cuts for optimal boundary and region segmentation of objects in N-D images. In: Proceedings of IEEE International Conference on Computer Vision (ICCV), vol. 1, pp. 105–112 (2001)
7. Bradski, G.: The OpenCV library. Dr. Dobb's J. Softw. Tools (2000)

8. Burgert, A., Dellen, B., Jaekel, U., Paulus, D.: Semi-supervised anomaly detection in skin-lesion images. In: Proceedings of the 20th International Joint Conference on Computer Vision, Imaging and Computer Graphics Theory and Applications, International Conference on Computer Vision Theory and Applications (Visapp 2025), vol. 2, pp. 535–541. SciTePress (2025)
9. Buslaev, A., Iglovikov, V.I., Khvedchenya, E., Parinov, A., Druzhinin, M., Kalinin, A.A.: Albumentations: fast and flexible image augmentations. Information **11**(2), 125 (2020). https://doi.org/10.3390/info11020125
10. Chan, S., Reddy, V., Myers, B., Thibodeaux, Q., Brownstone, N., Liao, W.: Machine learning in dermatology: current applications, opportunities, and limitations. Dermatol. Therapy **10**, 365–386 (2020)
11. Dhanachandra, N., Manglem, K., Chanu, Y.: Image segmentation using k-means clustering algorithm and subtractive clustering algorithm. Procedia Comput. Sci. **54**, 764–771 (2015)
12. He, K., Zhang, X., Ren, S., Sun, J.: Deep residual learning for image recognition. In: Proceedings of the IEEE Conference on Computer Vision and Pattern Recognition, pp. 770–778 (2016)
13. Khan, M.Z., Gajendran, M.K., Lee, Y., Khan, M.A.: Deep neural architectures for medical image semantic segmentation: review. IEEE Access **9**, 83002–83024 (2021). https://doi.org/10.1109/ACCESS.2021.3086530
14. Lei, T., Nandi, A.K.: Image Segmentation: Principles, Techniques, and Applications. Wiley (2022)
15. Lundberg, S.M., Lee, S.I.: A unified approach to interpreting model predictions. In: Proceedings of the 31st International Conference on Neural Information Processing Systems (NeurIPS), pp. 4765–4774 (2017)
16. Luuuyi: CBAM PyTorch implementation (ND). https://github.com/luuuyi/CBAM.PyTorch/tree/master. Accessed 11 Nov 2024
17. Molnar, C.: Interpretable machine learning. Lulu. com (2020)
18. Paszke, A., et al.: PyTorch: an imperative style, high-performance deep learning library. In: Advances in Neural Information Processing Systems (NeurIPS), vol. 32, pp. 8024–8035 (2019)
19. Rasmussen, C.: The infinite Gaussian mixture model. Adv. Neural Inf. Process. Syst. **12** (1999)
20. Real, E., Shlens, J., Mazzocchi, S., Pan, Y., Le, Q.V.: You might not need higher order penalties for semantic segmentation: revisiting optimal graph cut with object compatibility. In: Proceedings of the IEEE International Conference on Computer Vision (ICCV), pp. 2684–2693 (2017)
21. Ronneberger, O., Fischer, P., Brox, T.: U-Net: convolutional networks for biomedical image segmentation. In: Navab, N., Hornegger, J., Wells, W.M., Frangi, A.F. (eds.) MICCAI 2015. LNCS, vol. 9351, pp. 234–241. Springer, Cham (2015). https://doi.org/10.1007/978-3-319-24574-4_28
22. Selvaraju, R.R., Cogswell, M., Das, A., Vedantam, R., Parikh, D., Batra, D.: Grad-CAM: visual explanations from deep networks via gradient-based localization. In: Proceedings of the IEEE International Conference on Computer Vision (ICCV), pp. 618–626 (2017). https://doi.org/10.1109/ICCV.2017.74
23. Siddiqui, F.U., Yahya, A.: Clustering Techniques for Image Segmentation. Springer (2022)
24. Sundararajan, M., Najmi, A.: The many shapley values for model explanation. In: International Conference on Machine Learning, pp. 9269–9278. PMLR (2020)
25. Tang, Y., Li, Y., Zou, H., Zhang, X.: Interactive segmentation for medical images using spatial modeling mamba. Information **15**(10), 633 (2024)

26. Tschandl, P., Rosendahl, C., Kittler, H.: The HAM10000 dataset, a large collection of multi-source dermatoscopic images of common pigmented skin lesions. Sci. Data **5**(1), 1–9 (2018)
27. Woo, S., Park, J., Lee, J.Y., Kweon, I.S.: CBAM: convolutional block attention module. In: Proceedings of the European Conference on Computer Vision (ECCV), pp. 3–19 (2018)
28. Yakubovskiy, P.: Segmentation models for PyTorch (2024). https://github.com/qubvel-org/segmentation_models.pytorch
29. Zhang, Q., Yang, L.T., Chen, Z.: Deep computation model for unsupervised feature learning on big data. IEEE Trans. Serv. Comput. **9**(1), 161–171 (2016). https://doi.org/10.1109/TSC.2015.2497705
30. Zou, K.H., et al.: Statistical validation of image segmentation quality based on a spatial overlap index: scientific reports. Acad. Radiol. **11**(2), 178–189 (2004)

Accelerating Two-Dimensional k-Wave Ultrasound Simulations Through Pruned FFT: A Treatment Planning Optimisation

Ondrej Olsak[✉][iD], David Bayer[iD], and Jiri Jaros[iD]

Faculty of Information Technology, Brno University of Technology,
Brno, Czech Republic
iolsak@fit.vut.cz

Abstract. Wave propagation simulations are foundational tools across scientific and medical applications, yet their computational demands become significant for high-resolution simulations, particularly in medical applications where precise representation of different tissue geometries is crucial. This paper presents a novel approach to accelerate 2D wave propagation simulations in the k-Wave toolbox. Our method focuses on optimising Fourier transform computations through spectrum pruning. The Acoustic Field Propagator along with a bisection pruning algorithm to estimate the position of the spectral coefficients is used. Through these optimisations, our approach achieves significant performance gains, demonstrating speedups of up to 1.8x for large simulation domains. Experimental evaluation on medical ultrasound simulations demonstrates that the proposed method achieves focal point errors below 1% with minimal focus position shifts, while skipping up to 90% of spectral coefficients in large domains. This results in a significant simulation time reduction by half over the large simulation domains. Although the proposed method primarily focuses on accelerating k-Wave toolbox wave propagation simulation, it could be generally applied to wave propagation problems.

Keywords: Pruned Fast Fourier Transform · Ultrasound Simulation · Wave propagation simulation · k-Wave · Pseudo-spectral methods

1 Introduction

Wave propagation simulations play a crucial role in various fields of science and engineering. However, the time required for these simulations to complete can vary considerably, from a few seconds to days or even longer, depending on the complexity of the model and the level of resolution required. This paper focuses on accelerating 2D wave propagation simulations implemented by the k-Wave toolbox [19], which utilises k-space pseudo-spectral methods based on the Fourier Transform. The goal is to accelerate these simulations by optimising the computation of the Fourier Transform, which represents a significant part of the simulation. This optimisation is particularly crucial for simulations over high-resolution domains, where computational demands can be significant.

The most common approach to performing the Fourier transform is the Fast Fourier Transform (FFT) algorithm [2]. Since some applications only require a specific subset of frequency components from the FFT algorithm's output, computing the complete set of spectral coefficients may be unnecessary. In such cases, it may be beneficial to reduce the computational cost by implementing either a Sparse Fourier Transform (sparse FFT) [13] or Pruned Fourier Transform (pruned FFT) [7] algorithm. The sparse FFT is suitable for signals with few non-zero/significant coefficients compared to the size of the input signal, where their position in the spectrum is unknown. Sparse FFT algorithms are usually specially designed using domain-specific knowledge. Conversely, pruned FFT algorithms are optimised to compute spectral coefficients within known patterns that occur in the spectral domain. This leads to bypassing unnecessary computations in the standard Fast Fourier Transform [14].

For wave propagation simulations, spectral coefficients typically cluster near low frequencies [10]. This characteristic makes the pruned FFT particularly suitable as a replacement for the standard FFT currently employed in the k-Wave toolbox. By implementing this change, we aim to accelerate wave propagation simulations through reduced computational time in the Fourier transform phase and subsequent operations in the spectral domain.

Given that k-Wave is designed for time domain acoustic and ultrasound simulations in complex and tissue-realistic media, the evaluation of the proposed approach will focus on simulations used for non-invasive treatment pre-planning, such as the application of focused high-intensity ultrasound. This application exemplifies situations where precise representation of the media geometry can affect the accuracy of the simulation result [12,21]. Our evaluation will examine the method's impact on three critical aspects across various domain resolutions: simulation accuracy, focal point positioning and computational efficiency.

2 Mathematical Background

To compute the wave propagation, the k-Wave toolbox employs the pseudo-spectral method using Fourier basis functions. This technique involves representing the solution of the differential equation as a sum of specific basis functions. Unlike finite-difference time domain methods, which rely on local computations at neighbouring points, spectral methods use information from the entire domain, leading to higher accuracy [5]. The k-Wave toolbox runs simulations based on the following governing equations [19]:

$$\begin{aligned} \frac{\partial u}{\partial t} &= -\frac{1}{\rho_0}\nabla p \\ \frac{\partial \rho}{\partial t} &= -(2\rho + \rho_0)\nabla \cdot u - u \cdot \nabla \rho_0 \\ p &= c_0^2(\rho + d \cdot \nabla \rho_0 + \frac{B}{2A}\frac{\rho^2}{\rho_0} - L_\rho) \end{aligned} \qquad (1)$$

Where u is the acoustic particle velocity, p is the acoustic pressure, ρ is the acoustic density, ρ_0 is the ambient (equilibrium) density, c_0 is the isentropic sound speed, d is the acoustic particle displacement, B/A is the nonlinearity parameter characterizing finite-amplitude effects, and L_ρ is a loss operator that accounts for acoustic absorption and dispersion in the medium. Equation (1) can be written in a discrete form using the k-space pseudo-spectral method [16]. The following equations are part of the spatial gradient calculations based on the Fourier collocation spectral method:

$$\frac{\partial}{\partial \xi} p^n = \mathcal{F}^{-1}\{ik_\xi \kappa e^{ik_\xi \Delta \xi/2} \mathcal{F}\{p^n\}\}, \tag{2}$$

$$u_\xi^{n+\frac{1}{2}} = u_\xi^{n-\frac{1}{2}} - \frac{\Delta t}{\rho_0} \frac{\partial}{\partial \xi} p^n + \Delta t S_{\mathrm{F}\xi}^n, \tag{3}$$

$$\frac{\partial}{\partial \xi} u_\xi^{n+\frac{1}{2}} = \mathcal{F}^{-1}\{ik_\xi \kappa e^{-ik_\xi \Delta \xi/2} \mathcal{F}\{u_\xi^{n+\frac{1}{2}}\}\}, \tag{4}$$

$$\rho_\xi^{n+1} = \rho_\xi^n - \Delta t \rho_0 \frac{\partial}{\partial \xi} u_\xi^{n+\frac{1}{2}} + \Delta t S_{\mathrm{M}\xi}^{n+\frac{1}{2}} \tag{5}$$

For the Cartesian direction $\xi = x, y$ in R^2, $\mathcal{F}$ and $\mathcal{F}^{-1}$ denote the forward and inverse spatial Fourier transform, i is the imaginary unit, Δt is the time step, k_ξ represents the wave numbers in the ξ direction, and κ is the k-space operator defined as $\kappa = sinc(c_{ref} k \Delta t/2)$, where c_{ref} is a scalar reference sound speed. Equations (2) and (4) are spatial gradient calculations based on the Fourier collocation spectral method. Equations (5) and (3) represent update steps utilising a k-space corrected first-order accurate forward difference.

The Fast Fourier Transform algorithm is used to convert signals from the spatial domain to the spectral domain. Each simulation step of the 2D wave propagation simulation involves 11 FFTs. This computation consumes approximately 60% of the total simulation time, making it a significant part of the overall simulation [6].

3 Transducer Position

Since most simulations utilise narrow bandwidth sources, the spectral coefficients cluster around low frequencies, leading to sparsity in the spectral domain, particularly in high-resolution simulation domains [3]. Despite this apparent sparsity, sparse FFT approaches prove unsuitable as they discard small but crucial coefficients that contribute to simulation accuracy, while also introducing additional computational complexity through filtering processes. The spectrum is also not sparse enough to benefit from the usage of sparse FFT. In contrast, pruned FFT offers a more efficient solution by precisely computing the specified region within the spectral domain, preserving all coefficients regardless of their magnitude and eliminating the need for additional computational steps. This makes pruned FFT a more reliable choice for processing simulation data compared to sparse FFT methods [10].

One key factor that significantly impacts the position of computed spectral coefficients is the direction in which the wave propagates through the media [4]. Thus, the position of the coefficients in the spectrum (among other factors) depends on the orientation of the transducer that determines the direction of wave propagation. Figure 1 shows the wave propagation in water with different positions of an arc transducer, demonstrating different directions of ultrasound wave propagation. The direction of wave propagation clearly impacts the position of the coefficients in the spectrum (the zero-frequency components are shifted to the centre of the domain).

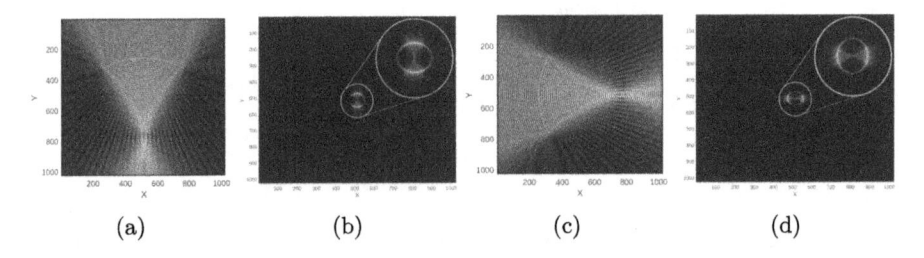

(a)   (b)   (c)   (d)

Fig. 1. Transducer position's effect on wave propagation and spectral domain: (a) pressure field distribution for Y-axis propagation in space domain, (b) frequency domain representation of Y-axis propagation showing spectral coefficient distribution, (c) pressure field distribution for X-axis propagation in space domain, and (d) frequency domain representation of X-axis propagation showing spectral coefficient distribution.

For the application of the pruned FFT in wave propagation simulation, the most suitable transducer positions are those aligned with one of the axes. This alignment enables reduction of the area computed by the pruned FFT algorithm [11]. If the transducer is not aligned with either axis, alignment can be achieved by rotating the domain around its centre. To obtain a suitable shape for the simulation domain, which is typically rectangular or square, the rotated domain can be filled with surrounding media. At the end of the simulation, if needed, the domain with the wave propagation result can be rotated back to its original position and cropped to its original size. This operation requires the ability to fill the domain with surrounding media and the presence of an absorbing layer, such as Perfect Match Layer (PML) [19] adjacent to the rotated domain. This layer absorbs the propagated wave around the simulation domain and prevents possible reflections that might affect the result of the original simulation. In our proof-of-concept implementation of the two-dimensional pruned FFT algorithm, the computation reduction is made only in the second dimension (columns - X). This means that over the first dimension (rows - Y), the full FFT is computed over each row. In the second dimension, only a given number of columns is computed. For our implementation, the position of the transducer aligned with the X-axis is the most suitable variant. However, the transducer can also be aligned with the Y-axis. In this case, the suitable reduced dimension would be Y (rows). This scenario is in actual implementation solved by rotating the domain by 90° to align the transducer with the X-axis.

4 Estimation of the Coefficient Area

An integral part of the pruned FFT algorithm is localising the area containing coefficients crucial for computing the wave propagation simulation. Most of these coefficients represent low frequencies, thus they are present at the corners of the spectral domain. Since the symmetry of the real-to-complex FFT in the Y dimension is used, only the first m columns of the X domain will be computed. To estimate the value of m, the Acoustic Field Propagator (AFP) [17] together with a bisection pruning algorithm is used.

The AFP enables the calculation of the wave field at all spatial positions at a given time in a single step. The advantage of AFP is its computational speed compared to a full wave propagation simulation. However, this method can only be used in homogeneous media with a single-frequency transducer and cannot compute the reflections and absorption of the propagated wave. Thus, it cannot replace a full ultrasound wave propagation simulation. When complex tissue interfaces create significant reflections or scattering, the AFP may fail to capture high-frequency components generated at these boundaries. Despite these limitations, it provides an acceptable estimation of spectral coefficient positions. The AFP is executed with the same transducer and homogeneous media, with a sound speed equal to the minimum value present in the original simulation. The lower the sound speed, the higher the frequencies that may occur in the spectral domain. The resulting spectrum of the propagated wave is used to estimate the position of spectral coefficients that will be computed by the pruned FFT in the wave propagation simulation.

To determine the first m columns for the pruned FFT, a bisection pruned algorithm is employed. In this algorithm, the dimension X is considered as an interval $[0, N/2]$, to find an optimal cutoff point. First, the AFP spectrum is shifted so that low frequencies are centred. Then, bisection is applied symmetrically to both halves along the X-axis. In each iteration, a midpoint $m = (lower + upper)/2$ is computed. Coefficients below m are temporarily set to zero, preserving only the spectral information in the interval $[m, N/2]$, after which an inverse FFT is performed. Due to the symmetrical properties of the FFT, this cutoff has a corresponding effect on the right half of the full domain $[N/2, N]$, creating a mirror image of the preserved region. The resulting spatial domain is compared to the original. Based on a user-defined error threshold, the iteration repeat with upper half of the interval if the error is below the threshold or lower half interval if it is above. The algorithm terminates when the border position stabilizes.

Three error thresholds were considered for the bisection pruning algorithm: Mean Absolute Percentage Error (MAPE), Root Mean Squared Percentage Error

(RMSPE), and Normalised Percentage L_∞ Error. However, RMSPE proved unsuitable due to its quadratic nature, which led to irregular error changes during the threshold search and limited spectral coefficient reduction. Table 1 compares the number of coefficients skipped by the pruned FFT algorithm using MAPE and Normalised Percentage L_∞ Error in a simulation with an arc transducer in water aligned along the X-axis. Note that the L_∞ error reflects the spatial domain error after applying the pruned FFT, excluding transducer-related discrepancies.

Table 1. Comparison of error and skipped rows/columns under different levels of Mean Absolute Percentage Error (MAPE) and Normalised Percentage L_∞ error in homogeneous domain with edge size of 1024.

	MAPE					Norm. Perc. L_∞				
	10%	20%	30%	40%	50%	1%	2%	3%	4%	5%
Rows skip	1	1	614	800	866	670	810	856	886	906
Rows skip [%]	0.10	0.10	59.96	78.12	84.57	65.43	79.10	83.59	86.52	88.48
L_∞ error [%]	0.003	0.003	0.005	0.054	0.312	0.005	0.072	0.240	0.498	0.768

The MAPE appears to yield better accuracy in the simulation results by computing significantly more coefficients than the Normalised Percentage L_∞ Error. However, the MAPE was found inadequate as it fails to accommodate zero values that may be encountered in spatial analysis. In contrast, the L_∞ Error provides a more reliable measurement by focusing on the maximum difference, without being influenced by the distribution of smaller errors. Additionally, this error metric is more intuitive for potential users, as it represents the maximum error occurring at a single grid point in the entire domain, making it easier to adjust based on specific needs. Thus, the Normalised Percentage L_∞ Error is more suitable for area estimation and will be used in all subsequent experiments.

The second method considered for area estimation involved computing the norm over each column of the spectrum and skipping the columns where the norm is below a given threshold. However, this approach has two main disadvantages compared to the bisection method. First, the resulting area of spectral coefficients may not be continuous. In contrast, when the estimated area is continuous, the pruned FFT result can be stored in a reduced-size matrix (with one dimension equal to the original size), which benefits element-wise operations and memory optimisation, especially for large domain simulations. Second, the preprocessing time is higher. The bisection pruning algorithm has a time complexity of $O(\log N)$, whereas the norm computation has a time complexity of $O(N)$ (with N representing half the number of columns, due to the spectral domain's symmetry). Moreover, when considering the computation of forward and backward FFTs to determine the threshold error for the AFP, the preprocessing time becomes significant. Figure 2 illustrates the experimental pipeline used to evaluate the proposed algorithm. The red dashed rectangle highlights the preprocessing stage, consisting of the AFP execution followed by the bisection pruning.

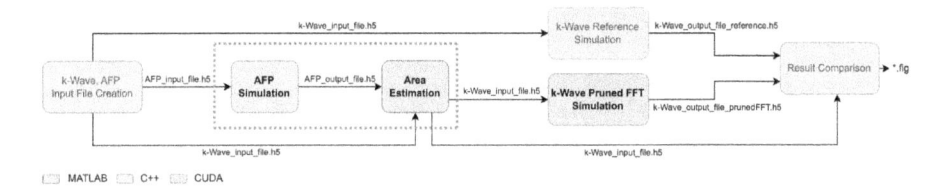

Fig. 2. The pipeline with preprocessing operations to estimate area in the spectral domain and the comparison of the reference and modified simulation.

5 Implementation

All principles described in the previous section result in a proof-of-concept version of the two-dimensional k-Wave wave propagation simulation. The original CUDA implementation of k-Wave was modified to incorporate the pruned FFT algorithm instead of the FFT.

The pruned FFT was implemented using the cuFFT [9] and VkFFT [15] libraries. The computation is divided into two stages: first, a full real-to-complex FFT along the X-axis using cuFFT, and second, a Y-axis FFT using VkFFT. Although cuFFT could perform both, VkFFT offers better flexibility for complex memory layouts—beneficial for future 3D extensions. Because spectral elements with an X coordinate greater than m are zero, only first m Y-axis FFTs are executed. The inverse FFT reverses this process: starting with a Z-axis transformation, then a Y-axis FFT for the first m X slices with VkFFT, followed by zeroing coefficients in the remaining X slices to eliminate artifacts from the initial transform, and finally an inverse X-axis transform with cuFFT.

To evaluate performance and error, the standard FFTs in the k-Wave toolbox were replaced by our pruned FFT in the wave propagation simulation. This approach accelerates computations for acoustic pressure, velocity gradients, and the absorption term during each simulation step. Additionally, by confining non-zero coefficients to a single spectral area, we can optimize element-wise matrix multiplications (highlighted in Eqs. 2 and 4), potentially further reducing simulation times, while leaving CUDA kernels for real domain operations unchanged.

6 Evaluation of the Method over Real Data

In this section, we evaluate the modified version of the k-Wave wave propagation algorithm that utilizes the pruned FFT. To simplify the experimental setup and focus on performance and computational error, all experiments align the wave

(a) (b) (c) (d)

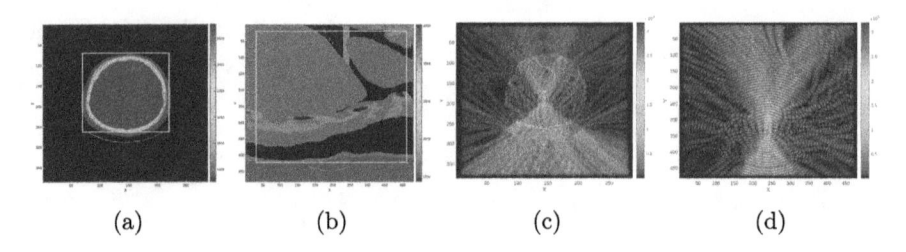

Fig. 3. Ultrasound simulation setups and results: (a) human skull configuration, (b) human liver configuration, (c) skull acoustic pressure distribution, and (d) liver acoustic pressure distribution. Orange rectangle shows error measurement area; green arc indicates transducer position. (Color figure online)

propagation direction with the X-axis, eliminating the need for domain rotation and additional preprocessing.

These examples represent practical clinical scenarios where precise targeting of specific locations within the human body with focused ultrasound waves is essential. Such applications are particularly relevant in therapeutic procedures that rely on ultrasound focusing techniques, allowing us to evaluate the algorithm's performance and precision in contexts closely mirroring real-world treatments.

In the search for the optimal focus position, multiple simulations may be required. To enhance computational efficiency, a two-stage approach is employed: initially, accelerated simulations with an acceptable error margin are performed to identify promising transducer positions. Once an approximate optimal position is determined, a full simulation is executed using that position to ensure accuracy and reliability. This hybrid strategy significantly reduces the overall computational time during the transducer position search while maintaining the necessary precision for medical applications [12].

Furthermore, high simulation resolution is crucial not only in medical applications but also in accurately representing diverse material geometries, especially for structures with large differences in material properties. A higher domain resolution helps prevent stair-casing artifacts and phase shifts [12,21]. Conse-

Table 2. Properties of the skull simulation across different domain resolutions.

	1x	2x	4x	8x	16x	32x
Nx	288	576	1152	2304	4608	9216
Ny	384	768	1536	3072	6144	12288
dx/dy [m]	9.375e−4	4.6875e−4	2.34375e−4	1.17187e−4	5.85937e−5	2.92969e−5
CLF	0.3	0.3	0.3	0.15	0.1	0.05
PPW (water)	5.36	10.73	21.46	42.92	85.85	171.69
Time steps	2798	5595	11189	44753	134258	537031

quently, high-resolution simulations tend to contain many more zero or negligible spectral coefficients, making them particularly suitable for the proposed spectrum pruning.

In all subsequent experiments, the domain properties—sound speed, density, and absorption coefficients—reflect those of real tissue. Figure 3a shows the transducer's position relative to the skull, while Fig. 3b illustrates its position relative to the liver. An additive transducer operating at 300 kHz with an amplitude of 100 kPa was used [20]. The same figures indicate the area within which acoustic pressure error is evaluated. Although the pruned FFT removes high frequencies—resulting in significant error at the transducer itself—the primary concern in ultrasound applications is the error within the tissue. This approach for assessing the accuracy of nonlinear wave propagation in layered, absorbing fluid media follows the method used in [8]. All experiments were executed on an NVIDIA RTX A5000.

Table 3. The measurements in the human skull with different bisection threshold.

	\multicolumn{6}{c}{1% Threshold}
	1x	2x	4x	8x	16x	32x
Skip [%]	22.92	43.40	63.02	77.78	87.41	93.51
L_∞ domain [%]	3.13	10.72	14.15	15.43	16.56	17.62
L_∞ focal point [%]	0.23	1.05	1.13	0.89	0.60	0.76
Focal point shift [mm]	0	0.469	0	0	0.059	0
Time original [s]	0.38	1.69	15.84	243.75	3636.13	54366.42
Time modified [s]	0.37	1.53	12.21	169.62	1916.76	29333.16
Step time original [ms]	0.136	0.296	1.386	5.325	26.48	98.97
Step time modified [ms]	0.132	0.268	1.069	3.706	13.96	53.40
AFP time [s]	0.23	0.74	2.62	9.10	32.34	42.34
Speedup	1.03	1.10	1.30	1.44	1.90	1.85
	\multicolumn{6}{c}{2% Threshold}
Skip [%]	39.58	61.11	77.43	87.59	93.27	96.44
L_∞ domain [%]	5.63	13.82	17.92	20.23	21.67	22.69
L_∞ focal point [%]	0.67	1.21	1.30	1.14	1.19	0.93
Focal point shift [mm]	0	0.469	0	0	0.059	0
Time original [s]	0.38	1.68	15.87	243.67	3654.35	54374.44
Time modified [s]	0.35	1.43	11.93	166.06	1847.03	28782.14
Step time original [ms]	0.136	0.294	1.389	5.324	26.61	98.98
Step time modified [ms]	0.125	0.250	1.044	3.628	13.45	52.39
AFP time [s]	0.26	0.62	2.66	9.47	33.25	174.20
Speedup	1.09	1.17	1.33	1.47	1.98	1.89

To measure the impact of the optimisation on the differently sized simulations, the resolution of the original simulation domains was upscaled using *nearest neighbour* approximation to maintain the domain in its segmented form. This ensures that no artificial material properties are introduced during the upscaling process. To show the impact of the bisection threshold on the simulation result, the measurements were made for 1% and 2% Normalised Percentage L_∞ bisection threshold. The simulation properties for both the skull and liver setup are presented in Tables 2 and 4 respectively.

Table 4. Properties of the liver simulation across different domain resolutions.

	1x	2x	4x	8x	16x
Nx	480	960	1920	3840	7680
Ny	480	960	1920	3840	7680
dx/dy [m]	3.33333e−4	1.66667e−4	8.33333e−5	4.16667e−5	2.08333e−5
CLF	0.3	0.3	0.3	0.3	0.15
PPW (water)	15.09	30.18	60.36	120.72	241.44
Time steps	5524	11047	22094	44187	176746

The original size of the simulation in skull was 288×384 with uniform grid spacing of 9.375×10^{-4}m. This simulation was upsampled up to 32 times while keeping the physical size the same as the original simulation. The original size of the liver simulation was 480×480 with uniform grid spacing of 3.333×10^{-4}m. This simulation was upsampled up to 16 times. Since absorption was present in all experiments, it was necessary to adjust the Courant-Friedrichs-Lewy (CFL) number to maintain simulation stability [18]. The CFL number affects the simulation time step, which can result in longer simulation times. The simulation time was chosen based on the time it takes the wave to travel from one corner of the grid to the geometrically opposite one.

When we examine the results in Table 3 (skull simulation) and Table 5 (liver simulation), we observe that the number of skipped spectral coefficients increases with the simulation domain resolution, reaching up to approximately 90%. The Fig. 4a and 4b show the error distribution of the normalized L_∞ error using a 1% bisection threshold and reveal that significant errors primarily occur at the tissue boundaries where sound speed and density change dramatically. Throughout most of the simulated media, errors remain manageable at just a few percentage points. As expected, increasing to a 2% bisection threshold leads to higher computation errors across all simulations, but also results in a more significant reduction in computed coefficients due to the greater loss of spectral information.

A crucial aspect for focused ultrasound procedures is the accuracy of the focal point and its position. With a 1% bisection threshold, the focal point error remains mostly below 1%. In skull simulations, the focal point shift is minimal, either zero or limited to a single grid point relative to the original domain size.

Table 5. The measurements in the human liver with different bisection threshold.

	1% Threshold				
	1x	2x	4x	8x	16x
Skip [%]	39.17	64.17	76.04	84.64	91.54
L_∞ domain [%]	10.84	20.49	25.25	29.78	33.53
L_∞ focal point [%]	0.47	0.99	0.86	0.05	2.07
Focal point shift [mm]	0	0.236	0.755	0.750	0.750
Time original [s]	1.03	6.75	72.35	537.40	9549.67
Time modified [s]	0.95	5.56	41.38	339.19	5106.51
Step time original [ms]	0.186	0.425	2.14	7.80	34.52
Step time modified [ms]	0.172	0.350	1.22	4.93	18.46
AFP time [s]	0.48	1.43	4.77	15.92	55.81
Speedup	1.08	1.21	1.75	1.58	1.87
	2% Threshold				
Skip [%]	59.58	75.42	85.00	91.41	95.47
L_∞ domain [%]	18.14	25.24	29.31	33.03	35.12
L_∞ focal point [%]	2.60	0.33	1.48	2.02	2.37
Focal point shift [mm]	0	0.850	0.755	0.750	0.750
Time original [s]	1.03	6.77	72.06	539.87	9554.37
Time modified [s]	0.92	5.40	40.13	323.92	4984.51
Step time original [ms]	0.186	0.426	2.13	7.84	34.54
Step time modified [ms]	0.167	0.340	1.19	4.70	18.02
AFP time [s]	0.48	1.32	4.60	16.69	94.15
Speedup	1.12	1.25	1.80	1.67	1.92

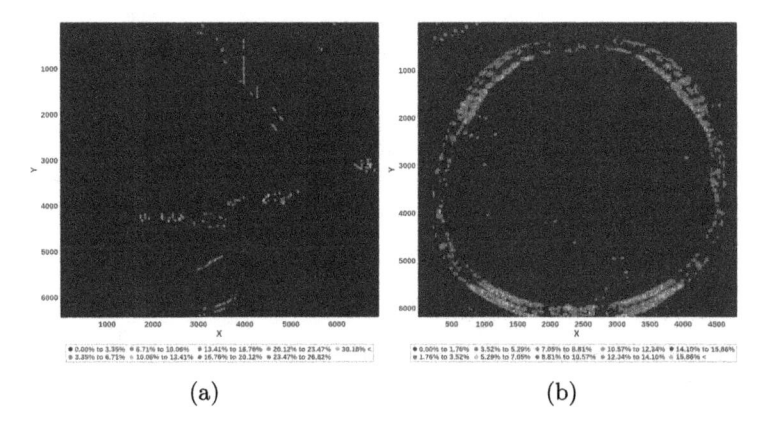

(a)   (b)

Fig. 4. The distribution of the Normalised L_∞ error over grid points of the final acoustic pressure distribution in (a) 16 times upscaled liver and (b) 32 times upscaled skull, grouped into 10 intervals.

In contrast, liver simulations can experience focal point shifts of several grid points due to the heterogeneous nature of bone tissue. Notably, simulation times improved significantly, especially for larger domains. With a 1% bisection error, a speed-up of up to 1.8 times was observed, reducing simulation time from roughly 15 to 8 h—a substantial saving when multiple simulations are required. However, when including the AFP preprocessing time, the pruned FFT's benefits diminish for small domain sizes due to overhead, while for large domains, AFP accounts for only a negligible portion of the total time saved.

Profiling

The profiling of the proof-of-concept pruned FFT implementation was performed on the same GPU used for the experiments, employing the identical input dataset utilized in the resolution evaluations described in Sect. 6.

Figure 5a compares the overall time spent on FFT computations in both directions between the original and modified implementations. The results indicate that acceleration was achieved for all simulation sizes, with larger simulations exhibiting greater speedup. The only exception is the largest simulation, where a decrease in speedup may be due to less efficient FFT algorithm selection. This trend is easily explained by the decreasing percentage of coefficients processed in the Y dimension. As noted in Sect. 2, FFTs account for approximately 60% of the total simulation time. According to Amdahl's law, with 60% of the computation optimisable, the maximum theoretical speedup is 2.5 times. Comparing this limit to the actual speedup shown in Fig. 5a highlights the high efficiency of the optimization.

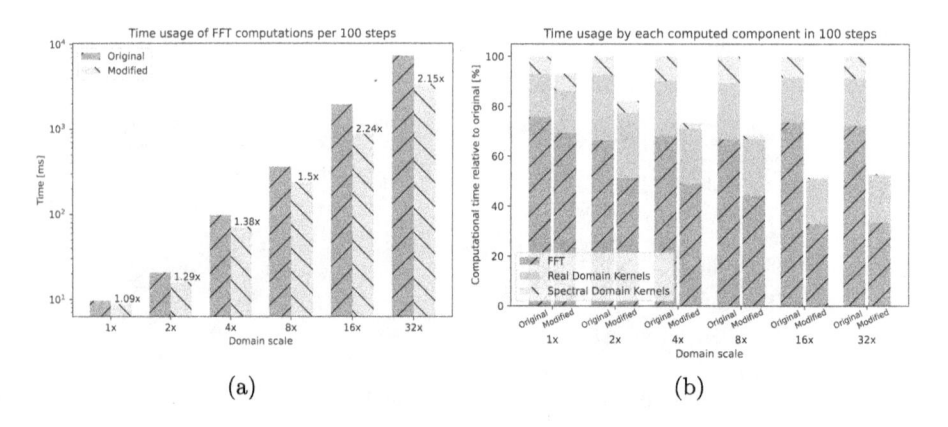

Fig. 5. Performance comparison between original and modified implementations: (a) FFT computation duration; (b) Time percentage breakdown of FFT and kernel computations.

Figure 5b illustrates the relative time usage of the original and modified implementations, divided into three segments: FFTs, real domain kernels, and spectral domain kernels, with the overall duration normalized to that of the original implementation. The significant reduction in computation time is observed in the FFT segment, which decreases in accordance with the domain scale. The time spent on spectral domain kernels is also greatly reduced; for the largest simulation, it drops from approximately 7% of the overall simulation time to less than 1%. As expected, no speedup was observed for the real domain kernels, since the optimizations were applied solely to the spectral part of the computation.

7 Discussion

Comparing our implementation with other acoustic pressure wave solvers [1], experiments at various resolutions show computation errors within acceptable cross-comparison ranges. For human head simulations similar to ours, relative L_∞ error ranges from 10% to 100% compared to k-Wave. Focal point shifts (0–2mm) and acoustic pressure errors (10^{-2}-10^1) also align with benchmarks. While direct comparison is challenging, the results achieved in the experiments presented here are promising and provide the first insight into the accuracy of this method over high-resolution simulations.

A limitation of this approach that requires further investigation is its performance in heterogeneous simulations where the difference between the properties of two media is so high that removing high frequencies from the spectrum prevents the wave from propagating or reflecting correctly. This can lead to significant errors at the boundaries of such media. Another limitation lies in the type of transducer used. For a piston transducer, the overall simulation error should be lower since most of the spectral coefficients representing this transducer are included in the computed part of the spectrum. However, if the piston transducer is unaligned, high computation errors may occur. In contrast, a point transducer is not suitable for this optimization, as accurately representing it requires the inclusion of high frequencies. The worst-case scenario represents the signals with widely distributed spectral content (sharp impulses or broadband noise). Similarly, simulation involving point source rather than distributed transducers may retain significant energy in high frequencies, limiting coefficient reduction.

There are several avenues for further improvement. Implementing the pruned FFT in the first dimension would reduce arithmetic operations and memory accesses. Applying the pruned FFT across all dimensions of the two-dimensional domain would allow storing the resulting spectral coefficients in reduced matrices that correspond to the computed area, potentially enhancing memory performance during simulation steps. Additionally, reducing the spectrum in both dimensions, rather than only along the X-axis, could be considered. However, this would likely increase computation error due to the additional removal of coefficients, and the performance gains may not justify losing the current advantage of maintaining a single continuous area of coefficients, which simplifies subsequent operations.

Despite FFT-pruning being well-established and the impact of spatial resolution on image quality being well understood, integrating these techniques for spectral methods in wave propagation simulations is novel. This approach bridges computational efficiency with high-resolution accuracy in complex wave models, opening new research opportunities in ultrasound simulation and acoustics.

8 Conclusion

This paper presented an approach to accelerate k-Wave's wave propagation simulation by replacing the standard FFT algorithm with a pruned FFT. The proposed method was demonstrated via a proof-of-concept implementation of the pruned FFT integrated into the k-Wave simulation framework. Experiments using simulation data from human skull and liver models showed significant improvements in computational time, particularly for high-resolution domains. Although transitions between media with significant differences in properties such as sound speed and density may lead to high computation errors at their boundaries, the overall impact on focal point accuracy was minimal, resulting in negligible focal shifts.

This approach shows promising potential for focused ultrasound procedures, especially in scenarios where multiple simulations are required to determine the optimal transducer position for targeting specific tissue areas. The improved computational efficiency makes it particularly suitable for treatment planning, where rapid iteration through various transducer configurations is necessary while maintaining acceptable accuracy. Notably, despite the challenges associated with ultrasound penetration through bone, our method achieved up to a 1.8x speedup in large simulation domains with a 1% bisection threshold, demonstrating its robustness in demanding scenarios.

Future research will focus on further improving the pruned FFT implementation and its integration into the k-Wave toolbox, including enhancements to the preprocessing phase, comprehensive evaluation and validation of the proposed approach, and potential extension to three-dimensional simulations. This work contributes to the ongoing effort to enhance the efficiency of spectral methods in wave propagation simulations, particularly for medical ultrasound applications. The promising results pave the way for more efficient high-resolution simulations, potentially enabling faster and more accurate treatment planning in clinical settings.

Acknowledgments. This work was supported by the Ministry of Education, Youth and Sports of the Czech Republic through the e-INFRA CZ (ID:90254). This project has received funding from the European Unions Horizon Europe research and innovation programme under grant agreement No 101071008. This work was supported by Brno University of Technology under project number FIT-S-23-8141.

References

1. Aubry, J.F., Bates, O., Boehm, E.A.: Benchmark problems for transcranial ultrasound simulation: intercomparison of compressional wave models. J. Acoust. Soc. Am. **152**(2), 1003–1019 (2022). https://doi.org/10.1121/10.0013426

2. Cooley, J.W., Tukey, J.W.: An algorithm for the machine calculation of complex fourier series. Math. Comput. **19**(90), 297–301 (1965). http://www.jstor.org/stable/2003354
3. Engholm, M., Stepinski, T.: Designing and evaluating transducers for narrowband ultrasonic spectroscopy. NDT & E Int. **40**(1), 49–56 (2007). https://doi.org/10.1016/j.ndteint.2006.07.006 https://www.sciencedirect.com/science/article/pii/S0963869506000673
4. Gircys, M., Ross, B.J.: Image evolution using 2D power spectra. Complexity **2019**(1), 7293193 (2019). https://doi.org/10.1155/2019/7293193
5. Gottlieb, S., Gottlieb, D.: Spectral methods. Scholarpedia **4**(9), 7504 (2009). https://doi.org/10.4249/scholarpedia.7504
6. Jaros, J., Treeby, B., Rendell, A.: Use of multiple GPUs on shared memory multiprocessors for ultrasound propagation simulations, vol. 127, pp. 43–52 (2012)
7. Markel, J.: FFT pruning. IEEE Trans. Audio Electroacoust. **19**(4), 305–311 (1971). https://doi.org/10.1109/TAU.1971.1162205
8. Martin, E., Jaros, J., Treeby, B.E.: Experimental validation of k-wave: nonlinear wave propagation in layered, absorbing fluid media. IEEE Trans. Ultrason. Ferroelectr. Freq. Control **67**(1), 81–91 (2020). https://doi.org/10.1109/TUFFC.2019.2941795
9. NVIDIA Corporation: cuFFT library. https://developer.nvidia.com/cufft. Accessed 25 Sept 2024
10. Olsak, O., Jaros, J.: Techniques for efficient fourier transform computation in ultrasound simulations. In: Proceedings of the 33nd International Symposium on High-Performance Parallel and Distributed Computing. HPDC '24, Association for Computing Machinery, New York, NY, USA (2024). https://doi.org/10.1145/3625549.3658825
11. Olsak, O., Jaros, J.: Accelerating ultrasound wave propagation simulations using pruned FFT. In: 2024 IEEE International Conference on High Performance Computing and Communications (HPCC), pp. 168–173. https://doi.org/10.1109/HPCC64274.2024.00032
12. Robertson, J.L.B., Cox, B.T., Jaros, J., Treeby, B.E.: Accurate simulation of transcranial ultrasound propagation for ultrasonic neuromodulation and stimulation. J. Acoust. Soc. Am. **141**(3), 1726–1738 (2017). https://doi.org/10.1121/1.4976339
13. Schumacher, J., Püschel, M.: High-performance sparse fast fourier transforms. In: 2014 IEEE Workshop on Signal Processing Systems (SiPS), pp. 1–6 (2014). https://doi.org/10.1109/SiPS.2014.6986055
14. Sorensen, H., Burrus, C.: Efficient computation of the DFT with only a subset of input or output points. IEEE Trans. Sig. Process. **41**(3), 1184–1200 (1993). https://doi.org/10.1109/78.205723
15. Tolmachev, D.: VkFFT-a performant, cross-platform and open-source GPU FFT library. IEEE Access **11**, 12039–12058 (2023). https://doi.org/10.1109/ACCESS.2023.3242240
16. Treeby, B., Cox, B., Jaros, J.: k-wave a matlab toolbox for the time domain simulation of acoustic wave fields user manual (2016). http://www.k-wave.org/manual/k-wave_user_manual_1.1.pdf
17. Treeby, B.E., Budisky, J., Wise, E.S., Jaros, J., Cox, B.T.: Rapid calculation of acoustic fields from arbitrary continuous-wave sources. J. Acoust. Soc. Am. **143**(1), 529–537 (2018). https://doi.org/10.1121/1.5021245

18. Treeby, B.E., Jaros, J., Rendell, A.P., Cox, B.T.: Modeling nonlinear ultrasound propagation in heterogeneous media with power law absorption using a k-space pseudospectral method. J. Acoust. Soc. Am. **131**(6), 4324–4336 (2012). https://doi.org/10.1121/1.4712021
19. Treeby, E.B., Jaros, J., Rendell, P.A., Cox, T.B.: Modeling nonlinear ultrasound propagation in heterogeneous media with power law absorption using a k-space pseudospectral method. J. Acoust. Soc. Am. **131**(6), 4324–4336 (2012). https://doi.org/10.1121/1.4712021
20. Ye, P.P., Brown, J.R., Pauly, K.B.: Frequency dependence of ultrasound neurostimulation in the mouse brain. Ultrasound Med. Biol. **42**(7), 1512–1530 (2016). https://doi.org/10.1016/j.ultrasmedbio.2016.02.012
21. Yoon, K., Lee, W., Croce, P., Cammalleri, A., Yoo, S.S.: Multi-resolution simulation of focused ultrasound propagation through ovine skull from a single-element transducer. Phys. Med. Biol. **63**(10), 105001 (2018). https://doi.org/10.1088/1361-6560/aabe37

A Computational Framework for Modelling Biomechanical Tumour Dynamics and Tissue Interactions: A Proof-of-Concept in Pleural Mesothelioma

Sacha Gijsbers[1,2], Valeria Krzhizhanovskaya[1], Stefano Trebeschi[2,3], and Vivek M. Sheraton[1(✉)]

[1] Computational Science Lab, Informatics Institute, University of Amsterdam, Amsterdam, The Netherlands
v.s.muniraj@uva.nl
[2] Department of Radiology, The Netherlands Cancer Institute – Antoni van Leeuwenhoek Hospital, Amsterdam, The Netherlands
[3] GROW – Research Institute for Oncology and Reproduction, Maastricht University, Maastricht, The Netherlands

Abstract. In Malignant Pleural Mesothelioma (MPM), solid stress and tissue deformation significantly impact tumour growth and invasion. This study presents a computational framework that integrates biomechanical tumour dynamics, tissue deformation, and force interactions within a realistic anatomical setting. Using the Finite Element Method, the framework is applied to a lung mesh reconstructed from CT scans, incorporating a synthetic mesothelioma tumour with defined material properties. Numerical results from the simulations closely match the analytical solution, with deviations within 5%, confirming the model's reliability and accuracy. Simulations of point compression and surface expansion effectively capture the localised tumour deformation and lung volume changes, replicating expected breathing mechanics under different conditions. The findings emphasize the role of mechanical interactions in tumour progression, demonstrating how increased tissue stiffness affects deformation patterns and respiratory dynamics. This study establishes a foundation for integrating computational biomechanics with predictive tumour modelling, offering potential applications in personalised medicine for MPM.

Keywords: Finite Element Method · Tumor biomechanics · Malignant Pleural Mesothelioma · Tissue deformation

1 Introduction

Cancer is one of the leading global health challenges, with growing mortality rates. In 2020, approximately 10 million deaths, or roughly 16% deaths worldwide, were attributed to cancer [31]. Cancer growth displaces and disrupts the normal functioning of healthy organs and tissues, contributing to organ failure and eventual death. One of the tumors that most exemplify this is Malignant Pleural Mesothelioma, also known as MPM. MPM is an aggressive lung cancer. MPM originates from the pleura, a thin membrane lining the lungs and thoracic cavity. It typically spreads in a crescent-shaped pattern, encasing the lung and exerting external pressure on both the lung and the mediastinum [16]. Overall, patient's survival after diagnosis is estimated to be between 4 to 12 months, with some multimodal therapies extending life expectancy to 5 years for 3% to 18% of patients [16]. Biomechanical modelling of tumour interactions with surrounding structures is essential but has often been overlooked, even in cases where its impact is evident, such as in malignant pleural mesothelioma. Therefore, it is critical to advance knowledge about the mechanisms involved in MPM tumour dynamics and the interplay with the surrounding environment in order to develop better treatment options and patient outcomes [23].

Multiple models have been developed to study tumour dynamics and treatment response. They can be divided into three types: discrete, continuum, and hybrid models. Discrete models focus on individual cells or groups of cells, used for cell signalling studies, while ignoring tissue mechanics [23]. Continuum models describe tumours as multi-phase systems using differential equations but lack single-cell resolution [12]. Lastly, the hybrid models combine both approaches, such as modelling cells individually while treating extracellular components as the continuum model [24]. While these models have advanced our understanding of tumour biology, they primarily focus on the microenvironment ignoring the macro-scale mechanical interactions between tumours and their surrounding tissues [26]. For example, tumours can deform adjacent organs or alter physiological functions (e.g., lung expansion during breathing), but this interaction is usually not modelled explicitly. Moreover, biomechanics plays a crucial role in tumour progression. In a realistic anatomical context, mechanical forces would stir the tumour into a growth path of least resistance [8], while causing tissue deformation, impairment of organ functioning and friction between the surrounding tissues [17]. This interplay of mechanical forces, alongside biochemical and biological factors, plays a crucial role in the onset, development, diagnosis, and treatment of cancer. Tumours, like other biological tissues, are subjected to various mechanical forces, which impact cellular function and behaviour [3]. These changes create a complex landscape where both cancer cells and surrounding tissues exhibit distinct physical abnormalities, directly impacting tumour behaviour and its response to treatments [17].

The transition from medical imaging to computational modelling follows a structured process of scanning, geometric reconstruction, and meshing. CT and MRI scans capture high-resolution tumour and lung structures, which are then processed for segmentation. Deep learning-based methods [30] have enhanced automation and accuracy in identifying anatomical features. Computational models automate meshing and predictive tasks, enhancing efficiency while preserving anatomical detail [32]. AI-driven models further support structural analysis, patient-specific simulations, and treatment planning, advancing personalized medicine [5].

One significant biomechanical factor is solid stress, an influential component within the tumour environment that plays a crucial role in cancer initiation, aggressiveness, and metastasis. While other mechanical stresses, such as interstitial fluid pressure (IFP), exist within the tumour microenvironment, solid stress is particularly important as it directly influences tumour growth and interactions with surrounding tissues [22]. Moreover, stresses like IFP and others, including fluid-induced shear stress and hydrostatic pressure, can be directly modelled within the current framework by extending the governing equations to incorporate fluid-structure interactions and pressure-driven deformations. Given its pivotal role in tumour progression, incorporating solid stress into predictive models is essential for improving diagnostic and therapeutic strategies [11].

In this study, this gap will be addressed by developing a computational framework to model tumour dynamics and tissue interactions, with a specific focus on mesothelioma. To investigate the deformation behaviours of lung tissue and tumours, the framework will be benchmarked using simple geometries subjected to compressive and tensile forces, allowing for an initial assessment of mechanical properties. These tests will be compared against analytical solutions to evaluate model performance. Building on this foundation, the model is extended to a realistic anatomical structure: a lung mesh derived from CT scans, with a synthetic mesothelioma tumour modelled as a sheet surrounding the lung. Point compression will be applied to the tumour to simulate forces from surrounding structures, while surface expansion within the lung will mimic breathing mechanics. By testing various deformation scenarios and implementing different elasticity models, this study aims to explore how mechanical forces impact the structural integrity and functionality of these biological structures.

2 Methods

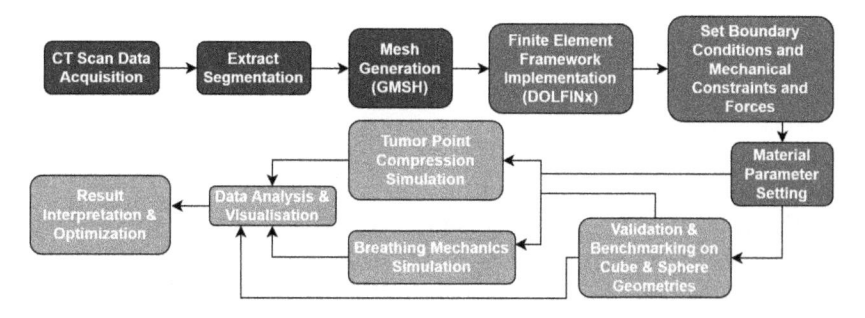

Fig. 1. Overview of the computational workflow for modelling tumour dynamics and tissue interactions. The process consists of data preprocessing, model implementation, simulations, and result analysis, with different colours indicating distinct stages.

Figure 1 illustrates the key steps of the computational framework, from data acquisition and mesh generation to finite element simulations and result analysis, highlighting the integration of different modelling approaches for benchmarking, validation, and biological scenario testing.

2.1 Mathematical Formulation of the Elasticity Model

In this section, we present the mathematical equations that form the basis of the computational framework, focusing on the elastic deformation of materials [6].

Linear Elasticity PDE: using the Navier-Cauchy equations for small deformations can be written as:

$$-\nabla \sigma = f \quad \text{in } \Omega \tag{1a}$$

$$\sigma = \lambda \operatorname{tr}(\varepsilon) I + 2\mu \varepsilon \tag{1b}$$

$$\varepsilon = \frac{1}{2}(\nabla u + \nabla u^T) \tag{1c}$$

where σ is the stress tensor, f is the body force per unit volume, λ and μ are Lamé's elasticity parameters for the material in Ω, I is the identity tensor, tr is the trace operator on a tensor, ε is the symmetric strain tensor (symmetric gradient), and u is the displacement vector field. Above we have assumed isotropic elastic conditions [7,21].

Variational Formulation: consists of forming the inner product of the Eq. 1. Which is gathered by deriving the weak form by integrating the linear elasticity equation against a test function and applying integration by parts [10]. This results in:

$$a = \int_\Omega \sigma(u) : \nabla v \, d\Omega \tag{2}$$

Analytical Solution: used in the simulation is based on uniaxial stress-strain relations from Hooke's Law, resulting in the following reaction force:

$$F_{\text{analytical}} = S \cdot A = \frac{E \cdot \Delta h}{H} \cdot A \tag{3}$$

where $F_{\text{analytical}}$ is the analytically computed force, S is the applied stress, A is the cross-sectional area, E is Young's modulus representing the material's stiffness, Δh is the applied deformation or step size, and H is the initial height of the material. This assumes a simple linear relationship between stress and strain in a homogeneous material under uniform loading, which aligns with the small deformation assumption in the finite element implementation [9].

2.2 Simulation Case Studies

Three unique cases were analysed and simulated in the study. This section summarises the methods utilised for simulating and analysing these cases.

Cube Compression and Stretching: In this simulation, the deformation of a cube under compression and stretching was analysed for varying amounts of total deformation, compared to an analytical solution to serve as a benchmark of the simulation. The deformation was applied in three steps: 10%, 50%, and approximately 100% of the cube's total height. The total deformation was applied uniformly on the cube. An analytical solution based on uniaxial stress-strain relations from Hooke's Law was compared to the resulting deformation. The results were then displayed in a corresponding load-deflection curve was recorded for both the numerical and analytical solution. The numerical and analytical solutions will closely match if the maximum margin of error is below the 5% for numerical approximations [33]. The deformed shapes of the cubes for each deformation level were also visualised.

Tumour Point Compression: The goal of this simulation is to investigate how point compression affects the deformation behaviour of a tumour mesh, later to be used to simulate the tumour-environment interactions. The sphere geometry was compressed at the top, to show how point compression works on simple geometries. After which, a bundle of points were selected, at extreme ends of the tumour mesh, as selecting one point is too small to have a visible impact. The deformation was applied at two levels: 10% and 20% deformation of the total height of the tumour mesh. Point compression was applied at the top and side of the mesh with different total deformation constraints. And at two points of the mesh, to show how multiple point compression can be performed simultaneously.

Lung Surface Extension: The third simulation involves modelling the stretching of surfaces on a cube to represent the surfaces that will also be selected in the lung mesh and the expected result of the expansion. In this simulation, 4 sides and the bottom of the cube and lung meshes are selected to be expanded. For the lung model, 20% of the lung width was selected for each side and bottom surfaces, with only a few of the highest points fixed to prevent displacement. Additionally, 30% of the lung width was selected for the side and bottom surfaces to observe the impact of surface selection on the deformation behaviour. The simulation was coupled with compression to model the breath cycle, and the corresponding changes in volume were plotted. To demonstrate the versatility of the model, the elasticity parameters were adjusted, and a non-linear, stiffer material model was introduced, in which E is scaled with strain. The more strain, the stiffer the model. This allowed for exploration of different material behaviours, showing how the simulation can be adapted for varying conditions. The volume rate of change during the breath cycle was also analysed.

2.3 Finite Element Simulation

The finite element method (FEM) is a numerical technique for solving partial differential equations (PDEs) over complex geometries by discretizing the domain into smaller elements. $FEniCSx$ an open-source platform designed for FEM simulations due to its automatic differentiation is used in this study [1]. $DOLFINx$, the $C++/Python$ interface and the back-end of $FEniCSx0.9$, were used for modelling. It provides the core functionality for mesh handling, function spaces, and solving PDEs [13,14]. A third-order Lagrange finite element space was employed to achieve higher accuracy in the numerical simulations.

Simulation Parameters and Material Properties: To perform the simulations, material properties for the tumour and lung tissues were selected based on reported values in the literature. These properties include the Young's modulus (E) and Poisson's ratio (ν), which were used to calculate the Lamé parameters (λ, μ) in Eq. 1b (Table 1).

Table 1. Material properties used in the simulations, including Young's modulus (E) and Poisson's ratio (ν) for tumour and lung tissues. The table also specifies the simulation domain size, mesh resolution, and relevant references.

Tissue Type	E [Pa]	ν	Domain Size (mm)	Mesh Cells	References
Tumour	10^5	0.3	$[161 \times 161 \times 273.52]$	$5,922,906$	[17,20,25]
Lung	10^4	0.3	$[147 \times 147 \times 267.89]$	$3,032,524$	[20,25]
Lung (Non-linear)	$10^4 - 10^8$	0.3	$[147 \times 147 \times 267.89]$	$3,032,524$	[20,25]

The table summarises the material properties used in Eq. 1, for the simulations, E and ν for tumour and lung tissues. For normal lung tissue, a constant E of 10 kPa was chosen [25]. For non-linear lung tissue, E is assumed to increase with stress, reflecting the strain-dependent stiffening typical of lung tissue, especially under larger deformations or pathological conditions. Tumour tissue is modelled with a higher E of 100 kPa, simulating the stiffness of fibrotic or advanced tumours [17]. A Poisson's ratio of 0.3 was used for both tissues, reflecting their near-incompressibility [20,25].

Mesh Generation

Mesh Definitions for Unit Cube and Sphere: A structured unit cube mesh was generated using $DOLFINx's$ built-in meshing capabilities as a benchmark model. The domain, $[0,1] \times [0,1] \times [0,1]$, was discretized into a hexahedral mesh, which consists of small cube-shaped elements. After which, the cube was uniformly divided into 12 elements along all axis, resulting in a total of 1,728 elements [14]. A sphere tetrahedral mesh was created using Gmsh, allowing for finer meshing control compared to FEniCSx's native tools and a complexer shape compared to the cube. The meshes were converted to XDMF format for use in FEniCSx simulations.

Tumour and Lung Mesh Definitions: Open-source CT scan data was used from the Cancer Imaging Archive [2], with *TotalSegmentator* [28] employed to extract the right lung instead of the left, as seen in Fig. 2a. The right lung was chosen due to its larger volume and three-lobe structure, making it a better candidate for studying tumour interactions. The segmented lung was processed to ensure accurate voxel spacing for further mesh generation. The segmented lung data was converted into a tetrahedral mesh using *Gmsh*, see Fig. 2b. Surface

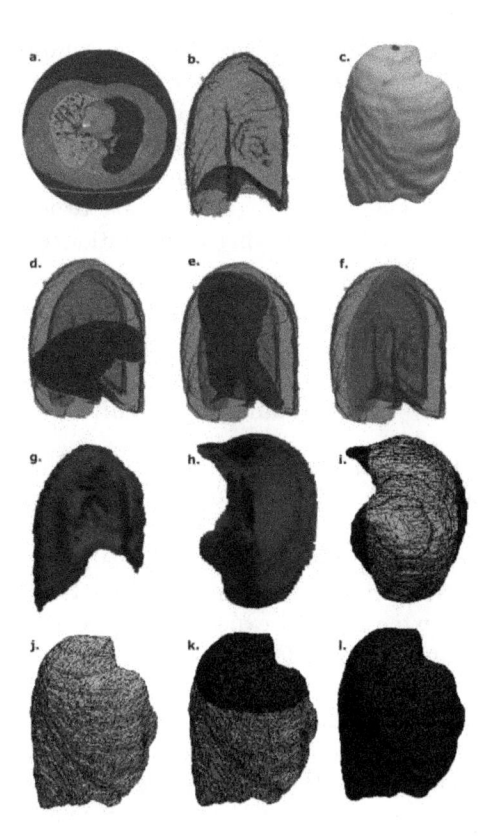

Fig. 2. (a) CT scan used to extract lung segmentation, with the right lung highlighted in magenta. (b) Tetrahedral mesh of the segmented right lung. (c) Synthetic tumour mesh, with the purple point indicating the top of the tumour used for compression simulations. (d)-(f) Cross-section images of the tumour, showing the empty space inside, representing the space occupied by the lung. (g)-(h) Cut surfaces of the tumour to visualise internal structures. (i) High-resolution sides of the tumour for surface expansion simulations. (j) Downsampled mesh with 5% of the original points for computational efficiency. (k) High-resolution region at the top of the tumour for point compression simulations. (l) Full-resolution tumour mesh.

extraction was performed using marching cubes, followed by mesh simplification and repair to ensure watertightness, meaning the mesh is completely enclosed without holes or gaps, making it suitable for simulations and further processing. The refined surface was then imported into *Gmsh*, where a surface loop and volume definition enabled high-quality mesh generation. Mesh parameters were adjusted to optimise curvature adaptation, and the final mesh was exported to *XDMF* format for use in *FEniCSx* simulations. A synthetic tumour was created encasing the segmented right lung based on the crescent-like mesothelioma growth patterns, see Fig. 2c. The tumour was generated using morphological dilation operations to simulate realistic spread, ensuring it remained confined to 2cm outside the lung, making it a significant sized tumour. See Fig. 2d-f for cross sections of the tumour and Fig. 2g & e, for the cut tumour, to see the inside structures. To optimise computational efficiency, the segmentation was downsampled to 5% of the original number of points (see Fig. 2j for the downsampled mesh and Fig. 2l for high resolution mesh) while maintaining high resolution in regions where pressure would be applied. The same *Gmsh*-based meshing process was applied to the tumour segmentation to ensure consistency in the tetrahedral mesh. The final meshes maintained high-resolution regions where mechanical stress would be applied while optimising computational efficiency, see Fig. 2i &k.

3 Results

3.1 Benchmark Simulation Results on a Unit Cube

The results of the cube compression and extension simulations are shown in Fig. 3, where the deformation patterns and load-deflection curves are compared to the analytical solutions 3. Figures 3(a-c) illustrate the normalised displacement for three different compression levels: 10%, 50%, and 100% of the total height. Figures 3(g-i) show the deformation patterns for cube stretching at 10%, 50%, and 100% total extension. As expected, the deformation increases proportionally with the applied load. Figures 3(d-f & j-l) show the corresponding load-deflection curves, comparing the numerical (Eq. 1) and analytical solutions (Eq. 3). Simulation results show a good agreement with the analytical solution, suggesting that the numerical method accurately captures the expected deformation behaviour. To quantify this, the maximum relative error between numerical linear elasticity solution and analytical solution of uniaxial stress-strain relations (from Hooke's Law) was calculated, revealing a consistent 4.68% error, well within the 5% requirement. All simulations exhibit the same error because the chosen linear elasticity model inherently scales proportionally with applied forces and deformations. The results demonstrate that the numerical model reliably predicts deformation under uniform loading and extension, ensuring sufficient mesh resolution and correctly applied boundary conditions, making it suitable for more complex simulations.

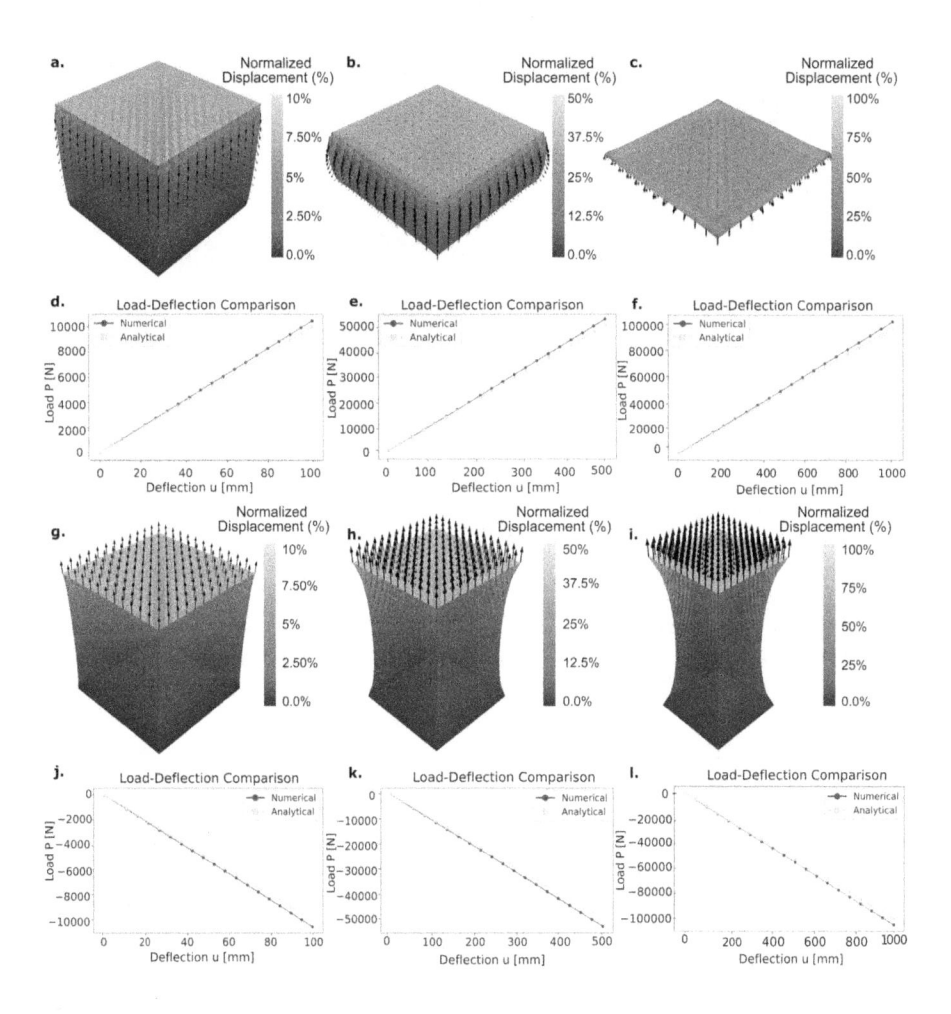

Fig. 3. (a-c) Normalised displacement for three different compression levels: 10%, 50%, and 100% of the total height. (g-i) Deformation patterns for cube stretching at 10%, 50%, and 100% total extension. The bottom of the cube remains fixed, and the displacement is visualised using a colour gradient, with higher displacement values at the top surface. Deformation increases proportionally with the applied load (d-f) and (j-l).

3.2 Tumour Point Compression Simulation Results

The point compression simulation was performed on a spherical geometry and a complex tumour mesh to analyse localised deformation. Applied forces and resulting displacements were recorded to assess the finite element model's accuracy. In the spherical case, top compression caused a localised indentation with a maximum normalised displacement of 10%, showing smooth, symmetric deformation consistent with theory.

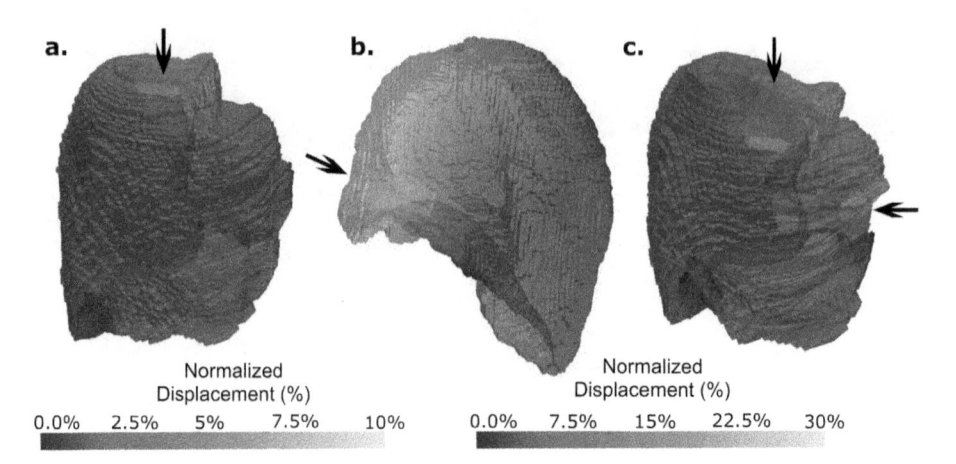

Fig. 4. (a) Tumour mesh under single-point compression at the top, producing a localised indentation with a maximum normalised displacement of 10% of the tumor's total height. (b) Increased deformation magnitude and shifted compression location, showing asymmetric displacement distribution. (c) Multiple-point compression applied at the top and side of the tumour mesh, simulating complex loading conditions with maximum normalised displacement reaching 30%. The displacement is visualized using a colour gradient, where higher displacement values are represented in yellow and lower displacement values in purple. (Color figure online)

The tumour mesh (Fig. 4b-d) was subjected to the single-point compression at the top, from one side and from 2 sides. The simulation tested different displacement magnitudes and locations. First, a small deformation (10% of the tumour's total height) was applied (Fig. 4b), followed by an increase in deformation magnitude and a shift in the compression location to evaluate the response (Fig. 4c). The results confirmed that the implemented FEM correctly captured the expected indentation behaviour, with the displacement distribution smoothly radiating from the compression point. Figure 4c shows the result of a single compression force at the side, testing the mesh's response under asymmetric loading. This preliminary test ensured that the code correctly handled irregular geometries and non-uniform stress distributions before introducing more complex loading conditions. Once verified, multiple points were compressed simultaneously, as seen in Fig. 4d, to simulate realistic interactions in a biological environment. This is particularly relevant as surrounding structures in the body exert different forces at various locations on the tumour, forcing it to deform and adapt accordingly. Compression at the top and side of the mesh resulted in highly localised deformations, with maximum normalised displacement reaching 30% in the most compressed regions. Across all test cases, the numerical model accurately captured the expected deformation patterns, confirming its reliability for simulating soft tissue mechanics under localised compression.

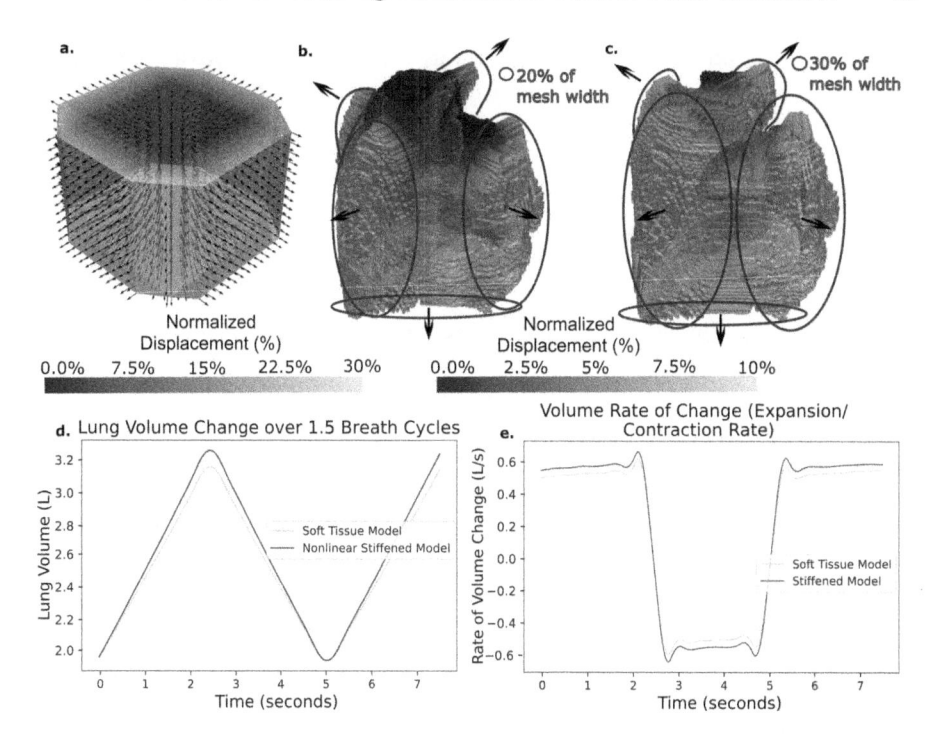

Fig. 5. (a) Expansion of selected surfaces in the lung mesh, illustrating the direction of expansion along four sides and the bottom surface, while the top surface remains fixed. (b, c) Applied expansion of 20% and 30% of the mesh surface width, respectively. The 20% expansion case shows localised deformation, whereas the 30% expansion leads to deformation across the entire lung. (d, e) Simulated lung volume and volume rate of change during breathing cycles.

3.3 Surface Expansion Simulation Results

The surface compression simulation simulated the expansion of surfaces in both the cube and lung models to analyse deformation behaviour under stretching forces. Figure 5a illustrates the expansion of selected surfaces in the lung mesh, showing the direction of expansion along four sides and the bottom surface, while the top surface remains fixed. The applied expansion was set to 20% and 30% of the mesh width, as shown in Fig. 5b & 5c, respectively. The 20% expansion case resulted in localised deformation, with some regions of the lung remaining nearly stationary. In contrast, the 30% expansion case produced deformation across the entire lung, suggesting a more realistic representation of breathing mechanics. Figures 5d & 5e show the corresponding lung volume and volume rate of change during simulated breathing cycles. The lung volume (Fig. 5d) follows a cyclic pattern with a sinusoidal-like curve, which is consistent with the expansion and contraction dynamics expected in real pulmonary function. The volume rate of change (Fig. 5e) further confirms this behaviour, showing a smooth transition between phases of inhalation and exhalation. The stiffened model, see Fig. 5d & e, which incorporates a non-linear elasticity response, resulted in a more gradual expansion at higher deformations compared to the soft tissue model. The ability to simulate both linear and non-linear deformation allows for a detailed evaluation of how different material properties influence lung mechanics under expansion forces.

4 Discussion

The results highlight the effectiveness of point compression in simulating diverse pressure distributions on the tumour. This method provides a foundation for future simulations where mechanical forces can be scaled based on the densities and stiffness of surrounding tissues, allowing for a more realistic representation of tumour-environment interactions. Similarly, the lung volume change simulations followed expected physiological patterns, capturing key characteristics of normal respiration and forced respiration. The volume rate of change reflects the known asymmetric behaviour of inspiration and expiration due to airway resistance and lung tissue viscoelasticity, consistent with experimental findings [18]. The observed reduction in peak lung volume and slower rate of volume change in the stiffened lung model aligns with clinical studies on pulmonary fibrosis, where increased tissue stiffness leads to restricted expansion and airflow dynamics [25].

Lung tissue is inherently non-linear, and experimental studies have demonstrated that lung parenchyma does not behave as a simple linear elastic material. Instead, strain-dependent stiffening occurs due to collagen fibre engagement during deep inhalation [25]. The stiffened model in this study successfully captures this effect, improving its biomechanical accuracy. Additionally, the cyclic breathing pattern observed in the simulations suggests energy-efficient breathing dynamics, where tissue resistance at higher lung expansion mirrors real-life pulmonary mechanics [29]. The damping effects observed in the volume rate of change further support this accuracy, as lung tissue naturally absorbs mechanical energy to prevent damage under large deformations. This aligns with previous studies highlighting how lung mechanics involve both elasticity and resistance to rapid expansion/contraction [25].

Unlike existing continuum or hybrid models that rely on homogenized tissue properties or fixed boundary conditions [23], this model provides spatially resolved, patient-specific mechanical feedback from actual lung and tumour geometries. This enhances its ability to simulate local tissue deformation and mechanical impedance, critical for mesothelioma's heterogeneous, surface-bound growth patterns. Future versions should incorporate time-dependent tumour growth, driven by proliferative pressure, evolving stiffness, and mechanical feedback from the surrounding lung parenchyma, allowing for more dynamic, realistic growth. Integrating growth kinetics or coupling with cellular automata models could better simulate the spatial and temporal progression of mesothelioma. This would not only help understand how increasing stiffness and tumour volume affect breathing mechanics but also identify the path of least growth resistance. By simulating tumour expansion, the model could pinpoint areas where the tumour is likely to grow with minimal resistance, aiding in treatment planning and surgical decision-making, including predicting tumour invasion and adaptation to mechanical changes.

A key application of this model lies in personalized medicine, where patient-specific tumour and lung geometries from medical imaging could be incorporated into individualized simulations [27]. Such models could improve predictions of tumour-tissue interactions, enabling more precise treatment strategies and surgical interventions tailored to a patient's unique biomechanical properties. Beyond oncology, the ability to simulate mechanical changes in diseased lungs has broader clinical implications. Conditions such as pulmonary fibrosis and chronic obstructive pulmonary disease (COPD) involve significant alterations in lung mechanics, which could be explored using similar modelling approaches. Understanding how stiffening or obstruction affects lung compliance may inform improved ventilation strategies in clinical settings [4,25].

Expanding beyond biomechanics, integrating this model with biochemical processes such as nutrient transport, oxygen diffusion, and metabolic activity would provide a more comprehensive understanding of tumour-environment interactions [19]. Additionally, incorporating blood flow dynamics could enhance realism by accounting for vascular adaptations, such as tumour-induced angiogenesis or perfusion deficits in diseased lung tissue [15].

5 Conclusions

This study aimed to develop a computational framework based on linear elasticity to model tumour dynamics and tissue interactions, with a specific focus on mesothelioma. The approach was first validated using simple geometries, such as a cube and a sphere, subjected to compressive and tensile forces. By comparing the numerical results with analytical solutions, the model demonstrated strong accuracy, with numerical deviations staying within the expected 5% error margin. This validation confirms the reliability of the framework for further simulations involving more complex geometries and biological structures.

Beyond simple benchmarking tests, the model was extended to simulate tumour and lung interactions using point compression and surface expansion analyses. The point compression simulations on both spherical geometry and tumour mesh successfully captured localised deformation behaviour, confirming the model's ability to handle asymmetric and multi-point loading conditions. Additionally, the surface expansion simulations on the lung mesh demonstrated realistic breathing dynamics, with lung volume changes following expected cyclic patterns. The results indicate that the framework can effectively simulate soft tissue deformation and mechanical interactions, providing a strong foundation for modelling tumour progression and biomechanical responses.

While the current framework effectively captures fundamental tissue mechanics, it remains a simplistic model and requires further refinement to more accurately represent complex biological systems. More advanced testing, including the use of more difficult force equations beyond linear elasticity, is necessary for improving the model's applicability to real-world scenarios. Future work will focus on incorporating additional mechanical properties, such as viscoelasticity and anisotropy, to better represent lung and tumour behaviour. The inclusion of patient-specific data from imaging techniques could further enhance the accuracy of simulations, making the model more relevant for MPM applications. This study sets a baseline from which these refinements can be made.

Acknowledgments. The authors acknowledge the National Cancer Institute and the Foundation for the National Institutes of Health, and their critical role in the creation of the free publicly available LIDC/IDRI Database used in this study. The authors would like to acknowledge the Research High Performance Computing (RHPC) facility of the Netherlands Cancer Institute (NKI).

References

1. Alnæs, M.S., Blechta, J., Hake, J.: The FEniCS project version 1.5. Arch. Num. Softw. **3** (2015). https://doi.org/10.11588/ans.2015.100.20553
2. Armato III, S.G., McLennan, G., Bidaut, L.: Data from LIDC-IDRI, the cancer imaging archive (2015). https://doi.org/10.7937/K9/TCIA.2015.LO9QL9SX
3. Bao, L., Kong, H., Ja, Y.: The relationship between cancer and biomechanics. Front. Oncol. **13** (2023). https://doi.org/10.3389/fonc.2023.1273154
4. Bates, J.H., Irvin, C.G.: Measuring lung function in mice: the phenotyping uncertainty principle. J. Appl. Physiol. **94**(4), 1297–1306 (2003)
5. Bharati, S., Mondal, M., Podder, P. Deep learning for medical image registration: a comprehensive review. arXiv preprint arXiv:2204.11341 (2022)
6. Bonet, J., Wood, R.D.: Nonlinear Continuum Mechanics for Finite Element Analysis. Cambridge University Press (1997)
7. Bower, A.F.: Applied Mechanics of Solids. CRC Press (2009)
8. Esmaeili, M., Stensjøen, A.L., Berntsen, E.M.: The direction of tumour growth in glioblastoma patients. Sci. Rep. **8**(1), 1199 (2018)
9. Huda, Z.: Mechanical Behavior of Materials: Fundamentals, Analysis, and Calculations. Springer (2021)
10. Hughes, T.J.: The Finite Element Method: Linear Static and Dynamic Finite Element Analysis. Courier Corporation (2003)
11. Islam, M.T., Righetti, R.: A novel finite element model to assess the effect of solid stress inside tumors on elastographic normal strains and fluid pressure. J. Eng. Sci. Med. Diagn. Therapy **2**(3), 031006 (2019)
12. Kaura, P., Mishra, T., Verma, N.: Effects of combined chemotherapeutic drugs on the growth and survival of cancerous tumours- an in-silico study. J. Comput. Sci. **54**, 101421 (2021). https://doi.org/10.1016/j.jocs.2021.101421
13. Logg, A., Wells, G.N.: DOLFIN: automated finite element computing. ACM Trans. Math. Softw. **37** (2010). https://doi.org/10.1145/1731022.1731030
14. Logg, A., Wells, G.N., Hake, J.: DOLFIN: a C++/Python finite element library. In: Logg, A., Mardal, K.A., Wells, G.N. (eds.) Automated Solution of Differential Equations by the Finite Element Method. LNCS, vol. 84. Springer (2012)
15. McDougall, S.R., Anderson, A., Chaplain, M.: Mathematical modelling of flow through vascular networks: implications for tumour-induced angiogenesis and chemotherapy strategies. Bull. Math. Biol. **64**(4), 673–702 (2002)

16. Napoli, F., Listì, A., Zambelli, V.: Pathological characterization of tumor immune microenvironment (time) in malignant pleural mesothelioma. Cancers **13** (2021). https://doi.org/10.3390/cancers13112564
17. Nia, H., Munn, L., Jain, R.: Physical traits of cancer. Science **370** (2020). https://doi.org/10.1126/science.aaz0868
18. Otis, A.B., McKerrow, C.B., Bartlett, R.A.: Mechanical factors in distribution of pulmonary ventilation. J. Appl. Physiol. **8**(4), 427–443 (1956)
19. Pries, A.R., Secomb, T.W.: Blood flow in microvascular networks. In: Microcirculation, pp. 3–36. Elsevier (2008)
20. Raveh Tilleman, T., Tilleman, M., Neumann, H.: The elastic properties of cancerous skin: Poisson's ratio and young's modulus. Optim. Incisions Cutan. Surg. Mohs' Micrograph. Surg. **105**(2) (2004)
21. Sadd, M.H.: Elasticity: Theory, Applications, and Numerics. Academic Press (2009)
22. Sarntinoranont, M., Rooney, F., Ferrari, M.: Interstitial stress and fluid pressure within a growing tumor. Ann. Biomed. Eng. **31**, 327–335 (2003)
23. Sciumè, G., Gray, W., Ferrari, M.: On computational modeling in tumor growth. Arch. Comput. Methods Eng. **20**, 327–352 (2013)
24. Sheraton, M.V., Chiew, G., Melnikov, V.: Emergence of spatio-temporal variations in chemotherapeutic drug efficacy: in-vitro and in-silico 3D tumour spheroid studies. BMC Cancer **20**, 1–16 (2020)
25. Suki, B., Bates, J.H.: Lung tissue mechanics as an emergent phenomenon. J. Appl. Physiol. **110**(4), 1111–1118 (2011)
26. Sun, Y., Yao, J., Yang, L.: Computational approach for deriving cancer progression roadmaps from static sample data. Nucleic Acids Res. **45**(9), e69–e69 (2017)
27. Viceconti, M., Henney, A., Morley-Fletcher, E.: In silico clinical trials: how computer simulation will transform the biomedical industry. Int. J. Clin. Trials **3**(2), 37–46 (2016)
28. Wasserthal, J., Breit, H.C., Meyer, M.T.: Totalsegmentator: robust segmentation of 104 anatomic structures in CT images. Radiol. Artif. Intell. **5**(5), e230024 (2023)
29. Weibel, E.R., Hsia, C.C., Ochs, M.: How much is there really? Why stereology is essential in lung morphometry. J. Appl. Physiol. **102**(1), 459–467 (2007)
30. Wilding, R., Sheraton, V.M., Soto, L.: Deep learning applied to breast imaging classification and segmentation with human expert intervention. J. Ultrasound, 1–8 (2022)
31. World Health Organization: Cancer (2020). https://www.who.int/en/news-room/fact-sheets/detail/cancer
32. Zhang, Z., Liu, Q., Wang, Y.: Road extraction by deep residual u-net. IEEE Geosci. Remote Sens. Lett. **15**(5), 749–753 (2018)
33. Zienkiewicz, O.C., Taylor, R.L.: The Finite Element Method: Volume 1, The Basis. Butterworth-Heinemann, 5th edn. Oxford, UK (2000)

Towards Sensitivity Analysis: 3D Venous Modelling in the Lower Limb

Magdalena Otta[1,2,3](✉), Karol Zając[1], Maciej Malawski[1,6],
Ian Halliday[2,3], Chung Lim[5], Janice Tsui[4,5], and Andrew Narracott[2,3]

[1] Sano Centre for Computational Medicine, Kraków, Poland
m.otta@sanoscience.org
[2] Division of Clinical Medicine and Population Health, University of Sheffield, Sheffield, UK
motta1@sheffield.ac.uk
[3] Insigneo Institute for in silico medicine, University of Sheffield, Sheffield, UK
[4] University College London, London, UK
[5] Royal Free London NHS Foundation Trust, London, UK
[6] Faculty of Computer Science, AGH University of Kraków, Kraków, Poland
https://sano.science/

Abstract. Deep vein thrombosis (DVT) of the lower extremity frequently leads to long-term complications known as post-thrombotic syndrome (PTS). The current clinical workflow for DVT and PTS treatment lacks sufficient evidence. The significance of the variation in the venous anatomy is yet to be understood. We report an analysis of a set of idealised 3D geometries of iliac vein unification to assess the importance of shape variability, inflow conditions, and viscosity on local haemodynamics - specifically on the wall shear stress metrics. Regions of low wall shear stress and high oscillating shear index have been associated with prothrombotic effects on the walls of blood vessels. A detailed steady state analysis focused on the wall shear stress distributions below three thresholds ($< 0.15 Pa, < 0.10 Pa < 0.05 [Pa]$). The preliminary work in the transient state focused on the oscillating shear index above three thresholds ($> 0.25, > 0.35, > 0.45$). We found that all the variations implemented had an effect on the size and shape of the absolute vein wall area subject to the shear metrics of choice under the assumed flow conditions. The results obtained in this research will serve as a basis for the interpretation of patient-specific geometries of the iliac vein unification affected by deep vein thrombosis.

Keywords: deep vein thrombosis · post-thrombotic syndrome · venous modelling · CFD · sensitivity analysis

1 Introduction

Post-thrombotic syndrome (PTS) is the most common long-term complication of deep vein thrombosis (DVT) of the lower extremity, a disease caused by abnormal blood clotting, often found in the iliofemoral region of the body [1]. Up to 100 in 100,000 experience the first episode of symptomatic DVT every year. Up to 50% will develop PTS in two years and about 33% will experience recurrent DVT within ten years [2]. Iliofemoral DVT usually presents the most severe symptoms and is more likely to cause PTS than other anatomical variants of the disease. The way DVT and PTS manifest can vary significantly between patients, making treatment challenging. In addition, current clinical workflows often lack strong evidence to support their effectiveness. More research is needed to understand the impact of variations in venous anatomy on the haemodynamics of the affected area. Blood flow changes, particularly stasis (prolonged blood residence) and recirculation, have been identified as a risk factor for thrombus development, but due to the potential variability of these changes in complex venous anatomy, this is not currently well understood.

Computational Fluid Dynamics (CFD) is an established approach for modelling blood flow in segments of the cardiovascular system. A short review of key concepts of cardiovascular modelling is provided by [3]. CFD analysis is typically based on solving the Navier-Stokes equations. Numerical solutions are obtained from the finite volume/element discretisation method. Depending on the purpose of the study, a numerical model can range from 0D to 3D in dimensionality. Representing and simulating all vessels in a numerical model would be impractical due to the high computational cost. Typically, vessels are lumped into several components that represent specific anatomical regions in reduced-order models (0D, 1D), sufficiently detailed to resolve the physiology of the real system [4], or localised segments of the cardiovascular system are modelled in detail (3D). Because DVT is a localised problem, 3D modelling could help to understand the local haemodynamics in the region of the thrombus. For any 3D model, the nature of the fluid dynamics depends on both the local 3D model geometry and the boundary conditions applied to this local 3D domain. Interpretation of the impact of flow changes on the biological response can be supported using parameters derived from the results of a CFD analysis. The local wall shear distribution has been associated with the behaviour of endothelial cells. The low wall shear stress region may cause endothelial dysfunction leading to an increased risk of thrombosis initiation and progression. Although the subject has been investigated in the context of arterial networks [5], the literature on wall shear stress that promotes thrombosis in the venous system remains scarce with a few exceptions, including [6] that used CFD to quantify haemodynamics in the portal venous system looking at the area of the wall regions exposed to low wall shear stress before and after splenectomy. A more relevant example, [7] investigated the impact of iliac vein stenting on blood flow, by simulating haemodynamics, including wall shear stress (WSS) and oscillatory shear index (OSI), in patient-specific geometries after stenting to assess the risk of restenosis. These studies focused on patient-specific geometries without investigating the relative influence of controlled variation on the calculated metrics.

This work focuses on the possibility of applying and personalising blood flow models to aid clinical decisions in the treatment of deep vein thrombosis. We investigated the significance of changes in vessel geometry, boundary conditions, and viscosity in predicted wall shear stress metrics in a set of idealised geometries of iliac vein unification.

2 Methods

In previous work, we developed a 0D model of the circulation in the lower extremities [8], including arterial inflow, capillaries, and venous outflow to investigate the influence of anatomical variability on global haemodynamics in the lower extremity; we achieved this by varying the radii of the vessels (±10%) from their reference values taken from the literature [9]. Through this analysis, we characterised the influence of such variability on the inflow and outflow in the iliofemoral region, particularly the unification of the internal and external iliac veins into a common iliac vein. In this study, these flow conditions were used to define boundary conditions for simulations using computational fluid dynamics (CFD) for a set of idealised 3D geometries of the iliac veins to investigate interactions between variability of these boundary conditions and other factors associated with the local definition of the 3D model on the nature of the detail of the flow in this region. Steady-state simulations were conducted in ANSYS Fluent. Preliminary sensitivity analysis of the 3D flow field was performed by varying the geometry, boundary conditions, and blood viscosity to investigate their relative influence on the distribution of WSS, a parameter related to blood coagulation and the development of thrombuses [10].

2.1 The Base State Simulation

All variations were considered relative to a single case, termed the *base case* or the *base state*. To reflect a realistic anatomical configuration, the geometry had two inlets and one outlet (Fig. 1). The parameters of each considered vessel were taken from [9] as provided in Table 1. The simulation was set as steady state *laminar* flow with constant density ρ (*incompressible*) and constant viscosity μ (*Newtonian*), with $\rho = 1050 \, kg \cdot m^{-3}$ and $\mu = 0.0035 \, Pa \cdot s$. To ensure developed flow profiles, velocity inlets were prescribed with parabolic profiles. The outlet reference pressure was set to zero. For the internal iliac vein (small inlet), the maximum velocity of the parabolic profile was $U_{max|int} = 0.52 \, m/s$ and for the external iliac vein (large inlet) $U_{max|ext} = 0.16 \, m/s$. These values were defined to match the volume flow rate obtained from previous simulations using the 0D model. To avoid a significant influence of the zero pressure outlet condition on the flow field within the region of interest, close to the unification of the veins, the outlet vessel was extended from its physiological length to ten times its diameter, that is, from $58 \, mm$ to $115 \, mm$.

2.2 Mesh Sensitivity

Mesh sensitivity was performed for the base state geometry, analysing pressure and velocity metrics, to obtain a mesh with sufficient resolution at an acceptable computational cost. Meshes of five different resolutions were considered, four

Table 1. Iliac veins - geometrical parameters

vein	d [mm]	L [mm]
external iliac	10.0	144.0
internal iliac	3.0	50.0
common iliac	11.5	58.0

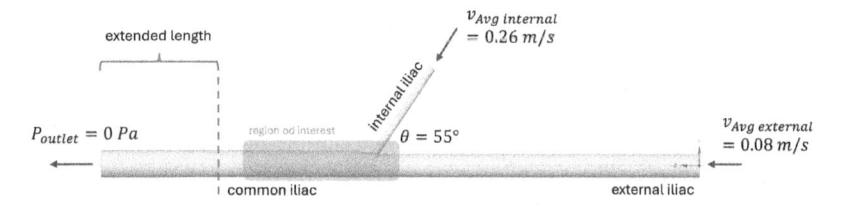

Fig. 1. Base case setup: base geometry with straight vessels and 55-degree angle between internal and external iliac veins; common iliac vein (outlet) extended from real length to ten times its diameter; inlets of fixed average velocity from the 0D model and zero-pressure outlet. Region of interest - expected low shear stress region highlighted close to the join of the vessels.

with global refinement of maximum element lengths of $1\,mm$, $0.5\,mm$, $0.25\,mm$ and $0.2\,mm$, and one with additional local refinement using a sphere of influence in the region around the unification with maximum element edge length of $0.25\,mm$ and refined edge length of $0.15\,mm$ (Fig. 2). The meshes were constructed from polyhedral elements. The finest refinement level was chosen considering the available computational resources. Each mesh had four inflation layers applied to the wall with a growth rate of 1.2. Figure 2 provides an overview of the five levels of mesh refinement that display the grid at the boundaries and on the cut plane through the long axis of the unification. To assess the quality of the mesh, we considered the variation in velocity and pressure metrics, as well as the area of the wall subject to low values of wall shear stress with mesh refinement level.

2.3 Variation in Vessel Geometry

The variation of vein geometry was considered by creating three subsets of idealised iliac veins. The first subset included a variation in the angle between the internal and external iliac vein $\pm 25°$ from the base state of $55°$ resulting in geometries with angles from $30°$ to $80°$. The second subset assumed a variation in the global curvature (simple, further referred to as type i) of the external to common iliac vein axis with the radius of curvature at $5\,mm$, $10\,mm$ and $15\,mm$. The third subset assumed variation in the curvature of each vein separately (complex, further referred to as type ii) with the curvature radii between $5\,mm$ and $15\,mm$. In both subsets of curvature changes, the choice of the bending direction was dictated by that observed in medical images of real iliac vein unifications.

Two types of vein connection were investigated for the base case and representative curvature variations: (1) the diameter was assumed constant along each vessel length and a tapered element was implemented to connect cross sections of different sizes, (2) the inlet and outlet diameters of the external and common iliac vein were set according to anatomical information from the literature, and a gradual transition in size was implemented along the vessel axes.

2.4 Variation in the Inflow Conditions

The variation in the inflow conditions to the internal and external iliac veins was considered by incrementally changing each inlet velocity over a range of $\pm 20\%$ from their average base values obtained from the 0D simulations, $0.26 m/s$ and $0.08 m/s$, respectively. This resulted in a range of $(0.208, 0.312)m/s$ for the internal iliac vein and $(0.064, 0.096)m/s$ for the external iliac vein. The parabolic velocity profiles for the base case and the variation extremes are shown in Fig. 4 A. In total, 25 steady-state simulations were performed to vary the inflow conditions for the geometry in the base state, 9 simulations changing inflow conditions for the simple (type i) change in curvature, and 9 for the complex (type ii) change in curvature.

2.5 Variation in Blood Viscosity

The viscosity of the blood was incrementally varied from the constant base value of $0.0035 Pa \cdot s$ to $0.0055 Pa \cdot s$ (the literature value for the maximum constant viscosity of the blood) for representative cases of each subset of variation of geometry. The use of a constant viscosity assumes that the blood behaves as a Newtonian fluid, which is true for large blood vessels (fluid as a continuum) and shear rates greater than $100 s^{-1}$, which is generally true in the arteries. Shear rates in the venous circulation can be much lower than this, and models such as the Carreau model [11] allow the shear-thinning properties of blood to be included, representing non-Newtonian behaviour. To investigate the influence of non-Newtonian blood viscosity on the flow field, we repeated analyses using a Carreau model given by Eq. (1). The effective viscosity μ_{eff} is a variable of the shear rate $\dot{\gamma}$ and determined by infinite-shear viscosity μ_∞, zero-shear viscosity μ_0, time constant λ and the power-law index n.

$$\mu_{eff}(\dot{\gamma}) = \mu_\infty + (\mu_0 - \mu_\infty)(1 + (\lambda\dot{\gamma})^2)^{\frac{n-1}{2}} \qquad (1)$$

At low shear rates, $\dot{\gamma} \ll 1/\lambda$, μ_{eff} approaches viscosity μ_0 and at high shear rates, viscosity μ_∞. At intermediate shear rates ($\dot{\gamma} \gtrsim 1/\lambda$), the model assumes a power-law fluid. For the purpose of this investigation, the infinite shear viscosity μ_∞ was set to $0.0035 Pa \cdot s$, zero-shear viscosity μ_0 was set at $0.056 Pa \cdot s$, the power law index n was set at 0.35, and the time constant λ, at $3.3s$ based on representative values from the literature [12]. The plot of the resulting effective viscosity versus the shear rate is shown in Fig. 4B.

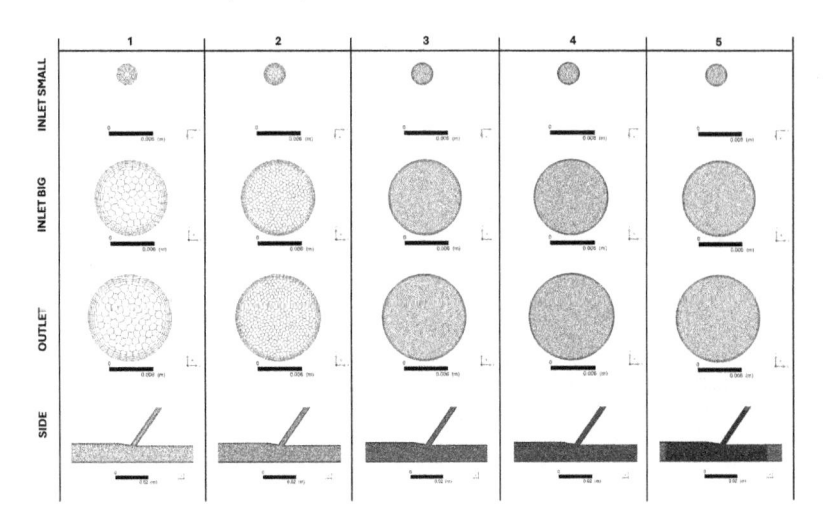

Fig. 2. Five levels of refinement of the investigated meshes.

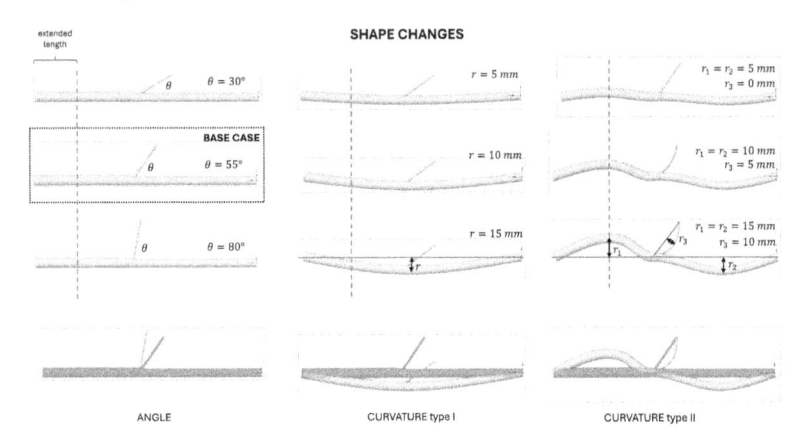

Fig. 3. Top three rows: variation in geometry - three subsets: (1) unification angle change, (2) curvature change (type i), (3) curvature change (type ii), in each column the cases are referred to as (0,1,2) from top to bottom; **bottom row:** the geometry change wrt base case.

2.6 Preliminary Transient Analysis

Preliminary work was performed to repeat all simulations (geometry, boundary conditions, and viscosity variation) in the transient state with oscillating inlet velocities defined by Eq. (2) assuming the parabolic inlets of the steady state $U_{profile|int}$ and $U_{profile|ext}$ for the internal and external iliac vein, respectively. The signal was assumed sinusoidal (justified by the results of Muller and Toro [9] in the considered region) with a frequency of $f = \frac{80}{60}$ equivalent to the heart rate of $80 bpm$.

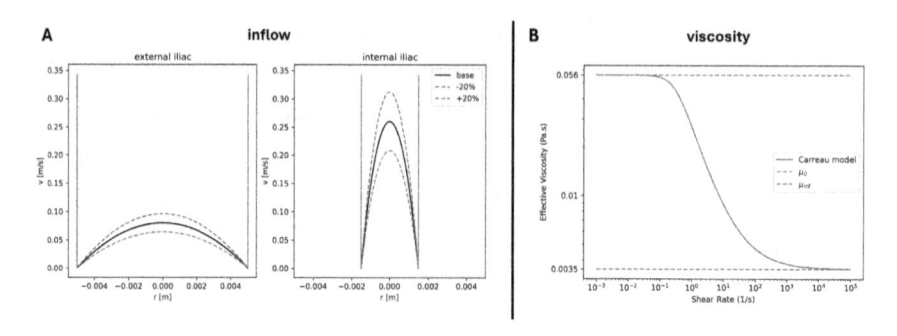

Fig. 4. (A) Parabolic profiles of inlet velocity for the base case and ±20% of maximum velocity for each vessel; (B) Carreau model for non-Newtonian viscosity.

$$U_{int} = U_{profile|int} \cdot (sin(2\pi f) + 1)$$
$$U_{ext} = U_{profile|ext} \cdot (sin(2\pi f) + 1) \qquad (2)$$

The oscillations in inlet velocities predicted by the 0D model were negligible, but the model did not assume any external sources of oscillation such as respiratory effects or the calf muscle pump. Therefore, the velocity values predicted by the 0D model were taken as average values of the oscillatory signal, and the amplitude of the oscillations was artificially increased. A total of 10 transient simulations were performed in this initial stage.

2.7 Metrics of Interest

In steady state, the impact of the described changes in vein geometry, boundary conditions, and flow properties on local haemodynamics was evaluated by computing the surface area of the vessel wall exposed to low WSS. Changes in distribution size and shape were analysed using three threshold values: ≤ 0.15 Pa, ≤ 0.10 Pa, and ≤ 0.05 Pa. In the transient analysis, we have looked at the oscillating shear index (OSI) defined by Eq. (3) where we approximate the integrals over one period of the inlet signals, T.

$$OSI = \frac{1}{2}\left(1 - \frac{\left|\int_0^T WSS\, dt\right|}{\int_0^T |WSS|\, dt}\right) \approx \frac{1}{2}\left(1 - \frac{|\overline{WSS}|}{\overline{|WSS|}}\right) \qquad (3)$$

with $|\overline{WSS}|$ being the magnitude of the time-averaged WSS and $\overline{|WSS|}$ - the time-averaged magnitude of WSS [13]. The OSI ranges between 0 and 0.5 where 0 indicates unidirectional flow and 0.5 fully reversible flow. The closer the OSI value to 0.5 the more oscillatory the WSS vector will be. The OSI distributions were analysed using three threshold values: $> 0.25, > 0.35, > 0.45$. All thresholds were set to values similar to those reported in other studies [6,7]. Both metrics are assumed to be associated with pro-thrombotic effect on the endothelial cells of the vein wall.

2.8 Computing Details

In total, we conducted 50 steady state simulations and 10 transient simulations. All were performed using ANSYS R2 2024 software, from the creation of idealised geometry in SpaceClaim 2024 to the meshing and simulation in Fluent R2 2024. The workflow was automated using the PyFluent library, enabling efficient script-based control of both meshing and solving processes. A batch processing approach was implemented to allow for parametrized execution through a CSV file. This setup enabled systematic variation of key parameters, including minimum and maximum mesh element sizes, maximum cell length, inlet velocities, blood viscosity, and solver-specific settings such as time-step size, number of time steps, and number of iterations. To run multiple simulations in parallel, batch jobs were submitted as an array job in the Slurm queuing system, with each instance running a specific combination of mesh configuration and solver settings. The implementation supported steady- and transient-simulations. Meshing was performed using 4CPUs, requiring approximately 10 min per case. The steady-state simulation on 4 CPUs took around an hour and required up to 50GB of RAM for highly refined global meshes. Transient simulations posed a greater challenge, each taking 4h on 192 CPUs with 90GB memory usage, and producing about 50GB of output data per case with 3s of simulated flow.

3 Results

This section describes the results of the performed simulation addressing each section of the Methods.

3.1 The Base State Simulation

In steady state, the metric of interest is the wall area subject to a low WSS below three different thresholds. Figure 5 shows an example of such a solution viewed in the XZ plane of the vessel geometry, where the area under each of the three thresholds considered is marked as a different-colour contour. The solution was obtained at mesh refinement level 4 (element length of $0.2\,mm$ as described in the Methods). The refinement was chosen on the basis of the mesh sensitivity investigation.

3.2 Mesh Sensitivity

To assess the necessary mesh refinement, the base case was simulated for the five mesh refinement levels described in Methods. The low wall shear stress distributions obtained from each simulation are shown in Fig. 6. Pressure and velocity metrics were also considered in this assessment, but tend to converge at smaller mesh refinements. In this context, the convergence means that there are little to no changes in the distribution with an increase in the mesh resolution. The results showed that refinement levels 1 and 2 were too coarse to capture the

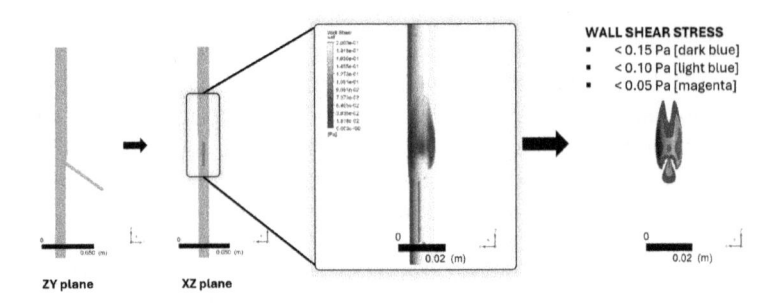

Fig. 5. Flow field obtained in the base state simulation.

features of the WSS distribution that only appeared at refinement level 3. The distributions from refinement levels 4 and 5 still showed some small changes, but the dominant features had already been resolved. Considering the computational cost associated with running each mesh type, we decided to run all steady-state simulations at mesh refinement level 4 (element size 0.2 mm).

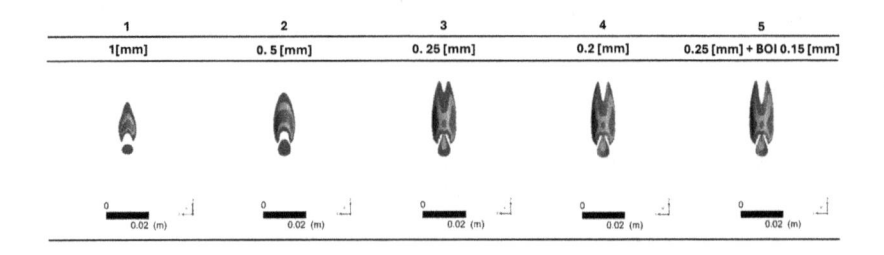

Fig. 6. The change in flow field with increasing mesh refinement.

3.3 Variation in Vessel Geometry

The variability in the geometry altered the absolute area and distribution of low WSS. The contours of the low WSS distributions obtained from simulations of different geometries are shown in Fig. 7A. The associated ranges for the change in the absolute area of the low WSS below the three thresholds considered are summarised in Table 2. For the change in angle from 30 to 80 degress, the WSS area ($< 0.05, < 0.10, < 0.15$ [Pa]) varied from (196, 428, 687) to (16, 123, 269) mm^2. For the change in simple curvature (type i) from the radius of curvature $r = 5\,mm$ to $r = 15\,mm$, the area varied from (29, 267, 514) to (21, 209, 568) mm^2. For the change in the complex curvature (type ii) from the radius of curvature $r = 5\,mm$ to $r = 15\,mm$ the area ranged from (74, 243, 448) to (53, 315, 452) mm^2. It is worth pointing out that in type i curvature, when the radius of curvature was increased, the absolute wall area of $WSS < 0.05 Pa$

and WSS $< 0.10 Pa$ decreased, while for WSS $< 0.15 Pa$ it increased. For type ii curvature, the area of WSS $< 0.05 Pa$ decreased, and for the other two thresholds it increased with increasing radii of curvature.

The effect of varying the type of connection between veins (tapered element vs. gradual transition) on the calculated WSS and OSI distributions was small compared to those caused by changes in the curvature and angle of unification.

3.4 Variation in the Inflow Conditions

The variability in the inflow conditions altered the absolute area and distribution of the low WSS for each geometry considered. Figure 7B shows a subset of low WSS areas obtained by varying the inflow velocities in the base geometry. The associated ranges for the absolute area of low WSS are summarised in Table 3. Both the area and the shape of the distribution change for each combination of the inlet conditions. Similar trends were observed for changes in the inflow velocity in the curved type i2 geometry and the curved type ii2 geometry.

3.5 Variation in Blood Viscosity

The variability in viscosity altered the absolute area and distribution of the low WSS for each case considered. The effect of varying constant viscosity on the range of the low WSS absolute area size for different vessel geometries is summarised in Table 4 and Fig. 8 shows the distributions for Newtonian viscosity at $0.0035 Pa \cdot s$, $0.0045 Pa \cdot s$ and $0.0055 Pa \cdot s$ vs. the non-Newtonian Carreau viscosity model in the base geometry. Increasing the constant viscosity generally leads to a decrease in the low WSS area. Considering a non-Newtonian model of viscosity proved to have an effect on the shape and size of the low WSS distributions.

Table 2. Change in the vessel wall area subject to the low wall shear stress due to geometry change

GEOMETRY CHANGE - WSS area $[mm^2]$			
	Angle	C Type i	C Type ii
WSS < 0.05 [Pa]	16 – 196	21 – 25	37 – 74
WSS < 0.10 [Pa]	123 – 428	209 – 266	243 – 315
WSS < 0.15 [Pa]	269 - 687	514 - 568	448 - 509

3.6 Preliminary Transient Analysis

The preliminary transient analysis identified regions of high OSI in approximately the same areas as regions of low WSS. Figure 9 shows an example comparison between the identified low WSS region and the high OSI region for the

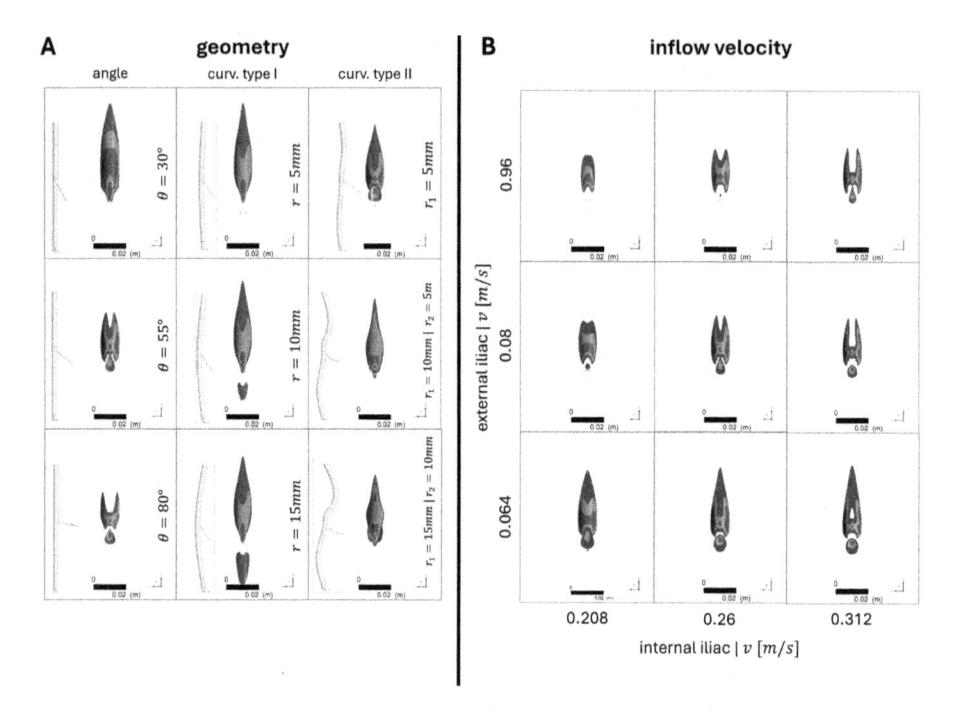

Fig. 7. Low WSS distributions for the three thresholds considered: (A) for different shapes of the iliac vein unification; (B) for different inflow velocities in the base geometry.

Table 3. Change in the vessel wall area subject to the low wall shear stress due to variation in the inlet velocity

INLET VELOCITY CHANGE - WSS area $[mm^2]$			
	BASE	C Type i 2	C Type ii 2
WSS < 0.05 [Pa]	28 – 114	9 – 293	38 – 129
WSS < 0.10 [Pa]	80 – 273	38 – 354	251 – 423
WSS < 0.15 [Pa]	171 - 620	216 - 2163	356 - 1448

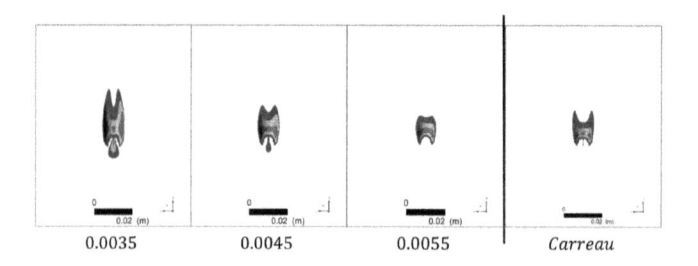

Fig. 8. Low WSS area identified for the base geometry for different values of constant viscosity and non-Newtonian Carreau model.

Table 4. Change in the vessel wall area subject to the low wall shear stress due to variation in viscosity

VISCOSITY CHANGE - WSS area [mm^2]			
	BASE	C Type i 2	C Type ii 2
WSS < 0.05 [Pa]	3 – 84	3 – 21	18 – 53
WSS < 0.10 [Pa]	97 – 233	20 – 209	116 – 315
WSS < 0.15 [Pa]	180 - 384	73 - 567	286 - 452

same case. The distributions share some characteristics for each simulated case and respond similarly to the applied changes, but also display unique features. The choice of thresholds is not correlated between steady-state and transient cases, so one should be careful when drawing conclusions from the comparison.

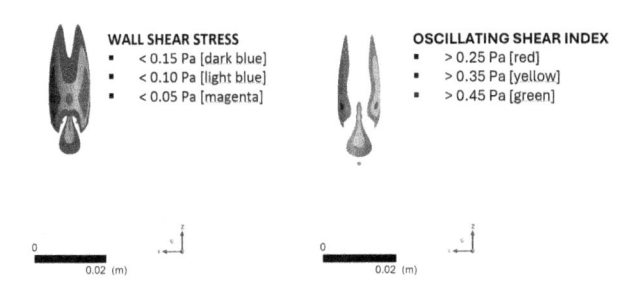

Fig. 9. Are subject to low WSS in the steady state (left) vs area subject to high OSI in transient state (right) for the base geometry. For the OSI: > 0.45 - green, > 0.35 - yellow, > 0.25 - red. (Color figure online)

4 Discussion

In this study, we simulated a set of idealised geometries of iliac vein unification to assess the relative importance of shape variation, inflow conditions, and viscosity in the predicted metrics of wall shear stress. The choice of changes was informed by the available clinical data for the geometries but was limited to simple changes in the angle between the vessels and the change in their curvature radii. A more thorough shape investigation would be challenging to perform with such a simple parametrisation. To achieve a more principled variation, methods such as statistical shape modelling could be used on a larger set of idealised and patient-specific geometries of the considered veins. It is part of ongoing work as a means of generating samples of variable geometry for uncertainty investigation, but it is beyond the scope of this paper.

The change in average velocity by ±20% was considered sensible given the values predicted by the 0D model in earlier analyses. It may not cover the true variability in this region between patients, but it provides a way to investigate the effect of flow changes in the absence of experimental data. The range of variation in constant viscosity was informed by the values reported in the available literature. Given the nature of the venous flow, non-Newtonian effects may be of significance, and a simple non-Newtonian model (Carreau) was used to check if it affects the metrics of interest. Although changes were observed, the parameters of the model were based on the available literature due to the lack of experimental data in the region of interest. Each of the investigated variations had an effect on the predicted wall shear distributions, and it is difficult to rank the effects according to their significance.

Flow parameter variation (inflow, viscosity) could benefit from a more advanced approach, for example, using formal Sobol analysis to assess the uncertainty in the predicted metrics, but it is beyond the scope of this paper. It would require a more in-depth investigation of the parameter space, including filtering out unphysiological parameter combinations. In addition, formal variance-based sensitivity analyses require a large number of input samples to obtain statistically significant outcomes. Depending on the number of input parameters, this could easily scale to thousands of simulations and would pose a greater computational challenge - if at all feasible - especially in the transient case. The approach presented in this work is more similar to local sensitivity analysis, and conducting it on very idealised models allows us to quantify the relative influence of variation in individual aspects of the flow problem that would be impossible to distil from patient-specific geometries, which will be analysed in the future work.

A mesh sensitivity study was performed to choose a grid that would resolve the critical features of predicted haemodynamics at an acceptable computational cost. Based on the steady-state simulation of the base geometry, we chose a refinement level with an element size of $0.2\,mm$. The global refinement to an element size of $0.15\,mm$ was computationally intractable with the available resources. Applying this refinement locally close to the unification substantially increased the computational cost without causing a significant change in the predicted haemodynamics. More computationally intensive transient analysis posed a greater challenge for the same mesh element size, consuming significantly more high-performance computing resources.

We assumed that wall shear stress metrics could serve as thrombosis risk metrics: low WSS in the steady state and high OSI in the transient state. This is an assumption based on the available literature. The choice of thresholds is subject to a biological and clinical interpretation of the specific location in the vasculature, but it does not undermine the findings of this research. An adjustment of the threshold to ones based on experimental data would be straightforward.

Vein compliance and fluid-structure interactions (FSI) were effectively ignored, which could be of significance in veins that are highly compliant compared to arteries. This assumption affects the flow dynamics and pressure distribution, which in turn may lead to inaccurate estimations of the wall shear stress. The choice was made for two reasons: (1) the complexity of the assumptions was incrementally increased to ensure that the importance of each step is assessed, (2) there are no clinical data collected on venous compliance of the iliac veins during standard clinical practice, and applying a generalised condition to all patients could result in more inaccurate predictions than with the rigid body assumption. It would only work if the vessel's environment was well known and other aspects such as respiratory effects were investigated.

The results obtained from the simulated cases will be used for the interpretation of patient-specific clinical data. Knowing the relative effect of the changes investigated on the shear metrics could help predict the expected effects of DVT treatment.

5 Conclusion

This work investigated the influence of geometry, inflow conditions, and viscosity on wall shear stress metrics – assumed to be related to prothrombotic responses – in 3D simulations of idealised iliac vein unifications with a detailed steady-state analysis and preliminary transient analysis. The results revealed substantial differences in the predicted metrics between each simulated case. Although further work is required to complete the analysis of the transient state, the investigations conducted provide a basis for interpretation of patient-specific vein geometries.

Acknowledgments. This project has received funding from the European Union's Horizon 2020 research and innovation programme under grant agreement No 857533. The publication was created within the project of the Minister of Science and Higher Education "Support for the activity of Centers of Excellence established in Poland under Horizon 2020" on the basis of the contract number MEiN/2023/DIR/3796 and is supported by Sano project carried out within the International Research Agendas programme of the Foundation for Polish Science, co-financed by the European Union under the European Regional Development Fund. The authors acknowledge the Polish high-performance computing infrastructure PLGrid (HPC Center: ACK Cyfronet AGH) for providing computer facilities and support within computational grant no. PLG/2024/017108.

Disclosure of Interest. The authors declare that there is no potential conflict of interest.

References

1. Baldwin, M.J., et al.: Post-thrombotic syndrome: a clinical review. J. Thromb. Haemost. **2**(11), 795–805 (2013). https://doi.org/10.1111/jth.12180
2. Kakkos, S.K., et al.: Guidelines on the management of venous thrombosis. Eur. J. Vasc. Endovasc. Surg. **2**(61), 9–82 (2021). https://doi.org/10.1016/j.ejvs.2020.09.023
3. Morris, P.D., et al.: Computational fluid dynamics modelling in cardiovascular medicine. Heart **2**(102), 18–28 (2016). https://doi.org/10.1136/heartjnl-2015-308044

4. Figueroa, C.A. et al.: Encyclopedia of Computational Mechanics, 2nd edn. Wiley (2017)
5. Belkacemi, D., et al.: Intraluminal thrombus characteristics in AAA patients: noninvasive diagnosis using CFD. Bioengineering 10, 540 (2023). https://doi.org/10.3390/bioengineering10050540
6. Wang, T., et al.: Predicting the risk of postsplenectomy thrombosis in patients with portal hypertension using computational hemodynamics models: a proof-of-concept study. Clin. Biomech. **98**, 105717 (2022). https://doi.org/10.1016/j.clinbiomech.2022.105717
7. Fan, Z., et al.: Insights from computational fluid dynamics and in vitro studies for stent protrusion in iliac vein: how far shall we go? Cardiovasc. Eng. Tech. **16**, 79–90 (2025). https://doi.org/10.1007/s13239-024-00758-7
8. Otta, M., Halliday, I., Tsui, J., Lim, C., Struzik, Z.R., Narracott, A.: Sensitivity analysis of a model of lower limb haemodynamics. In: Groen, D., de Mulatier, C., Paszynski, M., Krzhizhanovskaya, V.V., Dongarra, J.J., Sloot, P.M.A. (eds.) ICCS 2022. LNCS, vol. 13352, pp. 65–77. Springer, Cham (2022). https://doi.org/10.1007/978-3-031-08757-8_7
9. Möller, L.O., Toro, E.F.: A global multiscale mathematical model for the human circulation with emphasis on the venous system. Int. J. Num. Methods Biomed. Eng. **30**, 681–725 (2014). https://doi.org/10.1002/cnm.2622
10. Mukul, S.G., Scott, L.D.: Adhesion of normal erythrocytes at depressed venous shear rates to activated neutrophils, activated platelets, and fibrin polymerized from plasma. Blood **100**(10), 3797–3803 (2002). https://doi.org/10.1182/blood-2002-03-0712
11. Kannojiya, V., et al.: Simulation of blood as fluid: a review from rheological aspects. IEEE Rev. Biomed. Eng. **14**, 327–341 (2021). https://doi.org/10.1109/RBME.2020.3011182
12. Junaidi, A.R., et al.: Simulation of non-Newtonian flow of blood in a modified laparoscopic forceps used in minimally invasive surgery. Comput. Methods Biomech. Biomed. Engin. **24**(16), 1794–1806 (2021). https://doi.org/10.1080/10255842.2021.1919884
13. Soulis, J.V. et al.: Relative residence time and oscillatory shear index of non-Newtonian flow models in aorta. In: 2011 10th International Workshop on Biomedical Engineering. https://doi.org/10.1109/IWBE.2011.6079011

Cross-Scale Modeling of Healthcare Norms and Patient Features Dynamics with Interpretable Machine Learning

Chao Li[1]((✉)), Dutao Zhang[1], Fei Ren[2], and Sergey Kovalchuk[1]

[1] ITMO University, Saint Petersburg, Russia
316325@niuitmo.ru, kovalchuk@itmo.ru
[2] Independent Researcher, Shenzhen, China

Abstract. This study proposes an interpretable machine learning framework to model bidirectional dynamic interactions between macroscopic norms and microscopic features in clinical data. Leveraging real-world medical records from a specialized chest hospital (containing unstructured text, complex categorical variables, temporal indicators, and non-random missing patterns), we perform numerical processing through Latent Semantic Analysis and dimensionality reduction via Non-negative Matrix Factorization. Macroscopic therapeutic norms are identified using HDBSCAN clustering, while SHAP-XGBoost integration selects critical microscopic features, including multidrug-resistant tuberculosis diagnosis and liver function biomarkers. We integrate symbolic regression with the Peter-Clark Momentary Conditional Independence causal discovery method based on partial correlation, constructing cross-scale functional relationships with temporally rigorous constraints. Specifically, PySR derives nonlinear mapping equations, while partial correlation-based conditional independence tests establish time-lagged dynamic dependency networks. Guided by the Dynamic Maximum Entropy across Scales (DyMES) principle, multi-scale perturbation experiments reveal bidirectional mechanisms. Within our dataset and framework, DyMES reveals dynamic constraints' interplay driving statistical equilibrium between macroscopic clinical norms and microscopic patient characteristics through nonlinear coordination and threshold-triggered time-encoded mechanisms. Persistent constraint interactions induce novel steady states formation with dynamically preserved system memory.

Keywords: Medical Norms · Interpretable Machine Learning · Dynamic Maxent across Entwined Scales · Symbolic Regression · Distributed healthcare

1 Introduction

Medical norms, predominantly designed for human practitioners, are encoded in unstructured natural language with implicit references to clinical expertise, posing significant challenges for autonomous agents to dynamically adapt to evolving norms (e.g., treatment guidelines for drug-resistant tuberculosis) in distributed healthcare systems. Enabling autonomous agent systems to learn and perceive the norms among healthcare professionals is both intriguing and essential for their integration into real-world distributed healthcare environments [7,13]. The heterogeneous phenomena of medical norm propagation and adoption within autonomous multi-agent systems are intrinsically linked to the emergence of collective behaviors and the formation of organizational structures. These processes constitute fundamental manifestations of organized complexity [1,16]. One of the key challenges in current complex system modeling lies in the weak interpretability of emergent behaviors and insufficient formalized descriptions. Integrating machine learning into agent-based modeling is a highly promising research direction. Feature-based explanation approaches provide novel perspectives for understanding the complex emergent behaviors of multi-agent systems and linking micro- and macro-level characteristics [8,11].

Current research on norm propagation in medical autonomous multi-agent systems has achieved progress in formal modeling and validation with real-world clinical datasets [9]. However, two critical limitations persist. First, the absence of formalized cross-scale dynamic coupling mechanisms, particularly the lack of quantitative methodologies for macro-level norm and micro-level patient feature co-evolution. Second, existing models fail to effectively characterize the association between norm dynamic adaptation processes and system multi-scale characteristics [3,15]. For instance, the inability to quantify how microscopic feature variations (e.g., liver function test results) trigger macroscopic norm adjustments (e.g., medication plan revisions), and how such macroscopic adjustments subsequently influence treatment cycle timelines across diverse patients. This constitutes a core manifestation of organized complexity in healthcare systems. The central challenge lies in constructing a formal framework for dynamic constraints between microscopic individual features and macroscopic therapeutic norms.

Building upon the two limitations explored above, we extended the Dynamic Maxent across Entwined Scales (DyMES) theory [4] in complex systems to study medical norms in autonomous multi-agent systems. DyMES is a dynamic theory combining Top-Down information-theoretic inference with Bottom-Up state-variable-dependent mechanisms. In this framework, state variables influence microscale dynamics while being computed as averages over probability distributions of microvariables. This integration enables simultaneous prediction of time-evolving state variables and microvariable distributions. Central to DyMES is the notion of transition functions, which govern microvariable dynamics. Scale entwinement, and in particular, downward causation, is captured by explicit dependence of transition functions on state variables as well as on microvariables [4].

To address these gaps, we propose a three-stage computational framework synergizing interpretable machine learning with DyMES theory. First, XAI techniques disentangle macro-micro correlations from noisy EHR data. Second, symbolic regression distills these associations into cross-scale transition functions. Third, we integrate PCMCI-based causal discovery with partial correlation conditional independence tests to introduce temporally modulated functions for the derived transitions, optimizing their parameters through grid search based on these metrics. Finally, we construct a DyMES model with the optimized dynamic transition functions, conducting multiscale perturbation experiments on strictly monotonic temporal sequences to simulate bidirectional macro-norm-micro-feature interactions.

This study makes three core contributions: first, it establishes a DyMES framework that integrates maximum entropy principles with interpretable machine learning to formalize bidirectional cross-scale interactions; second, it develops a unified methodology combining symbolic regression-derived equations with PCMCI-validated bidirectional feature causality; finally, it introduces the first computational dynamical model for co-evolution between institutional medical norms and personalized patient characteristics.

2 Cross-Scale Dynamic Modeling Framework

We propose a general extensible framework comprising two categories of components. The first category corresponds to the colored sections in Fig. 1, specifically a Two-Stage Computational Architecture responsible for all XAI-related operations. The second category (white sections in Fig. 1) extends the DyMES model to healthcare datasets through simulation components, aiming to rigorously interpret macro-micro correlations within the dataset.

2.1 Two-Stage Computational Architecture

Fisrt Stage. The initial stage can be abstracted as: decoupling macro-micro correlations in datasets through diverse XAI tools based on their inherent characteristics and structural composition, where machine learning methods and feature processing approaches are selectively employed according to data properties and sparsity levels.

The fundamental principle of this first stage involves identifying crucial features from noisy data, uncovering strong dependencies between significant feature vectors, and subsequently distinguishing macro/micro features through integration with domain expertise and clinical knowledge.

For processing hybrid medical datasets containing clinical norms, the primary methodology involves unifying heterogeneous features into computable encodings, removing overly sparse and insignificant feature columns, followed by reasonable dimensionality reduction methods to prevent matrix oversizing. Our observations indicate these datasets typically exhibit semi-structured formats organized as structured tables with explicit headers corresponding to the categories documented in the 'Main Module' column of Table 1 (e.g., Basic Information, Clinical Process, Diagnostic Testing Modules). In this architecture, each patient sample comprehensively populates all categories defined under the 'Main Module' column of Table 1, forming a complete longitudinal record.

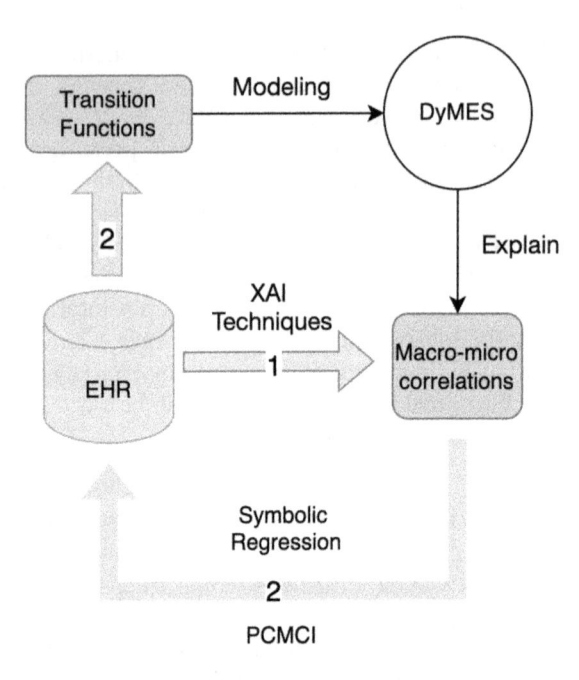

Fig. 1. Dynamic cross-scale norm interactions.

The identification of critical feature vectors within the processed feature matrix is primarily achieved through interpretable clustering approaches. Recommended clustering methods for handling hybrid feature matrices typically include HDBSCAN, Fuzzy C-Means Clustering, and Autoencoder KMeans. Subsequently, feature importance analysis is conducted based on the clustering results. Recommended static analysis methods applicable here include cluster persistence scores, density analysis of clusters, cluster membership probabilities, and cluster hierarchy trees. A more efficient dynamic approach involves training supervised classification models using cluster labels, followed by analyzing feature contributions to cluster label prediction through SHAP and LIME techniques. Based on the hybrid content formats of datasets, methods including association rule mining (Apriori/FP-Growth), dependency and correlation analysis, and rule induction (decision trees/RIPPER) can also serve as recommended alternative approaches for analyzing clustering results.

For encoded matrices derived from hybrid medical datasets containing clinical norms, analysis of clustering results typically yields a finite set of correlation combinations. Ultimately, through integration with medical domain knowledge and clinical expert validation, we can identify clinically significant macro-micro correlations within these combinatorial patterns.

Second Stage. Macro-micro correlations based on feature importance cannot be directly transformed into computational models. Therefore, the second-stage work involves mathematical formula mining through actual data of different feature columns corresponding to macro-micro correlations in the dataset. The core methodology here combines symbolic regression and PCMCI. The essence of symbolic regression lies in exploring vast function spaces using evolutionary algorithms or heuristic searches to identify mathematical expressions that optimally fit the data [17]. Symbolic regression achieves balance between "unknown candidate function forms in the library" and "required structural constraints for continuous dynamical systems." In multivariate, multiscale time series, causal structures often exhibit complexity. PCMCI (Peter-Clark Momentary Conditional Independence) is a statistical method that discovers statistically significant cross-scale causal mechanisms in medical temporal data through partial correlation conditional independence tests [14].

By integrating the two methodologies from Stage 2 performing function form searching across datasets and conducting multi-parameter optimization of identified functions on test sets we derive the formalized functional dependency between micro and macro features as expressed in Eq. 1. To streamline exposition, the complete mathematical definition of Eq. 1 is methodologically consolidated with the implementation framework for extending the DyMES to medical dataset simulations, thereby establishing an integrated analytical paradigm.

2.2 DyMES Model on Medical Datasets

We present key mathematical conventions based on the core DyMES framework. Detailed theoretical derivations are provided in the work of John Harte et al. [4].

First define m macroscale variables $X = (X_1, X_2, \cdots, X_m)$ with at most m corresponding microscale variables $\mathbf{x} = (x_1, x_2, \cdots, x_m)$. Here macroscale represent norms themselves while microscale variables manifest as salient features of individual patient samples. Both vector types X and $\mathbf{x}$ derive from previously mined macro-micro correlations within the dataset's feature vectors.

$R(\mathbf{x})$ represents the joint probability distribution of microscale variables. To determine $R(\mathbf{x})$, we maximize the Shannon information entropy of $R(\mathbf{x})$ under constraints imposed by X and dX/dt. We express these constraints as $F = (h_1(X), \cdots, h_m(X), \frac{dX_1}{dt}, \cdots, \frac{dX_m}{dt})$, where $h_\mu(X)$ are functions of macroscale variables. The average values of these functions over $R(\mathbf{x})$ yield the constraint conditions, denoted by $f_\mu(\mathbf{x}, X)$. For $\mu = 1, \cdots, m$, the functions f_μ depend solely on x_μ. In more complex cases when $\mu = m+1, \cdots, 2m$, f_μ may be functions of multiple microscale variables. Scale entwinement arises when f_μ serving as transfer functions can depend on both macroscale variables X and microscale variables $\mathbf{x}$ for $\mu = m+1, \cdots, 2m$. Therefore, we formulate all constraints as:

$$F_\mu = \sum_{\mathbf{x}} f_\mu(\mathbf{x}, X) R(\mathbf{x}|X) \tag{1}$$

where $\mu = 1, 2, \cdots, 2m$, the summation indicates integration over each microscale variable x_i, and explicitly denotes the conditional dependence of $R(\mathbf{x})$ on X.

Note that the transition function f_μ is methodologically extracted from the dataset through our Stage 1 and 2 XAI techniques.
By maximizing the Shannon information entropy of R: $H = -\sum_{\mathbf{x}} R \log(R)$, we obtain [5,6,12]:

$$R(\mathbf{x}|X) = \frac{e^{-\sum_\mu \lambda_\mu f_\mu(\mathbf{x},X)}}{Z} \qquad (2)$$

where $\lambda = (\lambda_1, \lambda_2, \cdots, \lambda_{2m})$ are Lagrange multipliers obtained by solving the constraint conditions [2,5]. Z is the normalization constant ensuring total probability sums to 1.

Equations 1 and 2 establish the foundational definitions for our DyMES framework. A core assumption of DyMES theory concerns the dynamic constraint updating process. When X and $\frac{dX}{dt}$ are known at time t, the Lagrange multipliers λ can be determined through maximum entropy conditions at time t. We omit the rigorous derivation process from John Harte et al. [4] and directly cite the key computational equations:

$$\sum_{v=1}^{2m} \mathrm{Cov}(f_{m+i}, f_v)\frac{d\lambda_v}{dt} = 0 \qquad (3)$$

where index i ranges from 1 to m, and $\mathrm{Cov}(A,B) = \langle AB \rangle - \langle A \rangle \langle B \rangle$ denotes covariance between A and B. Equation (3) provides m relationships among the $2m$ time derivatives of Lagrange multipliers.

$$\frac{dX_i}{dt} + \sum_{\mu=1}^{2m} \left(\mathrm{Cov}(f_i, f_\mu)\frac{d\lambda_\mu}{dt} \right) + \left(\mathrm{Cov}\left(f_i, \frac{df_\mu}{dt}\right) \lambda_\mu \right) = 0 \qquad (4)$$

Equations (3) and (4) can be efficiently solved through matrix inversion to determine the time derivatives of Lagrange multipliers, which are then iteratively updated. Equations (2), (3), and (4) formulate the theoretical foundation of DyMES. These equations characterize the intertwined dynamic relationships between macrostate variables and Lagrange multipliers within the system [4].

3 Experiments and Simulations

3.1 Dataset Processing

We conducted experiments using the dataset (collected in a specialized chest hospital), comprising longitudinal clinical data from the multidrug-resistant tuberculosis (MDR-TB) diagnosis and treatment database established and maintained by our research team. All enrolled patients underwent monthly follow-up assessments in strict accordance with therapeutic protocols developed by a multidisciplinary therapy group [10].

The dataset exhibits three primary characteristics: high sparsity with non-random missingness and heterogeneous medical data types. It comprises 31 major feature categories (see Table 1 Submodule Components) containing 1,245

Table 1. Main features in the dataset

Main Module	Submodule Components
Basic Information Module	Patient Identification Demographic Characteristics Clinical Baseline
Clinical Process Module	Clinical Examination Records Initial Diagnosis Documentation Follow-up Information Transfer Records
Diagnostic Testing Module	Hain GenoType MTBDRplus GeneXpert MTB/RIF Mycobacterial Speciation Chest Radiography Sputum Smear Microscopy Sputum Culture Liver Function Tests Conventional Drug Susceptibility (Selected Key Features)
Assessment & Monitoring Module	Evaluation Metrics Visit Assessment Protocols Adverse Events (AE/SAE Records) Therapeutic Outcome Documentation
Treatment Management Module	Therapeutic Regimen Specifications Treatment Protocol Documentation
Research Management Module	Case Enrollment Forms Longitudinal Follow-up Records Serial Number Identification

feature columns, where 23 categories demonstrate > 0.8 sparsity. The time span is from April 11, 2018 to December 14, 2023. These 1,245 columns incorporate temporal (follow-up dates, report dates, etc.); categorical (sputum smear results, conventional drug susceptibility testing, medication regimen codes, etc.); numerical (ALT levels from hepatic panels, serum creatinine values from renal profiles, etc.); binary (sputum culture submission flags, cavitation presence in chest imaging, etc.); and natural language data types (radiographic findings descriptions, etc.).

We take the zero-missing "follow-up date" column as the temporal axis. The datetime values are normalized to [0,1] with day granularity, followed by timestamp micro-adjustments for same-day samples to ensure strictly increasing time series aligned with dataset span; For categorical and binary feature columns, missing value indicator columns are appended before one-hot encoding, with subsequent NMF dimensionality reduction applied to high-dimensional features; Numerical columns with sparsity threshold <0.8 are filtered and retained; Textual description fields undergo TF-IDF vectorization extracting unigrams and bigrams as base features, accompanied by binary indicators for text missingness. Truncated SVD (i.e., LSA) reduces TF-IDF matrix dimensionality based on singular value decay curves. Semantic features are concatenated with missing indicators, forming final structured encoded features where each column has its own CSV file.

Pre/post-processing metadata including data types, encoding schemes, and notes are recorded in JSON files. All feature CSVs are merged into a 20872 (samples) × 387 (features) matrix, followed by Gower distance matrix computation.

3.2 Mining Macro-micro Correlations

For this high-dimensional dataset characterized by elevated sparsity, non-random missingness, and heterogeneous data types, we evaluated and implemented the four clustering methods detailed in Table 2. The experimental results demonstrate that HDBSCAN achieves optimal performance, attaining the highest Silhouette Score (0.5275) and lowest Davies-Bouldin Index (0.8192) among all evaluated approaches. While the hybrid Autoencoder+K-Means method exhibits potential competitiveness, its practical implementation faces challenges in architectural optimization of the deep neural network, which incurs significant engineering overhead and compromises computational efficiency.

Table 2. Clustering method performance Comparison

Clustering Method	Silhouette Score	Davies-Bouldin Index
HDBSCAN	**0.5275**	**0.8192**
Fuzzy C-Means	0.3489	0.9717
KMeans	0.3179	1.0144
Autoencoder + K-Means	0.4663	0.8754

The HDBSCAN clustering results show: Number of clusters = 55. Number of noise points = 7,980. These noise points represent specific cases that are not the current focus due to the high matrix dimensionality and sparsity. The final corrected valid samples shape is (12892, 12892). Initial static analysis reveals: the largest cluster contains 837 samples, the smallest cluster has 235 samples. The maximum persistence score is 0.6304, with 4 clusters exceeding 0.1 persistence score threshold.

To mine macro-micro correlations from the clustering model and results, we compared multiple methods.

Regarding core objectives: association rule mining primarily identifies frequent co-occurrence patterns among features (e.g., "feature A and feature B frequently co-occur"), dependency/correlation analysis focuses on detecting statistical relationships between features (using metrics like Pearson correlation coefficients and mutual information), rule induction aims to generate human-readable "if-then" rules (e.g., "age >60 AND complications > 3 ← high-risk cluster"), while SHAP analysis explains model decision logic for cluster assignments. Comparative evaluation reveals rule induction and SHAP methods demonstrate superior performance in medical data analysis (see Table 3).

Particularly, decision tree or RIPPER-based rule induction produces intuitive "if-then" rules that prove invaluable for clinical visualization and cross-domain expert collaboration. However, experiments on the considered dataset confront challenges from high-dimensional heterogeneous data (containing numerical, encoded, missing indicators, and NMF/SVD dimensionality-reduced features) - requiring noise control through feature selection, clinical binning, and pruning optimization. The feature type diversity and high dimensionality may lead to verbose rules with reduced interpretability, especially when features aren't rigorously refined. Single decision trees or RIPPER algorithms might generate complex logical structures with excessive branching, necessitating domain knowledge-guided secondary optimization.

Table 3. Comparative analysis of interpretation methods

Metric	Association Rules	Dependency Analysis	Rule Induction	SHAP
High-dim Support	Low	Medium	Low	High
Pattern Efficiency	Low	Medium	Medium	High
Interpretability	Medium	Low	High	High
TB Applicability	Low	Medium	High	High
Clinical Operability	Medium	Low	High	High

Here we selected XGBoost - a tree-based model demonstrating superior performance on tabular data. We trained the XGBoost model using 55 cluster labels as classification targets, then conducted global feature importance analysis on 387 features determining each cluster label, and generated their respective summary plots (bee swarm plots).

As shown in Fig. 2, this SHAP (SHapley Additive exPlanations) feature contribution diagram displays: The Y-axis lists semantically mapped feature names using clinically interpretable descriptions, sorted in descending order of global feature importance with the most discriminative key features positioned at the top. The X-axis represents the distribution range of SHAP values, which physically signifies the directional impact of features on sample assignment to specific clusters: Data points distributed on the right side (SHAP values > 0) indicate positive driving effects that enhance model confidence in assigning samples to corresponding clusters; points clustered on the left side (SHAP values < 0) reflect inhibitory effects on cluster membership. The color gradient (red-blue spectrum) encodes the magnitude of original feature values: Red spectrum indicates high feature values (e.g., abnormally elevated biomarker levels), while blue spectrum denotes relatively low-value states (e.g., physiological parameters at lower reference limits).

Cluster 30: Persistence = 0.6304

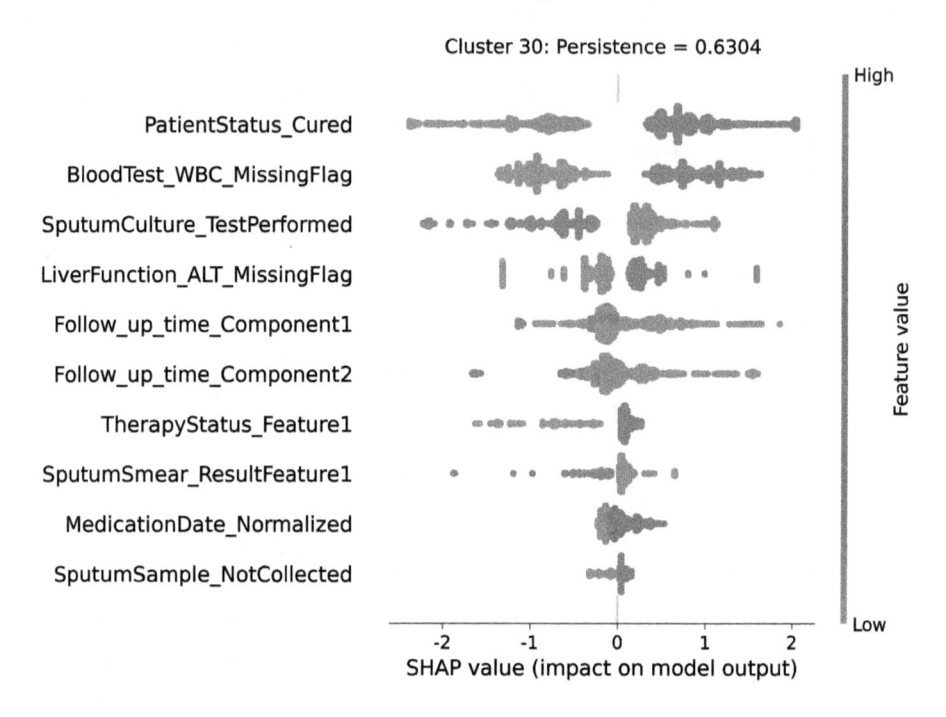

Fig. 2. Global feature importance analysis for Cluster 30.

Figure 2 displays Cluster 30 with the highest persistence score. We selected the TherapyStatus-Feature1 column, corresponding to the TherapyStatus variable containing eight distinct categories in the dataset, as a macroscale feature. These eight categories are: continuation of existing regimen, no treatment initiated, establishment/modification of treatment regimen, transfer-out, adverse drug reaction, voluntary discontinuation by patient, other, and exclusion of multidrug-resistant tuberculosis (MDR-TB) diagnosis.

The remaining features in the diagram predominantly represent patient-specific characteristics. To identify micro-level features exhibiting strong correlations with the TherapyStatus column, we retrained an XGBoost model using the eight macro-level states as classification targets and computed the predictive contribution of the remaining 386 features. The four most significant features are listed in Table 4.

3.3 Transition Function Search and Validation

The macroscale features encoded by the eight TherapyStatus categories were reduced to a single feature vector y via NMF dimensionality reduction. The four

Table 4. Top-4 micro-Level features contributing to macro-Level TherapyStatus classification

Feature	SHAP Value
x_0 =MxDataExt_MxName_NMF_2 (Follow up time-Component2)	0.8150031
x_1 =TbDiagnosis-Multidrug-Resistant Tuberculosis (MDR-TB)	0.6279506
x_2 =LiverFunc.Result.Dbil_DoubleValue (Direct Bilirubin)	0.19392538
x_3 =LiverFunc.Result.Alb_DoubleValue (Albumin)	0.18071306

features in Table 4 were sequentially designated as x_0, x_1, x_2, x_3 in descending order of importance.

Subsequent symbolic regression was implemented by utilizing the preprocessed globally monotonically increasing matrix as the search space. We employed PySR to execute the evolutionary algorithm with: 1,000 evolutionary rounds, 5 parallel populations, population size of 500 individuals (fundamental evolutionary units), symbolic parameter controlling maximum expression tree nodes $n_{\max} = 20$, and early stopping criteria Terminate if loss $<$ 1e−12 or complexity $C > 25$. The search results were autonomously logged with an evaluation rate of 2.710×10^3 expressions/second. Here the parameters of the evolutionary algorithm were configured solely based on computational resource availability, and comparative analysis across multiple independent experimental trials confirmed their negligible impact on search outcomes.

Subsequently, we conducted causal lag analysis on treatment state evolution patterns using the PCMCI (Peter-Clark Momentary Conditional Independence) method. The dynamic impacts of key variables exhibited the following characteristics: x_0 demonstrated negative regulatory effects at lag-1 (-0.0936) and lag-2 (-0.0310); x_1 showed positive driving effects at lag-1 (0.1066) and lag-2 (0.0646); x_2 revealed a positive association at lag-1 (0.0486).

We therefore introduced quadratic and exponential temporal modulation terms for cross-validation. The quadratic modulation form is expressed as $y_{\mathrm{mod}} = y_{\mathrm{base}} \times (1 + at + ct^2)$, where the linear term coefficient $a \in [0.01, 0.15]$ and quadratic term coefficient $c \in [0.001, 0.02]$. Through grid search on the training set (80% samples) with MSE as evaluation metric, we obtained optimal parameters $a = 0.01$, $c = 0.001$, achieving validation MSE 4.6×10^{-5} ($R^2 = 0.98$).

For exponential modulation $y_{\mathrm{mod}} = y_{\mathrm{base}} \times e^{bt}$, the growth rate $b \in [0.01, 0.1]$ was constrained to 2× the 0.05-level effect of x_2. Using 10-point uniform sampling, we determined optimal parameter $b = 0.1$, yielding validation MSE 4.8×10^{-4} ($R^2 = 0.79$). These results demonstrate effective capture of time-varying treatment state characteristics through our modulation functions.
The final transition functions we obtained are as follows:

$$Y(t) = \left[1.0001 - (x_0 + x_1^2)^{3.1569 \times 10^{-5}}\right] \times 0.13185 \cdot (1 + 0.01t + 0.001t^2) \quad (5)$$

$$Y(t) = (0.091956 - x_0)^{\exp(x_2)} \cdot e^{0.1t} \tag{6}$$

Interestingly, the fitting results of these transition functions on the dataset demonstrate that they embody computational formulations bridging macro-micro relationships. This connection manifests mathematically as expressions that remain computable for machines/models yet counterintuitive for human experts.

4 DyMES Simulation on the Dataset

We simulate the model using all 20,872 strictly increasing samples from the dataset by incorporating macro and micro feature columns contained in Eqs. (5) and (6). The DyMES framework is rigorously constructed following Eqs. (1)–(4). Our algorithmic innovation introduces precomputed acceleration matrices and covariance matrix-approximated Jacobians to significantly accelerate solving Eqs. (3) and (4), enabling whole-sample modeling without subsampling.

The precomputed acceleration matrix accelerates candidate function set $f_\nu(x, X, t)$ evaluations across all microstates $x \in x_\text{array}$ and macrovariables X. Conventional methods [4] require recomputing f_ν per iteration, yielding $\mathcal{O}(N_\text{iter} \cdot N_x \cdot N_\nu)$ complexity (N_x = microstate count, N_ν = constraint count). We pre-construct matrix $f_\text{matrix} \in \mathbb{R}^{N_x \times N_\nu}$ with elements: $f_\text{matrix}[i, j] = f_j(x_i, X, t)$. Matrix reuse strategy: Reusing f_matrix in probability distribution $R(x|X)$, constraint equations, and Jacobian computations reduces complexity to $\mathcal{O}(N_x \cdot N_\nu)$.

Traditional finite difference Jacobian calculation costs $\mathcal{O}(N_\nu^2 \cdot N_x \cdot N_\nu)$ with step-size sensitivity. Through leveraging f_matrix and distribution R via np.cov($f_\text{matrix}^\top$, aweights $= R$, bias $=$ True), we achieve secondary optimization: Eliminating extra function evaluations reduces complexity to $\mathcal{O}(N_\nu^2 \cdot N_x)$. Since we derived from Eq. 1:

$$J_{\mu\nu} = \frac{\partial (F_\mu - \mathbb{E}_R[f_\mu])}{\partial \lambda_\nu} = -\text{Cov}_R(f_\mu, f_\nu) \tag{7}$$

where $\mathbb{E}_R[f_\mu]$ denotes the expectation with respect to the probability distribution R.

Figure 3 demonstrates the fitting results across 100 time steps, where each grid unit on the horizontal axis encompasses 10 time steps. Here, λ_1 corresponds to the mean constraint of x_0 ($h_1(X) = \mathbb{E}[x_0]$), representing the follow-up time Component 2, while λ_2 corresponds to the mean constraint of x_1 ($h_2(X) = \mathbb{E}[x_1]$), associated with the MDR-TB diagnosis status. λ_3 and λ_4 encode the dynamic constraints governed by Eqs. 5 and 6, respectively. The experiment reveals that λ_2 remained constant, reflecting the stability of x_1's statistical distribution, which implies that the MDR-TB detection status maintains statistical equilibrium during microstate evolution, with the system preserving structural integrity through conserved mean values. The slight decline of λ_3 in later stages indicates temporal accumulation effects in Eq. 5's dynamic constraint (quadratic $0.001t^2$ term), requiring prolonged time modulation signal integration to trigger constraint adjustments.

The coupled dynamics of λ_1 and λ_4 - manifested through their exponential decay phase in the first 70 steps - arise from the nonlinear interaction between x_0's mean constraint (λ_1) and Eq. 6's dynamic constraint (λ_4) mediated by $\exp(x_2)$. The synergistic decay emerges from the coupling between x_0 and hepatic function indicators (x_2). The abrupt transition in later stages reveals a critical threshold (at $t \approx 70\Delta t$) where the time modulation factor $e^{0.1t}$ in Eq. 6 dominates the dynamical phase transition. The non-zero terminal states of λ_3/λ_4 signify the system's evolution toward a novel steady state incorporating time modulation terms, where persistent dynamic constraints maintain a "dynamic memory" encoded jointly by Eq. 5's quadratic temporal term and Eq. 6's exponential temporal driver.

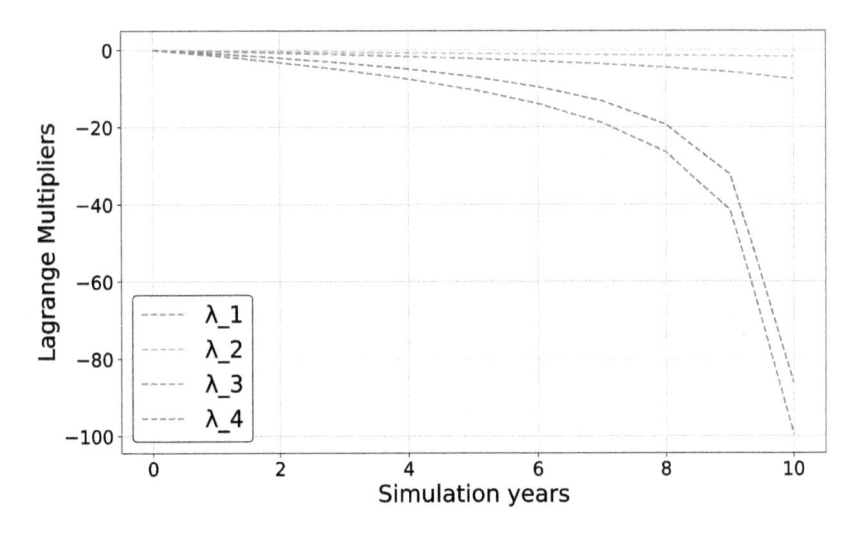

Fig. 3. Dynamic cross-scale norm interactions.

5 Discussion and Future Work

In most complex systems, causal relationships prove challenging to disentangle; DyMES may provide a quantitative methodology for determining both the directionality and magnitude of causal links [4]. Distinct from conventional top-down approaches [18], DyMES hybridizes mechanistic principles with Maximum Entropy (Maxent) theory, establishing an inferential framework that bridges fine-scale phenomena with coarse-grained outcomes, thereby enabling prediction of microscopic distributions from macroscopic knowledge. This methodology demonstrates capability in forecasting both the temporal evolution of state variables and probability distributions over microvariables.

These properties hold significant implications for investigating dynamic interactions between microscopic clinical practices/behaviors and macroscopic medical norms in distributed healthcare systems. Particularly, it facilitates modeling the dissemination and shared understanding of medical norms within autonomous multi-agent systems [9]. While conventional reinforcement learning paradigms employ reward-based mechanisms (e.g., reinforcement learning) to characterize and approximate agent behaviors at deeper levels, their utility remains limited for directly analyzing the complex scientific properties and inherent patterns within raw medical information datasets.

The application of DyMES, as a general mathematical framework in complexity science, to construct computable models for medical norm systems presents three principal challenges: (1) Transformation of information from unstructured multi-type non-random missing datasets into modelable microscopic features; (2) Formal definition and dynamic modeling of evolving clinical norms; (3) Reliable extraction of authentic transition function relationships between these elements from empirical data. Our current work systematically addresses these three fundamental issues.

Through implementation of an Explainable AI (XAI) framework, we propose a comprehensive three-phase mathematical modeling approach and empirically validate the effectiveness of extracted transition functions using clinical datasets. This investigation establishes a computational foundation for subsequent analyses of dynamic norm properties in healthcare environments.

Future research directions focus on three primary objectives: (1) Formalization and extraction of comprehensive composite microscopic features coupled with dynamic medical norms; (2) Investigation of bidirectional dynamic norm interactions under DyMES conditions in autonomous multi-agent systems; (3) Examination of micro-level agent practice (feature evolution) impacts on macroscopic norms and reciprocal constraint mechanisms. These explorations are anticipated to drive synergistic evolution of medical normative systems across theoretical and practical domains.

Acknowledgments. The research was supported by the Russian Science Foundation, agreement No. 24-11-00272, https://rscf.ru/project/24-11-00272/.

References

1. Anand, M., Gonzalez, A., Guichard, F., Kolasa, J., Parrott, L.: Ecological systems as complex systems: Challenges for an emerging science. Diversity **2**(3), 395–410 (2010). https://doi.org/10.3390/d2030395, http://dx.doi.org/10.3390/d2030395 systems as complex systems: Challenges for an emerging science. Diversity **2**(3), 395–410 (March 2010). https://doi.org/10.3390/d2030395, http://dx.doi.org/10.3390/d2030395
2. Arfken, G.B.: Mathematical methods for physicists (1967). https://api.semanticscholar.org/CorpusID:122141371
3. Boccara, N.: Modeling Complex Systems. Springer New York (2010). https://doi.org/10.1007/978-1-4419-6562-2

4. Harte, J., Brush, M., Umemura, K., Muralikrishnan, P., Newman, E.A.: Dynamical theory of complex systems with two-way micro–macro causation. Proc. Nat. Acad. Sci. U.S. Am. **121** (2024). https://api.semanticscholar.org/CorpusID:274566522
5. Jaynes, E.T.: Information theory and statistical mechanics. Phys. Rev. **106**, 620–630 (1957). https://api.semanticscholar.org/CorpusID:17870175
6. Jaynes, E.T.: On the rationale of maximum-entropy methods. Proc. IEEE **70**, 939–952 (1982). https://api.semanticscholar.org/CorpusID:42335268
7. Kovalchuk, S.V., Funkner, A.A., Metsker, O.G., Yakovlev, A.N.: Simulation of patient flow in multiple healthcare units using process and data mining techniques for model identification. J. Biomed. Inf. **82**, 128–142 (2017). https://api.semanticscholar.org/CorpusID:14429841
8. Kovalchuk, S.V., et al.: A conceptual approach to complex model management with generalized modelling patterns and evolutionary identification. Complexity **2018**, 1–15 (2018). https://doi.org/10.1155/2018/5870987 approach to complex model management with generalized modelling patterns and evolutionary identification. Complexity **2018**, 1–15 (2018). https://doi.org/10.1155/2018/5870987 N.O., Kalyuzhnaya, A.V., Vaganov, D.A., Bochenina, K.O.: A conceptual approach to complex model management with generalized modelling patterns and evolutionary identification. Complexity **2018**, 1–15 (Nov 2018). https://doi.org/10.1155/2018/5870987, http://dx.doi.org/10.1155/2018/5870987
9. Li, C., Petruchik, O., Grishanina, E., Kovalchuk, S.: Multi-agent norm perception and induction in distributed healthcare (2024). https://arxiv.org/abs/2412.18454
10. Ma, J., et al.: Treatment outcomes and risk factors of multidrug-resistant tuberculosis patients in Xi'an China, a retrospective cohort study. Infect. Drug Resist. **15**, 4947–4957 (2022). https://api.semanticscholar.org/CorpusID:251947734
11. Olsen, M., Kuhn, D.R., Raunak, M.: Explaining the impact of parameter combinations in agent-based models. J. Comput. Sci. **81**, 102342 (2024). https://doi.org/10.1016/j.jocs.2024.102342 combinations in agent-based models. J. Comput. Sci. **81**, 102342 (2024). https://doi.org/10.1016/j.jocs.2024.102342, http://dx.doi.org/10.1016/j.jocs.2024.102342 combinations in agent-based models. Journal of Computational Science **81**, 102342 (Sep 2024). https://doi.org/10.1016/j.jocs.2024.102342, http://dx.doi.org/10.1016/j.jocs.2024.102342
12. Pressé, S., Ghosh, K., Lee, J., Dill, K.A.: Principles of maximum entropy and maximum caliber in statistical physics. Rev. Mod. Phys. **85**, 1115–1141 (2013). https://api.semanticscholar.org/CorpusID:16150191
13. Rajpurkar, P., Chen, E., Banerjee, O., Topol, E.J.: Ai in health and medicine. Nat. Med. **28**, 31–38 (2022). https://api.semanticscholar.org/CorpusID:246098480
14. Runge, J.: Discovering contemporaneous and lagged causal relations in autocorrelated nonlinear time series datasets. arXiv:abs/2003.03685
15. Sayama, H.: Introduction to the modeling and analysis of Complex Systems. Open Suny Textbooks (2015)
16. Stallings, W.: Gerald m. weinberg. an introduction to general systems thinking. New York: Wiley, 1975, p. 279. Behavioral Science **21**(4), 289–290 (1976). https://doi.org/10.1002/bs.3830210409
17. Udrescu, S.M., Tegmark, M.: AI Feynman: a physics-inspired method for symbolic regression. Sci. Adv. **6** (2019). https://api.semanticscholar.org/CorpusID:167217655
18. Zhu, S.C., Zhang, R., Tu, Z.: Integrating bottom-up/top-down for object recognition by data driven Markov chain monte Carlo. In: Proceedings IEEE Conference on Computer Vision and Pattern Recognition. CVPR 2000 (Cat. No. PR00662). vol. 1, pp. 738–745. IEEE (2000)

Automatic Detection and Segmentation of Coronary Artery Stenosis in Coronary Angiography Images

Dmitrii Evtyukhov[1], Georgy Kopanitsa[1], Oleg Metsker[2], Aleksandr Mogilevskii[1], Alexey Yakovlev[2], and Sergey Kovalchuk[1(✉)]

[1] ITMO University, Saint Petersburg, Russia
{dmitrii.evtyukhov,kovalchuk}@itmo.ru
[2] Almazov National Medical Research Centre, Saint Petersburg, Russia

Abstract. In this paper, we present an approach for the detection, segmentation, and quantification of stenoses in coronary arteries using modern computer vision and deep learning techniques. Our system incorporates a detection model based on YOLOv8 and a segmentation model (DeepLabV3+) for precise localization and delineation of stenosis regions. In addition, a novel method is introduced to measure arterial thickness to support clinical decision-making. The experimental evaluation shows that the approach demonstrates high quality and performance in comparison to existing solutions. This work aims to improve diagnostic efficiency and reduce the reliance on expensive foreign-made equipment by providing an integrated solution that can operate on standard hardware.

Keywords: Coronary artery stenosis · Computer vision · Deep learning · YOLOv8 · DeepLabV3+ · Medical image analysis

1 Introduction

Cardiovascular disease remains one of the leading causes of death worldwide. Among various cardiovascular conditions, coronary artery stenosis stands out as a major concern—this narrowing of the arterial vessels that supply blood to the heart can lead to ischemic heart disease, myocardial infarction, and other severe complications. The timely diagnosis of stenosis plays a crucial role in preventing these outcomes and reducing mortality associated with cardiovascular diseases [9]. A modern diagnostic approach for stenosis is coronary angiography, a method that visualizes the heart's vessels using X-rays and contrast agents. However, this method requires the mandatory involvement of highly qualified specialists and substantial time for manual image analysis. Furthermore, the results of such analysis are susceptible to human error, especially under high workloads faced by medical personnel.

Automating the diagnosis of stenosis through computer vision methods has the potential to revolutionize the processing of medical images. Deep learning and computer vision algorithms enable rapid and accurate analysis of angiograms, allowing for the detection of abnormalities and the extraction of vessel geometry information. This approach not only accelerates the diagnostic process but also minimizes errors due to human factors. Moreover, automated analysis systems can be deployed in remote medical centers where access to highly specialized personnel is limited.

Previous studies have demonstrated the effectiveness of deep learning-based methods for coronary artery segmentation. For instance, in the study by Serrano-Anton et al. [7], a UNet-based model with transfer learning was proposed for coronary artery segmentation in CT angiography images. The findings indicate that transfer learning significantly improves segmentation accuracy, particularly when working with limited data. Additionally, Danilov et al. [2] introduces a fully automated approach for coronary angiogram interpretation using a sequence of deep neural networks, achieving high accuracy in stenosis detection. These studies highlight the potential of modern deep learning and computer vision techniques in automating coronary artery stenosis diagnosis, aligning with the objectives and methodologies proposed in our work.

2 Related Works

Advancements in artificial intelligence (AI), particularly deep learning, have significantly impacted the automated detection and quantification of coronary artery stenosis [8].

A meta-analysis published by Jie at al. [4] evaluates the diagnostic accuracy of AI-assisted CTA in detecting stenosis and characterizing plaque composition. The analysis, which included 11 studies with 1,484 patients, reported a pooled area under the receiver operating characteristic curve (AUROC) of 0.96 for assessing atherosclerotic plaque. For detecting $\geq 50\%$ stenosis, the AUROC was 0.95, and for $\geq 70\%$ stenosis, it was 0.96. The study concludes that AI-assisted CTA has high diagnostic accuracy but acknowledges substantial heterogeneity among studies and emphasizes the need for further research to standardize AI applications in clinical practice. Dundas et al. [3] evaluated an AI-based coronary stenosis quantification (AI-CSQ) tool and compared its performance with invasive quantitative coronary angiography (QCA). Their findings demonstrated high diagnostic accuracy, with the AI-CSQ model achieving an AUC of 0.92 for detecting stenosis $\geq 50\%$ and 0.93 for stenosis $\geq 70\%$. Additionally, the system exhibited a sensitivity of 80% and specificity of 88% for moderate stenosis ($\geq 50\%$) and sensitivity of 78% and specificity of 92% for severe stenosis ($\geq 70\%$). Li et al. [5] developed a deep learning model capable of automatically segmenting coronary arteries and diagnosing stenosis of $\geq 50\%$ severity. Utilizing a U-Net architecture for segmentation and a 3DNet for classification, the model achieved a mean Dice coefficient of 0.771 and an accuracy of 75% in diagnosing coronary artery disease (CAD). However, the study was limited by its single-center design, potentially affecting the model's generalizability. Danilov et al. [2] investigate the feasibility of real-time coronary artery stenosis detection using deep learning. The study evaluates eight neural network architectures, including MobileNet, ResNet-50, ResNet-101, Inception ResNet, and NASNet, using angiography data from 100 patients. The Faster-RCNN Inception ResNet V2 model achieves the highest accuracy (mAP = 0.95, F1-score = 0.96) but has a slow inference speed (3 fps). In contrast, SSD MobileNet V2 is the fastest (38 fps) but less accurate (mAP = 0.83, F1-score = 0.80). The RFCN ResNet-101 V2 model offers the best balance (mAP = 0.94, F1-score = 0.96, speed = 10 fps). The study confirms the potential of deep learning for real-time stenosis detection, improving diagnostic efficiency. However, the small dataset and lack of external validation limit generalizability.

Annotating stenotic regions in coronary angiograms and computed tomography angiography images requires expertise from cardiologists or radiologists. However, inter-observer variability is a significant issue, as different specialists may interpret and delineate stenotic lesions differently. This variability affects the consistency of ground truth labels, complicating model training and reducing generalizability. For instance, Zhang et al. [8] highlighted that machine learning and deep learning methods face challenges due to the lack of professional image annotations, which are manually added by experts.

There is no universal consensus on classifying stenosis severity (e.g., mild <50%, moderate 50–70%, severe >70%), leading to discrepancies in annotation protocols across datasets. Some studies use diameter reduction measurements, while others incorporate functional assessments like fractional flow reserve (FFR). This inconsistency impacts model robustness and comparability across different studies. A review by Aleksandric et al. [1] discussed the challenges, limitations, and future perspectives in the functional assessment of coronary stenosis severity, emphasizing the complexity of coronary physiology in the presence of valvular heart disease.

Despite these advancements, challenges persist, including the need for large, annotated datasets and the variability in imaging protocols across institutions. Future research should focus on developing models that are robust across diverse populations and imaging conditions. In conclusion, deep learning has significantly advanced the automated analysis of coronary artery stenosis, offering improved diagnostic accuracy and efficiency. However, addressing current limitations is essential for broader clinical implementation.

3 Deep Learning Model Development and Evaluation

The proposed algorithm for the automatic analysis of coronary artery stenosis comprises several distinct stages, each designed to enhance the robustness and precision of the diagnostic process.

The algorithm employed for detection, segmentation, and thickness measurement of coronary artery stenosis is structured as follows (see Fig. 1). Initially, input images undergo a preprocessing step aimed at normalizing and enhancing image quality. Subsequently, the processed images are directed into two parallel branches: one for detecting the stenotic regions and another for segmenting coronary arteries. The detection branch localizes areas suspected of stenosis, while the segmentation branch delineates the coronary artery structure. The outputs from these parallel processes, namely the detected stenosis regions and segmented coronary artery masks, are then mapped together to define precise areas of interest. These combined masks are subjected to a skeletonization process, enabling the extraction of vessel centerlines. Finally, the arterial thickness is determined by analyzing geometric properties derived from these skeletonized representations.

3.1 Annotation of Stenosis Data

Accurate annotation of coronary artery stenosis is a critical step in the development of robust computer vision models for automated diagnosis. In this work, experienced cardiologists manually annotated imaging data to delineate stenotic regions using the YOLO (You Only Look Once) format. This section outlines the methodology, guidelines, and quality control procedures employed during the annotation process.

Imaging data were sourced from standard clinical modalities, including coronary angiography and computed tomography angiography (CTA), reflecting real-world diagnostic practices. Annotations were performed using a specialized tool (e.g., LabelImg[1]) adapted for medical imaging. Cardiologists followed strict clinical guidelines to delineate the boundaries of stenotic lesions. Each lesion was annotated by drawing a bounding box that encapsulated the region of stenosis. The process was conducted by experts to capture even subtle variations in lesion morphology accurately.

The annotations were saved in the YOLO format, where each line in the annotation file represents a single object detection. The format includes the following normalized parameters:

- `class_id`: An integer representing the category (e.g., "0" for coronary artery stenosis).
- `x_center`, `y_center`: The normalized coordinates of the bounding box center relative to the image dimensions.
- `width`, `height`: The normalized width and height of the bounding box.

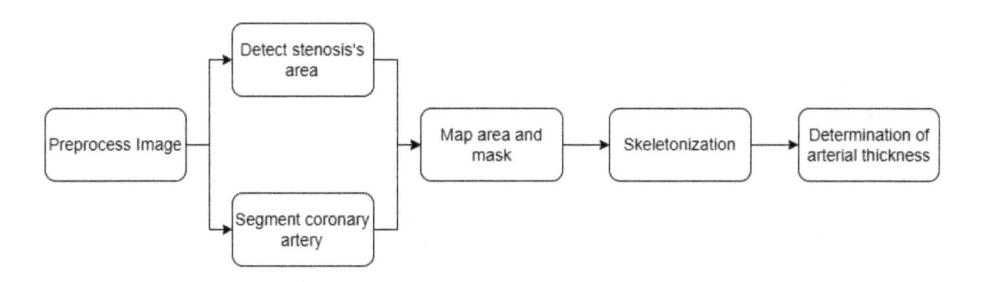

Fig. 1. General algorithm's scheme.

---
[1] https://github.com/HumanSignal/labelImg.

For example, an annotation line such as "0 0.450 0.550 0.200 0.150" indicates that the stenotic region has a center at 45% of the image width and 55% of the image height, with a bounding box spanning 20% of the image width and 15% of the image height.

To ensure consistency and accuracy, each image was independently annotated by at least two cardiologists. Discrepancies between annotations were resolved through consensus meetings. This dual-review process minimized inter-observer variability and ensured that the final annotated dataset accurately reflected the clinical characteristics of coronary stenosis.

3.2 Arterial Thickness Measurement

A reliable estimation of arterial thickness is crucial for quantifying the severity of coronary artery stenosis. In our approach, thickness is measured based on the segmented vessel mask by following a series of computational steps:

First, the segmented binary mask, representing the vessel region, is preprocessed to ensure a consistent data format. A skeletonization algorithm is then applied to this binary mask to extract the vessel's centerline, which serves as an approximation of the mid-curve running through the arterial lumen.

Subsequently, a Euclidean distance transform is computed on the binary mask. This transform assigns to each pixel a value corresponding to its shortest distance from the vessel boundary. For pixels that lie on the skeleton, the distance value effectively represents the approximate distance from the centerline to the edge of the vessel. Under the assumption that the vessel's full diameter is roughly twice this distance, the local arterial thickness is estimated by multiplying the distance value by two.

Finally, by aggregating the thickness estimates along the entire skeleton, key statistical metrics such as the minimum, maximum, mean, and median thickness are derived. These summary statistics provide a comprehensive quantitative description of arterial wall thickness, aiding in the assessment of stenosis severity and contributing to enhanced clinical decision-making.

3.3 Preprocessing

Medical imaging is often subject to challenges such as non-uniform illumination, contrast variability, and image noise. Consequently, a robust preprocessing pipeline is imperative to improve image fidelity and optimize model performance. The preprocessing pipeline consists of the following stages:

- **Standardization:** Normalization of pixel intensity values to mitigate inconsistencies in brightness and contrast across different angiographic images.
- **Contrast Enhancement:** Application of Contrast Limited Adaptive Histogram Equalization (CLAHE) to improve local contrast and enhance the visibility of vascular structures.
- **Gamma Correction:** Adjustment of image intensity using a power-law transformation to ensure a balanced brightness distribution.

- **Rescaling:** Standardization of image dimensions to maintain uniformity in spatial representation across different datasets.
- **Augmentation (for training purposes only):** Implementation of random transformations, such as rotations and flips, to enhance the generalizability and robustness of the trained models.

These preprocessing techniques collectively contribute to a more homogeneous dataset, reducing intra-class variations and facilitating the identification of salient anatomical features.

3.4 Detection Model

The identification of stenotic regions is performed using the YOLOv8m object detection model. This model was selected due to its optimal trade-off between detection accuracy and computational efficiency, making it well-suited for real-time clinical applications.

The performance of the detection model is assessed using established evaluation metrics:

- **Intersection over Union (IoU):** A metric quantifying the degree of overlap between the predicted bounding boxes and the ground-truth annotations, formally defined as:

$$IoU = \frac{|A \cap B|}{|A \cup B|} \quad (1)$$

where A and B represent the predicted and ground-truth bounding boxes, respectively.
- **Mean Average Precision (mAP@50):** Measures the detection accuracy at an IoU threshold of 0.5, providing an indication of model precision.
- **mAP@50-95:** Evaluates the detection performance over multiple IoU thresholds (ranging from 0.5 to 0.95 in increments of 0.05), offering a comprehensive assessment of model reliability.

The proposed model exhibits high precision while maintaining inference speeds conducive to real-time clinical deployment.

3.5 Segmentation Model

The delineation of stenotic regions is conducted using the DeepLabV3+ segmentation model, employing a ResNet-50 backbone. This architecture is particularly well-suited for high-resolution medical imaging tasks and utilizes atrous spatial pyramid pooling to capture multi-scale contextual information.

Segmentation accuracy is evaluated using the following quantitative metrics:

- **Intersection over Union (IoU):** Provides a measure of segmentation accuracy by computing the overlap between the predicted segmentation mask and the ground-truth annotation.
- **Dice Coefficient:** An alternative similarity measure defined as:

$$Dice = \frac{2\,A \cap B|}{|A| + |B|} \qquad (2)$$

where A and B denote the predicted and ground-truth segmentation masks, respectively. This metric is particularly sensitive to imbalances in class distributions.

The DeepLabV3+ model demonstrates superior capability in capturing fine-grained vascular structures, ensuring precise segmentation of stenotic regions.

3.6 Arterial Thickness Estimation

Following stenosis detection and segmentation, the arterial thickness is quantitatively assessed using a distance transform of the segmented vessel mask. This process enables an objective evaluation of vessel narrowing severity.

Key computational steps include:

- Skeletonization of the vessel structure to extract the centerline representation.
- Computation of the Euclidean distance from each skeleton pixel to the nearest vessel boundary.
- Estimation of arterial thickness as twice the computed distance, thereby providing an approximate measure of luminal diameter.

This methodology provides a rigorous quantitative assessment of stenosis severity, complementing traditional diagnostic approaches and enhancing clinical decision-making.

3.7 Dataset Description

The proposed system was evaluated on two distinct datasets: one for the detection task and one for the segmentation task.

Detection Dataset: A total of 9,000 coronary angiography images were used for the detection task. Of these, 10% (900 images) represent our proprietary data, while the remaining 90% (8,100 images) were sourced from the ARCADE dataset [6].

Segmentation Dataset: For segmentation, 250 high-resolution images with pixel-level annotations of coronary arteries were used. This dataset was similarly divided into training (175 samples), validation (50 samples), and test (25 samples) sets.

4 Model Training and Evaluation

The system was evaluated on both validation and test datasets using three different configurations.

The **baseline solution** (without preprocessing) utilized YOLO v5 for detection and U-Net for segmentation. In this setup, no image enhancement was applied, which resulted in a detection precision of 0.600 and mAP@50 of 0.880. The segmentation module achieved an IoU of 0.580 and a Dice coefficient of 0.740.

To improve performance, a **preprocessing pipeline** (PP) was introduced. This included standardization (to normalize pixel intensity values), CLAHE-based contrast enhancement, gamma correction for balanced brightness distribution, rescaling to standardize image dimensions, and data augmentation (random rotations and flips) for training. Applying PP to the YOLO v5 + U-Net configuration improved detection performance (precision = 0.672, mAP@50 = 0.930) and led to modest gains in segmentation (IoU ≈ 0.620, Dice ≈ 0.770).

The **optimal solution** aimed to further enhance performance by integrating a more advanced detection model (YOLO v8) and a more robust segmentation model (DeepLab v3+), while retaining the preprocessing pipeline. The YOLO v8 model significantly improved detection accuracy, achieving a precision of 0.966 and mAP@50 of 0.973. DeepLab v3+ further refined segmentation quality, with an IoU of 0.643 and a Dice coefficient of 0.781, surpassing the U-Net-based configurations.

This final configuration demonstrated the best balance between accuracy and computational efficiency. The combination of preprocessing, a superior detection model (YOLO v8), and an advanced segmentation model (DeepLab v3+) resulted in a system capable of high-speed processing while maintaining precise localization and detailed segmentation of stenotic regions.

Tables 1 and 2 summarize the key performance metrics for all three configurations.

Table 1. Detection Performance Metrics

Solution	Precision	mAP@50
Baseline (YOLO v5 + U-Net, no PP)	0.600	0.880
YOLO v5 with PP	0.672	0.930
Optimal (YOLO v8 + DeepLab v3+ with PP)	0.966	0.973

Table 2. Segmentation Performance Metrics

Solution	IoU	Dice
Baseline (YOLO v5 + U-Net, no PP)	0.580	0.740
YOLO v5 with PP	0.620	0.770
Optimal (YOLO v8 + DeepLab v3+ with PP)	0.643	0.781

The baseline configuration with U-Net yielded an IoU of 0.580 and a Dice coefficient of 0.740, indicating reasonable but suboptimal delineation of vascular structures, likely due to the absence of preprocessing and U-Net's limited contextual awareness in complex angiographic images.

With PP applied, the YOLOv5 + U-Net configuration improved to an IoU of 0.620 and a Dice coefficient of 0.770. These gains (7% in IoU, 4% in Dice) suggest that enhanced image quality facilitates better segmentation, particularly in capturing fine vessel edges. However, U-Net's performance plateaued, reflecting its architectural constraints in handling multi-scale features.

The optimal configuration excelled with a precision of 0.966, mAP@50 of 0.973, IoU of 0.643, and Dice of 0.781, outperforming the baseline by 61% in precision and 11% in mAP@50, and the intermediate setup by 44% and 5%, respectively. DeepLabV3+'s advanced feature extraction drove segmentation gains.

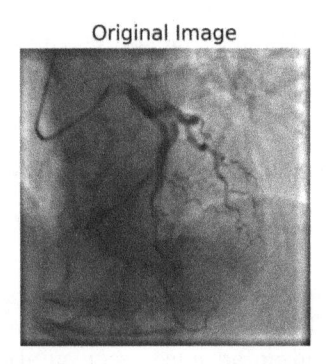

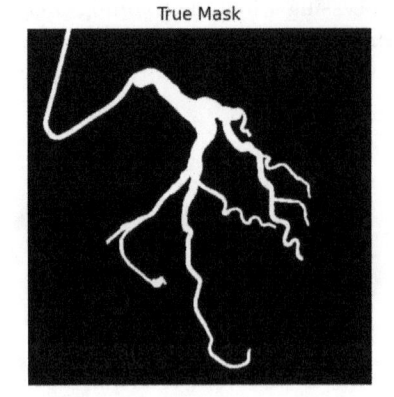

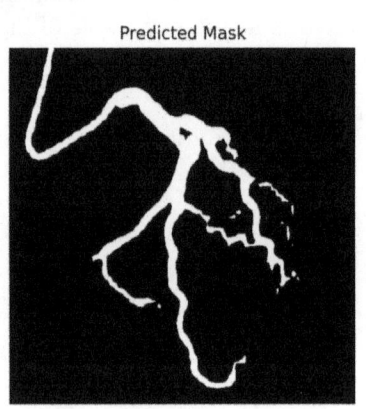

Fig. 2. Angiography segmentation steps.

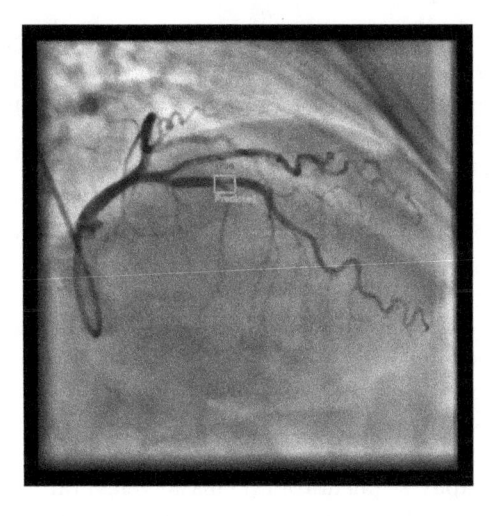

Fig. 3. Example of stenosis detection.

Arterial thickness estimates from the optimal setup ranged from 0.8 mm (severe stenosis) to 4.5 mm (healthy segments), with a mean of 2.4 mm, aligning with clinical norms and tied to segmentation accuracy. Inference speed was 18 fps on standard hardware, slower than the baseline (25 fps) but viable for real-time use. Qualitative results (Figs. 2 and 3) confirmed precise stenosis localization and delineation in the optimal setup.

We additionally evaluated the model on the ARCADE benchmark, where it achieved an F1-score of 0.524. Although this figure represents a solid level of performance, it falls short of the score obtained on our proprietary dataset—a gap we ascribe primarily to the divergent statistical properties and domain characteristics of the two data sources.

5 Discussion

The results of this investigation substantiate the efficacy of an integrated deep learning framework for the automated detection, segmentation, and quantification of coronary artery stenosis in coronary angiography images. The optimal configuration, employing YOLOv8 for detection and DeepLabV3+ for segmentation, complemented by a comprehensive preprocessing regimen, yielded a detection precision of 0.966, an mAP@50 of 0.973, a segmentation IoU of 0.643, and a Dice coefficient of 0.781. These outcomes surpass those of the baseline configuration (YOLO v5 + U-Net without preprocessing: precision = 0.600, mAP@50 = 0.880, IoU = 0.580, Dice = 0.740) and the intermediate configuration (YOLO v5 + U-Net with preprocessing: precision = 0.672, mAP@50 = 0.930, IoU = 0.620, Dice = 0.770), underscoring the synergistic effect of architectural choices and image preprocessing on diagnostic precision.

In comparison with prior research, the detection performance of this study aligns with, and in certain aspects exceeds, established benchmarks. Danilov et al. [2] reported an mAP of 0.950 and an F1-score of 0.960 using Faster-RCNN Inception ResNet V2 for real-time stenosis detection, albeit with a constrained inference rate of 3 fps. By contrast, the present YOLOv8-based model, with an mAP@50 of 0.973 and an inference speed of 18 fps, demonstrates enhanced precision and computational efficiency, rendering it more viable for real-time clinical implementation. Similarly, Dundas et al. [3] documented an AUROC of 0.920 for stenosis $\geq 50\%$ and 0.930 for stenosis $\geq 70\%$ using an AI-driven tool in CT angiography. Although AUROC and mAP are not directly equivalent, the precision of 0.966 achieved herein suggests a robust capacity for accurate stenosis identification within the angiography domain.

Regarding segmentation, the DeepLabV3+ model's Dice coefficient of 0.781 marginally exceeds the 0.771 reported by Li et al. [5] using a U-Net architecture for CT angiography segmentation. This incremental improvement may be attributable to DeepLabV3+'s incorporation of atrous spatial pyramid pooling, which facilitates superior multi-scale feature extraction relative to the convolutional framework of U-Net. Serrano-Antón et al. [7] also employed a U-Net model with transfer learning, though the absence of specific Dice metrics precludes direct comparison. The IoU of 0.643, while indicative of competent segmentation, suggests that further refinement in boundary delineation is warranted, particularly when juxtaposed with higher IoU values typical of non-medical imaging applications.

A distinctive contribution of this work lies in the development of an arterial thickness measurement technique, utilizing skeletonization and Euclidean distance transforms to quantify vessel narrowing. Thickness estimates ranged from 0.800 mm in severe stenosis to 4.500 mm in healthy segments, with a mean of 2.400 mm, consistent with clinical norms. This method provides a reproducible, non-invasive metric for stenosis severity assessment, distinct from invasive approaches such as fractional flow reserve (FFR) described by Aleksandric et al. [1]. However, its dependence on segmentation accuracy implies that enhancements to DeepLabV3+ could further bolster reliability.

Limitations of this study mirror challenges prevalent in the field. The reported performance metrics derive from controlled datasets, which may not fully encapsulate the heterogeneity of clinical imaging conditions, including variations in contrast agent distribution or equipment-specific artifacts—issues also noted by Danilov et al. [2] and Li et al. [5]. Although the dual-review annotation process mitigated inter-observer variability, as highlighted by Zhang et al. [8], the lack of a standardized classification for stenosis severity (e.g., mild <50%, moderate 50–70%, severe >70%) impedes consistent comparison with studies employing divergent criteria. Relative to Jie et al.'s [4] meta-analysis, which reported an AUROC of 0.950–0.960 for AI-assisted CTA, this study's emphasis on angiography extends the applicability of such techniques, though external validation across varied cohorts and modalities remains essential to address methodological disparities.

We additionally evaluated the model on the ARCADE benchmark, where it achieved an F1-score of 0.524. Although this figure represents a solid level of performance, it falls short of the score obtained on our proprietary dataset—a gap we ascribe primarily to the divergent statistical properties and domain characteristics of the two data sources. These results highlight that while the system performs well on proprietary data (F1-score = 0.681), domain shift remains a significant factor affecting generalization to external datasets such as ARCADE (F1-score = 0.524).

6 Conclusion

This study presents an integrated approach for automating coronary artery stenosis analysis using deep learning. By combining YOLOv8 for detection, DeepLabV3+ for segmentation, and a novel thickness measurement method, our system achieves a detection precision of 0.966, an mAP@50 of 0.973, a segmentation IoU of 0.643, and a Dice coefficient of 0.781. These consistent results across the evaluation phases demonstrate high performance and scalability for cardiovascular diagnostics. Future efforts will refine accuracy, expand datasets, and integrate the solution into clinical practice, potentially extending its utility to other cardiovascular conditions.

Acknowledgments. The research was supported by The Russian Science Foundation, agreement 24-11-00272, https://rscf.ru/project/24-11-00272/.

References

1. Aleksandric, S., Banovic, M., Beleslin, B.: Challenges in diagnosis and functional assessment of coronary artery disease in patients with severe aortic stenosis. Front. Cardiovasc. Med. **9** (2022). https://doi.org/10.3389/fcvm.2022.849032
2. Danilov, V.V., et al.: Real-time coronary artery stenosis detection based on modern neural networks. Sci. Rep. **11**(1) (2021). https://doi.org/10.1038/s41598-021-87174-2
3. Dundas, J., et al.: Artificial intelligence–based coronary stenosis quantification at coronary CT angiography versus quantitative coronary angiography. Radiol. Cardiothorac. Imaging **5**(6) (2023). https://doi.org/10.1148/ryct.230124
4. Jie, P., et al.: Diagnostic value of artificial intelligence-assisted CTA for the assessment of atherosclerosis plaque: a systematic review and meta-analysis. Front. Cardiovasc. Med. **11** (2024). https://doi.org/10.3389/fcvm.2024.1398963
5. Li, Y., et al.: Automatic coronary artery segmentation and diagnosis of stenosis by deep learning based on computed tomographic coronary angiography. Eur. Radiol. 1–9 (2022). https://doi.org/10.1007/s00330-022-08761-z
6. Popov, M., et al.: ARCADE: Automatic region-based coronary artery disease diagnostics using x-ray angiography images dataset (2023). https://doi.org/10.5281/ZENODO.8386059
7. Serrano-Antón, B., et al.: Optimal Coronary Artery Segmentation Based on Transfer Learning and UNet Architecture, pp. 55–64 (2023). https://doi.org/10.1007/978-3-031-46914-5_5

8. Zhang, X., Zhang, B., Zhang, F.: Stenosis detection and quantification of coronary artery using machine learning and deep learning. Angiology **75**(5), 405–416 (2023). https://doi.org/10.1177/00033197231187063
9. Zheng, J., Heidenreich, P.A., Kohsaka, S., Fearon, W.F., Sandhu, A.T.: Long-term outcomes of early coronary artery disease testing after new-onset heart failure. Circ. Heart Failure **16**(7) (2023). https://doi.org/10.1161/circheartfailure.122.010426

Explainable Artificial Intelligence for Doctors Decision Support in Diagnosing Spinal Pathologies

Aleksandra Vatian, Alexey Zubanenko(✉), Pavel Ulyanov, Alexander Golubev, Artem Beresnev, and Natalia Gusarova

1ITMO University, 49 Kronverksky Av, St. Petersburg, Russia
alexvatyan@gmail.com

Abstract. This paper presents an AI-based decision support system for spinal pathology diagnostics using MRI. The system incorporates an ensemble of neural networks and explainable AI (XAI) tools based on Grad-CAM. Our approach is aimed not only at enhancing the transparency of AI predictions, but also at improving clinical decisions in diagnostically complex cases. We experimentally show that (1) XAI can be used to restructure the training dataset to improve model performance, and (2) radiologists make more accurate diagnoses when provided with XAI maps alongside standard images. Our system shows promising results in detecting borderline cases of intervertebral disc protrusions, and lays the foundation for integrating XAI into clinical practice.

Keywords: Explainable AI · Grad-CAM · MRI · Spinal Pathology · Medical Imaging · Neural Networks

1 Introduction and Motivation

High-tech medical imaging, and in particular MRI images, are the first-line information sources in diagnosing spinal disorders [1–3]. The accuracy of human expert judgement of medical images is far below 100% [4, 5] and is determined not only by the qualifications and experience of radiologists, but also by their subjective differences, as well as by the degree of pathology manifestations.

This fully applies to degenerative spinal diseases [6, 7]. For example, intervertebral protrusions are often visually less pronounced compared to extrusions. Meanwhile, according to the Michigan State University classification [9], they represent significantly more complex diagnostic cases, where the boundaries between normal and pathological are blurred.

Today, artificial intelligence (AI) tools declare equal, if not better, accuracy in classifying spinal lesions than human experts [9–11]. However, they do not go beyond laboratory conditions to widespread clinical practice. This paradox is explained by a range of reasons, among which until recently priority belonged to the "black box" nature of AI. This challenge was expected to be addressed by the numerous means of explainable AI

(XAI) [12, 13], however, their inclusion in medical AI tools did not appear to change the situation much. This is largely because the European Medical Device Regulation (EU MDR) endorsed restrictions regarding transparency and other XAI features that have to be met before an AI based tool can be implemented in clinical practice [14].

In this regard, it would make sense to shift the focus of researchers from using XAI in medical domain for its "direct purpose", that is, as a means of explaining decisions made by an AI system, towards using XAI as an additional source of information for professionals, along with the image itself and its AI segmentation. The need for such a shift was clearly stated in [15, 16]. The inclusion of XAI tools in the structure of AI medical imaging systems has been widely reported [17–20]. However, we could detect only a few works containing methodically verified quantitative estimates of its feasibility.

Several studies [21–23] explored diagnostic support using Grad-CAM or segmentation overlays, but were limited in scope or dataset size.

[22] reported that using similar AI support in MRI diagnostics of brain disorders significantly, by almost 4%, decreased the number of cases erroneously classified as healthy (false negatives), even among experienced radiologists.

In [23], the radiologists of various level of qualification were consecutively presented with two series of images: MR images of vertebral bodies with degenerative changes separately, and then in combination with the results of segmentation according to the Modic scale [24] in a special AI system. The agreement between junior and senior neuroradiologists significantly improved in the latter case (from Cohen's kappa score of 0.52 to 0.58). It is noteworthy that both [22] and [23] cases did not raise any diagnostic doubts among experienced radiologists. In particular, [23] reported senior neuroradiologists having almost identical opinions of all the presented cases regardless of the experimental scenarios.

Of much greater research and practical interest are borderline cases, where the probability of error of both a human professional and an AI diagnostic system could be expected to be higher. For example, when diagnosing intervertebral hernias, the error rate even among experienced radiologists would reach 30% and more [25]. In addition to diagnostic errors themselves, this reduces the quality of the dataset markup used to train the AI system, and, accordingly, fundamentally limits its effectiveness.

Our paper aims to examine the effectiveness of XAI as an additional source of information for professionals to alleviate the above limitations in diagnostically complicated cases, using the example of spinal pathologies. We assume that the demonstration of XAI results will help the professional to discern inconspicuous details of the image and thereby increase the accuracy of diagnosis in complicated cases, including those regarded as borderline. To summarize, our contributions are as follows:

- We propose a novel AI system with ensemble architecture for spinal pathology segmentation, equipped with XAI tools in the form of Grad-CAM activation maps with the ability to select controlled neural network (NN) layers for ensemble elements. This would allow the medical practitioner to choose the most informative level of XAI granularity.
- We experimentally show that the results of XAI can be used as a means of modifying the training dataset, which would lead to an increase in the efficiency of the AI system in segmenting borderline cases.

– We experimentally show that demonstration of XAI results along with the original MRI image leads to increased diagnostic performance of professional radiologists in borderline cases.

The remainder of this paper is structured as follows. Section 2 presents the method, including architecture, preprocessing, and experimental setup. Section 3 discusses the results. Section 4 concludes the paper and outlines future directions.

2 Method and Materials

2.1 Neural Network Configuration

The developed NN for MRI image segmentation is based on the ensemble of models with SegResNet, UNETR, and Swin UNET architectures, respectively. The general structure of the developed NN is presented in Fig. 1.

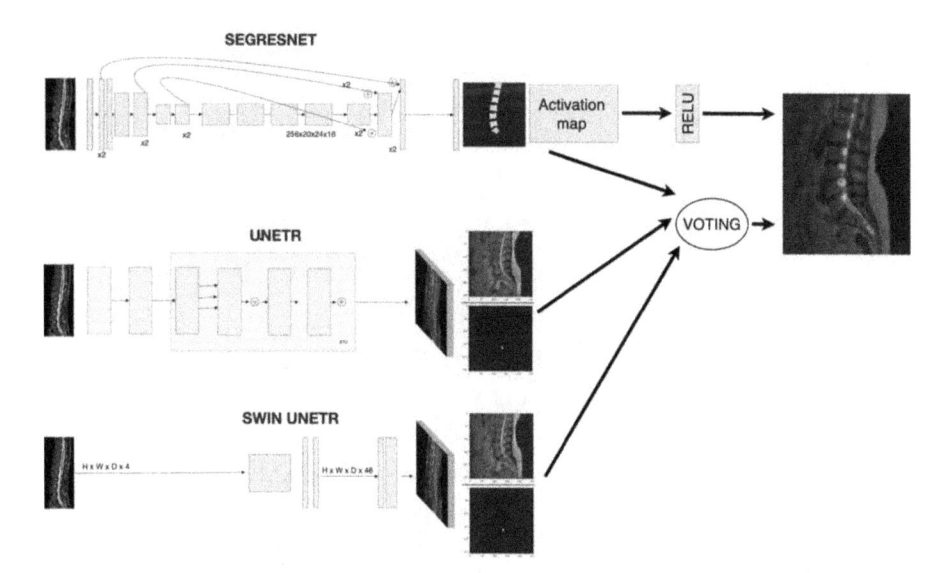

Fig. 1. Developed NN structure

SegResNet [26] from Monai [28] is a deep learning-based segmentation model optimized for medical imaging (e.g. MRI/CT) where edge accuracy and detail preservation are essential [29]. In the proposed NN model SegResNet serves as a backbone. UNETR [30] and Swin UNETR [31] are transformer-based architectures effective for 3D segmentation tasks. Their inclusion ensures a good balance between global context and spatial detail.

The final segmentation of the analyzed image is performed by pixel-by-pixel voting of predictions made by the models participating in the ensemble. The developed NN was trained using the Adam optimizer, an initial learning rate of 0.0001, and the loss function Dice + Focal Loss. The 5-fold cross-validation method was used for validation. The plateau criterion was used for early stop of training.

While other segmentation models such as Attention UNet, DeepLabV3+, and V-Net are also popular, they either lack support for volumetric data (e.g., DeepLabV3+) or show limitations in processing long-range dependencies in small datasets. Our selected architectures are complementary in design and offer a well-rounded ensemble suited to the task of detecting subtle spinal abnormalities.

2.2 XAI Implementation

The scheme of XAI implementation in the developed NN is shown in Fig. 2.

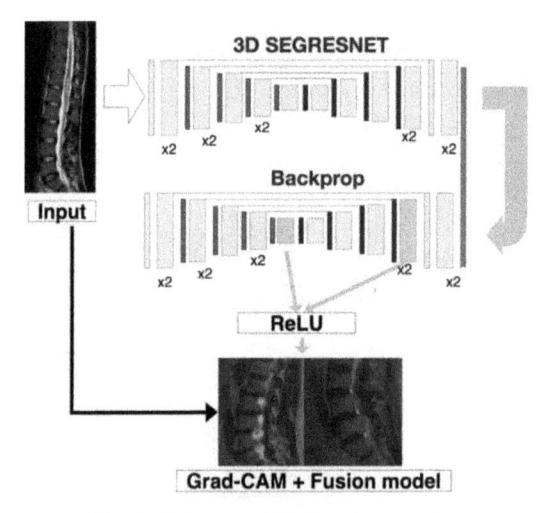

Fig. 2. Scheme of XAI implementation

The method chosen for implementing XAI was Grad-CAM [33], which captures activation maps of the selected NN layer and then overlays them on the input image as a heat map using gradient descent and appropriate spatial transformations. The Grad-CAM method has gained widespread acceptance in medical applications [21, 34, 35] as an intuitive means of visually demonstrating those areas of the input image that appeared to be the most important "from the point of view" of the AI model. Grad-CAM was chosen for its visual clarity and compatibility with CNN architectures, making it especially suitable for radiological interpretation.

Developing the Grad-CAM activation map begins with the selection of the NN target layer, which is shown in yellow in Fig. 2. A forward pass is then made through the 3D SEGRESNET architecture, where the input image proceeds through successive convolutional layers (light green blocks) and normalizations. This is followed by backpropagation of the error, calculating gradients in the target layer, which determines the relevance of individual features. The resulting activation map is scaled to the size of the original image and superimposed on the latter as a heat map, which visualizes the areas most highly relevant for the decision made by the model.

The last or penultimate layers of SEGRESNET were used as the target NN layer in our experiments.

2.3 Metrics

The model was evaluated using standard metrics, including Sensitivity, Specificity, Dice Score, and Cohen's Kappa.

2.4 Dataset

A dataset of 1500 axial T2-weighted lumbar spine MRIs (512×512 px, 3 mm slice thickness) was collected from scanners by Siemens, GE, and Philips to ensure imaging variability. All cases were reviewed and labeled by five radiologists in a cross-voting protocol, based on the Lumbar Disc Nomenclature 2.0 and the MSU classification. The dataset includes 200 normal images and 1300 pathological cases, all limited to Grade 1 intervertebral disc protrusions (MSU), deliberately excluding extrusions and sequestrations to focus on diagnostically ambiguous cases. Due to privacy restrictions, the dataset is currently unavailable but is planned for anonymized release.

2.5 Experimental Scenarios

We conducted experiments according to three scenarios.

Scenario 1. We also experimented with restructuring the training dataset based on Grad-CAM sensitivity rankings, which led to improved segmentation performance. Detailed results are omitted due to space constraints.

Scenario 2. was aimed at assessing the impact of Grad-CAM demonstration on the accuracy of diagnostics performed by medical practitioners. Five professional radiologists with different levels of experience (from 2 to 10 years of practical work) were involved in the experiment. A sample of 100 triplets "original medical image + result of segmentation performed by NN + its GradCAM map" was formed. A network trained on a restructured dataset (see Scenario 1) was used as the NN. Twenty triplets from the sample were randomly selected for the demonstration. The radiologists were assigned the task of diagnosing and segmenting the affected area. In the first experiment (**Scenario 2a**), they were demonstrated only the original MRI image, in the second experiment (**Scenario 2b**) - the entire above-described triplet. The randomly selected triplets made repeated demonstration of an image already seen by the doctor highly unlikely.

Scenario 3. Additionally, we evaluated the influence of Grad-CAM granularity levels on diagnostic decisions. These results will be presented in a future extended version.

3 Results and Discussion

Beyond the main diagnostic accuracy study (Scenario 2), we conducted auxiliary experiments involving dataset restructuring (Scenario 1) and varying Grad-CAM granularity (Scenario 3). Their results, while promising, are not shown here due to space constraints.

Table 1. Performance indicators of medical diagnostics when demonstrating original MRI images and triples of "MRI image + NN segmentation results + GradCAM markup results".

No of participian	Scenario 2a				Scenario 2b			
	Sensitivity	Specificity	Dice	κ	Sensitivity	Specificity	Dice	κ
1	0.87	0.83	0.85	0.81	0.92	0.94	0.87	0.85
2	0.82	0.86	0.80	0.75	0.91	0.92	0.85	0.83
3	0.84	0.82	0.82	0.78	0.90	0.93	0.88	0.86
4	0.86	0.85	0.84	0.80	0.93	0.94	0.86	0.85
5	0.85	0.84	0.84	0.81	0.89	0.92	0.84	0.82
Mean	0.85 ± 0.04	0.84 ± 0.03	0.83 ± 0.03	0.79 ± 0.05	0.91 ± 0.03	0.93 ± 0.03	0.86 ± 0.02	0.84 ± 0.04

Scenario 2. The results of the experiments for Scenario 2 are presented in Table 1.

Comparison of the results obtained in accordance with scenarios 2a and 2b shows that when demonstrating Grad-CAM maps along with the main MRI images, the efficiency of diagnostics performed by radiologists increases (Sensitivity increased from 0.85 to 0.91, and Specificity – from 0.84 to 0.93). This indicates that medical professional can more accurately determine the presence of pathologies, relying on the visualized attention zones of the model.

4 Conclusion and Future Works

The paper proposes an AI system for segmentation of spinal pathologies with ensemble architecture, equipped with an XAI tool in Grad-CAM form. The efficiency of the proposed system exceeds the SOTA model in diagnosing the most complex, borderline cases of intervertebral hernias.

Our experiments confirm that XAI improves model training and clinical performance in borderline cases. The ability to dynamically adjust XAI granularity enhances diagnostic clarity. Future work includes Active Learning based on XAI focus zones, and adaptive XAI interfaces for clinicians.

Acknowledgments. This work was supported by Russian Science Foundation, Grant № 23-11-00346.

Disclosure of Interests. The authors have no competing interests to declare that are relevant to the content of this article.

References

1. Merali, Z.A., Colak, E., Wilson, J.R.: Applications of machine learning to imaging of spinal disorders: current status and future directions. Global Spine J. **11**(1 Suppl), 23S-29S (2021). https://doi.org/10.1177/2192568220961353
2. Lee, A., Ong, W., Makmur, A., et al.: Applications of artificial intelligence and machine learning in spine MRI. Bioengineering **11**(9), 894 (2024). https://doi.org/10.3390/bioengineering11090894

3. Xuan J., Ke B., Ma W., et al.: Spinal disease diagnosis assistant based on MRI images using deep transfer learning methods. Front. Public Health. Sec. Digital Public Health (2023)
4. Fu, M.C., Buerba, R.A., Long, W.D., et al.: Interrater and intrarater agreements of magnetic resonance imaging findings in the lumbar spine: significant variability across degenerative conditions. Spine J. **14**(10), 2442–2448 (2014). https://doi.org/10.1016/j.spinee.2014.03.010. Epub 2014 Mar 15
5. Kim, J.-H., van Rijn, R.M., van Tulder, M.W.: Diagnostic accuracy of diagnostic imaging for lumbar disc herniation in adults with low back pain or sciatica is unknown; a systematic review. Chiropr. Man Therap. **21**(26), 37 (2018). https://doi.org/10.1186/s12998-018-0207-x
6. Azimi, P., Yazdanian, T., Benzel, E.C.: A review on the use of artificial intelligence in spinal diseases. Asian Spine J. **14**(4), 543–571 (2020). https://doi.org/10.31616/asj.2020.0147
7. Mbarki, W., Bouchouicha, M., Frizzi, S., et al.: Lumbar spine discs classification based on deep convolutional neural networks using axial view MRI. Interdisciplinary Neurosurgery **22**, 100837 (2020). https://doi.org/10.1016/j.inat.2020.100837
8. Fardon, D.F., Williams, A.L., Dohring, E.J., et al.: Lumbar disc nomenclature: version 2.0: Recommendations of the combined task forces of the North American spine society, the American society of spine radiology and the American society of neuroradiology. Spine J. **14**(11), 2525–45 (2014). https://doi.org/10.1016/j.spinee.2014.04.022
9. Mysliwiec, L.W., Cholewicki, J., Winkelpleck, M.D.: MSU classification for herniated lumbar discs on MRI: toward developing objective criteria for surgical selection. Eur. Spine J. **19**(7), 1087–1093 (2010). https://doi.org/10.1007/s00586-009-1274-4
10. Liawrungrueang, W., Park, J-B., Cholamjiak, W., et al.: Artificial intelligence-assisted MRI diagnosis in lumbar degenerative disc disease: a systematic review. Glob. Spine J. 21925682241274372LNCS Homepage (2024). http://www.springer.com/lncs. Accessed 25 Oct 2023
11. Qian, J., Su, G., Shu, X., et al.: Lumbar disc herniation diagnosis using deep learning on MRI. J. Radiat. Res. Appl. Sci. **17**(3), 100988 (2024)
12. Amisha, P., Malik, M.P., Rathaur, V.K.: Overview of artificial intelligence in medicine. J. Family Med. Prim. Care. **8**, 2328–2331 (2019). https://doi.org/10.4103/jfmpc.jfmpc_440_19
13. de Vries, B.M., Zwezerijnen, G.J.C., Burchell, G.L., et al.: Explainable artificial intelligence (XAI) in radiology and nuclear medicine: a literature review. Front. Med. 12 May 2023. Sec. Nuclear Med. **10** (2023). https://doi.org/10.3389/fmed.2023.1180773
14. Hafeez, Y., Memon, K., AL-Quraishi, M.S.: Explainable AI in diagnostic radiology for neurological disorders: a systematic review, and what doctors think about it. Diagnostics **15**(2), 168 (2025). https://doi.org/10.3390/diagnostics15020168
15. Beckers, R., Kwade, Z., Zanca, F.: The EU medical device regulation: implications for artificial intelligence-based medical device software in medical physics. Phys. Med. **83**, 1–8 (2021). https://doi.org/10.1016/j.ejmp.2021.02.011
16. Knapič, S., Malhi, A., Saluja, R.: Explainable artificial intelligence for human decision support system in the medical domain. Mach. Learn. Knowl. Extr. **3**(3), 740–770 (2021). https://doi.org/10.3390/make3030037
17. Chen, H., Gomez, C., Huang, C.M. et al. Explainable medical imaging AI needs human-centered design: guidelines and evidence from a systematic review. npj Digit. Med. **5**, 156 (2022). https://doi.org/10.1038/s41746-022-00699-2
18. van der Velden, B.H.M., Kuijf, H.J., Gilhuijs, K.G.A., Max, A., et al.: Explainable artificial intelligence (XAI) in deep learning-based medical image analysis. Med. Image Anal. **79**, 102470 (2022). https://doi.org/10.1016/j.media.2022.102470
19. Borys, K., Schmitt, Y.A., Nauta, M., et al.: Explainable AI in medical imaging: an overview for clinical practitioners – saliency-based XAI approaches. Eur. J. Radiol. **162**, 110787 (2023)

20. Fontes, M., De Almeida, J.D.S., Cunha, A.: Application of example-based explainable artificial intelligence (XAI) for analysis and interpretation of medical imaging: a systematic review. IEEE Access **12**, 26419–26427 (2024). https://doi.org/10.1109/ACCESS.2024.3367606
21. M.M.M., Manesh, T.R., V V.K., et al.: Enhancing brain tumor detection in MRI images through explainable AI using Grad-CAM with Resnet 50. BMC Med Imaging **24**, 107 (2024). https://doi.org/10.1186/s12880-024-01292-7
22. Chien, J.-C., Lee, J.-D., Hu, C.-S., et al.: The usefulness of gradient-weighted CAM in assisting medical diagnoses. Appl. Sci. **12**(15), 7748 (2022). https://doi.org/10.3390/app12157748
23. Finck, T., Moosbauer, J., Probst, M., et al.: Faster and better: how anomaly detection can accelerate and improve reporting of head computed tomography. Diagnostics **12**(2), 452 (2022). https://doi.org/10.3390/diagnostics12020452
24. Gao, K.T., Tibrewala, R., Hess, M.: Automatic detection and voxel-wise mapping of lumbar spine Modic changes with deep learning. JOR Spine. **5**, e1204. jorspine.com 1 of 10 (2022). https://doi.org/10.1002/jsp2.1204
25. Lange, M.B., Petersen, L.J., Lausen, M.: Influence of prior imaging information on diagnostic accuracy for focal skeletal processes—a retrospective analysis of the consistency between biopsy-verified imaging diagnoses. Diagnostics **12**(7), 1735 (2022). https://doi.org/10.3390/diagnostics12071735
26. Modic, M.T., Steinberg, P.M., Ross, J.S., et al.: Degenerative disk disease: assessment of changes in vertebral body marrow with MR imaging. Radiology **166**(1 Pt 1), 193–199 (1988)

Predicting Disease Transmission Rates for Hybrid Modeling of Epidemic Outbreaks: Statistical and Machine Learning Approaches

Maria Koshkareva, Elizabetty Guseva, Alyona Sharova, and Vasiliy Leonenko

ITMO University, Kronverksky Pr. 49A, 197101 St. Petersburg, Russia
{mpkoshkareva,vnleonenko}@itmo.ru

Abstract. Hybrid disease modeling is a perspective area of research that allows using detailed individual-based models for the outbreak onset phase and lightweight compartmental models to capture the general trend of the disease progression. In such a way, the method of hybrid modeling provides a good trade-off between the simulation speed and the accuracy of reproducing disease dynamics. One of the problems related to this approach is how to switch properly between the two models. That included detecting the right time moment to finish simulations with the detailed model and calculating correctly the input parameters for the compartmental model. In this paper, we propose an implementation of switching which relies on evaluation and prediction of disease transmission rate. Using an example with a network-based model and a discrete compartmental model, we demonstrate several methods of disease transmission prediction based on statistical models and machine learning approaches and analyze their advantages and disadvantages. The developed methods can be generalized to hybrid modeling of highly detailed demographic processes and propagation processes in general.

Keywords: network models · SEIR models · epidemics · compartmental models · machine learning · statistical models · disease outbreaks

1 Introduction

For epidemic outbreak modeling, compartmental models, such as SIR [6] and its modifications, are widely used. Their main drawback is a simplifying assumption of homogeneous mixing among individuals. Detailed individual-based models, like agent-based models (ABM) [7] or network-based models [10], can provide more realistic interactions of individuals allowing to capture localized transmission and super-spreading events. However, detailed models have higher computational costs and execution time, which is not desirable for situations requiring fast decision-making. As a result, selecting the best approach between the two for a particular use case is not an easy task [11]. Good news lies in the fact that the advantages of mentioned model techniques can be combined and drawbacks can be compensated by applying hybrid approaches.

Hybrid frameworks, such as those proposed by Bobashev et al. [3] and Hunter et al. [5], have shown that the usage of interacting simple and detailed models can improve both simulation accuracy and computational efficiency. The switching between the submodels occurs in a certain phase of the epidemic. It is assumed that in the early phase of the outbreak, when the number of infected individuals is small, detailed modeling of contacts plays a significant role, which makes an individual-based model a better option for simulation. Particularly, if transmission heterogeneity is high across a pathogen group, a small number of individuals plays a disproportionate role to the spread of a pathogen and targeting control measures towards those individuals can be very effective to reduce the epidemic burden [13]. Once the number of the infected is large, the population can be considered homogeneous, so transitioning to a compartmental model can be done. A key challenge is to choose the optimal switching moment. Early switching leads to information loss related to details of disease transmission, while late switching reduces the computational advantages of the hybrid approach. The success of hybrid simulation depends on an accurate switching moment, as well as on a proper alignment of submodels to ensure that the modeled disease prevalence curve won't show sudden surges or drops in the moment of switching [8].

Among the parameters of epidemic submodels, the disease transmission rate β plays a major role and requires special attention when switching between different model types. As it is noted in [3], thanks to the law of large numbers, the calculated cumulative average of β tends to stabilize as the number of the infected gets larger, which indicates a proper moment to use a compartmental model (a switch point) without altering epidemic dynamics. Hence, one can detect a switch point by waiting for the cumulative average of β to become constant and use this value further on in the simulation via the compartmental submodel. The authors of [3] pay attention to the fact that their hybrid model is built on top of a rather simple ABM (homogeneously mixed population, no community structure) which might lower the effectiveness of their approach for more complicated ABMs. Particularly, other works, such as [12], indicate that assuming time–dependent $\beta = \beta(t)$ throughout the whole simulation, rather than approximating it with a constant, is essential for accurately recreating epidemic trajectories in real settings.

In this research, we develop and assess methods of switch point detection and β estimation to ensure accurate switching between the submodels. As a baseline method, we regarded the switching of models using a last estimated value of β. The alternative proposed methods include various ways of dynamic β estimation along with β prediction based on the historically known epidemic waves. The methods are tested on a hybrid model which couples a detailed network submodel with a lightweight compartmental submodel. The goal of this work is to compare the computational efficiency (runtime) and accuracy (RMSE) of the analyzed methods. The source code implementing the methods proposed in this paper is available on GitHub [1].

2 Methods

2.1 Submodels for the Hybrid Approach

Compartmental Submodel. The baseline compartmental submodel we use in this study is a deterministic discrete SEIR model with a time step equal to one day. To address the stochasticity which is intrinsic to network and agent-based submodels of the hybrid framework, we also employ the stochastic version of the same model with probabilistic transitions using binomial sampling. In the experiments further on we demonstrate both the deterministic and the stochastic compartmental models. Let S be susceptible individuals, E—exposed individuals, I—infectious individuals, and R—recovered individuals. The dynamics of the groups' sizes over time for a discrete stochastic case are set by the following difference equations:

$$S_{t+1} = S_t - \xi_{SE}, \tag{1}$$

$$E_{t+1} = E_t + \xi_{SE} - \xi_{EI},$$

$$I_{t+1} = I_t + \xi_{EI} - \xi_{IR},$$

$$R_{t+1} = R_t + \xi_{IR},$$

$$S_0 \geq 0, E_0 \geq 0, I_0 \geq 0, R_0 \geq 0,$$

$$S_0 + E_0 + I_0 + R_0 = N, \tag{2}$$

where $\xi_{SE} \sim Bin(S_t, \beta I_t)$, $\xi_{EI} \sim Bin(E_t, \sigma)$, $\xi_{IR} \sim Bin(I_t, \gamma)$.

Network Submodel. The population is modeled as a network where each individual is a node, and interactions between the individuals are possible if and only if they are connected by an edge. In our simulations, we used a Barabasi-Albert network topology [2]. The Barabasi-Albert model generates networks through preferential attachment: when new nodes are added, they preferentially connect to existing nodes that already have a high degree (many connections). This process leads to the emergence of "hubs" - highly connected individuals - and a long-tailed degree distribution. In the context of disease spread, this implies that infections may spread more rapidly through these hubs. The network submodel does not have the β parameter as in the SEIR model. Instead, the disease transmission is characterized by τ, the probability for one person to infect the other via the common edge.

In the network submodel, the infection dynamics is defined by the contact network topology and the value of transmission probability τ. In a compartmental model, the infection transmission is solely governed by the transmission rate β. We assess β_t from a modeled output generated by the network model, using an approximate formula (Eq. 3):

$$\beta_t = -\frac{S_{t+1} - S_t}{S_t \cdot I_t}. \tag{3}$$

As stated earlier, the choice of β values for the compartmental submodel is important to ensure a proper switch. To address this, we generated baseline epidemic prevalence trajectories made solely by the network submodel, without switching. The values of σ and γ were fixed, we varied only 2 parameters – τ and I_0, the fraction of initially infected. Resulted epidemic curves, as will be described in detail later, constituted test and train datasets which were used to evaluate our methods of β analysis and prediction.

2.2 Beta Estimation Approaches

The regarded approaches were grouped into three categories: (a) those that rely on a current incomplete disease trajectory (which mimics the ongoing epidemic process), (b) those that are based on a train dataset with complete trajectories (which helps establish the form of β_t based on previously collected information), and (c) those that use both a train set and a current disease trajectory (which makes it possible to better adjust a form of β_t to an actual disease curve). Each group of approaches has their own baselines. In graphs and tables, we use the abbreviated names in the form of $M_{<type>}$, where <type> is the method name.

Estimation on Current Incomplete Data. To estimate the value of β for switching, we started with three baselines: choosing the last known β_t value, i.e. the closest to a switch point (M_{last_val}); taking the last value from the moving averages of β_t (M_{ma_val}), taking the last value from the cumulative averages of β_t (M_{ca_val}). The more advanced method of β estimation is to fit a function to incomplete β_t data before the day of switch and use it in a compartmental model (M_{biexp}). The function should have a similar shape with actually observed trends of β_t, i.e. bear similarity with a skewed bell. The chosen function is the biexponential decay function (Eq. 4):

$$\beta_t = a(e^{-bt} - e^{-ct}), \qquad (4)$$

where a, b and c are estimated through non-linear least squares.

Estimation on Train Set. The methods in this group obtain the form of β_t based on the train set, therefore the form of β_t is fixed and is not influenced by actual prevalence data related to the ongoing simulated outbreak. The baseline method (M_{median}) for this group is to use the trajectory of median values of β_t for each day t ($\tilde{\beta}_t$) from the train set. The alternative method is to use β_t in the form of third order polynomial regression (M_{regr}) with L2 regularization. The model takes t as input and outputs $log(\beta)$. Input values are modified by removing the mean and scaling to variance.

Estimation on Train Set and Incomplete Data. The methods from this group use both generated trajectories and the current sample's data to generate the forecast for the model switch. The baseline methods consist of taking the forecast β values from the previous group of methods and making them better comply with β values of current data.

The first baseline method (M_{median}^{shift}) is to shift modeled values of M_{median}, adding a scalar. The summand is calculated as the difference between an actual and estimated β at the switch point. The value of the actual β is chosen as a moving average at the switch point. Let β' denote β estimate from the baseline and t – the time of switch, then the shifted trajectory is set by the following equation:

$$\beta'_{shifted} = \beta' + (MA_t - \beta'_t), \tag{5}$$

where MA_t is a moving average for β_t.

The second baseline (M_{regr}^{shift}) is to shift a regression forecast (M_{regr}) in a similar manner. The third one (M_{regr}^{add}) is to additionally train a regression model for additional N epochs, using β values from the currently observed outbreak as target values.

The main methods are a regression model with an extended set of income features (M_{regr_ext}) and a Long Short-Term Memory (LSTM) network (M_{LSTM}) [4]. LSTM was chosen as it is often used for time series forecasting and for epidemic modeling in particular [9].

The regression model is with third degree polynomial features, i.e. all polynomial combinations of the features with degree 1 and 2. The regularization is L2. Input values are scaled by removing the mean and scaling to unit variance.

The LSTM network consists of 2 LSTM layers. The loss function is a mean squared error (MSE), the optimizer is RMSprop with an initial learning rate of 0.001 and a learning rate schedule: multiply the learning rate by 0.1 after 30 epochs. The training was conducted with a batch size of 64 for 100 epochs with early stopping: training was stopped if validation loss did not improve after 7 epochs. The model was trained to forecast based on 14 days prior. The chosen architecture (Fig. 1) is complex enough to catch the trends for β prediction on the train dataset.

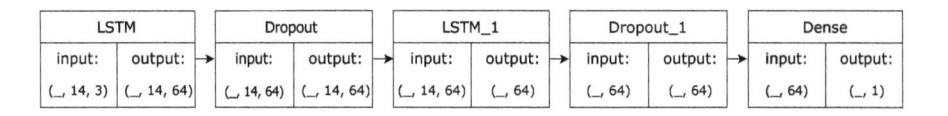

LSTM		Dropout		LSTM_1		Dropout_1		Dense	
input:	output:	input:	output:	input:	output:	input:	output:	input:	output:
(_, 14, 3)	(_, 14, 64)	(_, 14, 64)	(_, 14, 64)	(_, 14, 64)	(_, 64)	(_, 64)	(_, 64)	(_, 64)	(_, 1)

Fig. 1. LSTM architecture

2.3 Switch Point Detection Approaches

To determine the optimal time of switch from the network submodel to the compartmental submodel, we need to monitor β values. Initial days of an epidemic are characterized by a highly variable β with its further smoothing, therefore our criteria for switching is low β variance. This ensures that the compartmental model can accurately reproduce epidemic dynamics without significant loss of detail.

To compare dynamic methods described further, we also utilize static methods with constant switch points, later refered to in the form of $SP_{<n>}$, where n is the chosen day of switch.

Change in Variance. The first method (SP_{once}) involves calculating the moving variance of β, i.e. calculating variance over the previous k days. Let MV_t denote the calculated moving variance value for a given day t:

$$MV_t = Var(\beta_{t-k+1}, ..., \beta_t) \qquad (6)$$

To establish a switch condition regardless of the range of β values, we apply min-max scaling to get values in the range from 0 to 1 (Eq. 7):

$$MV_t' = \frac{MV_t - MV_{min}}{MV_{max} - MV_{min}} \qquad (7)$$

This scaling allows us to interpret $MV_t' = 0$ as the minimal observed variability. Therefore, when MV_t' falls below a chosen threshold (ε), we can consider β stable, which indicates the optimal time to switch.

Established Change in Variance. At the beginning of an epidemic the variance of β may accidentally decrease, not reflecting real stabilization. To account for such declines leading to early switching, we add the following condition: the value of MV_t' should stay below ε for a specified number of consecutive days. Therefore the second method (SP_{estab}) waits for an established change in variance.

Change in Variance After the Epidemic Situation. The third method (SP_{epi_sit}) introduces a new condition to avoid the early switch problem. The first condition, i.e. making sure that MV_t' is below ε, is accompanied by a second condition: the proportion of the infected population should reach a specified percent; this indicates the epidemic situation.

2.4 Evaluation Metrics

To assess the computational efficiency of the algorithms, time spent from the model initialization to the final prediction was measured. The observed period includes both training and inference phases. Forecast accuracy was measured by calculating RMSE between the actual and the predicted time series, both for I_t and for β_t.

Two additional metrics for I_t are peak height error (relative) and peak time error (absolute). Peak time error is negative when the model peak day is earlier than the real one, and positive otherwise. The value of peak height corresponds to maximal burden imposed on the healthcare units, whereas the value of peak time is used to assess the period of time remaining to prepare necessary resources. In case of positive peak time error, we will be caught by surprise unprepared, that is why negative peak time error is preferable.

3 Experiments and Results

In our simulations, we used a Barabasi-Albert graph network of 10^4 nodes, with each new node attached to 5 existing nodes. Epidemic trajectories were generated with $\sigma = 0.1$, $\gamma = 0.08$, $\tau \in \{0.01, 0.02, \ldots, 0.05\}$, $I_0 \in \{0.005, 0.006, \ldots, 0.011\}$. We regard these trajectories as outbreaks of some generic acute respiratory disease. The values of parameters σ and γ were chosen to be close to typical incubation and infection periods for ARIs. Values of τ and I_0 were selected to approximately match ARI outbreaks by the length of an epidemic and the maximal prevalence.

For each parameter combination, 50 simulation runs were performed with different seeds of a random number generator, thus reflecting the stochastic factor. A total of 1 500 unique epidemic trajectories were generated, with 20% of the data selected as a test dataset using a stratified approach based on the parameter values τ and I_0. All further experiments test the methods of β estimation and switch point detection on these 300 samples. The results for RMSE of I_t are presented in Table 1 and discussed further.

Table 1. RMSE of I_t for all methods of beta estimation and switch point detection; top-3 beta estimation methods for each switch point are highlighted in bold

β estimation method	Switch point detection method					
	SP_{20}	SP_{30}	SP_{40}	SP_{once}	SP_{estab}	SP_{epi_sit}
M_{last_val}	244.8 ± 116.0	154.7 ± 134.8	99.7 ± 134.8	265.2 ± 111.0	265.1 ± 109.1	287.5 ± 98.9
M_{ma_val}	273.7 ± 88.5	199.2 ± 124.6	111.5 ± 132.6	264.4 ± 84.6	262.4 ± 83.0	291.7 ± 86.9
M_{ca_val}	229.9 ± 59.9	221.0 ± 74.7	193.3 ± 95.3	244.8 ± 68.5	248.3 ± 69.7	243.3 ± 66.3
M_{biexp}	261.4 ± 152.8	205.9 ± 121.2	153.8 ± 84.8	324.8 ± 206.1	290.6 ± 202.4	325.1 ± 190.1
M_{median}	**74.2 ± 90.1**	**42.3 ± 56.9**	**28.6 ± 38.8**	**87.2 ± 76.2**	**83.8 ± 66.0**	**85.6 ± 71.6**
M_{regr}	109.6 ± 75.8	50.5 ± 65.5	42.7 ± 49.0	139.3 ± 77.4	131.5 ± 74.0	144.0 ± 73.5
M_{median}^{shift}	162.9 ± 117.0	70.9 ± 74.3	34.6 ± 61.5	178.1 ± 156.8	161.2 ± 159.6	183.6 ± 136.2
M_{regr}^{shift}	213.5 ± 67.0	112.5 ± 62.8	45.2 ± 71.6	209.1 ± 86.0	196.8 ± 87.9	217.2 ± 77.5
M_{regr}^{add}	83.4 ± 53.4	59.8 ± 48.9	48.2 ± 42.8	125.0 ± 61.1	120.0 ± 59.6	128.0 ± 58.1
M_{regr_ext}	**58.1 ± 48.2**	**37.2 ± 36.8**	**28.2 ± 28.5**	**66.9 ± 62.0**	**64.0 ± 60.2**	**67.1 ± 56.5**
M_{LSTM}	**80.9 ± 68.5**	**31.6 ± 26.8**	**20.9 ± 13.6**	**102.7 ± 85.6**	**94.6 ± 86.3**	**105.1 ± 85.5**

3.1 Beta Estimation Approaches

The aim of these experiment series is to compare the accuracy of β estimation approaches within each group, corresponding to their use cases.

Estimation on Current Incomplete Data. We applied the methods of β estimation from the first group (Sect. 2.2) to each test sample. Our additional subject of interest is to compare prediction results for constant and time-dependent β.

As can be seen in Fig. 2 and first 3 rows in Table 1, using constant β as input for the SEIR provides a poor fit to actual values of I_t. The β trajectory is highly variable, so using M_{last_val} leads to a big difference in RMSE based on the day of switch. The moving average (M_{ma_val}, the best results with a window size of 7 days) and the cumulative average (M_{ca_val}) approaches result in lower variance of RMSE compared to M_{last_val}. To further analyze limitations of constant β values, we took values from 0 to $4 \cdot 10^{-5}$ with a step size 10^{-6} as inputs to the SEIR model. The modeled I_t trajectories either are too wide or have a higher peak to match actual I_t (Fig. 3, left). This suggests that constant β gives poor results because the data requires a varying β trajectory, not because the constant value was chosen incorrectly.

The first time-dependent β estimation approach (M_{biexp}) does not estimate values similar to the initial β. The modeled β trajectory after the switch point does not match the curve of real values either, which leads to peak height underestimation. However, the method with time-dependent β has area of improvement, which will be showed further.

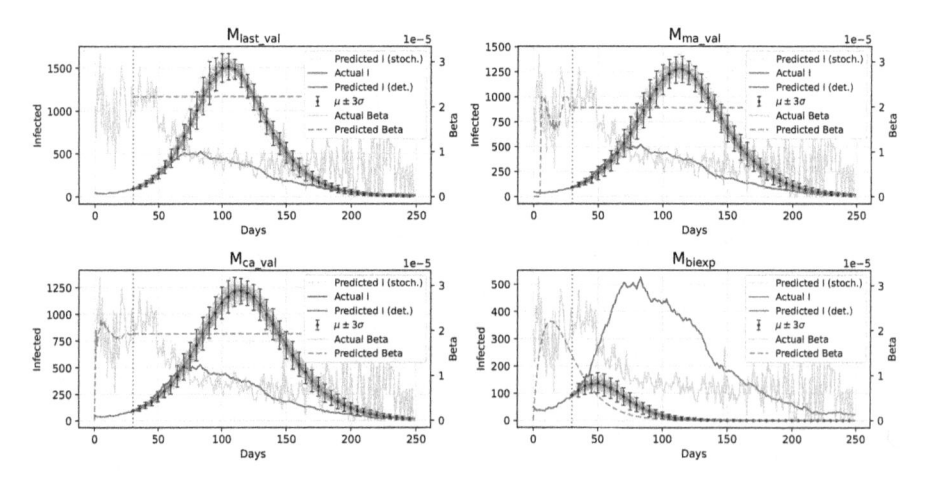

Fig. 2. Simulated prevalence curves for beta estimation methods based on current data: last value, moving average, cumulative average, biexponential decay

Estimation on Train Set. We applied the methods of β estimation from the second group (Sect. 2.2) to each test sample.

As can be seen in Fig. 3 (right), the maximum variation of actual values occurs at the beginning of an epidemic. Then, when an outbreak starts spreading rapidly and β values have the most effect on modeled I_t trajectories, the range of values is narrower. The methods M_{median} and M_{regr} have similar estimated trajectories after the 30th day when β values should be the most relevant for modeling, but the metrics vary significantly with M_{median} having lower errors.

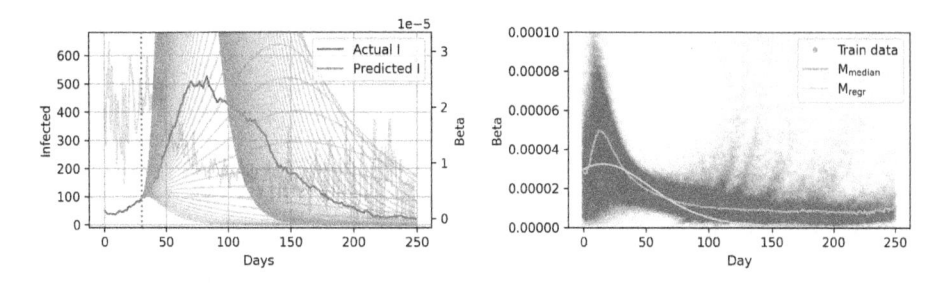

Fig. 3. Modeled trajectories for I_t with constant beta values (left); median β values of all generated trajectories and the output of the regression model fit on a day (right).

This implies that a slight mismatch even solely in first initial β values can nearly double the median RMSE of I_t.

Even a slight difference in β values results in a noticeable change in a number of infected people. This can be supported by Fig. 3 (left): each increment by 10^{-6} gives around 100 additional infected people at the modeled peak, i.e. 1% of the population.

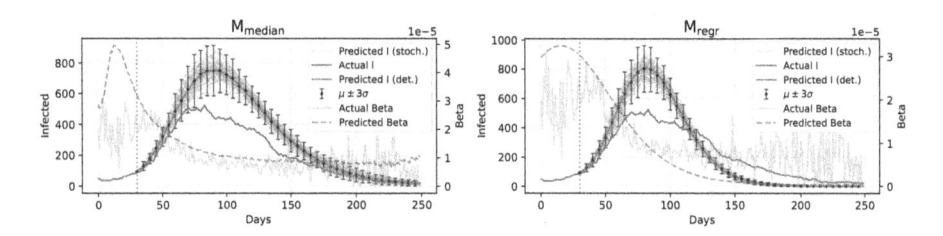

Fig. 4. Simulated prevalence curves for beta estimation methods based on train set: median of train set, regression on day

Estimation on Train Set and Incomplete Data. We applied the methods of β estimation from the third group (Sect. 2.2) to each test sample.

Both shift methods (the best results with a window size of 14 days) result in worse metrics than original methods (M_{median} and M_{regr}). One reason may be the shape of β values in simulated trajectories. While initial days have a high variability in values, shifting based on these days' values (even with rolling average) results in inflated β values at the end.

We can compare M_{regr}^{add} (the best results with 3 additional epochs) in Fig. 5 with M_{regr} in Fig. 4. The tail of estimated β from M_{regr}^{add} is the same, but the beta values near the switch point are closer to the actual β. This gives a lower peak, although the epidemic generally has the same duration.

M_{regr_ext} and M_{LSTM} methods are in top-3 for RMSE of I_t according to Table 1. Input features for M_{regr_ext} and M_{LSTM} were selected to achieve the least RMSE without redundant features. For M_{regr_ext}, the input features are: $t, S_t, E_t, I_t, R_t, I_{t-1}$. The input features for M_{LSTM} are: t, E_t, I_{t-2}. Feature values at switch point are compartment values from the network model. The next day is modeled with compartments from the discrete SEIR. We experimented with different values for I_{t-s}, where s is the shift; $s = 2$ gave the best results.

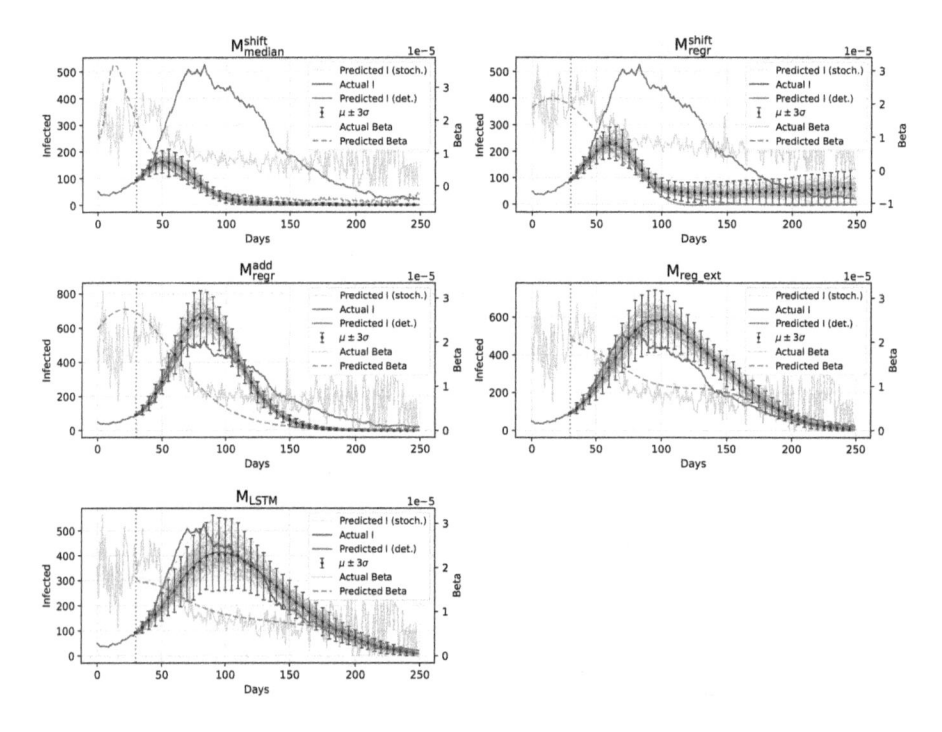

Fig. 5. Simulated prevalence curves for beta estimation methods based on current data and train set: shifted median, shifted regression, regression with additional learning, regression with extended input features, LSTM

3.2 Switch Point Detection Approaches

The aim of these experiment series is to assess switch point detection approaches and the stability of β estimation methods. To achieve this, we calculated the metrics for a fixed set of switch points and for variance-based detection approaches. RMSE of I_t for all β estimation methods and switch point detection methods can be found in Table 1.

The lowest accepted switch point was set to 14 days to not interfere with M_{LSTM}'s input shape. The most test samples visually have an epidemic start

(1% of the population is infected) around days 20, 30, and 40. We chose these three days as constant switch points. The choice of switch point does not affect the top-3 β estimation methods based on RMSE. This may suggest that even though SP_{20} may be too early to switch, or SP_{40} is too close to the peak, no method of β estimation gains any benefits to overcome the top-3 methods (M_{median}, M_{regr_ext} and M_{LSTM}).

Switch detection methods have a different optimal threshold ε. SP_{once} shows best results with $\varepsilon = 0.05$, i.e. 5%. SP_{estab} has best results with $\varepsilon = 0.05$ and 2 consecutive days. SP_{epi_sit} shows best results with $\varepsilon = 0.1$. The discrepancy comes from the methods' conditions: the last approach waits for 1% of the population to be infected, further β values are more stable, so we can switch with fewer risks, therefore a higher ε.

Constant switch points have limited usage for real-world applications because they do not consider the ongoing epidemic dynamics. Out of the three remaining methods, SP_{estab} has the lowest errors, according to Table 1. However, high accuracy may be a result of switching closer to the peak, which reduces the benefits of the hybrid approach. For instance, as it can be seen in Table 1, increasing the value of a constant switch point (from 20 to 30 and 40) results in decreasing the RMSE of I_t for the corresponding methods (SP_{20}, SP_{30}, SP_{40}). Thus, it is important to also consider the distance between the detected switch point and the actual peak time, i.e. the difference between their values in days. Figure 6 shows that the method with the largest median difference between the switch and the peak time is SP_{epi_sit}. Due to that reason, we prefer this method, although it has the largest errors among the 3 methods based on the variance of β.

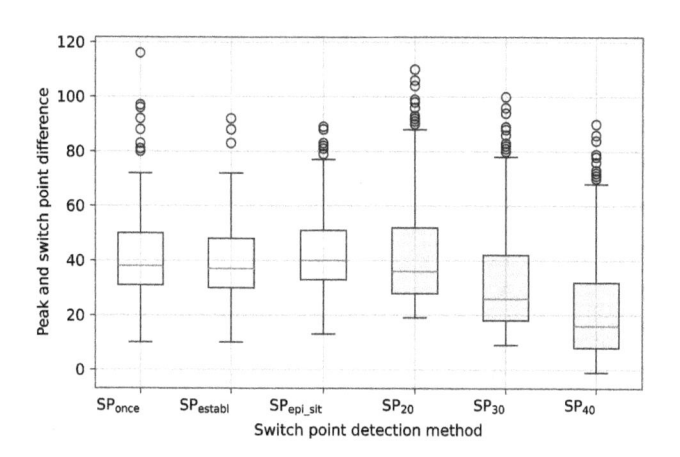

Fig. 6. Difference between the actual peak time and the detected switch point; boxplots for constant switch points are filled.

3.3 Evaluation Results

Accuracy (RMSE). All figures are presented for the best switch point detection approach – SP_{epi_sit}. For RMSE of I_t (Fig. 7), the methods with the best results are: M_{LSTM}, $\bar{M}_{median}$ and M_{regr_ext} with median errors 105.1, 85.5 and 67.1. Therefore, the error for top-3 methods is around 1% of the population. For RMSE of β values (Fig. 8), the best performing methods are: M_{LSTM}, M_{median} and M_{regr_ext} with median errors $1.09 \cdot 10^{-5}$, $1.08 \cdot 10^{-5}$ and $1.05 \cdot 10^{-5}$. The set of best methods is the same for both metrics. The purpose of a hybrid approach is to switch to a simpler model without the loss in accuracy for the predicted I_t trajectory, so RMSE of β is less relevant. As a result, the final top-3 methods are: M_{LSTM}, M_{median}, M_{regr_ext}.

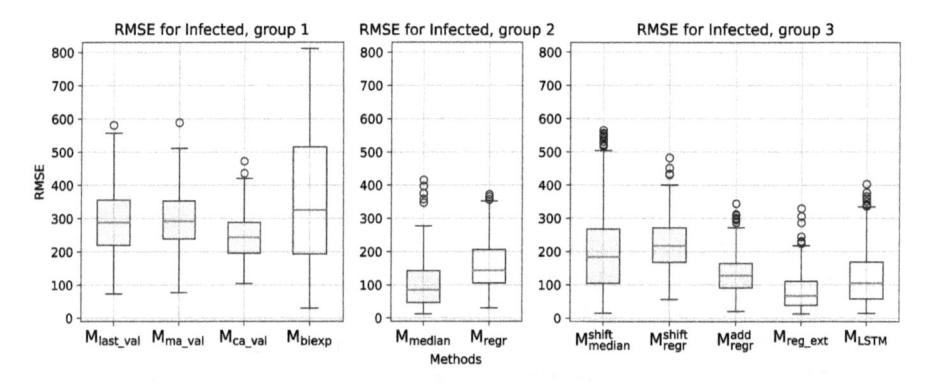

Fig. 7. RMSE for I_t. Beta estimation methods based on current data (group 1), train set (group 2), and combined (group 3). Boxplots for baseline methods are filled.

Table 2. Prediction time for the methods of β estimation

	M_{last_val}	M_{ma_val}	M_{ca_val}	M_{biexp}	M_{median}	M_{regr}
Time, sec	0.016 ± 0.006	0.005 ± 0.005	0.003 ± 0.005	0.003 ± 0.006	0.008 ± 0.006	0.005 ± 0.001

	M_{median}^{shift}	M_{regr}^{shift}	M_{regr}^{add}	M_{regr_ext}	M_{LSTM}
Time, sec	0.009 ± 0.007	0.005 ± 0.000	0.007 ± 0.001	0.362 ± 0.073	37.625 ± 14.519

Computational Efficiency. Time measurements for training and prediction were performed on a system equipped with an AMD Ryzen 5 5500U processor (up to 4.0 GHz), 16 GB of LPDDR4x RAM (4266 MHz), and integrated AMD Radeon Graphics. In a single-threaded configuration without parallelization, M_{LSTM} required 1228.4 s to finish the training, M_{regr_ext} took 2.86 s.

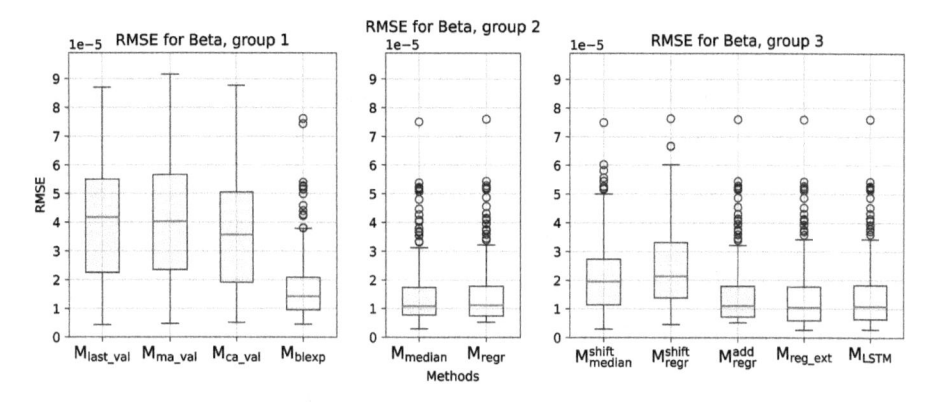

Fig. 8. RMSE for Beta. Beta estimation methods based on current data (group 1), train set (group 2), and combined (group 3). Boxplots for baseline methods are filled.

Time for predictions is presented in Table 2. M_{LSTM}, while demanding the most computational resources, is still considered top-3 due to its forecast accuracy. Two methods with higher RMSE and faster prediction time are M_{median} and M_{regr_ext}.

Accuracy (distance to Peaks). All figures are presented for the best switch point detection approach – SP_{epi_sit}.

For an easier interpretation, we present peak errors for each group of β estimation approach separately (Fig. 9). The x-axis corresponds to the difference between the predicted and actual peak time. The y-axis is the fraction of the predicted peak height to the actual peak height. The best case is at the point with coordinates (0, 1). Besides the best case, we are also interested in the second and third quadrants, which depict cases with peaks predicted earlier. All

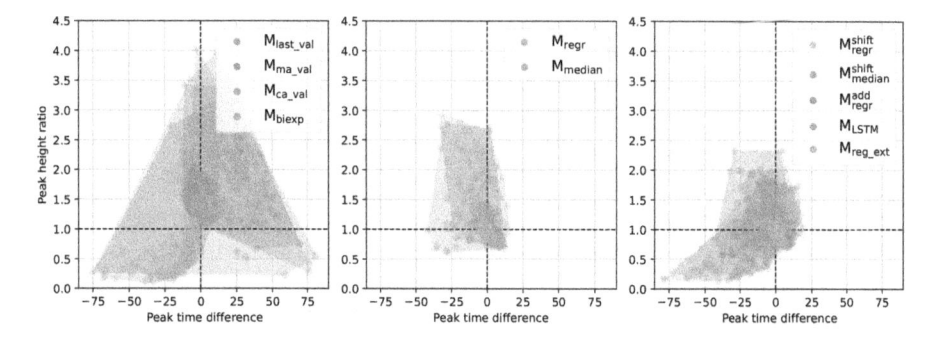

Fig. 9. Peak errors for beta estimation methods; x-axis is the difference between the predicted and actual peak time, y-axis is the fraction of the predicted peak height to the actual peak height

top-3 β estimation methods (M_{median}, M_{LSTM} and M_{regr_ext}) are close to the best case at (0, 1).

The remaining β estimation methods have larger areas of possible outcomes. Some additional conclusions may be drawn from methods based on regression (M_{regr}, M_{regr}^{shift}, M_{regr}^{add}). Shifted predictions by M_{regr}^{shift} skew the results closer to the best case for peak height ratio but with larger peak time underestimation. Additional learning for 3 epochs in M_{regr}^{add} makes the error in peak height slightly lower for almost each point.

4 Conclusions and Future Work

In this research, we presented the hybrid approach based on a network-based and a discrete SEIR submodels with dynamic switching. For the correct alignment of submodels, it is important to properly estimate the value of β (disease transmission rate). We conducted several experiments to analyze methods of β estimation and switch point detection. β estimation methods are divided into 3 groups based on their use conditions: only the current outbreak data are available, only historic data are available, historic and current data are available. Switch point detection methods are divided into static methods with constant switch points and dynamic detection methods based on β variance. Our work is based on the analysis of synthetic epidemic data; this is the first step towards the further application of the hybrid approach for modeling real-world epidemic processes.

The best switch point detection approach was concluded to be SP_{epi_sit}, i.e. switch point detection based on the declared epidemic. The method has two conditions: 1% of I_t and β variance lower than 10%. The best β estimation methods with the lowest RMSE in I_t and with the lowest peak errors are: M_{LSTM} (LSTM model), M_{median} (median of β values in train set) and M_{regr_ext} (regression model).

For future work, firstly, we plan to use interval estimates as opposed to point estimates to account for uncertainty. As some papers suggest, epidemic forecasts should always be done with deep uncertainty methods to enhance decision-making. Secondly, there is a separate field devoted to changepoint detection, which analyses when the change happens in the probability distribution of a certain signal. We can utilize state-of-the-art approaches and assess their applicability for our purposes. Thirdly, to formalize the assessment of switch point detection methods based on accuracy (distance to peak and RMSE of I_t) one can include the weighted metric. Finally, we also plan to generalize the methods of β estimation for more complex models such as ABM and real data. However, initial attempts showed more intricate β trajectories for ABM data, thus requiring changes in our approaches. ABM may also have an additional parameter, a fraction of immune individuals, to consider during modeling and β estimation.

Acknowledgement. This research was supported by The Russian Science Foundation, Agreement #22-71-10067.

References

1. https://github.com/vnleonenko/Network_hybrid
2. Albert, R., Barabási, A.L.: Statistical mechanics of complex networks. Rev. Mod. Phys. **74**(1), 47 (2002)
3. Bobashev, G.V., Goedecke, D.M., Yu, F., Epstein, J.M.: A hybrid epidemic model: combining the advantages of agent-based and equation-based approaches. In: 2007 Winter Simulation Conference, pp. 1532–1537 (2007). https://doi.org/10.1109/WSC.2007.4419767
4. Hochreiter, S., Schmidhuber, J.: Long short-term memory. Neural Comput. **9**, 1735–1780 (1997). https://doi.org/10.1162/neco.1997.9.8.1735
5. Hunter, E., Namee, B.M., Kelleher, J.D.: A hybrid agent-based and equation based model for the spread of infectious diseases (2020)
6. Kermack, W.O., McKendrick, A.G.: A contribution to the mathematical theory of epidemics. Proc. R. Soc. London. Ser. A **115**(772), 700–721 (1927). Containing papers of a mathematical and physical character
7. Kerr, C.C., et al.: Covasim: an agent-based model of COVID-19 dynamics and interventions. PLoS Comput. Biol. **17**(7), e1009149 (2021)
8. Leonenko, V.: A hybrid modeling framework for city-scale dynamics of multi-strain influenza epidemics. In: International Conference on Computational Science, pp. 164–177. Springer (2022)
9. Leonenko, V.N., Bochenina, K.O., Kesarev, S.A.: Influenza peaks forecasting in Russia: assessing the applicability of statistical methods. Procedia Comput. Sci. **108**, 2363–2367 (2017)
10. Pastor-Satorras, R., Vespignani, A.: Epidemic spreading in scale-free networks. Phys. Rev. Lett. **86**(14), 3200 (2001)
11. Rahmandad, H., Sterman, J.: Heterogeneity and network structure in the dynamics of diffusion: comparing agent-based and differential equation models. Manage. Sci. **54**(5), 998–1014 (2008)
12. Smirnova, A., deCamp, L., Chowell, G.: Forecasting epidemics through nonparametric estimation of time-dependent transmission rates using the seir model. Bull. Math. Biol. **81**(11), 4343–4365 (2019)
13. Tran-Kiem, C., Bedford, T.: Estimating the reproduction number and transmission heterogeneity from the size distribution of clusters of identical pathogen sequences. Proc. Natl. Acad. Sci. **121**(15), e2305299121 (2024)

Lightweight Heterogeneous SEIR Models for Epidemic Surveillance in Russian Cities: Turning Synthetic Populations Into Equations

Andrey Korzin and Vasiliy Leonenko[✉][iD]

ITMO University, 49 Kronverksky Pr., St. Petersburg 197101, Russia
vnleonenko@itmo.ru

Abstract. Influenza and other acute respiratory diseases pose a significant challenge to global health. The complexity of analyzing and mitigating influenza transmission is related to heterogeneity of contact network structures in modern cities. The need for effective public health strategies has driven the development of highly detailed network and agent-based models. To overcome a drawback of modeling multi-agent systems, which is their high demand for computational resources, approximate models can be employed. In our article, we present an approach that allows to convert heterogeneous synthetic populations into an input for the edge-based compartmental SEIR model. We demonstrate the method application by simulating influenza spread in a contact network of the synthetic population of Chelyabinsk, Russia. At a cost of neglecting some details in contact network structure, the proposed algorithm allows to greatly enhance simulation speed compared to multi-agent modeling, and at the same time to preserve population heterogeneity, which makes it a good choice for application in epidemic surveillance.

Keywords: epidemiology · influenza · synthetic population · complex networks · edge-based compartmental modeling

1 Introduction

Influenza, along with other acute respiratory diseases, continue to pose a significant challenge to global health. The complexity of influenza transmission, coupled with the need for effective public health strategies, has driven the development of a wide array of mathematical models. Mathematical models in epidemiology enable the prediction of epidemic dynamics and the evaluation of epidemic indicators, such as the reproductive number and the number of immune individuals. Furthermore, they allow computational experiments that facilitate the assessment of best vaccination strategies and other methods of disease control. The largest epidemics occur in major cities, where high population density and numerous contacts between individuals are prevalent. Modern cities can be considered as complex systems with heterogeneous network structure, which creates challenges for modeling the spread of diseases. Classical SIR-type compartmental models assume homogeneous mixing in the populations and are unable to account for effects associated with heterogeneity, such as existence of super-spreaders. The two most popular approaches that come to the rescue are multi-agent and network models.

The first method, multiagent modeling (MAM) based on detailed demographic data, is a powerful tool for modeling epidemic dynamics down to the level of a single individual. MAM allow us to consider effects connected with population heterogeneity, track the spatial spread of disease and chains of transmissions. However, high model detail results in long simulation time. This aspect is critical for epidemiological surveillance purposes, as the model calibration process requires numerous simulation runs. Another drawback is the need to build a synthetic population for every city of consideration. Collecting data on demographics, schools, urban development and workplaces of the city is time-consuming, and it is often difficult to verify the resulting datasets [12].

The second method, network modeling, represent a simplified approach, where contact networks are represented by random graphs and nodes are not distinguished by individual characteristics, enabling the conservation of contact tracing without the necessity of developing detailed synthetic populations. Barabási-Albert [4] and Erdős-Rényi [15] networks are commonly utilized for constructing contact networks. This method may not represent accurately the topology of contact networks in real populations.

To overcome a drawback of modeling multi-agent systems, which is their high demand for computational resources, approximate models can be employed. Kiss et al. showed that the computationally intensive SIR network model can be effectively approximated via a range of ordinary differential equation models [6]. Namely, edge-based compartmental modelling (EBCM), being the most exact of the approaches provided, approximates well the simulation results shown by network models and at the same time provides those results much faster than them [10]. While SIR models are partially applicable for the infections with short latent period, like influenza, their applicability to diseases with long incubation period, like COVID-19, is a matter of discussion. In order to fully take into account the latent period, it is more logical to consider the SEIR (Susceptible-Exposed-Infected-Recovered) model. The SEIR EBCM models were proposed and used in [3,13,14]. In these articles, the networks of standard topologies were used as an input, and the usage of synthetic populations in the models was not considered.

In this work, we propose a method that unites the usage of approximate heterogeneous compartmental models, namely, SEIR EBCM, with the generation of input contact network graphs based on synthetic populations. We demonstrate the usage of the method by constructing a contact network graph from the synthetic population of Chelyabinsk, Russia, and comparing the simulation of the disease outbreak using a SEIR EBCM with the simulation via a network model. The ability of the EBCM to replicate real disease incidence is demonstrated by calibration to 2022–2023 influenza incidence in Chelyabinsk.

2 Methods

2.1 Synthetic Population

A synthetic population is derived from diverse data sources reflecting the actual urban population. It is formatted into text files and serves as an input for the model. We have extensively employed multiagent modeling and creating datasets for synthetic populations in our previous research [7,8]. These datasets are compatible to the RTI synthetic populations standard [16], capturing individual attributes such as household residence, age, gender, and workplace or school identifiers. A detailed explanation of the data collection process and the synthetic population format can be found in a related publication [8]. An example of data from a synthetic population file `people.txt` describing the characteristics of individuals is given in Table 1. The variables `sp_id` and `hh_id` represent numerical identifiers of individuals and their households, while `work_id` identifies work office number. Other files that include data related to households, workplaces and schools have a similar format.

Table 1. Sample records from the file `people.txt` of a synthetic population

sp_id	age	gender	hh_id	work_id
1	25	M	784	14
2	10	F	294	83
3	74	M	33	X
⋮	⋮	⋮	⋮	⋮

Chelyabinsk, a major Russian city with a population of over one million, was chosen for this computational study. A synthetic population of the city was created using data current as of 2023. The spatial distribution of households, schools and workplaces is shown in Fig. 1. The data for workplaces is sourced from Yandex.Auditorii [2], the households are geocoded using the information from Open Street Map [1]. Workplaces are divided into offices, and households are divided into apartments. The homogeneous structure of the workplace arrangement depicted in Fig. 1 is related to the aggregated nature of office data collected by [2]. Each workplace point is associated with the aggregated information for workplaces in the corresponding area. The characteristics of the synthetic population are summarized in Table 2.

Table 2. Population statistics for Chelyabinsk

City	Population	Households	Workplaces	Schools
Chelyabinsk	1 189 000	436 000	56 800	114

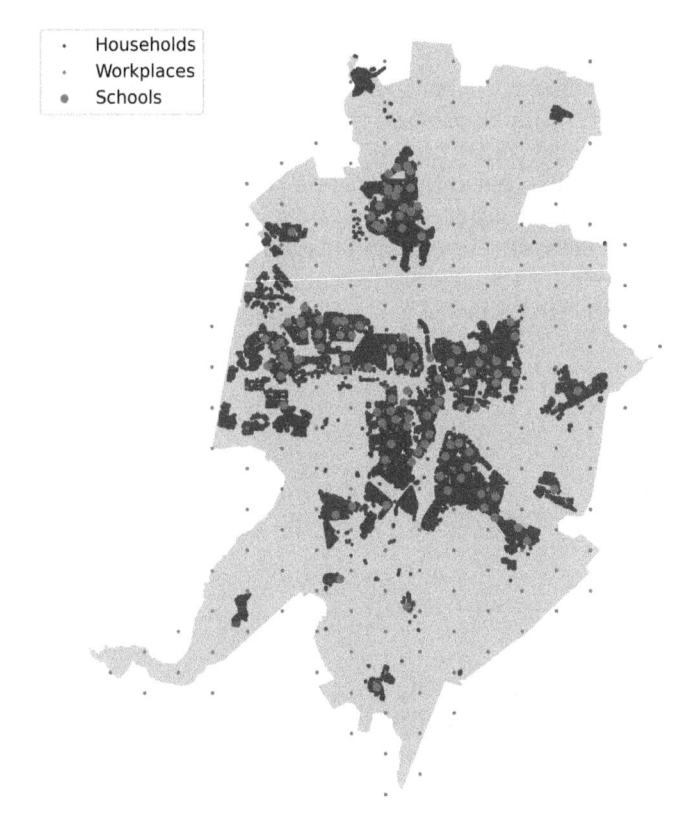

Fig. 1. Map of Chelyabinsk with spatial distribution of households, workplaces and schools from synthetic population data.

All computational experiments given in the article can be done on the full population of Chelyabinsk, however, converting a synthetic population to contact network graph of city with a population of 1 million people will require more than 30 GB of RAM and more than 2 h of calculations (using Intel Xeon Gold 6226R 2.9 GHz). To demonstrate the possibility of reducing computational cost, a sampling algorithm was employed on a full-scale synthetic population dataset. This algorithm involves reducing the population by leaving r percent of households in each district of the city. When the household is removed, the individuals attributed to it are also removed from the synthetic population.

To generate a contact network based on the synthetic population, the following algorithm was applied:

- Create a graph with N nodes where N is the size of the population. Each node corresponds to a specific individual.
- Create an edge between each pair of individuals that have the same `hh_id`, `work_id` or `school_id`.

2.2 SEIR EBCM

To explore the peculiarities of epidemic dynamics connected with unique topology of contact network and decrease the simulation time, we implemented an edge-based SEIR model (SEIR EBCM). A EBCM is an approximate model that incorporates probability generation function with degree distribution of input population network for constructing a system of equations that describes epidemic process on networks. Let the population consist of N individuals corresponding to N nodes of the network. According to works [13,14] the system of equations for SEIR EBCM model is the following:

$$\dot{\theta} = -\beta\psi,$$
$$\dot{\phi} = -(\alpha+\beta)\phi + (G''(\theta)\beta/G'(1))\psi,$$
$$\dot{\psi} = \beta\phi - \gamma\psi + (G''(\theta)\beta/G'(1))\psi,$$
$$S = G(\theta),$$
$$R = \gamma I,$$
$$\dot{E} = \beta SI - \alpha E.$$

Tables 3 and 4 show the description of the parameters, variables and their initial conditions, taking the notations for u and v as neighbor nodes, connected by edge in network. $G(x)$ stands for probability generating function (PGF):

$$G(x) = \sum_{k=0}^{\infty} p_k x^k,$$

where p_k is the probability that a randomly chosen node degree equals k. Variables $S(t), E(t), I(t)$ and $R(t)$ in the following equations refer to fractions of susceptible, exposed, infected and recovered individuals respectively.

The system of ordinary differential equations is solved numerically using `odeint` method from `scipy` library for `Python` language. Compartment sizes are multiplied by N to find the absolute value of individuals at each state.

2.3 Data

Influenza incidence data were sourced from Research Institute of Influenza, St. Petersburg, Russia. The number of individuals infected with each strain was assessed with the help of strain-specific laboratory diagnostic data. Details of the data processing methodology are outlined in [9]. In this research, we investigated 2022–2023 epidemic season in Chelyabinsk, Russia. The incidence data is shown in Fig. 2. Some time points do not have incidence values due to a lack of data for this period. During 2022–2023, the strain A(H1N1)pdm09 was the dominant strain, accompanied by strain B, which maintained lower incidence rates, not exceeding 2,000 new cases per week. For the purpose of simplifying the analysis of the epidemic dynamics, we used the aggregate number of new cases across all strains.

Table 3. Variables description for SEIR-EBCM model equations

Variable	Description	Initial value
$\theta(t)$	Probability that the node u did not transmit an infection to node v at a time step t	$1-\rho$
$\phi(t)$	Probability that the node u of an edge from u to v is exposed, and the edge did not transmit an infection at a time step t	$\varepsilon_\phi \ll 1$
$\psi(t)$	Probability that the node u of an edge from u to v is infectious, and the edge did not transmit an infection at a time step t	$\varepsilon_\psi \ll 1$
$S(t)$	Fraction of susceptible individuals in the population at time t	$1-\rho$
$E(t)$	Fraction of exposed individuals in the population at time t	0
$I(t)$	Fraction of infected individuals in the population at time t	ρ
$R(t)$	Fraction of recovered individuals in the population at time t	0

Table 4. Parameter description for SEIR EBCM equations

Variable	Description
β	Infection transmission rate over one edge
α	Rate for exposed nodes to become infectious
γ	Rate for infectious node to become recovered
ρ	Initial fraction of infectious nodes

3 Results

3.1 Synthetic Population Transformation

The synthetic population of Chelyabinsk was reduced using a sampling algorithm, so that only $r = 10\%$ of households were left. That resulted in approximately $N \approx 350000$ individuals compared to the full city population of 1189000. To evaluate the effect of the sampling algorithm on the contact network structure within the population, we generated histograms illustrating the distribution of households and workplaces sizes. The plots are presented in Fig. 3, showing that the distribution of household sizes moved to the right, while sizes of workplaces significantly decreased (from 8 to 2 individuals in average). Since in this study we do not calculate epidemic indicators, we consider these changes not crucial for fulfilling our goals, but in case of using the model in epidemic surveillance we consider using sampled populations with larger r, along with 'repopulating' workplaces to retain original average number of individuals in them.

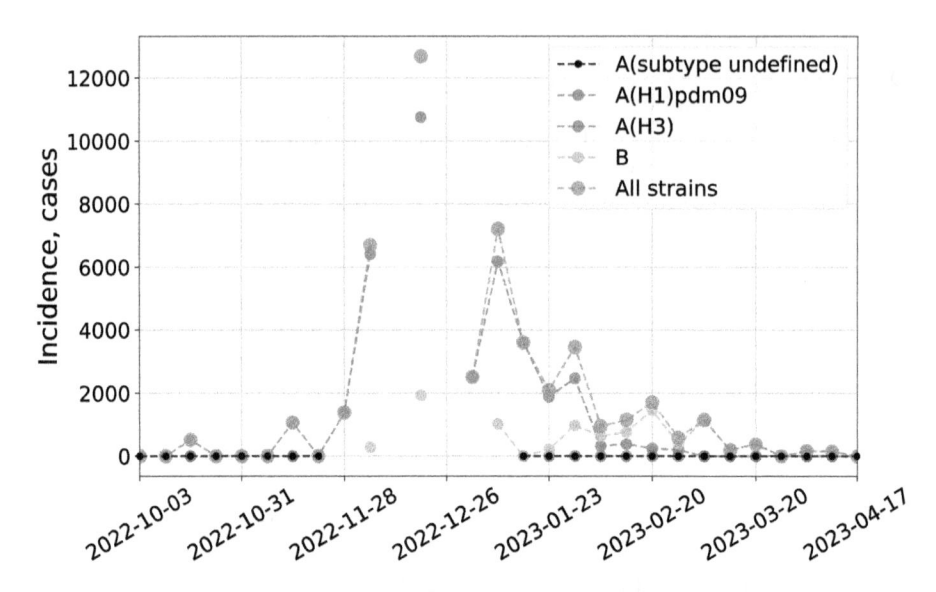

Fig. 2. Strain-specific influenza incidence data for 2022–2023 epidemic season in Chelyabinsk, Russia.

As a next step, a contact network was generated from the synthetic population, according to the aforementioned algorithm, with its software implementation in Python by means of `networkx` library. The resulting contact network degree distribution is shown in Fig. 4.

To demonstrate the usage of the contact network, we performed a simulation on it using a stochastic SEIR network model. The stochastic simulation was made by Gillespie algorithm [5] from EoN (Epidemics on Networks) library [11]. In a Gillespie simulation, the timing of the subsequent event is determined by computing the rate of all possible events at the current state. A waiting time is then sampled from an exponential distribution characterized by that rate. Subsequently, an additional random number is utilized to select which specific event among the possible options will occur. In our case, initially, most nodes are susceptible, with a certain fraction ρ of them set as infected. Each infected node may transmit infection to its susceptible neighbors according to a defined transmission rate. Exposed nodes transition to infected states, and infected nodes recover based on corresponding rates. The simulation output is time series equal to those of a compartmental SEIR model, i.e. $S(t), E(t), I(t), R(t)$.

3.2 Model Calibration

The second part of our method consists in feeding the created contact network to the edge-based SEIR model and launching calibration to data. The first 7 weeks on the graph with data were not considered for model calibration, as these data points represent only small fluctuations. We assume that a full-fledged epidemic

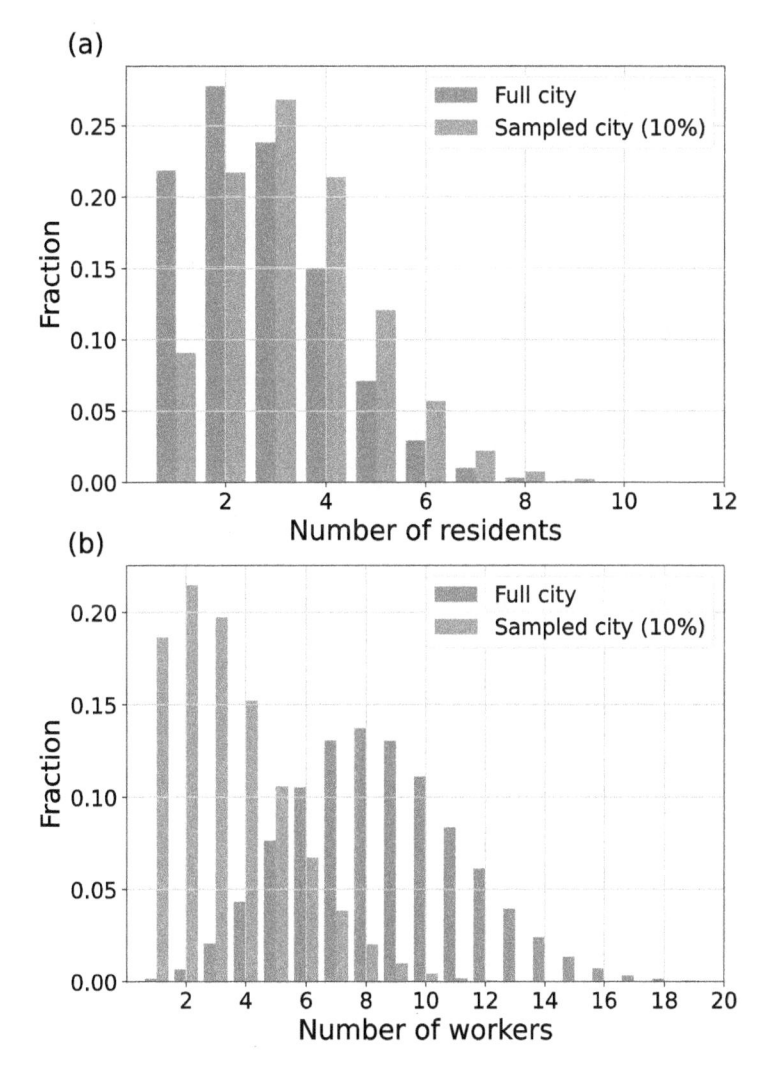

Fig. 3. Distribution of sizes of dwellings and offices according to synthetic population data.

begins from week 7, when the growth of number of new cases becomes apparent. The calibration was conducted using a combination of automatic approach via simulated annealing method and a manual parameter tuning. The resulting parameter values are presented in Table 4. To measure the accuracy of calibration, the R^2 metric was used. The calibration result is shown in Fig. 5 (Table 5).

The comparison of simulation time for different model types is presented in Table 6 and Fig. 6.

During the simulations, we have discovered an unpleasant effect which apparently takes place for some synthetic population structures. In Fig. 7, we

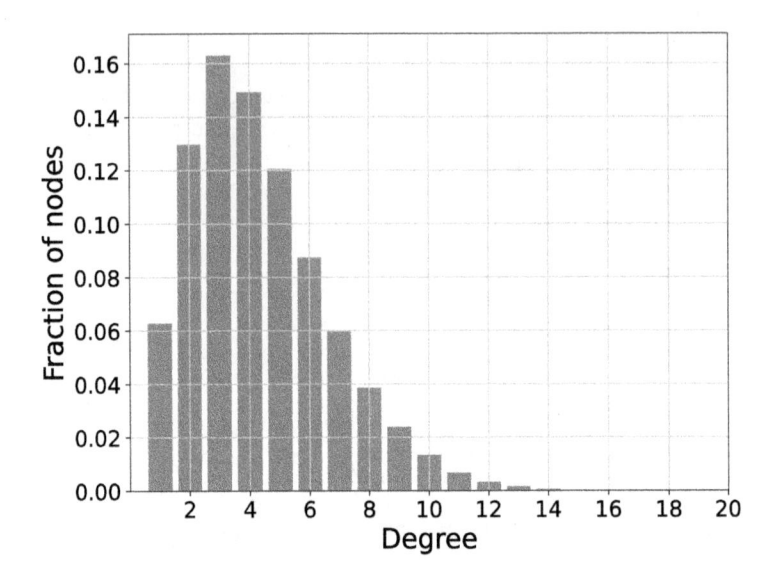

Fig. 4. Degree distribution of the contact network in a sampled synthetic population of Chelyabinsk.

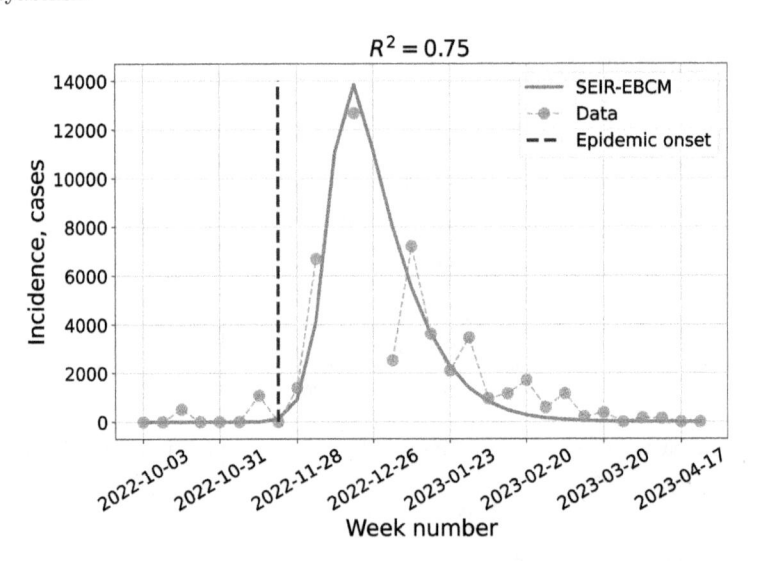

Fig. 5. SEIR EBCM calibration to influenza incidence data of 2022–2023 epidemic season in Chelyabinsk.

demonstrate the comparison of simulation curves for two different types of input contact networks. While the incidence curves for the EBCM and network models are well approximated when using a Barabási-Albert contact network, these incidence curves may have significant discrepancies when modeled on a contact

Table 5. Parameter values for SEIR EBCM obtained by calibration

Parameter	Value
β	0.0038
α	0.1
γ	0.14
ρ	0.0005

Table 6. Comparison of simulation time of different models.

Model	Avg. simulation time, sec	Complexity
SEIR ODE	$\approx 10^{-3}$	Low
SEIR EBCM	$\approx 10^{-1}$	Low
SEIR network model	$\approx 10^{2}$	Medium
Multiagent model	$\approx 10^{4}$	High

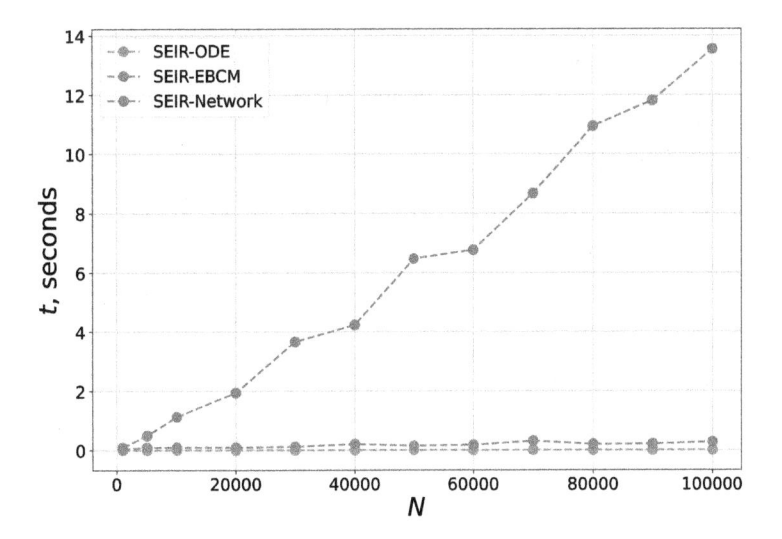

Fig. 6. Simulation time for different models depending on the number of nodes of the contact network graph. Experiments were conducted using Barabási-Albert network with $m = 5$.

network constructed from the synthetic population data. Particularly in this case, EBCM failed to reproduce the bimodality of the incidence curve. To our knowledge, this effect was not reported in the studies we relied upon. Consequentially, the usage of EBCM seems to be limited at least to unimodal curves, otherwise it changes the disease dynamics. Further studies are planned to quantify this limitation, exploring dependency of curves approximation quality on sampling ratio r and synthetic population structure of different cities.

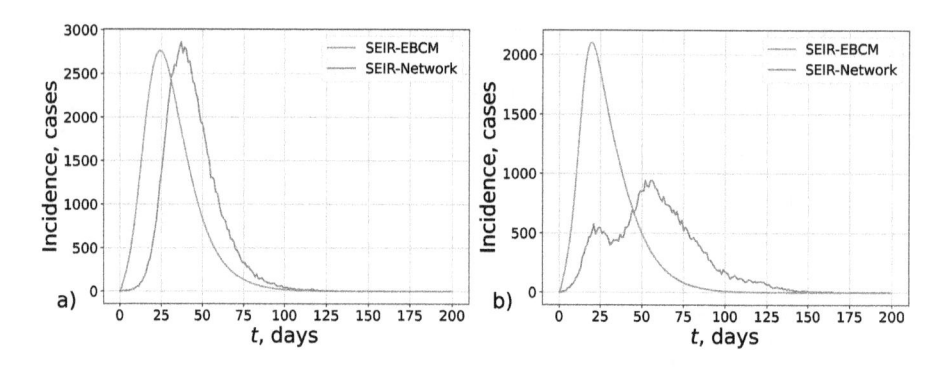

Fig. 7. Simulated daily incidence with different contact network graphs: a) simulation on Barabási-Albert network; b) simulation on a contact network graph based on synthetic population data.

4 Discussion

In this article, we proposed a method for fast and detailed modeling infection propagation which is based on combining synthetic population transformation and SEIR EBCM usage. To assess the applicability of the method, we compare different aspects of its usage with other modeling methods. Traditional methods, such as the SEIR ODE approach, offer fast simulation time and relatively straightforward calibration, making them appealing for rapid assessments of epidemic indicators and incidence prediction. However, these models fail to account for heterogeneous effects within populations. In contrast, multiagent modelling with synthetic populations provides a detailed view of epidemic spread at the individual level, capturing complex interactions and behavior patterns of individuals. Despite its advantages, multiagent modeling is hindered by high simulation time, which greatly complicates model calibration and data assimilation. Additionally, creating synthetic populations for these models is challenging and time-consuming, and these populations quickly become outdated due to demographic changes. In our method, using SEIR EBCM for modeling offers low simulation time and makes it possible to capture effects connected with population of concrete city using a transformed synthetic population. SEIR EBCM approach makes it an attractive choice for handling the complexities of epidemic modeling effectively. By integrating the benefits of different modeling techniques, SEIR-EBCM can provide more accurate predictions and better support public health strategies. However, to capture the effects connected with topology of each city the construction of synthetic populations is also needed.

In future studies, we plan to enhance our approach by upgrading the sampling techniques to preserve the distribution of apartment and work office sizes more accurately. Additionally, we aim to assess the feasibility of approximating contact networks based on synthetic populations using typical network models such as Barabási-Albert or Erdős-Rényi. Furthermore, we intend to employ this model for epidemic surveillance as part of a modeling framework that will allow ensemble forecasting and accurate assessing of epidemic indicators, such as R_t and fractions of immune population across different age groups.

Acknowledgments. This research was supported by The Russian Science Foundation, Agreement #22-71-10067.

References

1. Open Street Map. https://www.openstreetmap.org/. 2025/06/17 11:37:04
2. Yandex Audience. https://audience.yandex.ru/. 2025/06/17 11:37:04
3. Alota, C.P., Pilar-Arceo, C.P., de los Reyes V, A.A.: An edge-based model of seir epidemics on static random networks. Bull. Math. Biol. **82**(7), 96 (2020)
4. Barabási, A.L., Albert, R.: Emergence of scaling in random networks. Science **286**(5439), 509–512 (1999)
5. Gillespie, D.T.: Exact stochastic simulation of coupled chemical reactions. J. Phys. Chem. **81**(25), 2340–2361 (1977)
6. Kiss, I.Z., Miller, J.C., Simon, P.L., et al.: Mathematics of epidemics on networks. Cham: Springer **598**(2017), 31 (2017)
7. Korzin, A.I., Kaparulin, T.I., Leonenko, V.N.: Assessing the applicability of the multiagent modeling approach to the epidemic surveillance of COVID-19 in Russian cities. In: 2024 IEEE International Multi-Conference on Engineering, Computer and Information Sciences (SIBIRCON), pp. 237–242. IEEE (2024)
8. Leonenko, V., Arzamastsev, S., Bobashev, G.: Contact patterns and influenza outbreaks in Russian cities: a proof-of-concept study via agent-based modeling. J. Comput. Sci. **44**, 101156 (2020)
9. Leonenko, V.N.: Herd immunity levels and multi-strain influenza epidemics in Russia: a modelling study. Russ. J. Numer. Anal. Math. Model. **36**(5), 279–291 (2021)
10. Miller, J.C., Slim, A.C., Volz, E.M.: Edge-based compartmental modelling for infectious disease spread. J. R. Soc. Interface **9**(70), 890–906 (2012)
11. Miller, J.C., Ting, T.: EoN (epidemics on networks): a fast, flexible Python package for simulation, analytic approximation, and analysis of epidemics on networks. arXiv preprint arXiv:2001.02436 (2020)
12. Rineer, J., Kruskamp, N., Kery, C., Jones, K., Hilscher, R., Bobashev, G.: A national synthetic populations dataset for the United States. Sci. Data **12**(1), 144 (2025)
13. Shang, Y.: SEIR epidemic dynamics in random networks. Int. Sch. Res. Notices **2013**(1), 345618 (2013)
14. Wang, Y., Cao, J., Alsaedi, A., Ahmad, B.: Edge-based SEIR dynamics with or without infectious force in latent period on random networks. Commun. Nonlinear Sci. Numer. Simul. **45**, 35–54 (2017)
15. Watts, D.J., Strogatz, S.H.: Collective dynamics of 'small-world' networks. Nature **393**(6684), 440–442 (1998)
16. Wheaton, W.D., et al.: Synthesized population databases: a US geospatial database for agent-based models. Methods report (RTI Press) **2009**(10), 905 (2009)

Is Health Systems Sustainability Measurable? - Operationalizing SDG Targets Using SSP-TOPSIS Approach

Aleksandra Bączkiewicz[1(✉)], Jarosław Wątróbski[1,2(✉)], and Iga Rudawska[3]

[1] Institute of Management, University of Szczecin, ul. Cukrowa 8, 71-004 Szczecin, Poland
{aleksandra.baczkiewicz,jaroslaw.watrobski}@usz.edu.pl
[2] National Institute of Telecommunications, ul. Szachowa 1, 04-894 Warsaw, Poland
[3] Institute of Economics and Finance, University of Szczecin, ul. Mickiewicza 64, 71-101 Szczecin, Poland
iga.rudawska@usz.edu.pl

Abstract. The sustainability of health systems is examined through the lens of quality of care, emphasizing its long-term impact on public health. This paper presents research that extends traditional healthcare evaluation by integrating health outcomes with system resilience and environmental effects, in addition to financial costs. It uses a multi-criteria decision analysis (MCDA) approach using the Sustainable Development Goals (SDGs) for health. A key challenge with widely used MCDA methods, particularly those from American schools, is their compensatory nature. These methods allow poor performance in some criteria to be compensated by advantageous performance in others, which is inconsistent with the strong sustainability paradigm. To address this limitation, this paper presents the Strong Sustainability Paradigm based Technique for Order Preference by Similarity to Ideal Solution (SSP-TOPSIS) method, a novel approach designed to reduce compensatory effects in multi-criteria assessments. The contribution enhances understanding of sustainable health systems and offers a sophisticated tool for policymakers seeking to balance healthcare quality, resilience, and environmental impact in line with strong sustainability principles.

Keywords: Healthcare evaluation · Sustainability · Decision support system · Strong sustainability paradigm

1 Introduction

The domain of health systems sustainability is quality of care, taking into account the responsibility of health services on patients now, but also in the future. This long-term perspective accentuates the health system's impact on communities (public health domain) and on the environment, and consequently on the health of the entire population. Sustainability will therefore broaden the approach to the value of health care to estimate health outcomes in relation to the impact on the community (through system resilience) and the environment, in addition to the financial costs [20]. Therefore the purpose of this paper is to apply the Multi Criteria Decision Analysis (MCDA) approach to measure health systems sustainability using Health Targets in Sustainable Development Goals (SDG) as reference points [14]. To the authors' best knowledge, it is the first attempt of such kind. Previous works [8,10,29] concentrated on the triple bottom line and lacked direct application of SDG indicators. Therefore authors' study has the same potential to contribute to the state of the art in the field of the sustainability of health systems.

Despite the many advantages and applicability of MCDA methods in sustainability assessment, an important limitation among the popular and widely used group of multi-criteria methods derived from the American school is their compensatory character [27]. This implies that in assessments carried out using these methods, there is a possibility that weak values within certain criteria may be compensated for by outstanding values achieved against other criteria [24]. This phenomenon is undesirable from the point of view of the strongly sustainable development paradigm, which dictates the pursuit of favorable values within the widest possible range of criteria [31]. In the context of the limitations indicated, the purpose of this article is to present a multi-criteria SSP-TOPSIS method (the Strong Sustainability Paradigm based Technique for Order Preference by Similarity to Ideal Solution) for modeling compensation reduction to assess the sustainability of health systems in selected countries.

2 Literature Review

Health systems sustainability has become a key area of research, reflecting the need for healthcare organizations to balance quality care with public health (disease prevention, not just treatment), system resilience (for example, in the face of a pandemic), the environment and financial accountability. This initial concept synthesizes the approach proposed by The 2030 Agenda for Sustainable Development, adopted by all United Nations Member States in 2015 (https://sdgs.un.org/goals). It emphasizes the multifaceted nature of sustainability in healthcare. One of the fundamental aspects of sustainability in health systems is the integration of social (including economic) and environmental issues into a quality improvement framework. According to Veltman et al. (2020) [28] quality of care should include environmental impact. The authors suggest that healthcare organizations need to adopt a multi-directional approach to achieve sustainability. Baid & Damm (2021) [5] present similar findings, proposing the SusQI framework and incorporating sustainability into traditional quality domains such as performance and patient experience. The relationship between sustainability and quality improvement is also supported by Mortimer et al. (2018) [20] who argue that sustainability should be viewed as a quality domain in healthcare, extending the responsibility of healthcare services to future generations. Based on the literature review we propose a different approach incorporating SDG Health Targets and indicators.

Table 1. Structure model to measure the sustainability of health systems in the context of SDG health targets (reference to the SDG indicators in brackets).

Main Criteria	Sub-criteria	Relevant Studies	Proposed Measures (Relevant to SDG Target)
G_1 - Quality of Health Care	Patient Safety	Hurst & Jee-Hughes (2001) [13], Caunic (2019) [7], Kim & Jeon (2020) [15]	C_1 - Maternal mortality ratio (A.3.1.1), C_2 - Neonatal mortality rate (A.3.2.2)
	Effectiveness of Treatment		C_3 - Mortality rate attributed to cardiovascular disease, cancer, diabetes or chronic respiratory disease (A.3.4.1), C_4 - Suicide mortality rate (A.3.4.2)
	Transmission Stability		C_5 - Number of new HIV infections per 1,000 uninfected population (A.3.3.1), C_6 - Tuberculosis incidence per 100,000 population (A.3.3.2)
G_2 - Quality of Public Health	Effectiveness of Prevention of Harmful Stimulant Intake	Martin et al. (2024) [18], Arbour et al. (2023) [1], Livingood et al. (2018) [17]	C_7 - Harmful use of alcohol, defined according to the national context as alcohol per capita consumption (aged 15+) within a calendar year in liters of pure alcohol (A.3.5.2), C_8 - Age-standardized prevalence of current tobacco use among persons 15+ (A.a.1)
	Effectiveness of Prevention of Moral Hazard		C_9 - Death rate due to road traffic injuries (A.3.6.1)
G_3 - Financial Protection	Risk Protection	Murray & Frenk (2000) [21], Hurst & Jee-Hughes (2001) [13]	C_{10} - Coverage of essential health services – Index 2UN (A.3.8.1)
	Availability of Financial Resources		C_{11} - Proportion of total government spending on essential services i.e. health, social protection, education (B.1.a.2)
G_4 - System's Resilience	System's Robustness	Paschoalotto et al. (2023) [23], Fallah-Aliabadi et al. (2020) [11], Foroughi et al. (2022) [12]	C_{12} - Proportion of the target population covered by all vaccines included in their national program (3.b.1)
	System's Capacity		C_{13} - International Health Regulations (IHR) capacity (A.3.d.1)
	Availability of Human Resources		C_{14} - Health worker (doctors) density and distribution (A.3.c.1), C_{15} - Health worker (nurses) density and distribution (A.3.c.1)

We also apply MCDA as a robust method allowing multidimensional assessment of health systems sustainability. The preliminary framework is presented in Table 1. MCDA methods find application in the evaluation of problems requiring the consideration of multiple aspects, as they allow the consideration of many often conflicting criteria in the evaluation of multiple alternatives simultaneously [32].

However, many of the popular MCDA methods especially those originating from the American school, such as the Analytical Hierarchy Process (AHP), Technique for Order Preference by Similarity to Ideal Solution (TOPSIS),

Weighted Sum Method (WSM), or Multi-Attribute Utility Theory (MAUT) are compensatory in nature meaning that weaker performance of alternatives against a certain criterion can be offset by better performance against other criteria [27]. This phenomenon can cause an option that is very good against one criterion to be rated highly despite poorer performance for another criterion. This is not consistent with the strong sustainability paradigm, which assumes that certain criteria are not compensated by other criteria [24].

Multi-criteria approaches that support strong sustainability can be indicated. Among them are methods derived from the European school, like the Preference Ranking Organization METHod for Enrichment of Evaluation (PROMETHEE) and ELimination and Choice Expressing the Reality (ELECTRE) family of methods [30]. However, these methods are characterized by a more complicated algorithm compared to popular compensation methods, and not all of them produce quantitative results. The identified research gap became the motivation for the development of the SSP-TOPSIS approach, which allows for extending compensatory methods with the ability to model compensation reduction. The proposed approach makes it possible to maintain the simplicity of the known methods and reduce their shortcoming, i.e. compensability.

3 Methodology

This research aims to analyze the health systems sustainability of European Union countries, USA and Switzerland in terms of 15 evaluation criteria C_1-C_{15} belonging to four main dimensions G_1-G_4 listed in Table 1, where UK means United Kingdom. The conducted investigation uses the most recent and complete data obtained from United Nations, SDG Indicators Database (https://unstats.un.org/sdgs/dataportal/database). The particular data sources are included in a supplementary file named "data sources" provided on GitHub at the link https://github.com/energyinpython/SSP-TOPSIS-for-health-systems-assessment.

3.1 The SSP-TOPSIS Method

The SSP-TOPSIS method was developed on the basis of the principles of the widely used MCDA method called TOPSIS, which incorporates the distance of evaluated variants from two vectors representing ideal and anti-ideal solutions. The proposed method includes a novel stage that enables modeling criteria compensation reduction, which is a significant limitation of multi-criteria methods originating from American school represented by TOPSIS. The discussed stage incorporates compensation reduction by subtracting the Mean Deviation of the performance value MD calculated in relation to particular criteria. MD is the difference between the performance of the alternative for a considered criterion and the mean performance within the criterion across all alternatives multiplied by the value of the sustainability coefficient. The sustainability coefficient s is a component for modeling compensation reduction and yields real values from 0 to 1. By default, it can be adjusted to the value of the standard deviation from the normalized decision matrix within each criterion. Detailed steps of the presented SSP-TOPSIS method are given below.

To determine the weights of the criteria representing their relevance, the authors chose an objective weighting method called Criteria Importance Through Inter-criteria Correlation (CRITIC) [25]. This method determines the criteria weights based on the values in the decision matrix, taking into account their variability among the alternatives within each criterion. The choice of this weighting technique is justified by the fact that the weighting values determined using it are most evenly distributed among the criteria considered compared to other objective weighting methods, as shown in Fig. 1. This effect is in line with the assumptions of the investigation, according to which the significance of the criteria should be evenly distributed and no criterion should be overly favored or omitted.

Step 1. Compute the Mean Deviation MD_{ij} for each performance value incorporated in the decision matrix, subtracting the mean value $\overline{x_j}$ from performance values x_{ij} for each C_j criterion. Then, multiply the outcome by the value of the s_j coefficient. Coefficient s_j denotes the sustainability coefficient, which reflects the level of the compensation reduction of criteria performance. The sustainability coefficient takes real values from 0 to 1. Criteria are numbered by $j = 1, 2, \ldots, m$. High values of s_j represent relevant reductions in the compensation of a j-th criterion performance value. On the other hand, low values of s_j denote a low reduction of the compensation of a particular j-th criterion. The complete procedure of Mean Deviation calculation is carried out using Eq. (1).

$$MD_{ij} = (x_{ij} - \overline{x}_j)s_j \tag{1}$$

Step 2. Associate 0 values to these MD_{+ij} for profit criteria C_{+j} that are lower than 0. If MD_{+ij} is lower than 0 it means that x_{+ij} is lower than $\overline{x}_{+j}$. Assign 0 values for these MD_{-ij} for cost criteria C_{-j} that are higher than 0. It denotes that r_{-ij} are higher than $\overline{x}_{-j}$. The procedure described in this step is carried out as Eq. (2) demonstrates,

$$MD_{ij} = 0 \ \forall \ MD_{+ij} < 0 \ \lor \ MD_{-ij} > 0 \tag{2}$$

where MD_{ij} defines Mean Deviation values computed for criteria C_j. This stage is relevant because the purpose of it is to prevent unintended improvements in performance values outlying from the mean toward the worse.

Step 3. Subtract MD_{ij} values from performance values x_{ij} included in decision matrix x_{ij} according to Eq. (3).

$$t_{ij} = x_{ij} - MD_{ij} \tag{3}$$

The rest of the steps are analogous to the classic TOPSIS method.

Step 4. Perform the normalization of the decision matrix, which is demonstrated in Eq. (4) with chosen normalization technique, for example the Minimum-Maximum normalization or Vector normalization, which is the default normalization for the TOPSIS method. In Minimum-Maximum normalization, normalized values represented by r_{ij}^{+} for profit criteria and r_{ij}^{-} for cost criteria are achieved through the application of Eq. (5). After performing normalization, each criterion is already transformed to profit criteria.

$$T = [t_{ij}]_{m \times n} = \begin{bmatrix} t_{11} & t_{12} & \cdots & t_{1n} \\ t_{21} & t_{22} & \cdots & t_{2n} \\ \vdots & \vdots & \vdots & \vdots \\ t_{m1} & t_{m2} & \cdots & t_{mn} \end{bmatrix} \tag{4}$$

$$r_{ij}^{+} = \frac{t_{ij} - min_j(t_{ij})}{max_j(t_{ij}) - min_j(t_{ij})}, \ r_{ij}^{-} = \frac{max_j(t_{ij}) - t_{ij}}{max_j(t_{ij}) - min_j(t_{ij})} \tag{5}$$

Step 5. Compute the weighted normalized decision matrix. For this aim, multiply values in the normalized decision matrix by corresponding criteria weights w_j as Eq. (6) shows.

$$v_{ij} = r_{ij} w_j \tag{6}$$

Criteria weights were calculated using the CRITIC method.

Step 6. Determine the Positive Ideal Solution (PIS) using Eq. (7) and Negative Ideal Solution (NIS) using Eq. (8). PIS contains the maximum values of the weighted normalized decision matrix, while NIS contains its minimum values. Due to the previous normalization of the decision matrix, converting criteria into profit and cost is not required.

$$v_j^{+} = \{v_1^{+}, v_2^{+}, \ldots, v_n^{+}\} = \{max_j(v_{ij})\} \tag{7}$$

$$v_j^{-} = \{v_1^{-}, v_2^{-}, \ldots, v_n^{-}\} = \{min_j(v_{ij})\} \tag{8}$$

Step 7. Compute distance from PIS D_i^{+} and NIS D_i^{-} for each alternative according to Eq. (9). The default metric for distance computing in the TOPSIS method is Euclidean distance.

$$D_i^{+} = \sqrt{\sum_{j=1}^{n}(v_{ij} - v_j^{+})^2}, \ D_i^{-} = \sqrt{\sum_{j=1}^{n}(v_{ij} - v_j^{-})^2} \tag{9}$$

Step 8. Compute the score for each examined alternative according to Eq. (10). The C_i value is always within the range of 0 to 1. The alternative that has the highest C_i value is the ranking leader. The ranking is built by sorting alternatives according to preference values in descending order.

$$C_i = \frac{D_i^{-}}{D_i^{-} + D_i^{+}} \tag{10}$$

4 Results

The preliminary stage of the research was the selection of an objective weighting method for determining the significance values of evaluation criteria. The goal was to choose a technique that would return the most evenly distributed significance values since the authors did not intend to significantly favor or underestimate any criterion. The authors determined the criteria weights from a dataset employing six objective weighting methods including Entropy, Gini coefficient-based [4], IDOCRIW (Integrated Determination of Objective CRIteria Weights) [6], CILOS (Criterion Impact LOSs) [3], Angular [26], and CRITIC (CRiteria Importance Through Intercriteria Correlation) [25] weighting methods. The CRITIC method produced the desired result, as demonstrated in Fig. 1. As can be seen in the presented bar chart, neither value of the CRITIC criteria weights has a significant outlying value compared to the other criteria.

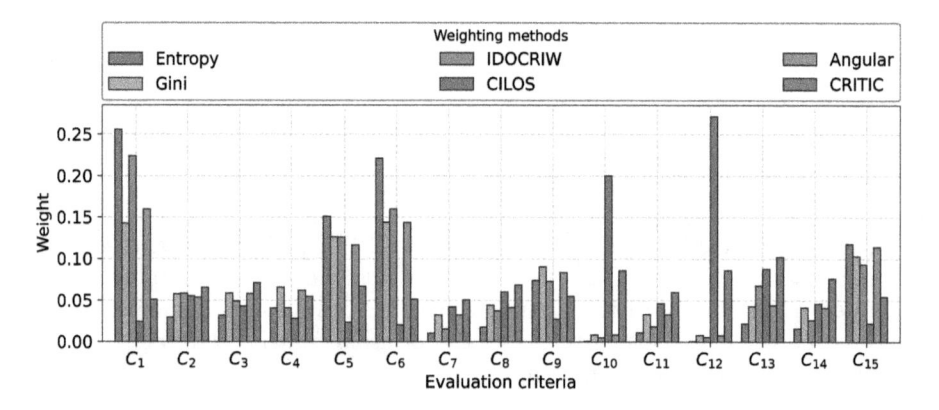

Fig. 1. Criteria weights determined with different objective weighting methods.

The first part of the research conducted using the proposed SSP-TOPSIS multi-criteria method involves comparing the results of the newly developed method with criterion compensation reduction and the classic TOPSIS compensation method. The sustainability coefficient in the SSP-TOPSIS method was set as the standard deviation of the data for each criterion (s = std). The results are shown in Table 2.

It can be noted that the leader of both rankings is Belgium. This indicates that this country has favorable and balanced values in relation to a wide number of evaluation criteria. In second place in both rankings came the USA, and in third place was France. Place 4 in the SSP-TOPSIS ranking was achieved by Lithuania, which was ranked 5th in the TOPSIS ranking. The more favorable score obtained in the reduced-compensation method than in the classical method testifies to balanced and favorable performance values in many of the evaluation dimensions represented by the individual criteria. An analogous situation is observed for Germany, which ranked 8th in the TOPSIS ranking and 10th

Table 2. Results of SSP-TOPSIS compared to TOPSIS results.

Country	TOPSIS score	TOPSIS rank	SSP-TOPSIS score s=std	SSP-TOPSIS rank s=std
Austria	0.4647	18	0.4836	19
Belgium	0.5651	1	0.5868	1
Bulgaria	0.4764	15	0.5008	12
Croatia	0.4617	19	0.4939	15
Cyprus	0.3834	29	0.4022	31
Czechia	0.5100	6	0.5349	6
Denmark	0.3805	30	0.4080	30
Estonia	0.3747	31	0.4095	29
Finland	0.4871	11	0.5087	11
France	0.5297	3	0.5514	3
Germany	0.4953	10	0.5172	8
Greece	0.4114	27	0.4269	28
Hungary	0.5219	4	0.5371	5
Ireland	0.4280	25	0.4577	25
Italy	0.4103	28	0.4456	26
Latvia	0.4789	13	0.4991	13
Lithuania	0.5175	5	0.5409	4
Luxembourg	0.4769	14	0.4951	14
Malta	0.4273	26	0.4451	27
Netherlands	0.4714	16	0.4930	17
Norway	0.4492	21	0.4699	23
Poland	0.4664	17	0.4908	18
Portugal	0.4609	20	0.4790	20
Romania	0.5037	8	0.5172	9
Slovakia	0.4441	22	0.4729	22
Slovenia	0.4412	23	0.4604	24
Spain	0.4366	24	0.4752	21
Sweden	0.5087	7	0.5266	7
Switzerland	0.4960	9	0.5140	10
UK	0.4800	12	0.4937	16
USA	0.5451	2	0.5759	2

in the SSP-TOPSIS ranking. Hungary, on the other hand, gained 4th place in the TOPSIS ranking while it was 5th in the SSP-TOPSIS ranking. This means that Hungary does not have enough favorable values in a sufficient range of evaluation criteria to remain in 4-th place for compensation reduction.

Czechia and Sweden were also among the well-rated countries. The countries that showed the weakest performance were Cyprus, Denmark and Estonia. These countries are at the bottom of both rankings, which denotes that they have much worse performance values within many criteria, which the favorable values achieved for other criteria cannot compensate for even in the lack of compensation reduction.

The next stage of the research involved conducting a sensitivity analysis with a stepwise increasing the sustainability coefficient value representing the degree of reduction in criteria compensation. Sustainability coefficient values were increased by 0.1 starting from 0.0 all the way up to 1.0 within the individual criteria groups of the model: G_1 displayed in Fig. 2, G_2 illustrated in Fig. 3, G_3 presented in Fig. 4, G_4 demonstrated in Fig. 5, and in the final step for all criteria shown in Fig. 6.

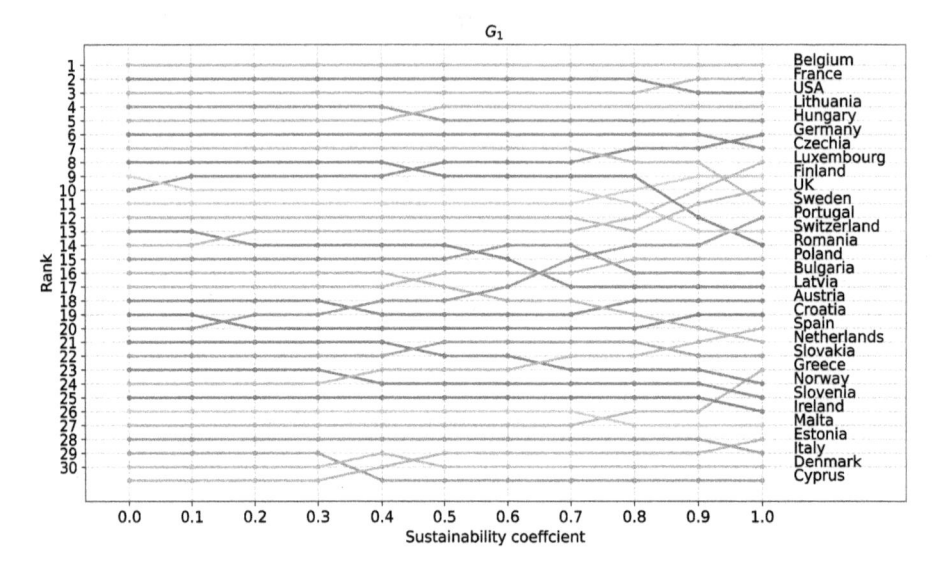

Fig. 2. Rank shifts caused by increasing criteria compensation reduction in G_1.

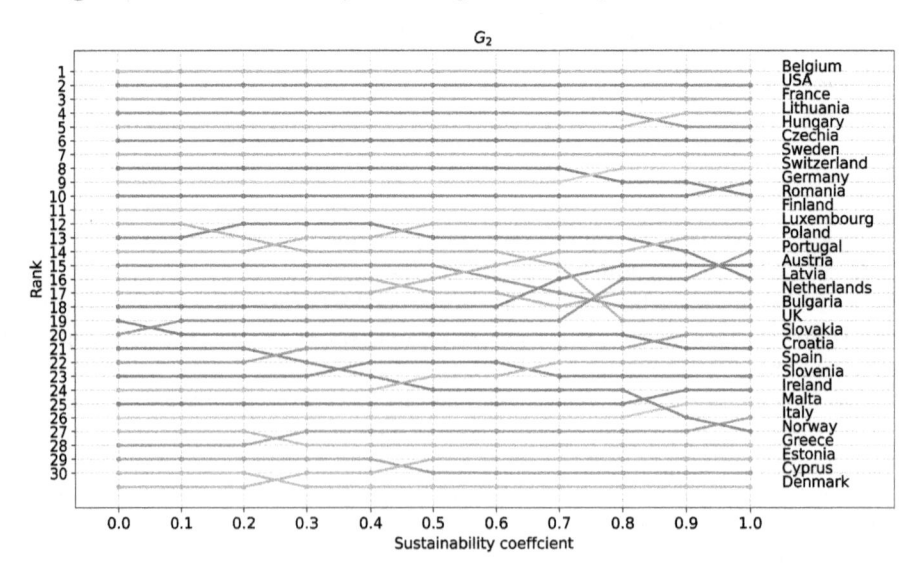

Fig. 3. Rank shifts caused by increasing criteria compensation reduction in G_2.

When, during increasing compensation reduction, a country advances or remains in a stable position it means that its performance values have favorable and balanced values within a wide range of criteria. On the other hand, if with increasing compensation reduction the country falls in the ranking it implies that it achieves favorable values within certain criteria that allow it to compensate for the weak values achieved for other criteria. In the case of increasing compensation reduction, the possibility of compensating for weak values is reduced, so the country achieves worse rankings. In the case of compensation reductions within the G_1 criteria, it can be observed that the largest decrease with an increase in compensation reduction was registered for Romania and the Netherlands. In contrast, the countries that advanced the most with the increase in compensation reduction were Portugal and Luxembourg. When reducing compensation in the G_2 criteria group, the largest decrease was observed for the United Kingdom and Norway, while the largest promotion was demonstrated by Portugal.

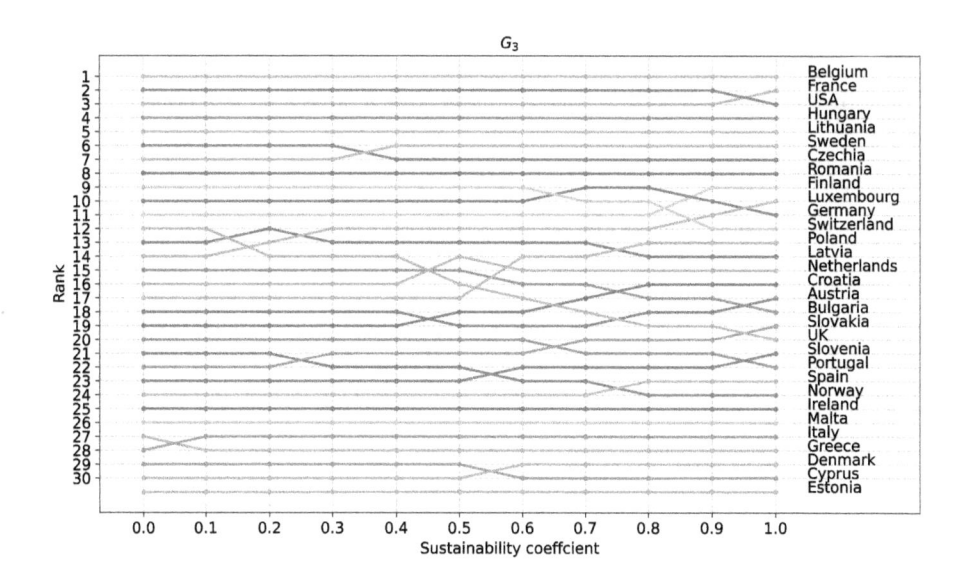

Fig. 4. Rank shifts caused by increasing criteria compensation reduction in G_3.

When reducing the compensation of the G_3 criteria group, the largest decrease was observed for the United Kingdom, and the largest promotion was demonstrated by Poland and Luxembourg. The increasing reduction in the compensation of the G_4 criteria group resulted in the greatest dynamics of changes in places among all criteria groups. The G_4 group had the most shifts with the largest range compared to the other groups. With increasing compensation reduction, the largest promotion was reported for Spain, Croatia, Bulgaria, Ireland, and Italy, and the largest decrease was shown by Hungary, Luxembourg, the United Kingdom, and Romania.

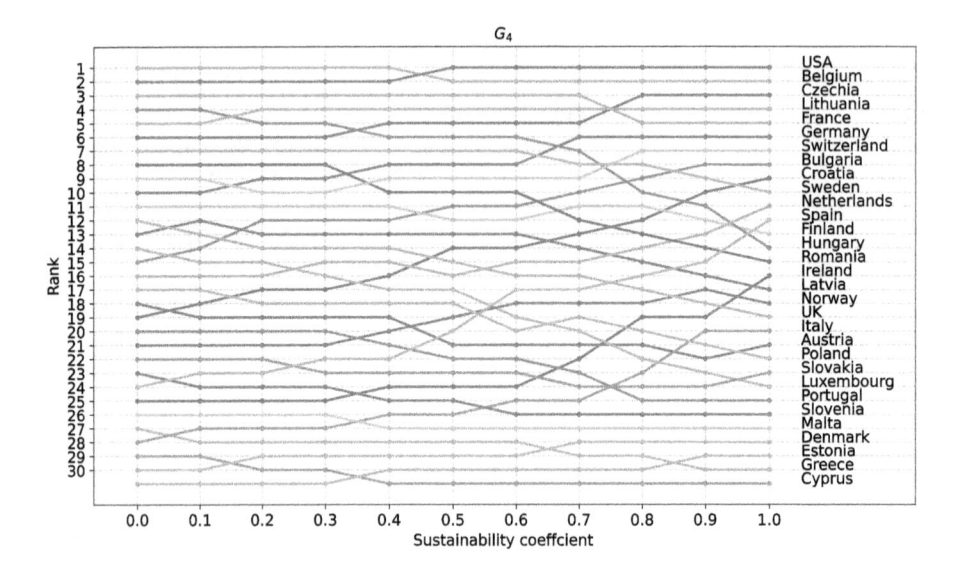

Fig. 5. Rank shifts caused by increasing criteria compensation reduction in G_4.

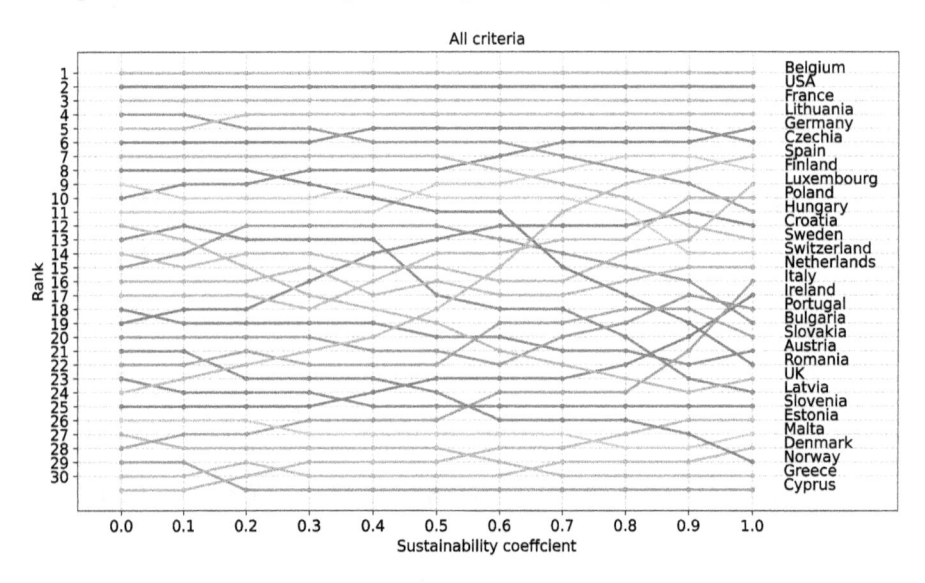

Fig. 6. Rank shifts caused by increasing compensation reduction of all criteria.

The high dynamics of the results when reducing the compensation of the G_4 criteria group indicates that there is a high sensitivity in the health system's resilience of the evaluated countries to the variability within this criteria group, which is determined by the large range of discrepancies in the performance values achieved by the evaluated countries against these criteria. For comparison, an analysis of the impact of compensation reductions on the SSP-TOPSIS ranking was also carried out with all criteria considered, resulting in the greatest dynamics of change with the widest range. In this case, Spain and Italy showed the greatest potential for advancement, while the largest decrease was recorded for Romania. The results obtained show that Belgium, the US, France, and Lithuania are among the countries ranked at the top regardless of the increasing reduction in compensation. The results received from the sensitivity analysis confirm the sustainability and stability of the performances achieved by these countries within a sufficiently broad group of criteria.

5 Discussion

Our results are consistent with those obtained by other authors [2,14,19] who emphasize the importance of sustainable development and adapting health policies to local needs and country health policies although our work is unique in terms of the applied method. Similar conclusions have been made by Konarzewska [16] who analyses the 12 indicators proposed by Eurostat to measure the achievement of Agenda 2030 Goal 3 on health and well-being. The study shows the dynamics of the values of these indicators between 2002 and 2017 and compares the situation in 28 EU countries in 2017, applying univariate and multivariate statistical analysis. The results show the varying situation of EU countries in the pursuit of healthy lives and well-being of citizens.

The European Commission's 'State of Health in the EU' [9] initiative aims to facilitate access to information on health systems, expertise and best practices for health policy makers. Our study can be used as a reference in this context and be helpful to prepare the reports, such as 'Health at a Glance: Europe' [22], which assess progress in building effective, accessible and resilient health systems in EU countries.

With our research, we open the scientific discussion on not only the need to tailor health goals to the specific conditions of each country, but first of all on the measurement of the SDG 3 health targets monitoring the progress toward sustainable health systems. Regular monitoring of progress and the use of available data and analysis, such as that provided by our work, are key to the effective setting and achievement of well-being and welfare in European countries. The methodical contribution is a proposal for solving this problem of sustainable health systems assessment of the SSP-TOPSIS method, which reduces criteria compensation, thus supporting the paradigm of strong sustainability.

In this research, criteria according to the WHO Impact Framework GPW 13 [33] were used to build the model, following the availability of data for selected countries. The few criteria for which Eurostat does not update data or countries do not report them were discarded. The model adopted criteria from group A, i.e. Health targets (SDG 3). In further work, the authors plan to expand the model to include criteria from group B, i.e. Health related SDG targets.

6 Conclusions

The paper presents a multi-criteria SSP-TOPSIS method for compensation reduction and confirms its applicability in decision support systems for assessing the sustainability of healthcare systems using the proposed author's evaluation model as an example. The modeling of compensation reduction in the proposed method provides broader analytical capabilities compared to the classical TOPSIS method and enables reliable sustainability assessment. Directions for future work include expanding the model to include additional evaluation criteria and extending other multi-criteria methods to include compensation reduction modeling capabilities. Benchmarking with more data is also required. Another interesting direction for further work is to analyze the long-term dynamics of the model's indicators using temporal MCDA methods.

Acknowledgments. This research was partially funded by National Science Centre, Poland 2022/45/B/HS4/02960, and Co-financed by the Minister of Science under the "Regional Excellence Initiative" Program RID/SP/0046/2024/01.

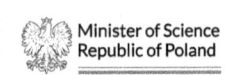

Disclosure of Interests. The authors have no competing interests to declare that are relevant to the content of this article.

References

1. Arbour, M., et al.: Sustaining and scaling a clinic-based approach to address health-related social needs. Front. Health Serv. **3**, 1040992 (2023). https://doi.org/10.3389/frhs.2023.1040992
2. Asma, S., et al.: Monitoring the health-related sustainable development goals: lessons learned and recommendations for improved measurement. The Lancet **395**(10219), 240–246 (2020). https://doi.org/10.1016/S0140-6736(19)32523-1
3. Ayan, B., Abacıoğlu, S., Basilio, M.P.: A comprehensive review of the novel weighting methods for multi-criteria decision-making. Information **14**(5), 285 (2023). https://doi.org/10.3390/info14050285
4. Bączkiewicz, A., Wątróbski, J.: Crispyn - a Python library for determining criteria significance with objective weighting methods. SoftwareX **19**, 101166 (2022). https://doi.org/10.1016/j.softx.2022.101166
5. Baid, H., Damm, E.: Reducing critical care's carbon footprint with financial and social co-benefits. Intensive Crit. Care Nurs. **64**, 103030 (2021). https://doi.org/10.1016/j.iccn.2021.103030
6. Bandyopadhyay, S., Mandal, I.: Proposing a novel MCDA technique based on dispersion and regression analysis. In: 2024 2nd International Conference on Computer, Communication and Control (IC4), pp. 1–4. IEEE (2024). https://doi.org/10.1109/IC457434.2024.10486734
7. Caunic, R.E., et al.: Frameworks and measures for health systems performance assessment. SEA-Practical Appl. Sci. **7**(21), 205–212 (2019)

8. Choi, B., Chen, C.L.: The triple bottom line and stabilization wedges: a framework for perioperative sustainability. Anesth. Analg. **134**(3), 475–485 (2022). https://doi.org/10.1213/ANE.0000000000005890
9. Commission, E.: State of Health in the EU (2025). https://health.ec.europa.eu/state-health-eu_en
10. De Smedt, D., Van Wilder, L., Boone, L., Dewulf, J.: HTA121 the triple bottom line in healthcare: a holistic sustainability assessment for decision support. Value Health **26**(12), S341–S342 (2023)
11. Fallah-Aliabadi, S., Ostadtaghizadeh, A., Ardalan, A., Fatemi, F., Khazai, B., Mirjalili, M.R.: Towards developing a model for the evaluation of hospital disaster resilience: a systematic review. BMC Health Serv. Res. **20**, 1–11 (2020). https://doi.org/10.1186/s12913-020-4915-2
12. Foroughi, Z., Ebrahimi, P., Aryankhesal, A., Maleki, M., Yazdani, S.: Toward a theory-led meta-framework for implementing health system resilience analysis studies: a systematic review and critical interpretive synthesis. BMC Public Health **22**(1), 287 (2022). https://doi.org/10.1186/s12889-022-12496-3
13. Hurst, J., Jee-Hughes, M.: Performance measurement and performance management in OECD health systems. Occasional Papers **47** (2001). https://www.oecd-ilibrary.org/docserver/788224073713.pdf?expires=1644004185&id=id&accname=guest&checksum=138BFDDA56718B93A4C4B456E1797D56
14. Ionescu, G.H., Firoiu, D., Tănasie, A., Sorin, T., Pîrvu, R., Manta, A.: Assessing the achievement of the SDG targets for health and well-being at EU level by 2030. Sustainability **12**(14), 5829 (2020). https://doi.org/10.3390/su12145829
15. Kim, H., Jeon, B.: Developing a framework for performance assessment of the public long-term care system in Korea: methodological and policy lessons. Health Res. Policy Syst. **18**(27), 1–10 (2020). https://doi.org/10.1186/s12961-020-0529-8
16. Konarzewska, I.: Realizacja celu zrównoważonego rozwoju: „Dobre zdrowie i jakość życia" w krajach Unii Europejskiej w roku 2017. Comp. Econ. Res. **23**(2), 53 (2020). https://doi.org/10.18778/1508-2008.23.12
17. Livingood, W.C., Bilello, L.A., Choe, U., Lukens-Bull, K.: Enhancing the science of discovery in public health systems and services research through participatory research methods. Popul. Health Manag. **21**(2), 155–162 (2018). https://doi.org/10.1089/pop.2017.0042
18. Martin, S., et al.: A scoping review of health equity interventions in governmental public health. J. Public Health Manag. Pract. **30**(4), 479–489 (2024). https://doi.org/10.1097/PHH.0000000000001947
19. Menne, B., et al.: Health and well-being for all: an approach to accelerating progress to achieve the Sustainable Development Goals (SDGs) in countries in the WHO European Region. Eur. J. Public Health **30**(Supplement_1), i3–i9 (2020). https://doi.org/10.1093/eurpub/ckaa026
20. Mortimer, F., Isherwood, J., Wilkinson, A., Vaux, E.: Sustainability in quality improvement: redefining value. Future Healthc. J. **5**(2), 88–93 (2018). https://doi.org/10.7861/futurehosp.5-2-88
21. Murray, C.J., Frenk, J.: A framework for assessing the performance of health systems. Bull. World Health Organ. **78**(6), 717–731 (2000)
22. OECD: Health at a Glance: Europe 2024 (2024). https://www.oecd.org/en/publications/health-at-a-glance-europe-2024_b3704e14-en.html
23. Paschoalotto, M., Lazzari, E.A., Rocha, R., Massuda, A., Castro, M.C.: Health systems resilience: is it time to revisit resilience after COVID-19? Soc. Sci. Med. **320**, 115716 (2023). https://doi.org/10.1016/j.socscimed.2023.115716

24. Sanatkumar, N., Berka-Harnmeijer, A.: Walking the tightrope: can integrated decision support transform business sustainability? J. Clean. Prod. **445**, 141366 (2024). https://doi.org/10.1016/j.jclepro.2024.141366
25. Saraji, M.K., Aliasgari, E., Streimikiene, D.: Assessment of the challenges to renewable energy technologies adoption in rural areas: a Fermatean CRITIC-VIKOR approach. Technol. Forecast. Soc. Chang. **189**, 122399 (2023). https://doi.org/10.1016/j.techfore.2023.122399
26. Shuai, D., Zongzhun, Z., Yongji, W., Lei, L.: A new angular method to determine the objective weights. In: 2012 24th Chinese Control and Decision Conference (CCDC), pp. 3889–3892. IEEE (2012). https://doi.org/10.1109/CCDC.2012.6244621
27. Silva, F., Lima, M.P., Corujo, D., Neto, A.J., Esposito, F.: A comprehensive step-wise survey of multiple attribute decision-making mobility approaches. IEEE Access (2024). https://doi.org/10.1109/ACCESS.2024.3436074
28. Veltman, L.M., Delnoij, D.M., Ossebaard, H.C.: Does quality of care entail environmental impact? A blind spot in our knowledge. Int. J. Healthc. **6**(2), 74 (2020). https://doi.org/10.5430/ijh.v6n2p74
29. Vergunst, F., Berry, H.L., Rugkåsa, J., Burns, T., Molodynski, A., Maughan, D.L.: Applying the triple bottom line of sustainability to healthcare research–a feasibility study. Int. J. Qual. Health Care **32**(1), 48–53 (2020). https://doi.org/10.1093/intqhc/mzz049
30. Wątróbski, J.: Temporal PROMETHEE II - New multi-criteria approach to sustainable management of alternative fuels consumption. J. Clean. Prod. **413**, 137445 (2023). https://doi.org/10.1016/j.jclepro.2023.137445
31. Wątróbski, J., Bączkiewicz, A., Rudawska, I.: A strong sustainability paradigm based analytical hierarchy process (SSP-AHP) method to evaluate sustainable healthcare systems. Ecol. Ind. **154**, 110493 (2023). https://doi.org/10.1016/j.ecolind.2023.110493
32. Wątróbski, J., Bączkiewicz, A., Sałabun, W.: New multi-criteria method for evaluation of sustainable RES management. Appl. Energy **324**, 119695 (2022). https://doi.org/10.1016/j.apenergy.2022.119695
33. WHO: GPW 13 WHO Impact Framework Programmatic targets and indicators: Mapping SDGs to GPW13 (2019). https://www.who.int/docs/default-source/documents/gpw/gpw13-wif-targets-and-indicators-en.pdf?sfvrsn=81cf3546_20

Computational Modeling and Artificial Intelligence for Social Systems

Automatic Detection and Identification of Causal Relationships in Polish Legal Texts

Łukasz Kurant[✉][iD]

University of Maria Curie-Sklodowska, Pl. M. Curie-Skłodowskiej 5, 20-031 Lublin, Poland
`lukasz.kurant@mail.umcs.pl`

Abstract. The paper focuses on the problem of detecting sentences containing causal relations in Polish legal texts. The identification of these relationships and their decomposition is a key factor in the effective analysis of legal texts and an important aspect in the extraction of parts of such relationships. This represents a contribution to the development of the field for languages other than English. The paper presents an analysis of the created dataset and based on it, classification was performed in nine different experiments using selected machine learning and deep learning algorithms (including several large BART-type models), taking into account the specifics of legal language. The experiments confirm the effectiveness of the proposed method, where the best model detected sentences containing both explicit and implicit causality with an accuracy of approximately 86%. These results lead to further questions and point to further directions for future development, especially in the field of reasoning from legal texts.

Keywords: causal relationships · argumentation · artificial intelligence

1 Introduction

Detecting cause-and-effect relationships is a challenge in natural language processing. This requires advanced cognitive processes, and the resulting data has wide applications in many scientific fields. However, there are several challenges to overcome [4], such as minimizing ambiguity and recognizing causal relationships that can exist in both explicit and implicit forms. Detecting relationships can be a significant obstacle, especially in the latter case. As with other Natural Language Processing (NLP) tasks, it is important to consider the impact of natural language, including its structural and semantic aspects such as vocabulary, sarcasm, and metaphors. Both the language itself and the specific domain can have an impact. Research on the influence of language on the process of formulating cause and effect should not be limited to specific domains or languages. Therefore, it is important to consider a wide range research in this subject. Detected causal relationships can be widely applied in many fields [2], including the field of law. Accurate causal reasoning is crucial for legal professionals in their daily work. For instance, judges use it to formulate sentences based on similar case law, while attorneys and prosecutors use it to determine the appropriate courtroom strategy.

Two main tasks can be distinguished in causality detection. The first task is to identify the locations (e.g. sentences) where causality occurs. Based on this information, the parts of the relationship can be extracted, determining the cause and effect, and filtering out irrelevant information. It is important to maintain a clear and logical flow of information with causal connections between statements.

2 Definitions

We can define causality as the relationship between two different events $event_1$ and $event_2$ in such a way that $event_2$ results from $event_1$. Various approaches are used to formalize the definition of causality, e.g. as an implication where the occurrence of cause c_1 implies the occurrence of effect e_1 ($c_1 \Rightarrow e_1$) or as a logical equivalent ($c_1 \Leftrightarrow e_2$), for the reason of avoiding ambiguity [12]. The choice of definition may thus depend on the specific problem that practitioners were to solve [3]. Therefore, due to the fact that during our experiments it does not matter whether an effect can occur for reasons other than those written out, we decided to use logical equivalence notation.

2.1 Division by Type

According to the definition of the type of relationship shown in [4], causal relationships can be divided into three categories as follows:

- $c_1 \Leftrightarrow e_1$, if c_1 exists then e_1 also exists, e.g. "The judge convicted him because the evidence was against him".
- $c_1 \Leftrightarrow \neg e_1$, if c_1 exists then e_1 does not exist, e.g. "It is clear from the witness's testimony that he could not do so".
- $\neg c_1 \Leftrightarrow \neg e_1$, if c_1 does not exist then e_1 does not exist either, e.g. "The witness was not allowed into the courtroom because he did not have a valid identity card".

2.2 Division by Complexity

We can also divide causality according to other criteria. Causality can be single-sentence or multi-sentence. In the case of the latter, the process of detection or extraction is much more complicated [25]. In some cases, causality may be more complex, such that the number of causes and effects may not be equal, i.e. many different causes may cause one effect, or vice versa: one cause may lead to many effects. Both causes and effects can be connected by conjunctions ($c_1 \Leftrightarrow e_1 \wedge e_2$) or by disjunctions, e.g. ($c_1 \Leftrightarrow e_1 \vee e_2$) in any way, including as combinations of these two. Because of this, causality can also lead to so-called causal chains, in the way that the effect of a cause, can be the cause of another effect, e.g. "The court did not allow a witness into the courtroom, due to an invalid identity document, which resulted in the person against whom the proceedings were taking place being found not guilty". We can therefore define such a chain as: "invalid document (c_1) ⇔ witness not allowed (e_1/c_2) ⇔ not guilty (e_2)".

2.3 Division by Form

We can also divide causality based on its form:

- *explicit causality*, occurring overtly in sentences, often with phrases or conjunctions indicating causality, e.g. "The judge convicted him because the evidence was against him",
- *implicit causality*, occurring implicitly, often with parts divided between different sentences, e.g. "The judge found him not guilty. No evidence of his guilt was found".

Within explicit causality, we can further distinguish: *marked causality*, when the text contains causal conjunctions, such as "because", "as"; and *unmarked causality*, when the text does not contain such a conjunct, but contains other phrases (e.g. verbs) that indicate the existence of causality. In addition, each linker, causal phrase can be divided into one of two categories, i.e. *ambiguous causal phrase*, when a phrase or word can be used in different contexts and only in some of them its use proves causality, and *non-ambiguous causal phrase*, when causality is inferred from almost every instance of the phrase's use.

2.4 Division by Order of Occurrence

The components of cause-effect relationships can be related to each other in different ways based on the timing of their occurrence. There are many different divisions based on such criteria, e.g. TimeML [18], which distinguishes as many as seven different types of temporal relationships, or CaTeRS [13], which distinguishes four types of them. Due to the complexity of the relationships of these types, we can simplify them into two groups:

- the cause occurred before the effect, e.g. "The witness failed to attend due to a car breakdown",
- the cause occurred together with the effect, e.g. "The witness spoke slowly because he had a speech defect".

In the second case, it does not matter whether the cause and effect occurred together throughout their existence if there was a concurrent time, the relation is qualified in this category.

3 Research Status

Among the methods currently used in the field of causality detection, we can distinguish between rule-based and pattern-based methods [6], statistical methods (including machine learning methods such as linear regression, Naive Bayes or decision trees), and more sophisticated methods using deep learning and neural networks. Among the latter, we can distinguish architectures such as CNNs, LSTMs, GRUs and Transformers such as BERT [4]. Each of these methods has its own advantages and disadvantages. Methods using patterns require domain expertise, while methods using machine learning need to be programmed and trained, which can result in the need for a large amount of resources. When detecting causality, the use of word embeddings also plays an important role and has a significant impact on improving performance [5].

The main problem during research is the lack of sufficient datasets, especially in the context of languages other than English. The rudimentary yet underdeveloped notion of causality in Polish legat texts was introduced in the [9]. In addition, the domain from which the samples in the collection are drawn is important. In this case, it is difficult to compare the chosen methods, due to the specificities of the language, in which certain features may be useless [8]. In particular, it is a difficult task to prepare a collection containing numerous instances of implicit causation, due to the difficulty for annotators to recognize it.

The process of detecting whether a relationship exists in a text is often the first step to extracting the components of a given causal relationship, but it is an important part of it. Hence, it is significant to get the best possible results in the first step, in order to avoid potential cascading errors in the future (when a sentence without causality is flagged by the model as having such a relationship, this can lead to further extraction errors [11]).

4 Dataset

The aim of our experiment was to perform causality detection at the sentence level. For this, it became necessary to prepare a suitable dataset, which would take into account not only whether a given sentence contains a cause-effect relation, but also the type of this relation, its complexity or form. For this purpose, 50 different court judgments in Polish were selected, from five different categories (10 judgments from each): animal protection, taxes, juvenile, infringement of privacy, international law.

The court judgments are taken from the publicly available Portal of Administrative Court Judgments [23]. Due to their nature, they have anonymised sensitive data, such as the names of individuals, place names, etc. Using the author's script (using the *beautifulsoup* library in Python [20]), the sentences were downloaded in HTML format and appropriately processed to split them into sentences (using the *NLTK* library [16]). The sentences prepared in this way were then imported into the *doccano* software [15] used to prepare their annotation. During the annotation process, the following tags were added for each sentence:

- *Causality/No Causality* if the sentence contains a causality relationship,
- *Not valid sentence* whether the sentence is a valid sentence. Legal texts sometimes contain referenced provisions or other phrases that should not be included in the detection process.

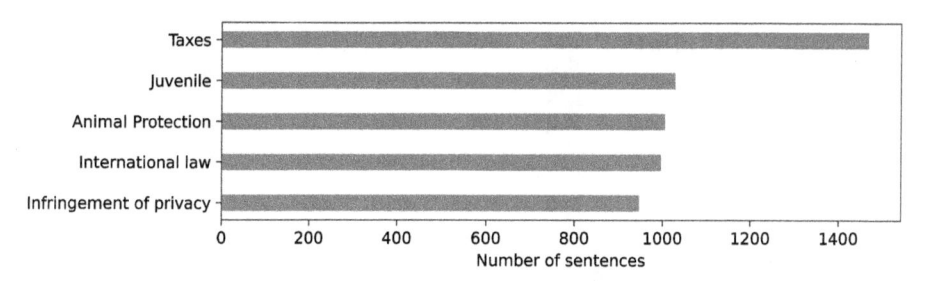

Fig. 1. Number of sentences by category of court judgement

If causality exists then:

- *Implicit causality/Explicit causality* whether causality is explicit or non-explicit,
- *Single cause/Multiple causes* when there is only one, or many different causes,
- *Single effect/Multiple effects* when there is only one, or many different effects,
- *Event chain* if sentence contains chain of causality,
- $c_1 \Leftrightarrow e_1\ relation/c_1 \Leftrightarrow \neg e_1\ relation/\neg c_1 \Leftrightarrow \neg e_1\ relation$ relation by type of constituent parts,
- *Cause before effect/cause together with effect* according to the timing of cause and effect.

If explicit causality exists then:

- *Marked/Unmarked* if the sentence contains phrase suggesting causality,
- *Ambiguous causal phrase/non ambiguous causal phrase* if a phrase suggesting causality does so only depending on the context or almost always.

The annotation process took place in two stages. First, we annotated the entire dataset, then we performed verification on the same dataset, but without access to the previously added tags. By comparing sentences whose tags differed between the two processes, we only added those that were more appropriate. In the process of preparing the collection for annotation, all the anonymised data referred to above were replaced by the tokens "ENTITY_0", "ENTITY_1", etc. depending on the number of unique abbreviations present in the sentence.

The sentences, which are the components of the judgments, have a specific, very formal structure in which words and phrases specific to the legal language are found (including often possessing phrases containing Latin). The detailed number of sentences in each category is shown in Fig. 1. As we can infer from it, this set is largely unbalanced in terms of the number of sentences with a causal link to those without. Nevertheless, this provides an indication that sentences that contain causality comprise a large proportion of all sentences. Among the items in the test set, sentences with a character count between 100 and 200 were the largest group (Fig. 2). Due to the fact that legal texts often contain very elaborate sentences, there is also a large representation of samples with a much higher number of characters. In each type of category, there is a certain number

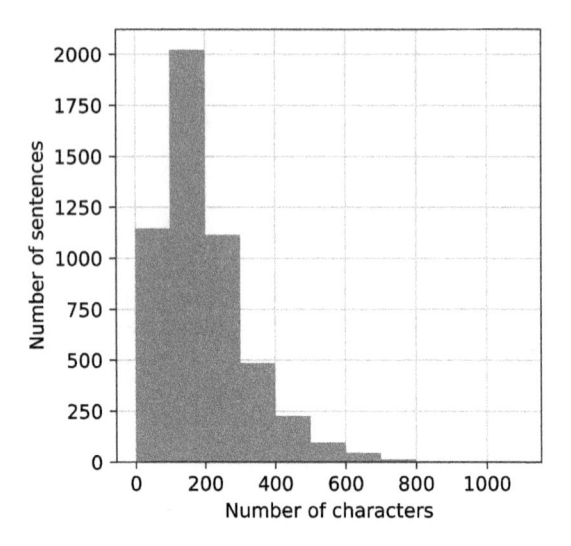

Fig. 2. Number of sentences with the selected character range

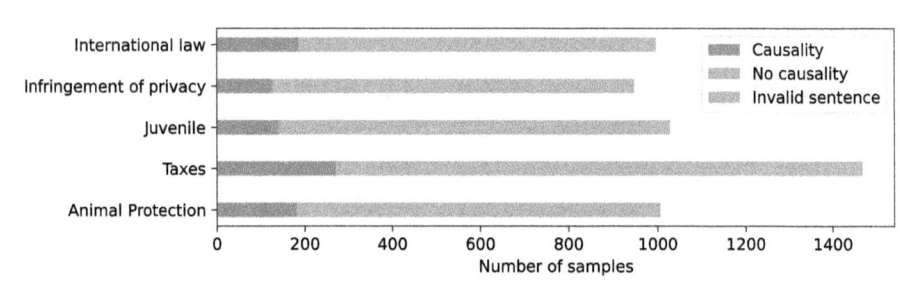

Fig. 3. Number of samples with Causality, No causality and Invalid sentence tags by category

of sentences with causality, without it, and those that are not correct sentences, so they will not be analysed further (such sentences in the entire set are about 5.25%). A detailed division by category is shown in Fig. 3.

In the dataset, each sentence was assigned appropriate tags as described above. The detailed number of elements of a given class, together with the percentage, is shown in Fig. 4. The collection for later use has been exported to JSON format, which is the input for the programmes using the selected machine learning models described in the next chapter.

Sentences with explicit causality, as described above, often have words or phrases that suggest the existence of such a relationship. Figure 5 shows a list of the most common words found in this type of sentence. These words, such as "gdyż" (because), "wynika" (follows) or "jeżeli" (if) are typical phrases from which causality is implied (other words, popular for a specific variety of language are also found).

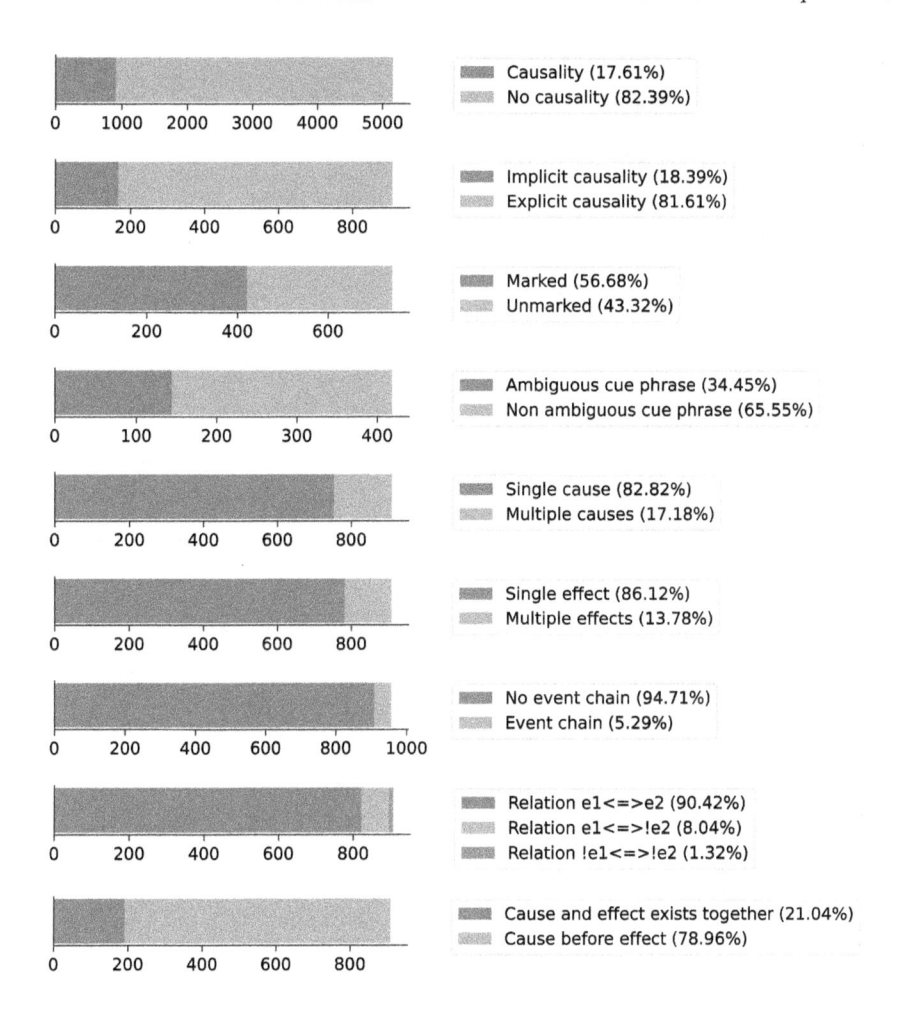

Fig. 4. Division of samples by class

5 Experiments

The main goal of the experiment was to detect sentences with a causal relationship, with a distinction between implicit and explicit causality. For this purpose, different classification methods were prepared and after a training process, they were appropriately evaluated using standard metrics such as Accuracy (ACC), Precision (P), Recall (R) and F1 measure. Classifiers from two categories were used for this purpose: machine learning-based and neural network-based. In each case, we dealt with binary classification, in the following classes (the first class is treated as positive and the second as negative):

- **(E1)** Causality/No Causality,
- **(E2)** Implicit causality/Explicit causality,

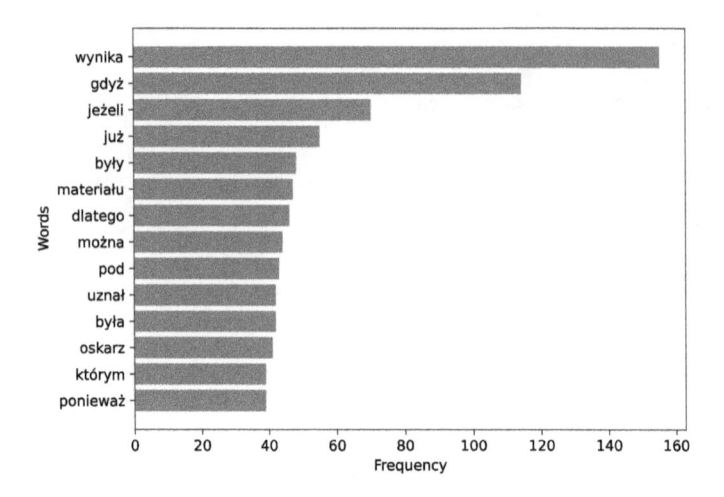

Fig. 5. The most popular words along with the number of occurrences in sentences containing causality relation.

- **(E3)** Marked/Unmarked,
- **(E4)** Ambiguous causal phrase/non ambiguous causal phrase,
- **(E6)** Single cause/Multiple causes,
- **(E6)** Single effect/Multiple effects,
- **(E7)** Event chain/No event chain,
- **(E8)** $c_1 \Leftrightarrow e_1$ relation/$c_1 \Leftrightarrow \neg e_1$, relation (due to insufficient number of samples, the class $\neg c_1 \Leftrightarrow \neg e_1$ was omitted),
- **(E9)** Cause before effect/cause together with effect.

Nine different experiments were therefore conducted, each with 16 different models. Because the sets were not balanced, to equalize the number of elements of the classes during training, the number of samples per class was equalized to the number of elements from the least numerous class during a given experiment. The equalization involved selecting random values from the dataset using default mechanisms contained in the Scikit-learn library [17].

5.1 Machine Learning-Based Methods

The first type of classifiers were machine learning models trained in a supervised manner based on a labeled dataset. The following models were used: Multinomial Naive Bayes (MNB), Support Vector Machine (SVM), K-Nearest Neighbour (KNN), Random Forest (RF) and XGBoost (XGB). The data were prepared in three formats: Bag of Words (BoW), Term Frequency-Inverse Document Frequency (TF-IDF) and in the form of dense vectors (word embeddings, WE) preprepared FestText vector sets for Polish, trained using the CBOW technique, stored in 300 dimensions using n-grams of characters of length 5 and windows size equals 10 negatives [7]. A detailed list of hyperparameters is presented in Table 1.

Table 1. Hyperparameters of the models used in the experiments

Model	Input	Hyperparameters
MNB	BoW, TF-IDF	alpha: 1, fit_prior: True
SVM	BoW, TF-IDF	c: 1, kernel: rbf, degree: 3, gamma: scale, tol: 1e-3
KNN	BoW	n neighbors: 20, algorithm: ball_tree, leaf_size: 30
RF	BoW	n estimators: 100, criterion: gini, max features: sqrt
XGB	BoW, WE	n estimators: 1000, subsample: 0.8, early stopping rounds: 10
SN	WE	optimizer: adam, loss: Binary Crossentropy
CNN	WE	optimizer: RMSprop (lr: 1e-4), loss: Binary Crossentropy
BiLSTM	WE	optimizer: adam, loss: Binary Crossentropy, lstm units: 32
Transformer	WE	embed_dim: 32, num_heads: 2, ff_dim: 32, maxlen: 300, optimizer: adam, learning_rate: 1e-4, loss: Binary Crossentropy
DistilBERT	BERT tokens	104 languages, embed_dim: 768, hidden_layers: 7, num_heads: 12
RoBERTa	BERT tokens	Polish only, embed_dim: 768, hidden_layers: 12, num_heads: 12
HerBERT	BERT tokens	Polish only, embed_dim: 768, hidden_layers: 12, num_heads: 12
Polbert	BERT tokens	Polish only, embed_dim: 768, hidden_layers: 12, num_heads: 12

5.2 Neural Network-Based Methods

The second type of classifiers are methods based on selected neural networks. Several popular architectures based on such networks were chosen for the experiment, such as: Shallow Neural Network (SN, only with one hidden layer), Convolutional Neural Network (CNN), Bidirectional Long Short-Term Memory Network (BiLSTM), Transformer Network (TN). In each of these cases, the input data had the format of word embeddings vectors (the same as described in the section on machine learning methods).

In addition, other models based on Bidirectional Encoder Representations from Transformers (BERT), prepared for the Polish language on which the finetuning process was carried out, were also used: DistilBERT (base, multilingual, cased) [22], Polish RoBERTa v2 (large) [1], HerBERT (base, cased) [14], Polbert (base, uncased) [10]. All these models were used with the *Simple Transformers* library [19]. The models were selected based on the results of the KLEJ benchmark [21] (the GLUE equivalent for English models [24]). As for the previous models, the hyperparameters of the models are detailed in Table 1.

Table 2. Results of the Experiment 1 (E1)

Model	AC	Macro P	Macro R	Macro F1	Causality P	Causality R	Causality F1	No causality P	No causality R	No causality F1
NB (BoW)	0.68	0.71	0.68	0.67	0.63	0.86	0.73	0.78	0.51	0.61
NB (TF-IDF)	0.66	0.69	0.66	0.64	0.61	0.87	0.72	0.78	0.44	0.56
SVM (BoW)	0.74	0.74	0.74	0.74	0.74	0.73	0.73	0.73	0.75	0.74
SVM (TF-IDF)	0.75	0.75	0.75	0.75	0.78	0.69	0.73	0.72	0.80	0.76
KNN	0.54	0.71	0.54	0.43	**0.91**	0.10	0.17	0.52	**0.99**	0.68
RF	0.77	0.77	0.77	0.77	0.81	0.71	0.76	0.74	0.83	0.78
XGB (BoW)	0.79	0.79	0.79	0.78	0.83	0.72	0.77	0.75	0.85	0.80
XGB (WE)	0.85	0.86	0.85	0.85	0.81	**0.93**	0.86	**0.91**	0.78	0.84
SN	0.73	0.73	0.73	0.73	0.73	0.73	0.73	0.73	0.74	0.73
CNN	0.70	0.74	0.70	0.69	0.65	0.89	0.75	0.82	0.51	0.63
BiLSTM	0.73	0.73	0.73	0.73	0.74	0.71	0.72	0.72	0.75	0.74
TN	0.72	0.74	0.72	0.72	0.68	0.85	0.75	0.79	0.59	0.68
DistilBERT	0.82	0.82	0.82	0.82	0.80	0.85	0.82	0.84	0.79	0.81
RoBERTa	**0.86**	**0.87**	**0.86**	**0.86**	0.83	0.92	**0.87**	0.91	0.81	**0.86**
HerBERT	0.85	0.85	0.85	0.85	0.81	0.91	0.86	0.89	0.79	0.84
Polbert	0.79	0.79	0.79	0.78	0.77	0.82	0.79	0.80	0.75	0.78

Causal words or phrases are a great help in identifying obvious causation. In the case of legal texts, such words differ to some extent from words used in informal language. For example, the word "albowiem" (because, but very formal) may or may not be used in causal contexts. Models can also fail when a sentence contains a cause-effect relationship, but either the cause or effect is split across multiple sentences. This can result in a lack of relationship at the sentence level, highlighting the need to study such links at a larger level than just one sentence. Considering the complex sentence structures commonly found in legal documents, it is important to keep in mind that these constructions can often be lengthy due to the presence of multiple subordinate clauses.

6 Results

The main experiment (E1) was the detection of sentences with cause-and-effect relationships. Table 2 shows the detailed results for this task by model. Depending on the model, accuracy scores range from 0.54 to 0.87. The best performer

Table 3. Summary results of accuracy in experiments E2-E9

Model	E2	E3	E4	E5	E6	E7	E8	E9
NB (BoW)	0.65	0.67	0.65	0.64	0.62	0.66	0.64	0.65
NB (TF-IDF)	0.66	0.65	0.61	0.63	0.60	0.66	0.64	0.65
SVM (BoW)	0.77	0.80	0.75	0.74	0.71	0.88	0.73	0.72
SVM (TF-IDF)	0.76	0.77	0.77	0.71	0.70	0.78	0.72	0.73
KNN	0.52	0.54	0.52	0.54	0.53	0.56	0.53	0.50
RF	0.78	0.83	0.80	0.75	0.74	0.88	0.75	0.75
XGB (BoW)	0.83	**0.93**	**0.90**	0.76	0.75	**0.91**	0.76	0.72
XGB (WE)	0.84	0.84	0.89	0.80	0.79	0.91	0.83	**0.83**
SN	0.76	0.77	0.50	0.71	0.68	0.50	0.73	0.50
CNN	0.70	0.65	0.61	0.67	0.68	0.88	0.72	0.63
BiLSTM	0.73	0.77	0.50	0.74	0.68	0.50	0.72	0.50
TN	0.68	0.75	0.50	0.71	0.56	0.50	0.71	0.50
DistilBERT	0.85	0.84	0.72	0.76	0.80	0.72	0.84	0.65
RoBERTa	**0.90**	0.91	0.84	**0.86**	**0.86**	0.84	**0.88**	0.76
HerBERT	0.86	0.91	0.81	0.85	0.81	0.56	0.85	0.65
Polbert	0.80	0.78	0.80	0.74	0.79	0.75	0.80	0.65

was the RoBERTa model, which presented good results for both the class with and without causality. Similar results were obtained by the XGBoost model based on word embeddings vectors. Detecting causality in legal texts is therefore a possible task, although quite difficult in the case of implicit causality. In the case of the latter experiment (E2), the results are also solid, but also in this case, the RoBERTa model was the best, achieving an accuracy of 0.90. In the results, we can also notice a regularity that models based on BERT trained using only Polish texts perform noticeably better than the multilingual model. In the case of models based on machine learning, there is not much difference between the results of models based on BoW or TF-IDF. When evaluating such models, we should keep in mind that if we wanted to use this type of data in further experiments (e.g. extraction of such compounds), we should focus on better Recall results than Precision, due to the cascade errors described earlier. The situation is similar for the other experiments (E3-9). The results for accuracy for the experiments are shown in Table 3. Detailed results for both classes are shown in Table 4.

Table 4. Detailed results of experiments E2-E9

Experiment 2	Macro P	R	F1	Positive P	R	F1	Negative P	R	F1	Experiment 3	Macro P	R	F1	Positive P	R	F1	Negative P	R	F1
NB (BoW)	0.70	0.65	0.62	0.60	0.91	0.72	0.81	0.39	0.52	NB (BoW)	0.70	0.67	0.66	0.62	0.86	0.72	0.77	0.48	0.59
NB (TF-IDF)	0.73	0.66	0.63	0.60	0.93	0.73	0.85	0.38	0.52	NB (TF-IDF)	0.70	0.65	0.63	0.60	0.90	0.72	0.80	0.41	0.54
SVM (BoW)	0.77	0.77	0.77	0.77	0.78	0.77	0.77	0.76	0.77	SVM (BoW)	0.80	0.80	0.80	0.78	0.83	0.80	0.82	0.77	0.79
SVM (TF-IDF)	0.76	0.76	0.76	0.77	0.75	0.76	0.75	0.78	0.77	SVM (TF-IDF)	0.77	0.77	0.77	0.78	0.77	0.77	0.77	0.78	0.77
KNN	0.72	0.52	0.39	**0.93**	0.05	0.10	0.51	**1.00**	0.68	KNN	0.73	0.54	0.43	**0.93**	0.09	0.17	0.52	**0.99**	0.68
RF	0.78	0.78	0.78	0.80	0.75	0.77	0.77	0.81	0.79	RF	0.83	0.83	0.83	0.82	0.84	0.83	0.84	0.82	0.83
XGB (BoW)	0.84	0.83	0.83	0.85	0.82	0.83	0.82	0.85	0.84	XGB (BoW)	**0.93**	**0.93**	**0.93**	0.93	0.93	**0.93**	0.93	0.93	**0.93**
XGB (WE)	0.85	0.84	0.84	0.81	0.91	0.85	0.90	0.78	0.83	XGB (WE)	0.86	0.84	0.84	0.78	0.94	0.86	0.93	0.74	0.82
SN	0.76	0.76	0.76	0.76	0.75	0.76	0.76	0.77	0.76	SN	0.77	0.77	0.77	0.78	0.76	0.77	0.77	0.78	0.78
CNN	0.74	0.70	0.68	0.64	0.92	0.75	0.85	0.47	0.61	CNN	0.76	0.65	0.60	0.59	**0.98**	0.74	0.94	0.32	0.47
BiLSTM	0.77	0.73	0.73	0.67	0.91	0.77	0.86	0.56	0.68	BiLSTM	0.77	0.77	0.77	0.81	0.71	0.75	0.74	0.83	0.78
TN	0.71	0.68	0.66	0.79	0.48	0.60	0.63	0.87	0.73	TN	0.75	0.75	0.75	0.75	0.76	0.75	0.75	0.75	0.75
DistilBERT	0.85	0.85	0.85	0.83	0.88	0.86	0.87	0.82	0.85	DistilBERT	0.84	0.84	0.84	0.82	0.87	0.85	0.86	0.81	0.84
RoBERTa	**0.91**	**0.90**	**0.90**	0.86	**0.96**	**0.91**	**0.96**	0.84	**0.89**	RoBERTa	0.91	0.91	0.91	0.87	0.96	0.91	**0.96**	0.86	0.90
HerBERT	0.87	0.86	0.86	0.80	0.96	0.87	0.94	0.77	0.85	HerBERT	0.91	0.91	0.91	0.91	0.91	0.91	0.91	0.91	0.91
Polbert	0.80	0.80	0.80	0.78	0.84	0.81	0.83	0.76	0.79	Polbert	0.78	0.78	0.78	0.80	0.74	0.77	0.76	0.82	0.79
Experiment 4	Macro P	R	F1	Positive P	R	F1	Negative P	R	F1	Experiment 5	Macro P	R	F1	Positive P	R	F1	Negative P	R	F1
NB (BoW)	0.68	0.65	0.63	0.60	0.85	0.71	0.75	0.44	0.55	NB (BoW)	0.66	0.64	0.64	0.71	0.49	0.58	0.61	0.80	0.69
NB (TF-IDF)	0.66	0.61	0.59	0.58	0.88	0.69	0.74	0.35	0.48	NB (TF-IDF)	0.66	0.63	0.61	0.73	0.41	0.52	0.59	**0.85**	0.69
SVM (BoW)	0.76	0.75	0.75	0.71	0.83	0.77	0.80	0.67	0.73	SVM (BoW)	0.74	0.74	0.74	0.76	0.71	0.74	0.73	0.77	0.75
SVM (TF-IDF)	0.77	0.77	0.77	0.77	0.77	0.77	0.77	0.77	0.77	SVM (TF-IDF)	0.71	0.71	0.71	0.71	0.73	0.72	0.72	0.70	0.71
KNN	0.76	0.52	0.38	**1.00**	0.04	0.08	0.51	**1.00**	0.68	KNN	0.76	0.53	0.41	0.52	**1.00**	0.68	**1.00**	0.07	0.13
RF	0.81	0.80	0.80	0.84	0.75	0.79	0.77	0.85	0.81	RF	0.75	0.75	0.75	0.72	0.83	0.77	0.79	0.67	0.73
XGB (BoW)	**0.90**	**0.90**	**0.90**	0.93	0.85	**0.89**	0.87	0.94	**0.90**	XGB (BoW)	0.78	0.76	0.76	0.72	0.88	0.79	0.84	0.65	0.73
XGB (WE)	0.89	0.89	0.89	0.86	0.92	0.89	0.91	0.85	0.88	XGB (WE)	0.80	0.80	0.80	0.78	0.84	0.81	0.82	0.76	0.79
SN	0.25	0.50	0.33	0.50	**1.00**	0.67	0.00	0.00	0.00	SN	0.71	0.71	0.71	0.69	0.78	0.73	0.74	0.64	0.69
CNN	0.78	0.61	0.55	0.56	1.00	0.72	**1.00**	0.23	0.37	CNN	0.68	0.67	0.66	0.74	0.53	0.61	0.63	0.81	0.71
BiLSTM	0.25	0.50	0.33	0.50	1.00	0.67	0.00	0.00	0.00	BiLSTM	0.75	0.74	0.74	0.78	0.67	0.72	0.71	0.81	0.76
TN	0.25	0.50	0.33	0.50	1.00	0.67	0.00	0.00	0.00	TN	0.71	0.71	0.71	0.72	0.71	0.71	0.71	0.72	0.71
DistilBERT	0.72	0.72	0.72	0.72	0.71	0.72	0.71	0.73	0.72	DistilBERT	0.76	0.76	0.76	0.74	0.79	0.77	0.78	0.72	0.75
RoBERTa	0.85	0.84	0.84	0.81	0.90	0.85	0.88	0.79	0.84	RoBERTa	**0.87**	**0.86**	**0.86**	0.84	0.90	**0.87**	0.90	0.82	**0.86**
HerBERT	0.81	0.81	0.81	0.81	0.81	0.81	0.81	0.81	0.81	HerBERT	0.85	0.85	0.84	0.83	0.88	0.85	0.87	0.82	0.84
Polbert	0.80	0.80	0.80	0.81	0.79	0.80	0.80	0.81	0.80	Polbert	0.74	0.74	0.74	0.74	0.76	0.75	0.75	0.73	0.74
Experiment 6	Macro P	R	F1	Positive P	R	F1	Negative P	R	F1	Experiment 7	Macro P	R	F1	Positive P	R	F1	Negative P	R	F1
NB (BoW)	0.64	0.62	0.60	0.70	0.42	0.52	0.58	0.82	0.68	NB (BoW)	0.69	0.66	0.64	0.61	0.88	0.72	0.78	0.44	0.56
NB (TF-IDF)	0.63	0.60	0.57	0.70	0.35	0.47	0.57	0.85	0.68	NB (TF-IDF)	0.80	0.66	0.61	0.59	**1.00**	0.74	**1.00**	0.31	0.48
SVM (BoW)	0.71	0.71	0.71	0.73	0.67	0.70	0.69	0.75	0.72	SVM (BoW)	0.88	0.88	0.88	0.88	0.88	0.88	0.88	0.88	0.88
SVM (TF-IDF)	0.70	0.70	0.70	0.70	0.70	0.70	0.70	0.69	0.70	SVM (TF-IDF)	0.79	0.78	0.78	0.85	0.69	0.76	0.74	0.88	0.80
KNN	0.66	0.53	0.40	0.51	**0.98**	0.68	0.81	0.07	0.12	KNN	0.77	0.56	0.46	**1.00**	0.13	0.22	0.53	**1.00**	0.70
RF	0.74	0.73	0.73	0.72	0.77	0.74	0.75	0.70	0.72	RF	0.88	0.88	0.87	0.93	0.81	0.87	0.83	0.94	0.88
XGB (BoW)	0.75	0.75	0.75	0.73	0.79	0.76	0.77	0.71	0.74	XGB (BoW)	**0.92**	**0.91**	**0.91**	0.84	1.00	**0.91**	1.00	0.81	**0.90**
XGB (WE)	0.80	0.79	0.79	0.78	0.82	0.80	0.81	0.77	0.79	XGB (WE)	0.92	0.91	0.91	0.84	1.00	0.91	1.00	0.81	0.90
SN	0.69	0.68	0.68	0.67	0.72	0.70	0.70	0.65	0.67	SN	0.25	0.50	0.33	0.50	1.00	0.67	0.00	0.00	0.00
CNN	0.69	0.68	0.68	0.73	0.58	0.64	0.65	0.79	0.71	CNN	0.90	0.88	0.87	0.80	1.00	0.89	1.00	0.75	0.86
BiLSTM	0.69	0.68	0.68	0.66	0.76	0.70	0.71	0.60	0.65	BiLSTM	0.25	0.50	0.33	0.50	1.00	0.67	0.00	0.00	0.00
TN	0.61	0.55	0.49	0.53	0.91	0.67	0.68	0.20	0.31	TN	0.50	0.50	0.38	0.50	0.06	0.11	0.50	0.94	0.65
DistilBERT	0.80	0.80	0.80	0.80	0.80	0.80	0.80	0.80	0.80	DistilBERT	0.72	0.72	0.72	0.73	0.69	0.71	0.71	0.75	0.73
RoBERTa	**0.86**	**0.86**	**0.86**	0.86	0.87	**0.87**	0.87	0.86	**0.86**	RoBERTa	0.88	0.84	0.84	0.76	1.00	0.86	1.00	0.69	0.81
HerBERT	0.81	0.81	0.81	0.78	0.85	0.82	0.84	0.77	0.80	HerBERT	0.77	0.56	0.46	1.00	0.13	0.22	0.53	1.00	0.70
Polbert	0.79	0.79	0.79	0.79	0.79	0.79	0.79	0.79	0.79	Polbert	0.83	0.75	0.73	1.00	0.50	0.67	0.67	1.00	0.80
Experiment 8	Macro P	R	F1	Positive P	R	F1	Negative P	R	F1	Experiment 9	Macro P	R	F1	Positive P	R	F1	Negative P	R	F1
NB (BoW)	0.67	0.64	0.63	0.74	0.45	0.56	0.60	0.84	0.70	NB (BoW)	0.66	0.64	0.63	0.61	0.81	0.70	0.71	0.48	0.57
NB (TF-IDF)	0.68	0.64	0.61	0.77	0.39	0.52	0.59	**0.89**	0.71	NB (TF-IDF)	0.67	0.64	0.63	0.61	0.83	0.70	0.72	0.46	0.56
SVM (BoW)	0.73	0.73	0.73	0.74	0.72	0.73	0.73	0.75	0.74	SVM (BoW)	0.72	0.72	0.72	0.73	0.73	0.73	0.73	0.71	0.72
SVM (TF-IDF)	0.72	0.72	0.72	0.70	0.75	0.73	0.73	0.68	0.70	SVM (TF-IDF)	0.73	0.73	0.73	0.73	0.73	0.73	0.73	0.73	0.73
KNN	0.73	0.53	0.41	0.52	**1.00**	0.68	**0.95**	0.07	0.13	KNN	0.75	0.51	0.35	**1.00**	0.02	0.03	0.50	**1.00**	0.67
RF	0.75	0.75	0.75	0.74	0.78	0.76	0.77	0.72	0.74	RF	0.76	0.75	0.74	0.83	0.63	0.71	0.70	0.87	0.77
XGB (BoW)	0.76	0.76	0.76	0.76	0.77	0.76	0.76	0.76	0.76	XGB (BoW)	0.72	0.72	0.72	0.72	0.73	0.73	0.73	0.71	0.72
XGB (WE)	0.83	0.83	0.83	0.84	0.80	0.82	0.81	0.85	0.83	XGB (WE)	**0.84**	0.83	**0.83**	0.79	**0.91**	0.85	0.89	0.76	**0.82**
SN	0.73	0.73	0.73	0.73	0.73	0.73	0.73	0.73	0.73	SN	0.25	0.50	0.33	0.00	0.00	0.00	0.50	1.00	0.66
CNN	0.73	0.72	0.72	0.78	0.61	0.69	0.68	0.83	0.75	CNN	0.72	0.63	0.59	0.87	0.31	0.46	0.58	0.95	0.72
BiLSTM	0.73	0.72	0.72	0.75	0.66	0.71	0.70	0.78	0.74	BiLSTM	0.25	0.50	0.33	0.00	0.00	0.00	0.50	1.00	0.66
TN	0.72	0.71	0.71	0.75	0.65	0.69	0.69	0.78	0.73	TN	0.25	0.50	0.33	0.00	0.00	0.00	0.50	1.00	0.66
DistilBERT	0.84	0.84	0.84	0.84	0.83	0.83	0.83	0.84	0.84	DistilBERT	0.65	0.65	0.65	0.63	0.68	0.66	0.66	0.61	0.63
RoBERTa	**0.88**	**0.88**	**0.88**	0.86	0.91	**0.88**	0.90	0.86	**0.88**	RoBERTa	0.77	0.76	0.76	0.73	0.84	0.78	0.81	0.69	0.75
HerBERT	0.85	0.85	0.85	0.83	0.89	0.86	0.88	0.81	0.84	HerBERT	0.65	0.65	0.64	0.67	0.57	0.62	0.63	0.72	0.67
Polbert	0.80	0.80	0.80	0.80	0.80	0.80	0.80	0.80	0.80	Polbert	0.66	0.65	0.65	0.63	0.75	0.68	0.69	0.56	0.62

7 Conclusions

The main issue with this type of analysis, in terms of causality, is the absence of a suitable dataset, particularly for languages other than English. Therefore, it is essential to create such a collection yourself, which can be a time-consuming task, given the chosen domain of texts, such as legal texts.

Excluding legal texts from analysis can facilitate the process, but it can also create issues when attempting to generalize methods. It is important to note that this collection contains various types of documents, including statutory texts and court judgments, which may differ significantly in their use of formal language. However, court judgments often share a similar structure, typically including the same components, such as the operative part, grounds, and referenced provisions.

Upon analysis of the collection, it can be concluded that although causality sentences make up less than 20% of the collection, they convey crucial information for future legal analysis. Additionally, determining the type of causality can provide significant information, particularly in extracting the constituent parts of such compounds. Explicit causality is the main form in which causality occurs. However, detecting implicit causality can be challenging due to the ambiguity involved. Our empirical findings confirm that detecting causality in sentences from real-world data, such as court judgments, can be achieved with a satisfactory F1 index score of 86%.

8 Future Work

After a thorough analysis of the relationships present in legal texts, it is worth considering the extraction of the components of these relationships. The extracted parts, including causes and effects, can provide important data for further analysis, however, it is necessary to construct a specific models to identify these components. The prevailing view in current research is that it is important to develop general methods that can be applied to different fields. However, it is worth noting that in some fields, such as medicine or law, the languages used are so specific that they can pose a significant challenge for this type of analysis. It is therefore crucial to carry out research dedicated to specific fields. Another important step is to identify causality relations between a wider spectrum of sentences. Often sentences in close neighborhood exhibit relationships that are not discernible through single-sentence analysis. Models using the attention mechanism can therefore be used for this purpose, in such a way as to find connections between fragments of text located further apart.

Disclosure of Interests. The authors have no competing interests to declare that are relevant to the content of this article.

References

1. Dadas, S., Perełkiewicz, M., Poświata, R.: Pre-training polish transformer-based language models at scale. In: Artificial Intelligence and Soft Computing, pp. 301–314. Springer International Publishing (2020). https://doi.org/10.48550/arXiv.2006.04229
2. Dasgupta, T., Saha, R., Dey, L., Naskar, A.: Automatic extraction of causal relations from text using linguistically informed deep neural networks. In: Komatani, K., Litman, D., Yu, K., Papangelis, A., Cavedon, L., Nakano, M. (eds.) Proceedings of the 19th Annual SIGdial Meeting on Discourse and Dialogue, pp. 306–316. Association for Computational Linguistics, Melbourne, Australia (2018). https://doi.org/10.18653/v1/W18-5035
3. Fischbach, J., Frattini, J., Méndez, D., Unterkalmsteiner, M., Femmer, H., Vogelsang, A.: How do practitioners interpret conditionals in requirements? CoRR arxiv:abs/2109.02063 (2021). https://doi.org/10.48550/arXiv.2109.02063
4. Frattini, J., Fischbach, J., Mendez, D., Unterkalmsteiner, M., Vogelsang, A., Wnuk, K.: Causality in requirements artifacts: prevalence, detection, and impact. Requirements Eng. **28**(1), 49–74 (2023). https://doi.org/10.1007/s00766-022-00371-x
5. Girju, R.: Automatic detection of causal relations for question answering. In: Proceedings of the ACL 2003 Workshop on Multilingual Summarization and Question Answering, pp. 76–83. Association for Computational Linguistics, Sapporo, Japan (2003). https://doi.org/10.3115/1119312.1119322
6. Girju, R., Moldovan, D.: Text mining for causal relations. In: Proceedings of the Fifteenth International Florida Artificial Intelligence Research Society Conference, pp. 360–364. The Florida AI Research Society (2002)
7. Grave, E., Bojanowski, P., Gupta, P., Joulin, A., Mikolov, T.: Learning word vectors for 157 languages. In: Proceedings of the International Conference on Language Resources and Evaluation (LREC 2018) (2018). https://doi.org/10.48550/arXiv.1802.06893
8. Keskes, I., Zitoune, F.B., Belguith, L.H.: Learning explicit and implicit Arabic discourse relations. J. King Saud Univ. Comput. Inf. Sci. **26**(4), 398–416 (2014). https://doi.org/10.1016/j.jksuci.2014.06.001
9. Kurant, Ł.: Mechanism for detecting cause-and-effect relationships in court judgments. In: Proceedings of the 18th Conference on Computer Science and Intelligence Systems. vol. 35, pp. 1041–1046. ACSIS (2023). https://doi.org/10.15439/2023F4827
10. Kłeczek, D.: PolBERT - polish BERT (2020). https://huggingface.co/dkleczek/bert-base-polish-uncased-v1
11. Li, Z., Li, Q., Zou, X., Ren, J.: Causality extraction based on self-attentive BiLSTM-CRF with transferred embeddings. CoRR arxiv:abs/1904.07629 (2019). http://arxiv.org/abs/1904.07629
12. Mavin, A., Wilkinson, P., Harwood, A., Novak, M.: Easy approach to requirements syntax (ears). In: Requirements Engineering Conference, 2009. RE '09. 17th IEEE International, pp. 317–322 (2009). https://doi.org/10.1109/RE.2009.9
13. Mostafazadeh, N., Grealish, A., Chambers, N., Allen, J., Vanderwende, L.: CaTeRS: causal and temporal relation scheme for semantic annotation of event structures. In: Palmer, M., Hovy, E., Mitamura, T., O'Gorman, T. (eds.) Proceedings of the Fourth Workshop on Events, pp. 51–61. Association for Computational Linguistics, San Diego, California (2016). https://doi.org/10.18653/v1/W16-1007

14. Mroczkowski, R., Rybak, P., Wróblewska, A., Gawlik, I.: HerBERT: efficiently pretrained transformer-based language model for Polish. In: Proceedings of the 8th Workshop on Balto-Slavic Natural Language Processing, pp. 1–10. Association for Computational Linguistics, Kiyv, Ukraine (2021). https://www.aclweb.org/anthology/2021.bsnlp-1.1
15. Nakayama, H., Kubo, T., Kamura, J., Taniguchi, Y., Liang, X.: Doccano: Text annotation tool for human (2018). https://github.com/doccano/doccano
16. NLTK: Natural language toolkit (2023). https://www.nltk.org. Accessed 12 Jan 2024
17. Pedregosa, F., et al.: Scikit-learn: machine learning in Python. J. Mach. Learn. Res. **12**, 2825–2830 (2011)
18. Pustejovsky, J., et al.: TimeML: Robust specification of event and temporal expressions in text, pp. 28–34 (2003)
19. Rajapakse, T.C.: Simple transformers. https://github.com/ThilinaRajapakse/simpletransformers (2019). Accessed 12 Jan 2024
20. Richardson, L.: Beautiful soup python library (2004-2023). https://www.crummy.com/software/BeautifulSoup/bs4/doc/. Accessed 12 Jan 2024
21. Rybak, P., Mroczkowski, R., Tracz, J., Gawlik, I.: KLEJ: comprehensive benchmark for polish language understanding. CoRR arxiv:abs/2005.00630 (2020). https://doi.org/10.48550/arXiv.2005.00630
22. Sanh, V., Debut, L., Chaumond, J., Wolf, T.: DistilBERT, a distilled version of BERT: smaller, faster, cheaper and lighter. ArXiv arxiv:abs/1910.01108 (2019). https://doi.org/10.48550/arXiv.1910.01108
23. Sprawiedliwości, M.: Portal orzeczeń sądów powszechnych (2012-2024). https://orzeczenia.ms.gov.pl. Accessed 04 Nov 2023
24. Wang, A., Singh, A., Michael, J., Hill, F., Levy, O., Bowman, S.R.: GLUE: A multi-task benchmark and analysis platform for natural language understanding (2018)
25. Yang, J., Han, S.C., Poon, J.: A survey on extraction of causal relations from natural language text. CoRR arxiv:abs/2101.06426 (2021). https://doi.org/10.48550/arXiv.2101.06426

A Parameter-Free Model for the Online Spread of Far-Right Messages: Combining Agent-Based Models with Large-Language Models

Stephen Zhong[1], Nathalie Japkowicz[1], Frédéric Amblard[2], and Philippe J. Giabbanelli[3](✉)

[1] Department of Computer Science, American University, Washington, DC 20016, USA
{sz8367a,japkowic}@american.edu
[2] IRIT, Universite de Toulouse, Toulouse, France
frederic.amblard@ut-capitole.fr
[3] Virginia Modeling, Analysis, and Simulation Center (VMASC), Old Dominion University, Suffolk, VA 23435, USA
pgiabban@odu.edu

Abstract. Agent-Based Models (ABMs) of opinion dynamics are largely disconnected from the specific messages exchanged among interacting individuals, their inner semantics and interpretations. Rather, ABMs often abstract this aspect through corresponding numerical values (e.g., -1 as against and $+1$ as totally in favor). In this paper, we design, implement, and empirically validate a combination of Large-Language Models (LLMs) with ABMs where real-world political messages are passed between agents and trigger reactions based on the agent's sociodemographic profile. Our computational experiments combine real-world social network structures, posting frequencies, and extreme-right messages with nationally representative demographics for the U.S. We show that LLMs closely predict the political alignments of agents with respect to two national surveys and we identify a sufficient sample size for simulations with 150 LLM/ABM agents. Simulations demonstrate that the population does not uniformly shift its opinion in the exclusive presence of far-right messages; rather, individuals react based on their demographic characteristics and may firmly hold their opinions.

Keywords: Belief spread · Hybrid Model · Online social network

1 Introduction

The rise of extreme right ideologies on online social media platforms, such as X (formerly Twitter), has become an important phenomenon with profound social and political implications. These ideologies, often characterized by hate speech, misinformation, and polarizing rhetoric, have found fertile ground in the digital age. Algorithms amplify the natural tendency of individuals to prefer more extreme views within their political group [63], particularly by amplifying the political right [27]. Amplification contributes to creating echo chambers that foster radicalization [43], even if the concept of an echo chamber is approached differently across studies [35]. Characterizing the dynamics for the spread of far-right online messages is critical, as the influence of 'e-extremism' [61] extends beyond the virtual realm, contributing to real-world violence [9], the erosion of democratic norms, and the marginalization of vulnerable communities [56].

Modeling the spread of beliefs in online social networks continues to be a fertile area of research, as exemplified by multiple empirical studies at the International Conference on Computational Science [21,32]. Models specialized in the spread of hate speech need to account for several characteristics. First, although the far-right may share some narratives (e.g., collective victimhood), it is composed of extremely heterogeneous organizations [18] and individuals with different motives. Online hate thus varies substantially across users [5]. Given this heterogeneity, several models explicitly represent each individual instead of grouping them into aggregates assumed to behave identically. A commonly used technique is Agent-Based Modeling (ABM) has been particularly used, where each entity has its own attributes and/or rules and interacts with others in a local environment that may be digital or physical [11,39,52]. Second, ideas do not exactly spread like viruses: instead, there is a *gradual* build up in a person's beliefs and attitudes [46]. From a simulation standpoint, models thus often track extremism among individual agents using a numerical scale rather than through categorical states (e.g., susceptible or 'infected' with extreme ideas). Although complex ABMs may not allow us to identify the analytical solutions afforded by simpler mass action models of political extremism [12], they are helpful in identifying tailored solutions [57,58] (e.g., for different user profiles, behaviors, locations), improve accuracy by incorporating spatial and network effects [48], and they can estimate uncertainty by capturing stochasticity at the individual level. Despite these advantages, current ABMs for the spread of far-right messages have two important limitations, summarized as follows.

First, in current models, *agents do not exchange actual texts*; rather, their interactions are abstracted as a stochastic process such as the probability of passing a type of message, or an agent gradually aligns itself on the state most commonly encountered among its peers [42]. A model is a simplification, and this longstanding abstraction of text evades the complexity of text processing while answering important questions. But without the text, we miss an important marker for detecting and preventing violence [15]. Furthermore, without knowing how agents react to specific messages, we cannot estimate the effect of campaigns to debunk specific arguments, such as COVID or election conspiracies [36].

Second, several parameters were created to keep the ABMs simple, such as the 'ease' or 'volatility' at which agents would change opinions [11], the strength at which they would 'influence' others [8], or their 'tolerance' threshold to other opinions [59]. These are called *free parameters* [30] as they *are very difficult to calibrate empirically* [8,11,59] since they do not directly map to a real-world characteristic. Their combined values are calibrated by comparing aggregate outcomes for the overall ABM with expectations, but their individual values cannot be known [30]. For example, there does not exist a general 'ease' at which somebody changes their mind: it depends on the person and the message, among other aspects. Other fields have stressed the importance of empirical grounding for ABMs of social spreads. For instance, a review on innovation diffusion emphasized that the ability to calibrate ABMs from data is instrumental to shift their use from a learning tool onto guiding policy decisions [60]. We thus need ABMs with minimal reliance on parameters that cannot be individually calibrated.

Our main contribution is to address both limitations by avoiding the use of parameters and by supporting the spread of actual messages. This is achieved by combining ABMs with Large-Language Models (LLMs), such that each agent uses a LLM to model its reactions with respect to a specific message based on the agent's key demographics (age, sex, race and ethnicity, educational attainments). In this paper, we design, implement, and assess the hybrid ABM/LLM model on a sample case study using empirical data that includes U.S. demographics, political opinions, and messages from the Truth Social platform.

The remainder of this paper is organized as follows. In Sect. 2, we succinctly cover the design of ABMs for the spread of far-right messages or ideas that marginalize vulnerable social groups, along with the emerging practice of hybrid ABM/LLM models for computational social science. In Sect. 3, we present the design of our model and explain its empirical grounding in representative data sources. Experiments in Sect. 4 show how agents change in reaction to different messages, at several population scales. Finally, we summarize the core limitations of the present model in Sect. 5 and provide directions for extensions. To support replicability, our model and experiments are accessible on a third-party repository at https://osf.io/h8zme/.

2 Background

2.1 Design of ABMs for the Spread of Far-Right Ideas

Building upon foundational work in opinion dynamics from the 1960s [1] and 1970s [14], the past two decades have witnessed a surge in ABM applications to this field, exemplified by seminal contributions such as those by Deffuant et al. [13] and Hegselmann & Krause [47], among many others (see [33] for a comprehensive overview). This rich literature reveals key distinctions among models, which we organize in the five following critical characteristics.

First, most models represent opinions numerically (binary, discrete, or continuous) facilitating straightforward measurement of opinion distances and thus, the quantification of influence processes among agents. Interestingly, this numerical encoding parallels methods used in political science to represent political actors within multidimensional ideological spaces [31, 54].

Second, the number of actors involved in each influence event also varies. While many models focus on peer-to-peer interactions, involving only two agents, other configurations have been investigated. Many-to-one influence, where a single individual is influenced by multiple agents (e.g., averaging opinions in the immediate social environment [47]), and one-to-many influence, better suited for information diffusion [51], are notable examples.

Third, the core process of influence itself forms another significant point of divergence among ABM models. For a considerable period, mimetic influence– the tendency for agents to adopt opinions similar to those observed in their social environment– dominated the literature [16]. However, the inclusion of contrarian dynamics [19] allows to model behaviors such as radicalization or the deliberate distancing of opinions. More sophisticated models integrate both mimetic and contrarian dynamics, making the influence process dependent on the opinion distance between interacting individuals. Agents with similar views converge, while those with dissimilar views diverge even further [28].

Fourth, the *substratum* of social influence–the underlying social structure within which interactions occur– represents another key variable. Early models often employed assumptions of random mixing within the population. More recent work range from abstract network models (e.g., small-world or scale-free networks) to the incorporation of empirically derived social network data [10], often producing more nuanced and realistic results.

Finally, the lack of longitudinal data on individual opinions, often due to sensitivity concerns, limits the evaluation of accuracy and predictive power of ABMs. While macroscopic-level snapshots of opinion distributions are available, the absence of detailed individual-level data restricts comprehensive validation efforts. Experimental data from social psychology and ethology provide partial micro-validation of specific influence mechanisms, but often fall short of capturing complex real-world influence processes, such as the role of media [16].

2.2 Combining ABMs with Large Language Models

Combining ABMs with LLMs creates a *hybrid* (systems) model since it uses a simulation technique along with a technique from another domain [40]. As this specific type of hybrid is relatively new, it goes by different names such as 'LLM-based agents' [20] or 'Generative Agent-Based Modeling' [23]. Several works have either proposed or demonstrated that describing the sociodemographic characteristics [4,45] of agents (i.e., 'conditioning' a prompt or creating a 'persona') can then "leverage the vast data within LLMs to capture human behavior and decision-making [instead of] relying on modelers' assumptions" [23].

A conceptual framework for disinformation research and LLMs proposed to power agents within a social network via LLMs [45]. GPT-4 was viewed as a promising tool to suggest the evolution of opinions given a user profile and exposure to an idea such as electoral fraud. Several of these ideas were realized in a study by Zheng and Tang, released in November 2024 [62], who created a small model where agents interact on Twitter (post, retweet, reply, like). Operating without empirical data, the model illustrated changes in attitudes on an abstraction of the Roe v. Wade case on abortion. Although no statistical analyses were conducted, visualizations suggested that average attitudes may fluctuate over time without ever stabilizing, depending on the synthetic network topology employed (small-world vs. scale-free networks). Opinion diversity (measured as the number of unique opinions) also depended on the topology by increasing in one case (small-world) and decreasing in the other (scale-free) [62].

3 Methods

3.1 Design of the Hybrid ABM/LLM Model

Overview. We initialize each agent in our population with a set of demographic characteristics (age, sex, race and ethnicity, educational attainments) that are partially predictive of their initial opinion score. Agents are connected to mimic the follower relationships on mainstream social media such as Twitter (before becoming X). A set of right-leaning agents post specific messages based on an empirical frequency that accounts for the relation between the amount of posts and the number of followers (i.e., agents with fewer followers post less). In this one-to-many influence model (Sect. 2.1), all followers of the posting agents will read their posts. Followers react immediately upon reading by using a LLM that accounts for the content of the messages and the reader's demographic characteristics The LLM is tasked with *suggesting* a new plausible opinion score, which may become more conservative or more liberal as the LLM integrates contrarian and mimetic dynamics (Sect. 2.1). Since humans do not widely change opinions by reading a single post, the suggested score is compared with the agent's current opinion and leads to a moderate update that follows the empirical literature on gradual changes in opinions. Previous models have shown that dynamics may embrace a chaotic or an oscillatory trajectory [62], thus we end the simulation after a set number of steps rather than stabilization.

Formal Description. The population is modeled as a graph $G = (V, E)$ consisting of a set of users V and the users whom they follow via directed edges E. Each agent $i \in V$ has a constant set of demographic characteristics i_{dem} over the duration of the simulation, and a variable opinion score $i^t_{opi} \in [-10, 10]$ that is updated over discrete time ticks t. At $t = 0$, we use a LLM to initialize the scores based on each agent's demographics, that is, $i^0_{opi} \leftarrow LLM(i_{dem}) \forall i \in V$. Positive scores indicate left-leaning agents and negative scores indicate right-leaning agents. The simulation proceeds for a duration specified by the user.

At each tick $t > 0$, we perform an *asynchronous update* in three steps. First, right-leaning agents $\{i | i \in V, i^t_{opi} > 0\}$ have a probability $P(i) = f(d_{in}(i))$ of posting based on their number of followers, that is, the number of incoming edges $d_{in}(i) = |\{u \in V | (u, i) \in E\}|$. A specific message is chosen at random from a set of far-right messages $\mathbb{M}$. Second, for all agents who post, their followers $\{j | i, j \in V, i^t_{opi} > 0, (j, i) \in E\}$ read and react to the messages. That is, the LLM is tasked with *suggesting* a *plausible numerical* score based on the reader's demographics and the message content, $LLM(j_{dem}, m \in \mathbb{M})$. Due to stochastic variations in the LLM, it may not deliver a numerical score, it may not be within the target range, or it may be an implausibly wide departure from the agent's current opinion. We thus treat $LLM(j_{dem}, m \in \mathbb{M})$ as a random variable F that starts by drawing a sample F' and calls itself again if criteria are not yet met:

$$\begin{cases} F'(j_{dem}, m) & \text{if } F'(j_{dem}, m) \in [-10, 10] \text{ and } |j^t_{\text{opi}} - F'(j_{dem}, m)| \leq 2, \\ F(j_{dem}, m) & \text{otherwise} \end{cases} \quad (1)$$

Unlike in a synchronous update where agents' values are buffered until all have been visited, the asynchronous update computes and updates values in the order in which agents are visited. This order may change at each time tick, as agents are updated in reaction to a stochastic event (following an account that posted). This mechanism thus uses an asynchronous update with random order.

Implementation Considerations. Engineering the prompt is an essential component of a LLM-based system. The prompt used to update an agent is shown in Box 1. We experimented several versions of this wording. Suggesting that the post was sent *by a friend* or *by a peer* had a risk of biasing the LLM in trusting the message, thus we removed any description of the sender. This illustrates the well-documented notion that less can be better in a prompt [37]. We had to engage in 'roleplay' by including *"pretend you are"* in the prompt, otherwise the LLM would state that it is a machine learning model that cannot give an opinion. As it is stochastic (even with a temperature of 0), the LLM may occasionally return an invalid response, such as a variation of "I cannot answer this" that lacks an opinion score. We thus loop queries to the LLM until a satisfactory answer is obtained, in line with other recent works on prompting [53].

Box 1. Prompt to GPT to suggest a new opinion score for an agent.

Role prompt: Pretend you are a «age»-year old «race» «sex» who has completed «education»

Main prompt: Pretend you have a political opinion score «opinion» where -10 is far-right Republican and 10 is far-left Democrat. What is your new opinion score after you see «message» sent to you on social media? Do not explain your reasoning.

For replicability, note that a simulation primarily depends on OpenAI 1.58.1 (for the GPT API), Numpy 1.26.4 to store the agents' attributes and connections (scaling-up since computations with arrays are faster than Python primitive data structures), and re 2.2.1 (to parse the LLM response using regular expression).

3.2 Empirical Data

Social Network. The far right heavily relies on social media, with an established presence on all well-known platforms including Twitter [29]. The use of such a mainstream platform is suitable for our case study, since we simulate the potential spread of far-right ideas among a *general population* rather than only among right-leaning groups who may already endorse some of these ideas (e.g., on the Parler or Truth Social platforms). To support replicability of our simulations, we use a public domain sample of Twitter data consisting of 11,316,811 nodes (users) and 85,331,846 edges (representing a follower relationship) [2].

While using the *whole sample* for simulations is sometimes unnecessary to observe representative trends, it is potentially very expensive [20] and wasteful in terms of computations, and costly when using paid LLMs such as GPT. Simulations commonly right-size the computations by using a *sufficient sample* size for the experiments [34] (see Subsect. 4.2). Each sample should be representative of the data. Using a python library for graph sampling [49], we noted that a random sampling of node was unsuitable (the network consists of scattered users lacking connectivity for message spread) and sampling by PageRank had poor scaling (memory needs exceed 32GB). We use a *degree-based sampling* strategy [26], which preserves characteristics of the degree distribution (e.g., heavily skewed) but creates changes as evidenced by a Jensen-Shannon distance of 0.64 between a sample of 200 nodes and the whole network (using normalized degree distributions). As noted by Moran-Tovar and colleagues, "while the degree distribution ignores the specific topology of the network, it captures the effect of largely connected nodes or hubs on the transmission statistics" [38]. It is thus appropriate to study the dynamics of spreading phenomena, but it may not support a more fine-grained analysis (e.g., detecting communities). The sampling library returns an *undirected* graph but Twitter data is *directed*: a user A follows B and that is not necessarily reciprocal. We restored directionality by checking for all sample nodes whether they appeared as a pair the edge list of the original data, then we added the corresponding edges.

Messages. While Twitter shows us the *reaction* of a wide population to extreme ideas, the diversity of topics and tones encountered on this platform limits its usability as a *source* for such ideas. The partisan tone of alternative media websites make them valuable sources to retrieve far-right ideas. In particular, Truth Social (launched by Donald Trump) is the most right-leaning alternative platform [17]. Per the Pew Research Center, most prominent accounts (94%) were individuals rather than organizations [17]. It is thus a suitable source to collect far-right messages as expressed by individuals. For replicability, we use an open Truth Social dataset crawled from February 21st, 2022 to October 15th, 2022. The authors sampled posts and accounts using Trump's account as the seed, then spread to his followers and other popular accounts. The dataset has over 823,000 posts and over 454,000 accounts [22].

Not all messages can be used to spread ideas via a simulation, since some may be too short or are only interpretable in the context of a discussion. We thus undertook three typical steps for pre-processing political online messages [50]. First, we used the well established `TextBlob` Python library 0.18.0 [3] to assign a sentiment polarity to each message from -1 (very negative) to 1 (very positive). In order to study reactions regarding strongly worded messages, we retained posts with absolute polarity above 0.9. Then, we removed posts with fewer than 25 characters (since they are either too context-dependent or lack content). Finally, we removed all posts that contained links because the LLM may attempt and fail to retrieve the URL's content thus causing simulation errors. This technical limitation may impact the results since real-world social media users may be more affected by posts that include references. Following our three steps, we obtained 13,239 messages to spread in our simulations (exemplified in Box 2).

> **Box 2. Sample of Truth Social posts**
>
> - Insurrection my ass. Video after video now showing the Capitol Police encouraging and inviting people through! We have all seen the videos and most of us have them saved!!!
> - Ilhan Omar is the perfect example of why we need an immigration moratorium in America.
> - Ignorant idiots! Y'all have no idea what the CDC, WHO, and this government is getting you to do! Wake up now!
> - If the U.S can afford to send 40 Billion dollars to Ukraine, then We can afford to put armed security in all 131,000 schools in America!! Protect the Children!! Evil people do not care about laws!!
> - Yup! Fauci is evil and does need to be locked up!

Post Frequency. Social media accounts do not continuously produce content; rather, they post at a given frequency. To seed their simulated Twitter network, Ben Sliman and Kohli analyzed the empirical distribution of the average number of tweets per day as a function of the number of followers [7]. We extracted the numerical data from the plotted distribution in their article using https://automeris.io Since the distribution is discrete (e.g., 260 followers, 270, 280...), we used a linear interpolation to obtain a continuous distribution that allows us to quantify the frequency for all user accounts (e.g., 263 followers). The authors' plotted distribution starts at 24 followers, thus we used an extrapolation to estimate the posting frequency of users with fewer followers, under the assumption that agents without followers would have no posts.

Demographics. Reviews on conspiracies and politically divisive decisions in the U.S. (e.g., whether to vaccinate for COVID-19) have shown that key determinants include age, sex, educational attainment, race and ethnicity. As we previously detailed, *these four determinants should be initialized together when creating virtual agents due to their dependencies* [6]. Otherwise, we would erroneously create agents with plausible age distribution and (separately) valid educational attainments, but their joined distributions may not match the data. We use the U.S. Census CPS Basic Monthly Data from October 2024 as this nationally representative survey provides tables for the joint distributions of the four social determinants [55].

4 Results

Evaluations of LLM agents can be performed at two levels [20]. At the *micro-level*, simulated decisions from the agents (particularly through the prism of the LLM) must align with real-world data (Subsect. 4.1). At the *macro-level* we assess dynamics over the entire population, which may differ from the sum of the individuals given that ABMs often show *emerging* properties. Prior LLM/ABM models for the spread of political opinions used 15 agents over 10 simulation steps and presented macro-level findings through visuals [62]. However, there is a risk for such results to be an artifact of the limited model's size, or that findings lack statistical significance. We thus use a statistical approach to identify a suitable model size (Subsect. 4.2) then we analyze the results (Subsect. 4.3).

4.1 Validating the Use of LLMs for Political Opinions

We use a LLM to quantify the political opinion of an agent based on four demographic features. Assessing the accuracy of the LLM in performing this complex task contributes to validating its use to initialize our agents and informs us of the confidence margin associated with the simulation results. Since the demographics of voters change over time, we compared the LLM results with the most recent 2022 data from the Pew Research Center [25] as well as the 2024 post-voting polls from NBC News [41] (Table 1). Surveys have limited generalizability since many eligible voters do not vote. *On average, the prediction of GPT are 6% points away from either of the two surveys*, which makes it suitable for our application. The predictions are more accurate with respect to race (maximum error of 6%), followed by age (overestimating elderly as conservative and young adults as liberal), educational attainments (overestimating liberals among college

Table 1. Prevalence of right-leaning voters in two surveys vs. prediction of GPT. For GPT, we create a complete population and we aggregate to obtain the target feature value. For example, for 'male', we aggregate all male agents with weights corresponding to the prevalence of race, educational attainments, and age category among males. We also generate American Indians and Alaskan Natives (as one group) but they were omitted from the racial breakdown due to their low prevalence.

Demographic feature	Group	Pew 2022	NBC 2024	GPT-4
Sex	Male	54%	55%	65%
	Female	48%	45%	32%
Race	White	57%	57%	56%
	Black	5%	13%	10%
	Asian	32%	40%	34%
Educational attainments	Postgraduate	37%	38%	45%
	College graduate	48%	50%	36%
	Some college	54%	51%	49%
	High school or less	59%	62%	60%
Age category	18–29	31%	43%	32%
	30–49	45%	47%	43%
	50–64	55%	54%	54%
	65+	56%	50%	62%

graduates) and sex (with the highest error of 16% on female voters). Our results based on GPT-4 confirm previous reports based on GPT-3 [4]: it could not be distinguished from humans when associating keywords with political parties; it was also highly correlated with political votes when agents profiles were provided based on race, age, sex, and seven other characteristics.

4.2 Right-Sizing the Model: Effects of Scaling

We aim to identify a *sufficient* population size so that results reflect the dynamics of the model instead of being an artifact of the model's size (see Sect. 3.2–**Social network**). As in previous works, we identify a sufficient size by starting with a small population, gradually increasing its size, and measuring whether the outputs depend on the population size (cf. Figures 7–9 in [24]). As expected, the standard deviation decreases as the population size increases (Table 2). A one-way ANOVA between the simulation outputs for each population size shows no statistically significant differences for 11 of the 13 demographic groups. In the case of black agents (ANOVA p-value = 0.03), a post hoc Tukey HSD revealed that a population size of 100 was statistically different from 150 (Q = 4.57, p = 0.02); there were no other differences. In the case of college graduates (ANOVA p-value = 0.01), a population size of 100 was statistically different from 150 agents (Q = 5.23, p = 0.009) and 50 agents (Q = 4.12, p = .04); again, there were no differences between other sizes. In summary, a population size of 100 is insufficient as its results differ from other sizes. There is no difference between

Table 2. For each population size, we report the average ± standard deviation over 5 runs of the political opinion score, which ranges from -10 (far-right) to 10 (far-left). We also perform an ANOVA on outputs of the 5 runs across population sizes.

Demog.	Group	50	100	150	200	ANOVA p-value
Sex	Male	−4.04 ± 1.09	−3.20 ± 1.03	−3.72 ± 0.44	3.42 ± 0.32	0.39
	Female	−1.24 ± 0.75	−0.20 ± 1.26	−1.48 ± 0.39	1.17 ± 0.52	0.09
Race	White	−3.33 ± 0.88	−2.31 ± 0.92	−3.37 ± 0.48	3.04 ± 0.25	0.09
	Black	1.40 ± 0.70	1.97 ± 0.60	0.63 ± 0.77	−1.14 ± 0.54	0.03
	Asian	−2.18 ± 2.17	−1.08 ± 0.42	−1.62 ± 1.79	0.38 ± 1.30	0.33
Educational attainments	Postgrad.	−0.56 ± 1.62	−0.56 ± 1.46	−0.60 ± 0.91	0.55 ± 0.52	0.99
	College grad.	−1.69 ± 0.90	−0.53 ± 0.66	−2.00 ± 0.46	1.36 ± 0.32	0.01
	Some college	−2.28 ± 1.85	−0.83 ± 1.56	−2.67 ± 0.64	2.36 ± 0.85	0.16
	≤high school	−4.29 ± 0.98	−3.65 ± 0.90	−3.86 ± 0.56	3.66 ± 0.25	0.48
Age cat.	18–29	−0.90 ± 0.83	0.07 ± 1.02	−0.56 ± 0.26	0.28 ± 0.45	0.21
	30–49	−1.34 ± 1.45	−1.50 ± 1.15	−2.30 ± 0.72	1.93 ± 0.17	0.44
	50–64	−3.21 ± 0.78	−2.56 ± 0.65	−3.16 ± 0.57	2.95 ± 0.84	0.47
	65+	−4.07 ± 2.07	−2.66 ± 1.15	−4.24 ± 0.88	3.96 ± 0.43	0.23
Standard deviation		1.51	1.23	0.86	0.60	

218   S. Zhong et al.

population sizes of 150 and 200 so either can be employed. In the remainder of this paper, we use 200 agents as it yields a narrower standard deviation.

4.3 Dynamics of the Population

Prior works measured the diversity of opinions as the *number* of different opinions among the agents [62], but this does not account for the *frequency* at which these opinions hold. We thus measure the diversity of opinions using Shannon entropy, where a higher entropy means more diversity. Figure 1-a shows that the diversity initially rises modestly from 3.70 and plateaus at 3.84. The average opinion value starts almost neutral in the population and steadily becomes more right leaning, oscillating at -2.25 (Fig. 1-b). Together, these results suggest that being *exclusively* exposed to far-right messages produces a change in the population. As shown in Fig. 2, this change is not merely a *shift* where every individuals experience the same decrease in opinion score. Rather, individuals react based

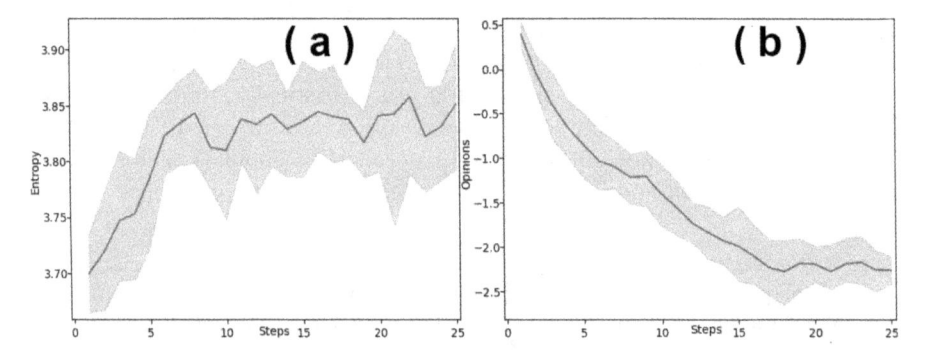

Fig. 1. The diversity of opinions quickly plateaus (a) while the average opinion plateaus after 20 steps (b). Standard deviations (blue bands) are based on 5 simulation runs. (Color figure online)

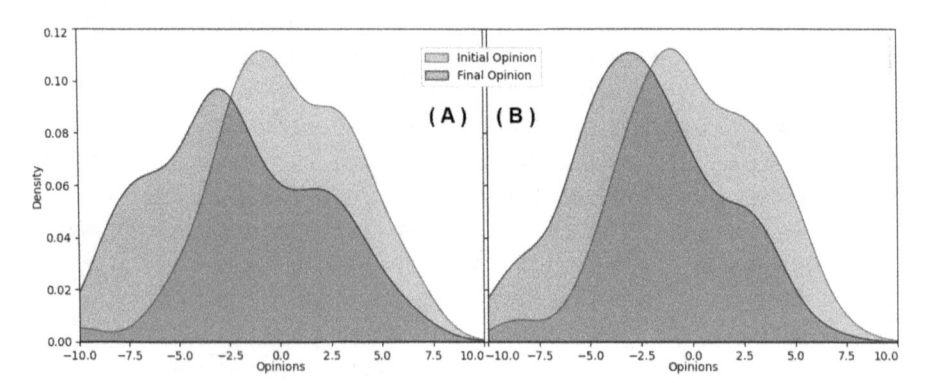

Fig. 2. The initial distribution of opinions at $t = 0$ morphs into its final configuration at $t = 25$ across two simulation runs (A, B) for 200 agents.

on their demographics, with some holding firmly to their opinions (as the distribution continues to go up to 10 – far left) and others having a stronger reaction (as shown by the increased weight in the first half of the distribution).

5 Discussion

Given prior works on using GPT to emulate voting patterns [4] or key political debates such as abortion [62], we have shown the feasibility of simulating *changes* in opinions due to exposure to specific political messages. Our work confirms the potential of combining LLMs with ABMs to to develop models that represent human behavior and decision-making [23]. As stated by Park and colleagues, a model *generates* behaviors in social media according to certain specifications [44]. The goal of a model is not always to merely witness a phenomenon (that we already knew was happening); rather, it serves as a virtual laboratory to test the consequences of possible interventions. The reddit simulation from Park *et al.* thus paired a generating model with a what-if component to study scenarios such as moderator interventions. By following our process or directly reusing our open-source implementation, researchers can test strategies such as combining the model with detection algorithms (e.g., for hate speech, incitement to violence, or misinformation) to delete posts, ban their authors, or algorithmically deprioritize posts (i.e., reduce their visibility in a reader's feed).

As we are only in the infancy of generative agent-based modeling, there are several interesting avenues to extend the model. Pastor-Galindo *et al.* stressed that it is "imperative to simulate and model realistic social networks" by modeling three aspects [45]. In this paper, we focus on the first aspect of *direct communications* as agents write posts to which their followers react. Our model did not account for *information sharing* by mining links and other dynamic contents shared on a network (which can be achieved by the LLM), and we did not represent how *user engagement* varies depending on the type of content (which needs a change in the model and prompt). Dynamicity can be important depending on the simulated time window: at the scale of a few days, we can assume that the network is static (as in our study), but over longer durations, there would be changes since individuals unfollow others or create links.

While our study used several real-world datasets for empirical grounding, there are two limitations in data availability regarding individual opinions in general (see Sect. 2.1) and for certain political groups in particular. First, our agents have a representative and internally consistent set of demographics but given the paucity of data that *associates such features with social media accounts*, agents were assigned at random to Twitter accounts. This makes it possible for an extremely left-leaning agent to follow extreme-right accounts. As a result, the changes observed in our simulation are an over-estimate of changes happening in real-world networks characterized by assortativity and echo chambers. Second, we examined changes due to exposure to far-right posts from Truth Social, while noting that social media platforms such as Twitter/X contain a *variety* of posts. At present, there is no left-wing equivalent to Truth Social that allows for the same large-scale data collection. As new platforms (e.g., Bluesky) emerge, it may become possible to simulate complete exposure to left- and right-leaning posts.

Acknowledgments. We gratefully acknowledge the financial support of American University's Signature Research Initiative Project. SZ wishes to thank Eric Schuler for support in using American University's High Performance Computing Cluster Zorro.

Disclosure of Interests. The authors have no competing interests to declare that are relevant to the content of this article.

References

1. Abelson, R.P.: Mathematical models of the distribution of attitudes under controversy. Contributions to mathematical psychology (1964)
2. Aché, M.: Twitter edge nodes. https://www.kaggle.com/datasets/mathurinache/twitter-edge-nodes/data. Accessed 31 Dec 2024
3. Aljedaani, W., Rustam, F., Mkaouer, M.W., et al.: Sentiment analysis on twitter data integrating TextBlob and deep learning models: the case of us airline industry. Knowl. Based Syst. **255**, 109780 (2022)
4. Argyle, L.P., Busby, E.C., Fulda, N., et al.: Out of one, many: using language models to simulate human samples. Polit. Anal. **31**(3), 337–351 (2023)
5. Baele, S.J., et al.: Uncovering the far-right online ecosystem: an analytical framework and research agenda. Stud. Confl. Terror. **46**(9), 1599–1623 (2023)
6. Beerman, J.T., Beaumont, G.G., Giabbanelli, P.J.: A scoping review of three dimensions for long-term Covid-19 vaccination models: hybrid immunity, individual drivers of vaccinal choice, and human errors. Vaccines **10**(10), 1716 (2022)
7. Ben Sliman, M., Kohli, R.: Asymmetric relations and the friendship paradox. Columbia Business School Research Paper (18-73) (2018)
8. von Briesen, E.M., Bacaksizlar, N.G., Hadzikadic, M.: Modeling genocide at the system and agent levels. J. Policy Complex Syst. **3**(2), 31–48 (2017)
9. Brown, O., Smith, L.G., Davidson, B.I., et al.: Online signals of extremist mobilization. Personal. Soc. Psychol. Bull., 01461672241266866 (2024)
10. Cointet, J.P., Roth, C.: How realistic should knowledge diffusion models be? JASSS **10**(3), 5 (2007)
11. Coscia, M., Rossi, L.: How minimizing conflicts could lead to polarization on social media: an agent-based model investigation. PLoS ONE **17**(1), e0263184 (2022)
12. Crokidakis, N.: Recent violent political extremist events in brazil and epidemic modeling: the role of a sis-like model on the understanding of spreading and control of radicalism. Int. J. Mod. Phys. C **35**(02), 2450015 (2024)
13. Deffuant, G., Neau, D., Amblard, F., Weisbuch, G.: Mixing beliefs among interacting agents. Adv. Complex Syst. **3**(01n04), 87–98 (2000)
14. DeGroot, M.H.: Reaching a consensus. J. Am. Stat. Assoc. **69**(345), 118–121 (1974)
15. Ebner, J., Kavanagh, C., Whitehouse, H.: Assessing violence risk among far-right extremists: a new role for natural language processing. Terrorism Polit. Violence **36**(7), 944–961 (2024)
16. Flache, A., Mäs, M., Feliciani, T., et al.: Models of social influence: towards the next frontiers. JASSS **20**(4), 2 (2017)
17. Forman-Katz, N., Stocking, G.: Key facts about truth social (2022)

18. Froio, C., Ganesh, B.: The transnationalisation of far right discourse on twitter: issues and actors that cross borders in western European democracies. Eur. Soc. **21**(4), 513–539 (2019)
19. Gambaro, J.P., Crokidakis, N.: The influence of contrarians in the dynamics of opinion formation. Phys. A **486**, 465–472 (2017)
20. Gao, C., et al.: Large language models empowered agent-based modeling and simulation: a survey and perspectives. Humanit. Soc. Sci. Commun. **11**(1), 1–24 (2024)
21. Geller, M., Vasconcelos, V.V., Pinheiro, F.L.: Toxicity in evolving twitter topics. In: In: Mikyška, J., de Mulatier, C., Paszynski, M., Krzhizhanovskaya, V.V., Dongarra, J.J., Sloot, P.M. (eds.) International Conference on Computational Science, pp. 40–54. Springer (2023). https://doi.org/10.1007/978-3-031-36027-5_4
22. Gerard, P., Botzer, N., Weninger, T.: Truth social dataset. In: Proceedings of the International AAAI Conference on Web and Social Media, vol. 17, pp. 1034–1040 (2023)
23. Ghaffarzadegan, N., Majumdar, A., et al.: Generative agent-based modeling: an introduction and tutorial. Syst. Dyn. Rev. **40**(1), e1761 (2024)
24. Gibson, M., Portugal Pereira, J., et al.: Agent-based modelling of future dairy and plant-based milk consumption for UK climate targets. JASSS **25**(2) (2022)
25. Hartig, H., Daniller, A., Keeter, S., Green, T.V.: (2023). https://www.pewresearch.org/politics/2023/07/12/voting-patterns-in-the-2022-elections/
26. Hu, P., Lau, W.C.: A survey and taxonomy of graph sampling. arXiv preprint arXiv:1308.5865 (2013)
27. Huszár, F., Ktena, S.I., O'Brien, C., Belli, L., Schlaikjer, A., Hardt, M.: Algorithmic amplification of politics on twitter. PNAS **119**(1), e2025334119 (2022)
28. Jager, W., Amblard, F.: Uniformity, bipolarization and pluriformity captured as generic stylized behavior with an agent-based simulation model of attitude change. Comput. Math. Organ. Theory **10**, 295–303 (2005)
29. Kakavand, A.E.: Far-right social media communication in the light of technology affordances: a systematic literature review. Ann. Int. Commun. Assoc. **48**(1), 37–56 (2024)
30. Kasaie, P., Kelton, W.D.: Guidelines for design and analysis in agent-based simulation studies. In: Winter Simulation Conference, pp. 183–193 (2015)
31. Laver, M., Sergenti, E.: Party Competition: An Agent-Based Model. Princeton University Press (2012)
32. Lipiecki, A.: Strategic promotional campaigns for sustainable behaviors: Maximizing influence in competitive complex contagions. In: International Conference on Computational Science, pp. 62–70. Springer (2024). https://doi.org/10.1007/978-3-031-63759-9_
33. Lorenz, J.: Continuous opinion dynamics under bounded confidence: a survey. Int. J. Mod. Phys. C **18**(12), 1819–1838 (2007)
34. Lutz, C.B., Giabbanelli, P.J.: When do we need massive computations to perform detailed Covid-19 simulations? Adv. Theory Simul. **5**(2), 2100343 (2022)
35. Mahmoudi, A., Jemielniak, D., Ciechanowski, L.: Echo chambers in online social networks: a systematic literature review. IEEE Access (2024)
36. Mead, E.L., McNerney, H.W., Agarwal, N.: Text mining domestic extremism topics on multiple social media platforms. In: International Conference Computational Linguistics and Natural Language Processing, pp. 104–111. IEEE (2024)
37. Memmert, L., Cvetkovic, I., Bittner, E.: The more is not the merrier: effects of prompt engineering on the quality of ideas generated by GPT-3. In: Proceedings 57th Hawaii International Conference on System Sciences, pp. 7520–7529 (2024)

38. Morán-Tovar, R., Gruell, H., et al.: Stochasticity of infectious outbreaks and consequences for optimal interventions. J. Phys. A **55**(38), 384008 (2022)
39. Müller, A., Lopez-Sanchez, M.: Countering negative effects of hate speech in a multi-agent society. In: Artificial Intelligence R&D, pp. 103–112 (2021)
40. Mustafee, N., Powell, J.H.: From hybrid simulation to hybrid systems modelling. In: 2018 Winter Simulation Conference (WSC), pp. 1430–1439. IEEE (2018)
41. NBC News: Exit polls (2024). https://www.nbcnews.com/politics/2024-elections/exit-polls
42. Negahban, A., Giabbanelli, P.J.: Hybrid agent-based simulation of adoption behavior and social interactions: alternatives, opportunities, and pitfalls. IEEE Trans. Comput. Soc. Syst. **9**(3), 770–780 (2021)
43. O'Hara, K., Stevens, D.: Echo chambers and online radicalism: assessing the internet's complicity in violent extremism. Policy Internet **7**(4), 401–422 (2015)
44. Park, J.S., Popowski, L., Cai, C., et al.: Social simulacra: creating populated prototypes for social computing systems. In: Proceedings 35th Annual ACM Symposium on User Interface Software and Technology, pp. 1–18 (2022)
45. Pastor-Galindo, J., Nespoli, P., Ruipérez-Valiente, J.A.: Large-language-model-powered agent-based framework for misinformation and disinformation research: Opportunities and open challenges. IEEE Secur. Priv. **22**, 24–36 (2024)
46. Popa-Wyatt, M.: Online hate: is hate an infectious disease? is social media a promoter? J. Appl. Philos. **40**(5), 788–812 (2023)
47. Rainer, H., Krause, U.: Opinion dynamics and bounded confidence: models, analysis and simulation. JASSS **5**(3) (2002)
48. Rothut, S., Schulze, H., et al.: Ambassadors of ideology: a conceptualization and computational investigation of far-right influencers, their networking structures, and communication practices. New Media Soc. **26**(12), 7120–7147 (2024)
49. Rozemberczki, B., Kiss, O., Sarkar, R.: Little ball of fur: a Python library for graph sampling. In: Proceedings 29th ACM International Conference Information & Knowledge Management, pp. 3133–3140 (2020)
50. Sandhu, M., Vinson, C.D., Mago, V.K., Giabbanelli, P.J.: From associations to sarcasm: mining the shift of opinions regarding the supreme court on twitter. Online Soc. Netw. Media **14**, 100054 (2019)
51. Serrano, E., Iglesias, C.Á., Garijo, M.: A novel agent-based rumor spreading model in Twitter. In: Proceedings 24th International Conference on World Wide Web, pp. 811–814 (2015)
52. Stokes, B.M., Jackson, S.E., Garnett, P., Luo, J.: Extremism, segregation and oscillatory states emerge through collective opinion dynamics in a novel agent-based model. J. Math. Sociol. **48**(1), 42–80 (2024)
53. Tao, K., et al.: GPT-4 performance on querying scientific publications: reproducibility, accuracy, and impact of an instruction sheet. BMC Med. Res. Methodol. **24**(1), 139 (2024)
54. Ting, M.M., Bendor, J., Diermeier, D., Siegel, D.A.: A Behavioral Theory of Elections. Princeton University Press (2011)
55. United States Census Bureau: CPS basic monthly (2024 OCT) Custom Table on PESEX, PTDTRACE, PEEDUCA, PRTAGE (2024). https://data.census.gov/app/mdat/CPSBASIC202410
56. Unlu, A., Kotonen, T.: Online polarization and identity politics: an analysis of Facebook discourse on Muslim and LGBTQ+ communities in finland. Scand. Polit. Stud. **47**(2), 199–231 (2024)
57. Waldherr, A., et al.: Worlds of agents: prospects of agent-based modeling for communication research. Commun. Methods Meas. **15**(4), 243–254 (2021)

58. Weisburd, D., Wolfowicz, M., et al.: Using agent based modelling to advance evaluation research in radicalization and recruitment to terrorism: prospects and problems. Studies in Conflict & Terrorism, pp. 1–24 (2024)
59. Westermann, C.J., Coscia, M.: A potential mechanism for low tolerance feedback loops in social media flagging systems. PLoS ONE **17**(5), e0268270 (2022)
60. Zhang, H., Vorobeychik, Y.: Empirically grounded agent-based models of innovation diffusion: a critical review. Artif. Intell. Rev. **52**, 707–741 (2019)
61. Zhang, X., Davis, M.: E-extremism: a conceptual framework for studying the online far right. New Media Soc. **26**(5), 2954–2970 (2024)
62. Zheng, W., Tang, X.: Simulating social network with LLM agents: an analysis of information propagation and echo chambers. In: Tang, X., Huynh, V.N., Xia, H., Bai, Q. (eds.) International Symposium on Knowledge and Systems Sciences, pp. 63–77. Springer, Singapore (2024). https://doi.org/10.1007/978-981-96-0178-3_5
63. Zimmerman, F., Bailey, D.D., Muric, G., et al.: Attraction to politically extreme users on social media. PNAS nexus **3**(10), pgae395 (2024)

Accelerated Approximation of Bellman Equation Solutions: Agent Policy Optimization With a Feedforward Neural Network

Victoria M. Garibay

University of Amsterdam, Amsterdam, Netherlands
Victoria.Garibay@gmail.com

Abstract. Solving recursive equations through iteration can be a computationally expensive endeavour, and the time required to reach an optimal solution delays the progress of any dependent processes. To address this issue for a specific use case of decision-making in an agent-based model, a method of replacing the iterative function used in said model, a Bellman equation, with a feedforward multilayer perceptron was developed. A hyperparameter grid search was performed to determine the combination of architecture, learning rate, and batch size which produced results deviating the least from those of the original iterative method. With the resulting neural network, accepting four inputs and yielding two outputs, the time required to compute outputs scales sublinearly with the number of agents. Excluding training time, for a set of 1,000 agents, the selected neural network produces output at over 66,000 times the speed of the original function. It achieves this acceleration while maintaining a 99.3% accuracy in adaptation strategy selection and 0.10 mean absolute error in consumption, leading to its ready adoption as an acceptable replacement for the original method.

Keywords: Dynamic Programming · Recursive Function · Surrogate Model · Multilayer Perceptron

1 Introduction

Modelling autonomous agents interacting in an environment is an increasingly computationally intensive task. As new technologies develop improving the efficiency and capacity of agent based models (ABMs), new levels of complexity become realizable [3,9]. Rising towards this higher potential can sometimes involve reconsidering the approach to certain traditional model aspects. The particular case addressed in this study is an ABM which follows the evolution of stock capital for heterogeneous household agents in a society subjected to exogenous shocks. Loosely founded on the Boltzmann wealth model, within a timestep, agents exchange capital and decide how to optimally consume and invest their resources depending on their attributes and perceptions of their changing environment. The decision on how to partition their capital is based on a Bellman equation [2], which has a recursive structure accommodating sequential processes such as forming an optimal consumption policy based on its expected future values. The form the Bellman equation took in the case model did not have a known analytical solution. Solutions to the Bellman equation have been approximated in a variety of creative ways in the past [13, 15, 22, 23][1]. One such calculation applicable to the custom Bellman equation arrangement required by the ABM is policy iteration, a numerical method which approximates an optimal policy by alternately evaluating and optimizing the result of a value function until convergence is reached [22]. This method in its established form was not a feasible option for the case model due to the desired scale (1,000,000+ agents), which a newly developed tensor-based framework (DGL-ABM) was otherwise capable of supporting. Due to the imperfect nature of the process of human decision making, there was some flexibility in the performance of a replacement and, conveniently, the capacity to generate samples as needed with the existing method, limited only by time and memory dedicated to their production. To overcome the barrier posed by iterative calculation for a problem set of sufficient magnitude to serve a million potentially unique agents, several options for interpolation from known results to form a reference landscape map or response surface were considered [12, 24][2]. However, the dimensionality issues rising from a four-input, two-output structure and the irregularity of the custom Bellman equation made traditional interpolation methods poorly suited to the task. Instead, attention shifted to the development of a surrogate model of the equation. The judgment was reached that a Gaussian process regression was inappropriate due to the piecewise discontinuity of certain specialized aspects of the equation, moderately high dimensionality of the problem, and the potentially large sample size required to accommodate this dimensionality. With these restrictions in mind, it was decided to pursue a neural network approximation for the equation. There are many examples of neural networks being used to model equations, particularly in engineering and earth sciences [4, 6, 16]. The potential applicability of neural networks to agent behavior is now also being appreciated [14, 15]. However, approaches used in existing research differ from that taken for the task in this study, as they primarily focus on using recursive neural networks to capture the learning process of agents. For the proposed use case, instead of considering agents to be themselves neural networks, the equation which describes their decision-making is being approximated with a multilayer perceptron (MLP) trained on computed results. A similar technique has been applied to economic models in the past, primarily seeking proof of concept, but literature on practical application attempts, successful or not, is limited [17]. The contents of this manuscript are a written record of the motivation, exploration, and formation of a static mapping technique for the results of a custom Bellman equation computation. The primary objectives of this research were to 1) identify a superior

---

[1] Based on methods currently found in [7].
[2] Based on methods from [1].

combination of hyperparmeters with which to train a neural network for the outlined computational task and 2) assess whether the resulting MLP could be a worthy replacement for the iterative method—specifically, if it could achieve the substantial speed gains required by the ABM setup while maintaining, at minimum, a 90% fidelity in investment strategy and, at maximum, an average absolute error of 0.25 in consumption. The following methods were developed out of need, naive to the specific science and standards surrounding neural network design. Through the process, the conductor of this research has gained knowledge, become more aware of limitations in the field, and come to appreciate that there is much remaining for all researchers to understand and discover in this vast frontier. As such, it is hoped that the methods documented and ensuing discussion can contribute as precedent for improved, streamlined efforts to design similar approximation tools in the future.

2 Methods

Subsequent sections describe the methods followed in the process of training an MLP as a potential replacement for an existing solver used as a decision model for agents in an ABM.

2.1 Problem Context

The ABM itself is not the focus of this manuscript, thus it will not be explained in great detail. However, readers may find it helpful to know a summary of relevant points regarding the agent decision process. Agent decisions are described by a constrained Bellman equation satisfied by a custom value function maximizing the value of the state of stock capital, k, resulting from a particular consumption, c, and investment in adaptation i_a (Eq. 1). Maximization is constrained by the rule that consumption and investment do not exceed available funds at the following time step. Investment in adaptation is associated with a multiplier which reduces the impact of a perceived exogenous shock, rendering an adjusted shock factor $\theta_{m,t}$. The function u is a standard isoelastic utility function with exact value dependent on the risk aversion of an agent, σ, while f_{income}—a function of aptitude, α, and stock capital—is the maximum value from the results of a capital-only Cobb-Douglas production function evaluated for available capital output elasticity exponents, γ, minus the *cost* assigned to their use (u and f_{income} are described further in Algorithm 1 in the Appendix). The depreciation, δ, and discount factor, β, were considered model constants, homogeneous for all agents.

$$V(k_t) \equiv \max_{c_t, i_{a,t}} \{u(c_t, \sigma) + \beta V(\theta_{m,t}[f_{income}(\alpha, k_t) + (1-\delta)k_t - c_t - i_{a,t}])\} \quad (1)$$

2.2 Bellman Equation Sample Data Generation

Before attempting to model the Bellman equation (Eq. 1) with a neural network, it was necessary to develop a training and testing dataset. As mentioned, the solution to the equation was being computed iteratively relying on techniques documented by Stachurski [22]. For additional context, this iterative optimization was occurring up to three times per set of agent properties as conceptually represented in Algorithm 1, found in the Appendix[3]; this was because three variations of the Bellman equation, one for each adaptation option, needed to be evaluated to determine which produced the highest value for the agent. The inputs to the algorithm were the agent properties k_t, stock capital, θ_t, shock perception, σ, risk aversion, and α, aptitude for income generation, as well as the information on adaptation options, technology coefficients, and their associated costs. The outputs were the optimal consumption and choice of adaptation investment, i_a, between options of a N(one), L(ow), and H(igh) protection multiplier, m. The multiplier affects Eq. 1 with $\theta_{m,t} = \theta_t + m(1 - \theta_t)$.

A Sobol' sequence-based Saltelli sampling matrix (N = 8,192) was formed from the distributions specified in Table 1 [20,21]. Duplicate agents were removed, leaving 47,935 unique agents. All agents for which the Bellman solution resulted in an error or did not converge were removed from the dataset, resulting in 47,281 remaining samples (a 98.8% retention rate). Prior to training and testing the model, each input variable was scaled according to its distribution:

– For uniformly distributed:

$$\text{input} = \frac{\text{input} - \text{minimum}}{\text{maximum}}$$

– For normally distributed:

$$\text{input} = \frac{\text{input} - \text{mean}}{\text{standard deviation}}$$

Likewise, the output variables were scaled according to their maximum values:

$$\text{output} = \frac{\text{output}}{\text{maximum}}$$

---

[3] The code used to generate samples is also available as a notebook [5].

Table 1. Distributions used in Saltelli sampling of agent attributes.

Variable	Distribution Type	Distribution Parameters
θ	uniform	[0,1)
k	uniform	[0.1,10)
α	normal	loc = 1.08; scale = 0.074
σ	uniform[*]	[0.1,2)

[*] Value rounded to nearest tenth.

2.3 Hyperparameter Grid Search

An exhaustive grid search was performed to determine the most appropriate model architecture and hyperparameters for the equation mapping task. In the process of defining ranges and points for the search, formal and informal guidelines were consulted (e.g., Philipp [18], Goodfellow et al. [8], Richetti et al. [19]), but they were—by their own admission in some cases—fairly vague and supported the idea that hyperparameter selection can be highly application-dependent. The MLP neural network for which the search was conducted had an input layer of four nodes, output layer of two nodes, and between two and five hidden layers (Fig. 1). The relative shape of the network was generalized into five architectural variations, or ratios of layer widths, as specified in Table 3. The n_{max} referenced in that table is the maximum number of nodes (i.e., the number of nodes in the widest layer of the network). All layers used a basic ReLU (rectified linear unit) activation function. Model training used the Adam optimizer and L2 loss (i.e., mean squared error, MSE, Eq. 3) to minimize the difference between the predicted values and the training dataset consisting of 80% of the agent sample data. The remaining 20% of the sample was reserved for validation. Training ran for 200 epochs with an early stop triggered after 20 epochs with no improvement in validation MSE. Other considerations for the grid search were batch size and learning rate. Due primarily to storage constraints, three random seeds were utilized for each unique combination of architecture type, learning rate, n_{max}, and batch size listed in Table 2. The code used for the model grid search can be found publicly [5]. The foundational network training and testing scripts followed the Pytorch Template Project [11].

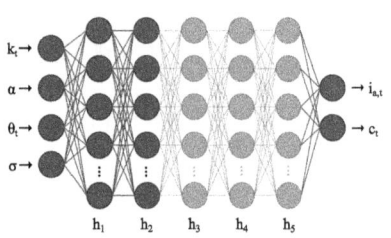

Fig. 1. Highly generalized diagram of the multilayer perceptron used to map the iterative Bellman equation input to output. The number of hidden layers and their respective widths were varied according to Table 3 to execute a grid search for the best architecture and hyperparameter combination.

Table 2. Grid search space for which combinations of one item of each column determined the model and hyperparameters for training.

Architecture Type	n_{max}	Learning Rate	Batch Size
ThreeLayer (3L)	512	0.001	64
FourLayer (4L)	1024	0.01	128
FiveLayer (5L)	2048	0.0001	512
PudgeFiveLayer (P5L)	-	-	-
PudgeSixLayer (P6L)	-	-	-

Table 3. Layer widths specified in relation to maximum number of nodes (n_{max}) for each architecture type. The column titles $h_1...h_5$ correspond to the hidden layers as labeled in the stylized illustration of the network provided by Fig. 1.

Architecture Type	h_1	h_2	h_3	h_4	h_5
ThreeLayer (3L)	$n_{\max}$	$n_{\max}/2$	-	-	-
FourLayer (4L)	$n_{\max}$	$n_{\max}/2$	$n_{\max}/4$	-	-
FiveLayer (5L)	$n_{\max}$	$n_{\max}/2$	$n_{\max}/4$	$n_{\max}/8$	-
PudgeFiveLayer (P5L)	$n_{\max}/2$	$n_{\max}$	$n_{\max}/2$	$n_{\max}/4$	-
PudgeSixLayer (P6L)	$n_{\max}/4$	$n_{\max}/2$	$n_{\max}$	$n_{\max}/2$	$n_{\max}/4$

2.4 Metrics and Model Selection

The metrics recorded for the best model of each combination of parameters included mean absolute error (MAE, Eq. 2) for the model, percentage of incorrect guesses for i_a, and MAE for c_t of the validation data. For each of these three metrics, the average value produced from the three seeds was considered to represent the performance of each unique architecture and hyperparameter combination. It was decided that the primary metric of consideration would be the combined MAE, provided that the same model also fell within the top ten performers in the other two metrics. In the event of failure to meet this secondary requirement, the next best model would be evaluated, loosening the top n standard by one, and so on down the rankings.

Beyond selecting the best model for the intended purpose, additional information was collected on differences in model performances to investigate any potential trends for variables in the grid search space. For this aspect, model MSE and MAE were the chosen metrics of comparison.

$$MAE = \frac{1}{n}\sum_{i=1}^{n}|y_i - \hat{y}_i| \text{ and} \qquad (2)$$

$$MSE = \frac{1}{n}\sum_{i=1}^{n}(y_i - \hat{y}_i)^2, \qquad (3)$$

where y_i is the basis point, $\hat{y}_i$ is the predicted value, and n is the sample size.

3 Results and Discussion

With respect to execution time alone, using an MLP as opposed to the, admittedly poorly optimized, iterative computation method was a tremendous success. In Fig. 2, it is possible to observe the linear scaling of the iterative version and the sublinear scaling of the MLP version with increasing sample size. It should be taken into account that the iterative computation method was applied to agents in series, implying that time efficiency could have been increased through parallelization.

While an improvement in computation time was anticipated with the MLP method, the sheer magnitude of the improvement, emphasized by the identical log scales of the boxplots, is extremely advantageous in the context of developing a population-scale ABM and far beyond any improvement that could currently be achieved through simple parallelization of the iterative computation method (Fig. 2). For 10 runs of a 1,000-agent sample, the average execution time was 38,957 s (standard deviation 956 s) for the iterative method and 0.53 s (standard deviation 0.23 s) for the MLP method [5]. The following subsections contain more information on the MLP itself and how its performance compared to the original iterative computation.

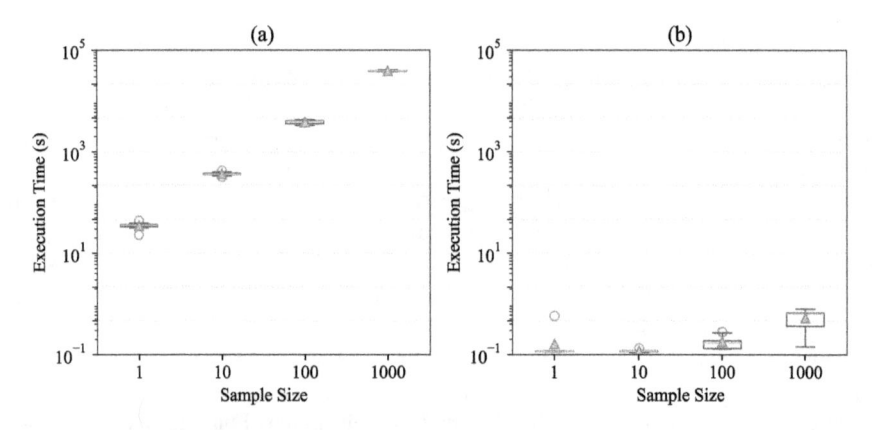

Fig. 2. Boxplots of execution time in seconds on a log scale for (a) the original iterative solution method and (b) the neural network solution method for increasing sample size demand. Triangles indicate the mean of 10 runs. The y-axis scale is kept constant for easier comparison.

3.1 Model Selection

In Fig. 3, the best and worst performers in the metrics chosen as a basis of model selection are shown. The highest ranked model for overall combined MAE was the P6L architectural ratio with a maximum layer width of 2,048 nodes, a learning rate of 0.001, and a batch size of 64. Note that, while the same model was also ranked first for the metric pertaining to i_a, it placed tenth for c_t MAE. The closeness of the former and discrepancy of the later may be due to the order of magnitude difference between the scales for the outputs; the error in i_a (maximum 0.5) had finer granularity in its influence on MSE loss during training than that of c_t (maximum 40). The precise model applied to the ABM used the previously stated hyperparameters with a seed of 84 and took 954 s to train.

Other findings well highlighted by the plots are the great similarity in performance among the better models, and the poor performance of 3L models with 0.0001 learning rates in terms of overall MAE (Fig. 3a). Also, visible in Fig. 3c, the higher learning rate, 0.01, produced results with greater variance in response to seed change. This tendency was also generally true of the middle values not shown in the figure. The trend is sensible since a high rate leads to coarser network updates, making the initialization more crucial and potentially pushing the model into suboptimal regions of the loss landscape [8]. It should also be noted that neither batch normalization nor dropout were used in the network model.

3.2 Trends in Hyperparameter Optimization

While conducting hyperparameter grid search without optimization in the search itself may be deemed inefficient, it was elected primarily as a straightforward, thorough way of scientifically selecting a model and in response to the lack of empirical studies reporting architecture and hyperparameter selection in the context of MLPs for equation mapping. As computational budgets for this task permitted, the exhaustive search was conducted to determine if any general trends could be identified to direct future attempts at similar mapping problems. It may be speculated that while the findings in this research cannot be generalized per se, they may perhaps serve as a way to justify targeting a particular region of the search space.

The differences in performance by controlled hyperparameter averaged over the remaining three are shown in Fig. 4. For the maximum layer width, n_{max}, the downward trend in medians is very subtle and not observable in the mean at all, indicating that after 1,024 nodes there is little improvement; however, there is a reduction in variance as number of nodes increases within the studied range (Fig. 4a). The response of changing the learning rate is more complex, with more of the combinations performing their worst with the lowest learning rate and their best at a moderate learning rate of 0.001 (Fig. 4b). This is consistent with the prior observation about high variance with high learning rate and for the lowest learning rate may be speculatively attributed to the limitation to 200 epochs [8]. For the architectural shape, while the improvement from adding a fourth layer is substantial, it appears that adding layers beyond four results in very small marginal improvements in mean performance (Fig. 4c). Another note of interest is the small gain achieved by ramping up to the widest hidden layer as demonstrated in the differences in median and mean between 5L and P5L (Fig. 4c, Tab. 3). As documented in literature, an increase in batch size is one technique sometimes used to decrease model training time; this choice is, at some

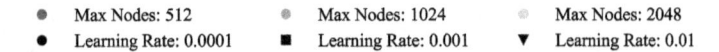

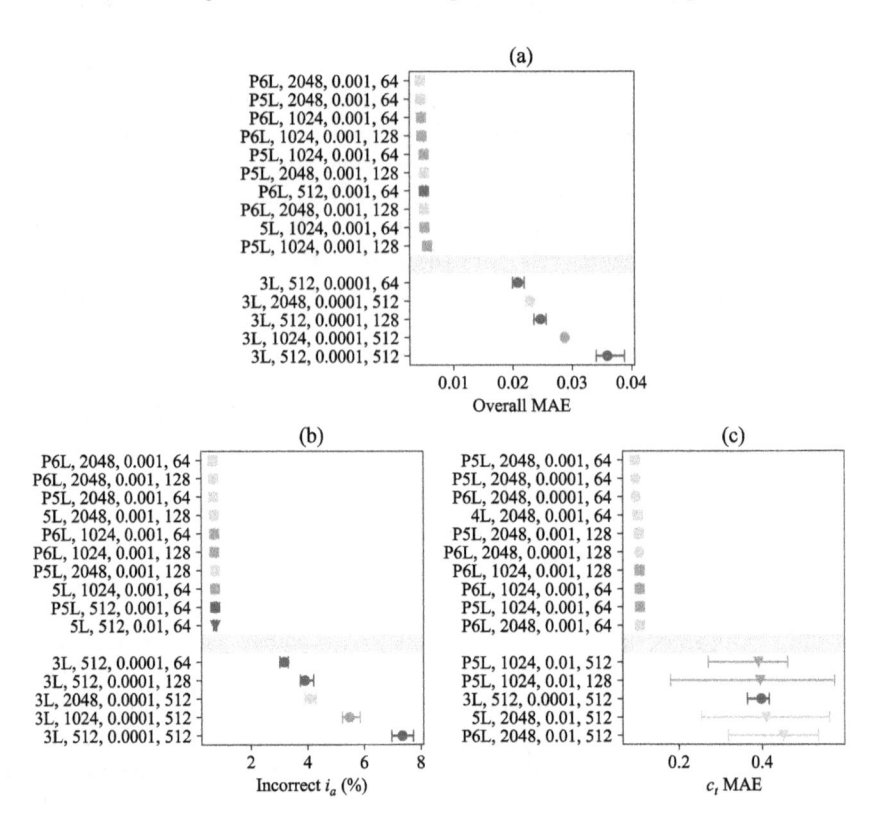

Fig. 3. The ten highest- and five lowest-ranked hyperparameter combinations performance in (a) overall mean absolute error, (b) percent of incorrect guesses for adaptation investment, and (c) mean absolute error in consumption for the validation sample in comparison with the original iterative Bellman equation method. The combination labels follow the convention: Architecture Type, n_{max}, Learning Rate, and Batch Size as specified in Table 2. For each point marking the mean value of the metric, the error bars represent the minimum and maximum value, the point shape indicates the learning rate, and the point color is indicative of n_{max}.

point, to the detriment of the model performance [25]. Within the parameters of this grid search there was an upwards trend in error with increasing batch size (Fig. 4d). This suggested that, for the tested learning rates, most of the batch sizes were larger than the critical size, beyond which gradient noise becomes too low, decreasing the generalizability of the trained model to testing data. All of the observations stated for the MAE aligned closely with results for MSE.

For all hyperparameters, the variance in performance suggested the importance of their interaction effects. The heatmaps in Fig. 5 facilitate the investigation of relationships between hyperparameters as well as successful and poor

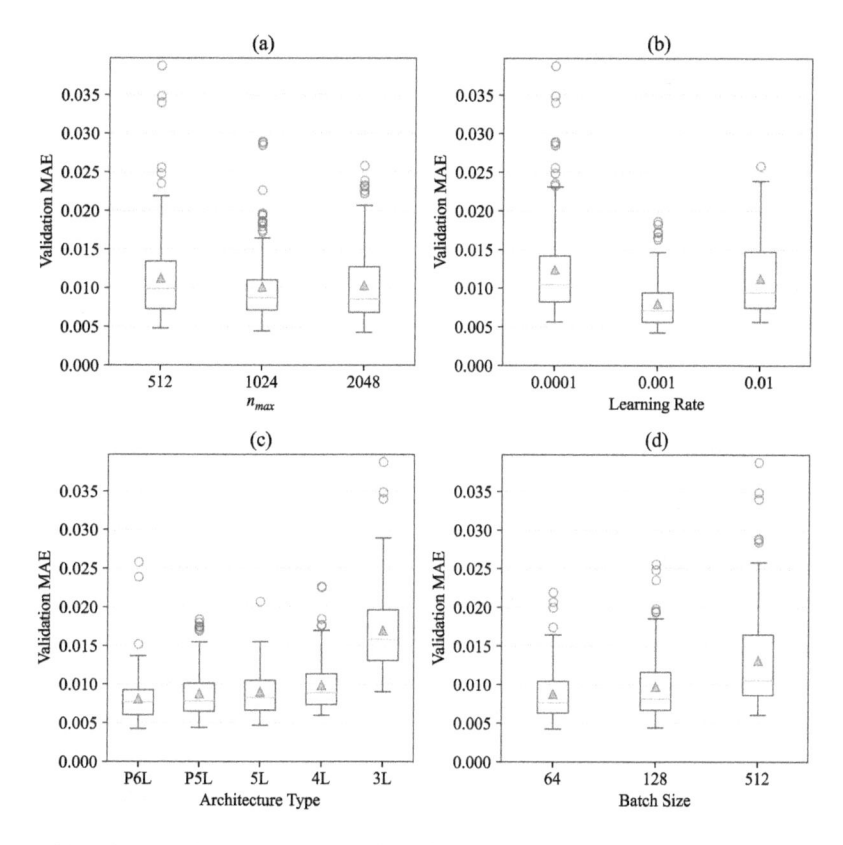

Fig. 4. Boxplots marking the spread of performance in overall mean absolute error for the validation sample categorized by (a) maximum number of nodes, (b) learning rate, (c) architecture type, and (d) batch size. Triangles indicate the mean value.

combinations. Similarly to the boxplots, it can be observed that learning rate had a strong influence, but in the heatmap, it is more clear that the width of the network had an impact on the magnitude of that effect (Fig. 5f). As mentioned before, the average performance decreased with increase in batch size and improved with the addition of layers, however the learning rate appears to modulate these effects (Fig. 5d,e). From these plots, general combinations to embrace or avoid in future training exercises can be identified; e.g., a model with batch size of 64, n_{max} of 1024, learning rate of 0.001, and P6L structure will likely perform well, while a model with batch size of 512, n_{max} of 512, learning rate of 0.0001, and 3L structure will likely perform very poorly (Fig. 5).

3.3 Limitations, Notes, and Future Work

A weak point in making generalizations, even in the very specific context of this case study task, is the small model seed sample size of three. Although the variance for points of the same combination was fairly low, particularly for

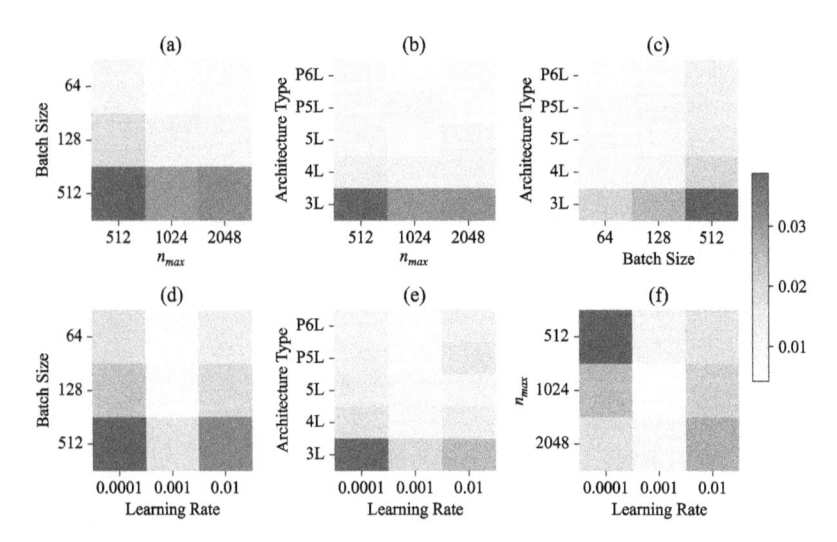

Fig. 5. Heat maps showing the combined effects of two hyperparameters on the overall mean absolute error for the validation sample. The hyperparmeters compared include (a) batch size to maximum nodes, (b) architecture type to maximum nodes, (c) architecture type to batch size, (d) batch size to learning rate, (e) architecture type to learning rate, and (f) maximum nodes to learning rate. Dark blue indicates relatively high error while light green indicates relatively low error. (Color figure online)

the best-performing models, uncertainty in performance and ranking could be reduced by training more models with each hyperparameter combination. Upon further research into general guidelines surrounding MLPs, it was surmised that 2,048 nodes is considered very wide for a network with only four inputs and that such an overabundance of nodes may potentially lead to memorization of training sets [8]. For models with an n_{max} of 2,048, the average MSE of the validation dataset was approximately 93% higher than that of the training dataset, strongly suggesting that these models were subject to overfitting; however, as the average absolute difference was ∼0.00044, the practical consequences were considered negligible, and the initially selected model was retained for use with the ABM. This knowledge will be considered for the next iteration of hyperparameter search and training, which will incorporate the adaptation option information as input variables so that more flexibility can be gained in the ABM implementation. Future efforts will also use larger training sample sizes to improve generalizability and reduce overfitting. Expanding the model inputs would provide some answers as to whether it might be appropriate to generalize the research to other Bellman equations. Further comparison against optimized solvers of other equation forms would be useful in determining the robustness of the method and findings, but this type of testing was far beyond the scope and aspiration of this research.

The evidence that the shape of the architecture had relevance, more specifically that ramping the width up then down versus abruptly up was better for results, deserves further experimentation. It is unclear whether the total number of nodes being slightly higher was the driver of this outcome or the shape itself was responsible. Although only tangentially related, a similar observation was made in a study of convolutional neural networks [10]. Other experiments which were conducted outside of the scope of this manuscript included splitting the task into two MLPs (one for each output with i_a as categorical), experimenting with alternate activation functions (Swish and Mish), and experimenting with scaling inputs and outputs. With regard to splitting the model, preliminary results did indicate that there was minor improvement, particularly in results for consumption, which is to be expected given the current arrangement in which it is effectively underrepresented in the loss calculation during training. By using Swish and Mish in place of ReLU, there was improvement in the average metrics for incorrect i_a and c_t MAE, particularly with Mish. However, in both the split case and the alternate activation function case, the improvement was so marginal, that it was deemed unnecessary to replace the equation in the ABM. Still, these aspects may be explored further in future training exploits. The experiments on scaling included omitting the scaling process for inputs, outputs, or both. While the outcome of these experiments will not be fully summarized here, the most telling statistics were that, in an average performance, using both unscaled inputs and outputs yielded a 10,940% increase in model MSE, 1,259% increase in overall MAE, and 18% increase in training time over fully scaled counterparts.

4 Conclusion and Takeaway

As ABMs and other models which rely on computationally complex subfunctions grow in scale, occasions where their implementation is inhibited by techniques which were previously acceptable become more prevalent. The case study solved as documented in this manuscript is but a single example to set a precedent for the use of MLPs to overcome this situation. While the journey taken in the process of conducting this research could definitely be termed a learning experience, the results met the need that was established well enough to allow the modelling of agent behavior with a Bellman equation to continue without a major overhaul in fundamental basis. While perhaps not a perfect replacement, the neural network performs with sufficient accuracy, at a far faster rate, and scales sublinearly with the addition of more agents. This provides hope that other models may benefit from similar replacement strategies in the future, perhaps incorporating some of the caveats against overfitting, observations on thresholds of diminishing return for node and layer additions, and notes on architectural performance differences.

Acknowledgments. The author acknowledges and is grateful for constructive academic discussions with Debraj Roy and Alex Gabel. This research was conducted with support from the Dutch Research Council (NWO) under contract 27020G08, titled "Computing societal dynamics of climate change adaptation in cities".

Disclosure of Interests. The author is aware of no competing interests regarding the content of this article.

A Appendix

Algorithm 1: Computation of optimal consumption and adaptation

Input: Dictionary of agent parameters with k_t, θ_t, σ, α and $AdapTable$ (array of i_a and m pairs); $TechTable$ (array of γ and $cost$ pairs)
Output: $i_{a,t} \in \{N, L, H\}$ and c_t

1 $feasible \leftarrow []$
2 **for** $i_a \in AdapTable$ **do**
3 $\quad$ **if** $i_a < \max_{i=1,...,n} [\mathrm{f}(\alpha, k_t, TechTable) + \theta_t (1-\delta) k_t]$ **then**
4 $\quad\quad$ $(c, v, \text{convergence}) \leftarrow$ solve_bellman(BellmanEquation(u, f, $k_t, \theta_t, \sigma, \alpha, i_a, m$))
5 $\quad\quad$ $feasible.\mathrm{append}([c, v, \text{convergence}])$

6 $index \leftarrow "NaN"$, $\ converged \leftarrow []$, $\ values \leftarrow []$, $\ \mathrm{labels}=[\mathrm{N,L,H}]$
7 **for** $result \in feasible$ **do**
8 $\quad$ $converged.\mathrm{append}(result[2])$
9 $\quad$ $values.\mathrm{append}(result[1])$

10 **if** $False \in converged$ **then**
11 $\quad$ **return** "NaN", "NaN"
12 **else**
13 $\quad$ $index \leftarrow values.\mathrm{argmax}()$
14 $\quad$ $i_a \leftarrow \mathrm{labels}[index]$ max $\leftarrow feasible[index]$ $c_t \leftarrow \max[0]$
15 $\quad$ **return** i_a, c_t

16 **Function** solve_bellman(BellmanEquation):
17 $\quad$ Function solve_bellman(BellmanEquation) is primarily a while loop of update_bellman(v_{grid}, BellmanEquation) based on the method described in Stachurski [22]. If errors for the policy solution are not within tolerances at loop exit, convergence $\leftarrow$ False.
18 $\quad$ **return** $c, v, convergence$

19 **Function** update_bellman(v_{grid}, BellmanEquation):
20 $\quad$ Function update_bellman(v_{grid}, BellmanEquation) iterates over an array of grid points to maximize and update their value function results, v_{grid}, through optimization of consumption, c_{grid}, very similar to the method described by Stachursky [22].
21 $\quad$ **return** c_{grid}, v_{grid}

22 **Function** u(c, σ):
23 $\quad$ Function u(c, σ) returns utility of given consumption.
24 $\quad$ **if** $\sigma \neq 1$ **then**
25 $\quad\quad$ **return** $\frac{c^{1-\sigma}-1}{1-\sigma}$
26 $\quad$ **else**
27 $\quad\quad$ **return** $\ln(c)$

28 **Function** f(k_t, α, $TechTable$):
29 $\quad$ Function f($k_t, \alpha, TechTable$) returns agent income.
30 $\quad$ $income \leftarrow []$
31 $\quad$ **for** $i \in TechTable.\mathrm{keys}()$ **do**
32 $\quad\quad$ $entry \leftarrow \alpha * k^{TechTable[i][0]} - TechTable[i][1]$
33 $\quad\quad$ $income.\mathrm{append}(entry)$
34 $\quad$ **return** $\max(income)$

References

1. Barber, C.B., Dobkin, D.P., Huhdanpaa, H.: The quickhull algorithm for convex hulls. ACM Trans. Math. Softw. **22**(4), 469–483 (1996). https://doi.org/10.1145/235815.235821
2. Bellman, R.: On the theory of dynamic programming. Proc. Natl. Acad. Sci. **38**(8), 716–719 (1952). https://doi.org/10.1073/pnas.38.8.716
3. DeAngelis, D.L., Diaz, S.G.: Decision-making in agent-based modeling: a current review and future prospectus. Front. Ecol. Evol. **6**, 237 (2019). https://doi.org/10.3389/fevo.2018.00237
4. Espinosa Barcenas, O.U., Quijada Pioquinto, J.G., Kurkina, E., Lukyanov, O.: Surrogate aerodynamic wing modeling based on a multilayer perceptron. Aerospace **10**(2), 149 (2023). https://doi.org/10.3390/aerospace10020149
5. Garibay, V.M.: nnbellman (2025). https://github.com/vmgaribay/nnbellman
6. Garzón, A., Kapelan, Z., Langeveld, J., Taormina, R.: Machine learning-based surrogate modeling for urban water networks: review and future research directions. Water Resour. Res. **58**(5), e2021WR031808 (2022). https://doi.org/10.1029/2021WR031808
7. Garín, J., Lester, R., Simms, E.: Intermediate macroeconomics (2021). https://sites.nd.edu/esims/textbook/
8. Goodfellow, I., Bengio, Y., Courville, A.: Deep Learning. MIT Press (2016). http://www.deeplearningbook.org
9. Grignard, A., Taillandier, P., Gaudou, B., Vo, D.A., Huynh, N.Q., Drogoul, A.: GAMA 1.6: advancing the art of complex agent-based modeling and simulation. In: Boella, G., Elkind, E., Savarimuthu, B., Dignum, F., Purvis, M.K. (eds.) PRIMA 2013. LNCS (LNAI), vol. 8291, pp. 117–131. Springer, Heidelberg (2013). https://doi.org/10.1007/978-3-642-44927-7_9
10. Han, D., Kim, J., Kim, J.: Deep pyramidal residual networks. In: Proceedings of the IEEE Conference on Computer Vision and Pattern Recognition, pp. 5927–5935 (2017). https://doi.org/10.48550/arXiv.1610.02915
11. Huang, V.: Pytorch-template (2020). https://github.com/victoresque/pytorch-template. Accessed 4 Mar 2024
12. Jones, D.R., Schonlau, M., Welch, W.J.: Efficient global optimization of expensive black-box functions. J. Global Optim. **13**, 455–492 (1998). https://doi.org/10.1023/A:1008306431147
13. Joppan, N.T.: Modelling poverty alleviation strategies using resilience thinking. Master's thesis, University of Amsterdam (2021). https://scripties.uba.uva.nl/search?id=record_30354
14. Jäger, G.: Replacing rules by neural networks a framework for agent-based modelling. Big Data Cognit. Comput. **3**(4) (2019). ISSN 2504-2289, https://doi.org/10.3390/bdcc3040051
15. Kuriksha, A.: An economy of neural networks: learning from heterogeneous experiences. PIER Working Paper No. 21-027 (2021). https://doi.org/10.2139/ssrn.3973697
16. Ma, C., Zhu, B., Xu, X.Q., Wang, W.: Machine learning surrogate models for landau fluid closure. Phys. Plasmas **27**(4) (2020). https://doi.org/10.1063/1.5129158
17. Maliar, L., Maliar, S., Winant, P.: Will artificial intelligence replace computational economists any time soon? CEPR Discussion Paper No. DP14024 (2019). https://ssrn.com/abstract=3464569

18. Philipp, G.: The nonlinearity coefficient - A practical guide to neural architecture design. ArXiv (2021). https://doi.org/10.48550/arXiv.2105.12210
19. Richetti, J., Diakogianis, F.I., Bender, A., Colaço, A.F., Lawes, R.A.: A methods guideline for deep learning for tabular data in agriculture with a case study to forecast cereal yield. Comput. Electron. Agri. **205**, 107642 (2023). ISSN 0168-1699, https://doi.org/10.1016/j.compag.2023.107642
20. Saltelli, A.: Making best use of model evaluations to compute sensitivity indices. Comput. Phys. Commun. **145**(2), 280–297 (2002). ISSN 0010-4655, https://doi.org/10.1016/S0010-4655(02)00280-1
21. Sobol', I.M.: Global sensitivity indices for nonlinear mathematical models and their monte carlo estimates. Math. Comput. Simul. **55**(1), 271–280 (2001). ISSN 0378-4754, https://doi.org/10.1016/S0378-4754(00)00270-6, the Second IMACS Seminar on Monte Carlo Methods
22. Stachurski, J.: Economic Dynamics, second edition: Theory and Computation. MIT Press (2022). ISBN 9780262544771
23. Su, J., Cheng, H., Guo, H., Huang, R., Peng, Z.: An approximate quadratic programming for efficient bellman equation solution. IEEE Access **7**, 126077–126087 (2019). https://doi.org/10.1109/ACCESS.2019.2939161
24. Virtanen, P., et al.: SciPy 1.0: fundamental algorithms for scientific computing in Python. Nature Methods **17**, 261–272 (2020). https://doi.org/10.1038/s41592-019-0686-2
25. Zhang, G., et al.: Which algorithmic choices matter at which batch sizes? Insights from a noisy quadratic model (2019). https://doi.org/10.48550/arXiv.1907.04164

Simulation-Based Inference in Agent-Based Models Using Spatio-Temporal Summary Statistics

Eric Dignum[✉][iD], Harshita Choudhary, and Mike Lees

University of Amsterdam, Science Park 900, 1098 XH Amsterdam, The Netherlands
e.p.n.dignum@uva.nl

Abstract. In agent-based models (ABMs), traditional statistical inference faces challenges due to intractable likelihoods and computational costs. This study evaluates neural posterior estimation (NPE) and neural ratio estimation (NRE) for parameter inference in ABMs and compares them with approximate Bayesian computation (ABC). NPE and NRE are argued to be more efficient than traditional methods such as ABC and circumvent some of their limitations. The assessment of the methods focuses on the satisfaction threshold in Schelling's model of residential segregation, including regions of high variance and non-equilibrium dynamics. As these simulation-based methods still require summary statistics as high-level descriptions of the ABM, we propose a general approach to construct them based on spatial and/or temporal information and evaluate how the different summary statistics affect performance. Both NPE and NRE generally outperform ABC regardless of summary statistics. Most notably, NRE excels when employing the most detailed spatio-temporal information, but adding spatial or temporal information alone is not always beneficial for NPE, NRE and ABC. This holds true for different training budgets and when estimating multiple parameters. Hence, the study underscores the importance of spatio-temporal information for accurate parameter inference in this ABM, but information redundancy can degrade performance as well. Therefore, finding optimal high-level descriptions to capture fundamental emergent patterns in the model through summary statistics might prove crucial in cases where the systems are governed by more complex behaviour.

Keywords: simulation-based inference · agent-based modelling · neural posterior estimation · neural ratio estimation · approximated bayesian computation

1 Introduction

Traditional statistical methods for parameter inference heavily rely on likelihood functions. However, in many cases, models reach a level of complexity where the likelihood becomes impossible to derive exactly or even sample from. In such cases, simulation-based inference (SBI) methods have emerged as a powerful tool for parameter inference when it is still possible to generate data from the model under study. Recently, there has been an intersection of deep learning with the field of SBI, resulting in novel neural network-based inference approaches [2]. One modelling paradigm that might benefit from these novel methods is agent-based modelling. Agent-based models (ABMs) often consist of many (heterogeneous) agents with rule-based actions that interact with each other and their environment. This can result in macro-scale emergent outcomes that are often not expected when studying agents in isolation [3]. Using mathematical modelling and computational algorithms, one can simulate the behaviour of agents, their interactions, and consequential actions, all of which influence the overall dynamics of the system [1]. While a flexible modelling paradigm, due to interdependent and complicated behavioural rules, non-linear dynamics, and inherent stochasticity, it is often very difficult or impossible to perform likelihood-based inference on. As these new SBI techniques can identify complex relationships efficiently, they seem to be very suitable for calibrating ABMs on data [5].

Approximate Bayesian computation (ABC) [16] is a SBI method that tries to approximate the posterior, the values of the parameters that are most likely to have generated the observed data, based on the discrepancy between the simulated (model-generated) data and the observed data. It is common practice in SBI to use summary statistics which provide a condensed representation of model behaviour, capturing essential information while reducing the dimensionality. ABC in particular requires choosing a distance metric and a threshold; a parameter value is accepted to come from the posterior distribution if the discrepancy between simulated and observed summary statistics falls within the threshold or filter those with the smallest distances to the observed data. However, deciding the distance metric and the threshold value can greatly affect performance [5]. Moreover, ABC aims to estimate an approximate posterior, the accuracy of which depends on the chosen threshold, and inference is only valid for the specific observation used; hence, the procedure has to be repeated for different (empirical) observations of the same system. The newly proposed neural network-based approaches aim to learn the relationship between the parameter values and their corresponding summary statistics. This addresses some of the limitations encountered in traditional ABC techniques. They eliminate the need to choose a distance metric and threshold, and they are amortised, meaning they can estimate posterior probabilities for any new observation not seen by the network. Lastly, they do not throw away samples as in ABC, but they use all available data. This can lead to a more efficient use of simulations, which is especially important when these are computationally expensive, which is typically the case for ABMs. However, this new generation of inference techniques has yet to be extensively tested for parameter inference in ABMs, nor has it been simultaneously compared with ABC. In addition, summary statistics still have to be selected carefully for the calibration methods, and a general approach to do this for ABMs is currently missing.

In this paper, we propose a general approach to construct summary statistics for spatio-temporal ABMs, while comparing the performance of neural posterior estimation (NPE) [6,12] and neural ratio estimation (NRE) [7,11] with ABC. NPE and NRE have shown promising results for economic ABMs, producing significantly more accurate parameter estimations while requiring fewer simulations [5], but only using a specific configuration of the models tested. We demonstrate our approach by calibrating the entire range of the satisfaction threshold parameter of Schelling's model of residential segregation [13]. This model is chosen as a test case as it has only one parameter that governs segregation dynamics and its underlying mechanisms are well described. Depending on the value of the threshold, one can observe emergent behaviour, non-linear patterns, non-equilibrium dynamics and/or regions with a large variance in outcomes. These can provide challenging cases for the calibration methods while still having relatively easy-to-understand dynamics. Our approach is general enough that it can be applied to other (spatial) ABMs, and our results provide some indication of how the inference methods may perform in other scenarios. Both NPE and NRE generally outperform ABC regardless of summary statistics. Most notably, NRE excels when employing the most detailed spatio-temporal information, but adding spatial or temporal information alone is not always beneficial for the methods. Hence, the study underscores the importance of spatio-temporal information for accurate parameter inference in this ABM, but information redundancy can degrade performance as well. Therefore, finding optimal high-level descriptions to capture fundamental emergent patterns in the model through summary statistics might prove crucial in cases where the systems are governed by more complex behaviour.

2 Background and Related Work

In this section, only three approaches for calibrating simulation models are discussed. For a detailed overview of existing methods, we refer to [2]. In SBI the aim is to estimate the posterior probability distribution $P(\theta|X)$, i.e., the posterior distribution of the model parameters (θ) conditional on both the simulated data from the model (X^{sim}) and potentially the observations of the real system (X^{obs}). When the likelihood $P(X|\theta)$ is intractable, but it is possible to simulate X from a generative model given a set of parameter values (θ), one can perform SBI. One of the most commonly used techniques in this area is approximate Bayesian computation (ABC).

2.1 Approximate Bayesian Computation

ABC has been applied to the calibration of ABMs in various fields, including economics [8], epidemiology [18], and cancer research [14]. In this section, only the most basic version of ABC will be described. However, for more details, [9] provide an elaborate study on the various improvements and extensions of ABC algorithms. The general idea of ABC methods is to generate samples from the posterior distribution by simulating data $X^{sim} \sim p(X|\theta)$ and to assess whether the simulated data are close to the observed data. Specifically, if the discrepancy between X^{sim} and X^{obs} according to a distance metric d falls within a certain acceptance threshold ϵ. Note that this makes posterior inference using ABC only valid for the specific observation (X^{obs}) used, which is also called: non-amortised. Rejection ABC is the most basic algorithm of the ABC methods [16]. Here, θ_i is randomly sampled from a prior distribution $P(\theta)$. Those θ_i for which the distance between X^{sim} and X^{obs} is less than ϵ are accepted as samples from the posterior distribution. General ABC rejection for sampling one parameter from the posterior is as follows:

1. Sample $\theta_i \sim P(\theta)$
2. Run the generative model with θ_i as input and save summary statistics X_i^{sim} that are a description of the model behaviour
3. Accept θ_i as a sample coming from the posterior $P(\theta|X)$ if $d(X_i^{sim}, X^{obs}) \leq \epsilon$
4. Repeat this until a specific number of accepted samples or total simulations is reached

This technique draws independent samples from the approximate posterior. However, there are some drawbacks to this approach. First, one has to choose an appropriate distance metric and a value for ϵ. Second, the acceptance rate is typically low, specifically when the posterior is much narrower than the prior distribution, and third, it can be computationally very expensive for a small value of ϵ. In this study, 10% of the simulations with the smallest Euclidean distances to the observed data are kept to estimate the posterior distribution.

2.2 Neural Density Estimation

To avoid the dependence on ϵ, [12] proposed NPE, that frames parameter inference as a conditional density estimation problem. The method takes the input X and produces the posterior distribution $P(\theta|X)$ by training a neural network (NN) on simulated data X^{sim} and the corresponding parameter vector θ. The NN tries to learn the probabilistic relationship between X^{sim} and θ through this training. Note that the empirical data has not been utilised yet and the NN essentially acts as a surrogate for the generative model. This approach amortises the inference process, involves training a neural conditional density estimator once, using training data consisting of data-parameter pairs (X_i^{sim}, θ_i) where $X_i^{sim} \sim P(X|\theta = \theta_i)$, and then condition the posterior distribution for any X^{obs}. This is an improvement over ABC, as ABC's inference is only valid for the specific X^{obs} used. Various variants were developed, but we use the NPE-C variant of [6] as it is the most flexible and best performing approach.

NRE takes a different approach and uses supervised classification [11]. In its simplest form, it works as follows: create a random dataset of independent pairs by shuffling the data-parameter pairs as described above. A neural network is then trained to classify which combinations belong to the dependent dataset and which ones to the independent one. Specifically, NRE uses a neural network as a classifier to distinguish between dependent data points $(X_i^{sim}, \theta_i) \sim P(X, \theta)$ and independent data points $(X_i^{sim}, \theta_j) \sim P(X)P(\theta)$. The dependent pairs are generated from the simulator, while the independent pairs can be obtained by shuffling the (X_i^{sim}, θ_i) pairs. This destroys the dependencies, thus associating X_i^{sim} with a random θ_j. Using the likelihood-to-evidence ratio, which represents the likelihood that the pair belongs to the dependent dataset, one can obtain the posterior distribution with the help of MCMC sampling. Similarly to NPE, NRE is an ϵ-free inference technique that does not require the acceptance-rejection step. It is also simulation-efficient (it does not reject simulations) and does not rely on a distance function. However, unlike NPE, direct sampling from the posterior is not feasible, as likelihood ratios are calculated. This necessitates the use of a sampling technique such as MCMC, introducing an additional, more computationally expensive step in the process. The variant of [11] is employed here (NRE-C) as it performs better than other variants in their experiments.

Both NPE and NRE have been used for parameter inference in ABM by [5] where they show similar or even improved performance compared to kernel density estimation techniques with fewer simulations in an economic ABM. However, they do not systematically sweep the parameter space. Hence, there might be parts where the techniques struggle to infer the parameters of the models. Moreover, they employ one set of self-constructed summary statistics and use a NN to learn a set of summary statistics from the simulations, but it is not clear how varying summary statistics more systematically affect the methods.

3 Methodology

This section provides a detailed explanation of our specific implementation of the Schelling model, as it is important to relate it to the performance of the calibration methods. Subsequently, the setup of several experiments with different summary statistics for the calibration methods is described, followed by the calibration pipelines for ABC, NPE and NRE to estimate the true value of the satisfaction threshold.

3.1 Schelling's ABM of Residential Segregation

To investigate the phenomenon of racial segregation, Schelling [13] developed one of the first ABMs. He showed that even a mild preference for having people of the same group in your local neighbourhood could result in a highly segregated society. Hence, the model yielded counter-intuitive findings, demonstrating that the outcome of the collective behaviour of agents could differ from the intentions of the isolated individuals, due to the non-linearity caused by the interacting agents. Although not completely similar to the original, below we describe the version used in this study.

- **Initialisation:** Agents are randomly placed on an 80 × 80 grid and their total number is such that the density is 90% to allow for relocation. With a probability 0.5, an agent belongs to either the blue (B) or orange (O) group.

- **Movement:** At each time step, 15% of agents are allowed to move. An agent is satisfied in their current location if the fraction of individuals of the same group in their 8 surrounding cells (Moore neighbourhood) is above a certain threshold (μ_h). If an agent is not satisfied, they move to a randomly chosen vacant cell, otherwise they stay in their location. The tolerance level of each household (μ_h) is sampled from a truncated normal distribution with mean μ and standard deviation of 0.05.
- **Stopping:** The simulation is stopped after 100 time steps (to reduce computation time) or when every agent is satisfied.

The relocation process in the model occurs iteratively over time, where the movement of dissatisfied agents can subsequently affect the satisfaction levels of agents in their new neighbourhood. This potentially triggers a cascade of relocations. Segregation is measured as the average fraction of similar neighbours. For every household, the fraction of similar neighbours (the eight surrounding cells) is calculated, itself excluded, and averaged over all households. Although numerous metrics have been proposed to quantify segregation [10], the average fraction of similar neighbours is selected because of straightforward use and interpretation.

This simple but powerful model demonstrates that segregation can be much higher than would be needed to satisfy individual preferences. Figure 1 illustrates the results of the Schelling model with different tolerance levels, denoted by μ. These tolerance values indicate the threshold required for agent satisfaction. When $\mu = 0.3$, an agent is considered satisfied if at least 30% of its immediate neighbours, here considering a neighbourhood size of 8, belong to the same group. Note that satisfaction threshold, tolerance parameter, and similar terms are used interchangeably and refer to the parameter μ.

3.2 Experimental Design

ABMs are time-driven models and often have an explicit spatial context [1]. We thus propose and base our experiments on a generic approach of taking summary statistics at different points (or scales) in time and space. With the Schelling model, a series of experiments is undertaken across the entire range of the tolerance parameter $\mu \in \{0.1, 0.2, ..., 0.8\}$. Both 0 and 0.9–1 are excluded as they are very similar to those of 0.1 and 0.8 respectively.

This assessment allows us to analyse the effectiveness of both methods in various regimes of model behaviour, including stable regions, non-equilibrium dynamics, and regions with substantial variance in observations to test the accuracy of ABC, NPE and NRE in different circumstances and compare them. In this context, the summary statistics are expected to describe the dynamics of the model sufficiently, such that one can infer the true values of the parameters that generated the data. Relatively low-dimensional summary statistics are often comparatively easy to interpret and can serve as a good indicator of the overall dynamics of the model. However, it might also lead to the loss of important information and affect the performance of the calibration methods. Therefore, several summary statistics are tested to see how this affects performance of ABC, NPE and NRE.

- **Scalar**: A scalar value that quantifies the level of segregation (average fraction of similar neighbours) observed at the end of one simulation ($t = 100$).
- **Temporal**: A 5-dimensional vector that represents the segregation evaluated in time steps $t = 20, t = 40, \ldots, t = 100$.
- **Spatial**: The average fraction of similar neighbours is calculated with a Moore neighbourhood of sizes 1, 2, 3, 4 and 5 at $t = 100$. The relocation rule for the agents is still based on a radius of 1.
- **Spatio-temporal**: Combines both spatial and temporal dimensions. A 25-dimensional vector that includes segregation observed at time steps (20, 40, ..., 100) and within various radii values (1, 2, 3, 4, and 5).

The training budget for the methods is varied ($N \in \{500, 1000, 2000\}$) to see how this affects the performance for the different summary statistics. The spatio-temporal summary statistics, containing 25 values, might benefit most from increasing the training budget. With lower sample sizes, inference methods could struggle to approximate posteriors in this high-dimensional space. As the implemented ABC-method retains 10% of the samples, the number of samples drawn from the posterior estimations of NPE and NRE are changed accordingly, to make it a fair comparison (i.e., $M \in \{50, 100, 200\}$). Moreover, a heterogeneous version of Schelling's model, where the two groups have different μ values, is tested. In this case, the methods have to infer two parameters and possibly a more complicated relationship between the tolerance parameters and the observed level of segregation. Note that we calculate the level of segregation at different spatio-temporal scales, but this can be any other metric that is of interest. For example, in an epidemiological model, the percentage of infections on the different scales could be used instead of segregation.

In summary, ABC, NPE and NRE will be used to estimate the satisfaction threshold (μ) of the Schelling model described in Sect. 3.1. The parameter represents the preference to have at least a certain fraction (μ) of similar neighbours. Its value ranges between 0.1 and 0.8, where 0.8 signifies a high degree of intolerance toward other groups, while 0.1 indicates impartiality or (almost) lack of preference. All three methods try to approximate $P(\mu|X)$ using the simulated data (X_i^{sim}, μ_i) generated from the model and an observation X^{obs}. In our case, X^{obs} is also generated by the model and setting $\mu = \mu_{true}$, but in real settings μ_{true} is generally not known. The SBI Python package is used to implement ABC, NPE and NRE with their default architectural specifications [15]. The inference procedure for ABC can be found in Sect. 2.1 and for NPE/NRE it can be summarised as follows:

1. Randomly sample $N \in \{500, 1000, 2000\}$ values from $P(\mu) \sim \mathcal{U}(0,1)$.
2. Run the Schelling model with the input parameters μ_i for each $i \in \{1, 2, ..., N\}$ and save the summary statistics of each run.
3. Train the NN for NPE and NRE on the simulated data pairs (X_i^{sim}, μ_i).

4. Sample $M \in \{50, 100, 200\}$ times from the approximated posterior distribution and calculate the root mean squared error (RMSE) given μ_{true}: $RMSE = \frac{1}{M}\sqrt{\sum_{i=1}^{M}(\hat{\mu}_i - \mu_{true})^2}$.

4 Results

Given the model specifications and parameter settings in Sect. 3.1, the Schelling model is simulated for μ ranging from 0.1 to 0.8, in increments of 0.1 with 10 replications for each parameter value. Figure 1 shows the impact of changing the tolerance parameter on the resulting level of segregation as measured by the average fraction of similar neighbours. For low intolerance values ($\mu = 0.1$), everyone is satisfied with their initial placement, resulting in very low segregation. However, even a slight increase to $\mu = 0.3$ leads to high levels of segregation of more than 0.7. Furthermore, as μ approaches 0.6, the level of segregation increases significantly, resulting in a fully segregated system. For satisfaction thresholds of 0.7 and higher, a huge decrease in segregation can be seen. This is because agents have difficulty finding satisfactory locations. This results in agents continually moving to new locations. At a certain point, there is no solution that can satisfy all (or a sufficient number of) agents. It is difficult to achieve a convergence with such a high threshold, and only a few of the conditions may satisfy all agents. This means that the resulting level of segregation is close to 0.5 as they stay in a different, but close to random configuration every time step (non-equilibrium dynamics). For 0.1–0.7 the model eventually reaches a stable state and for 0.7–0.76 there is a steep decrease with high variance.

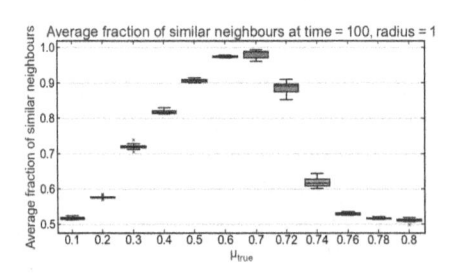

Fig. 1. Average fraction of similar neighbours at $t = 100$ as a function of the tolerance parameter (μ_{true}) and 10 replications for each μ_{true} value. Note that between 0.7–0.8, more values are added to include the high-variance region.

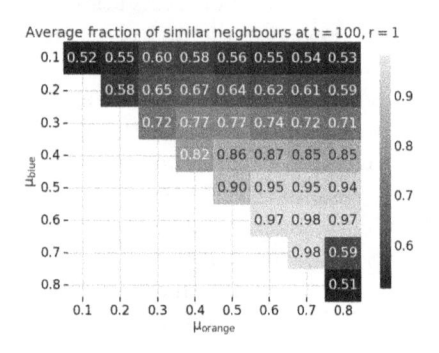

Fig. 2. Average fraction of similar neighbours at the end of a Schelling model run as a function of the satisfaction thresholds (μ_{blue} and μ_{orange}). The numbers are based on the average of 10 model runs for each combination.

4.1 Empirical Calibration with Spatio-Temporal Summary Statistics

Figure 3 displays the posterior samples obtained using the NPE and NRE methods for different summary statistics. Ideally, NPE and NRE draw posterior samples close to the true tolerance value μ_{true} denoted by the dashed line. Although all posteriors assign at least some probability mass around the true value, numerous posteriors are bimodal. This is most evident when μ_{true} is 0.1 or 0.74 and higher. In these cases, some posterior mass is centred around the true value, but most have two modes, one close to 0–0.1 and the other close to 0.8. This seems especially to be the case for the scalar summary statistics. This makes sense given its limited information, since the average fraction of similar neighbours is the same for a satisfaction threshold of 0.1 and 0.8 (Fig. 1). Adding spatial or temporal information to the summary statistics makes this problem less severe, but not necessarily in all cases. For $\mu_{true} \in \{0.1, 0.8\}$, adding spatial information seems to lead to a more bimodal posterior than for the scalar summary statistic, for example. However, when adding spatial and temporal information simultaneously, the bi-modality disappears, and all posterior mass is centred around the true values.

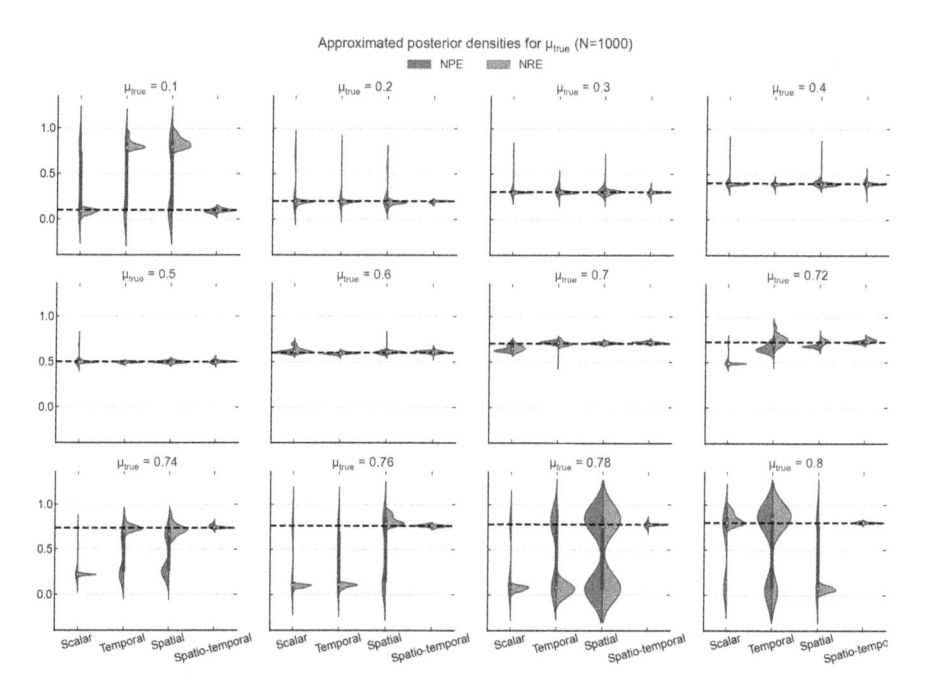

Fig. 3. Approximated posterior distributions for different summary statistics. Plots are based on 100 samples, dashed lines are the μ_{true} values.

To provide a more quantitative assessment of the methods, the different summary statistics and how they compare to ABC, the RMSE is reported in Fig. 4. Moreover, because the Schelling model and the training of the neural networks contain stochastic elements, the calibration procedure is performed 10 times to calculate an average RMSE. Calculations are based on sampling 10% of the training budget, i.e., $M = 100$ samples for $N = 1000$.

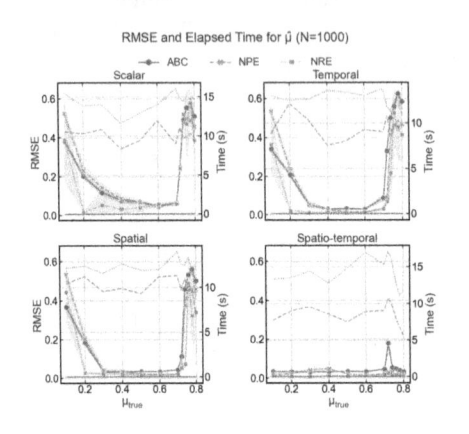

Fig. 4. RMSE for ABC, NPE, NRE and different summary statistics. For every μ value, the methods are trained 10 times, with a different training set consisting of 1000 samples. Hence, the reported RMSE is an average of 10 RMSEs, each based on 100 samples from the approximated posterior. Average elapsed time (lines without markers) is on the axis on the right.

Fig. 5. RMSE when μ_{blue} is varied along the x-axis and averaged over the different values for μ_{orange}. The numbers are based on the average of 10 RMSEs for each combination.

In Fig. 4 it can be seen that in the region $\mu_{true} \in [0.3, 0.7]$, the methods are able to estimate the value of μ_{true} quite accurately regardless of the summary statistic used, with most RMSEs below 0.10. The scalar summary statistic which contains only the average fraction of similar neighbours, seems to perform marginally worse than the other summary statistics in the easy-to-infer region (0.3–0.7). Hence, spatial and temporal trajectories provide important information on the true value of μ. Moreover, in most cases, NRE performs better than NPE and ABC. Using spatio-temporal information increases performance substantially, and NRE does better than NPE, while both outperform ABC. However, for NPE and $\mu_{true} \in \{0.3, 0.4, 0.5\}$ there is an increase in RMSE. In the analysis of Fig. 3, the posteriors are wider than for NRE, but from only this figure it is not clear why this performance decrease is observed for NPE.

Moving towards the more difficult to infer regions of $\mu_{true} \leq 0.2$ and $\mu_{true} > 0.7$, one can see that the methods experience a significant increase in RMSE (Fig. 4) and thus a decrease in performance. Using Fig. 3, the larger RMSEs can be explained because the methods have difficulty distinguishing the regions around 0.1 and 0.8. For the scalar summary statistic, this makes sense as they lead to similar values and adding spatial or temporal information does not seem to change this. Interestingly, the addition of both leads to a substantial improvement, resulting in very low RMSEs. When $\mu_{true} = 0.2$, NRE performs significantly better than NPE and ABC for all summary statistics. As mentioned when describing the dynamics of the Schelling model, both regions are rather extreme cases. For low values, everyone is satisfied immediately, and for high intolerance, agents keep moving around randomly but are unsatisfied. Interestingly, only adding spatio-temporal information simultaneously seems to be able to grasp this correctly.

Between 0.7 and 0.8, a region with a high variance in the average fraction of similar neighbours can be observed (Fig. 1). Here, all methods have the greatest difficulty in terms of RMSE. Not much can be said on the differences in performance of the methods for scalar, spatial, or temporal summary statistics. However, adding spatio-temporal information leads to a significant decrease in RMSE, and this extra information benefits NRE and NPE even more so than ABC. Here, the usefulness of the neural-inspired methods might stand out, as they are better able to learn the (more) complex relationship between the input values and summary statistics in this region of the parameter space. However, this comes at a computational cost. NPE needs roughly 10 s to train and sample (with no clear difference between summary statistics, see Fig. 4), while ABC runs in an instant. Moreover, NRE needs between 10 and 15 s, which is likely due to the extra MCMC step.

As these results can depend on the number of training samples used, the same experiment is repeated for different training budgets ($N \in \{500, 2000\}$). Figures 7 and 8 show results that are very similar to those using $N = 1000$, but there are some noticeable differences. Firstly, in terms of computational cost, NPE and NRE take a couple of seconds less for $N = 500$, but significantly more time for $N = 2000$ (between 200 and 600 s). Moreover, using spatial summary statistics instead of scalar is not always beneficial. In the case of 2000 training samples, NRE performs even better for the scalar summary statistic than for temporal and spatial statistics separately in the case of $\mu_{true} \leq 0.2$. Hence, information redundancy can also hurt performance. Lastly, it seems that increasing or decreasing the training budget causes a slight general drop ($N = 2000$) or an increase ($N = 500$) in RMSE.

4.2 Multi-parameter Problems: Two Groups with Different Thresholds

To test the methods on potentially more challenging and realistic problems, the Schelling model is extended with additional parameters. In this extension, groups have independent and possibly different satisfaction thresholds (μ_{blue}, μ_{orange}), where one group can be tolerant while the other is not, for example. In this case, calibration methods need to approximate the joint posterior distribution and a potentially more complex interaction pattern.

Figure 2 shows the average fraction of similar neighbours for different combinations of the two parameters. Almost complete segregation is the result if both groups have a threshold value between 0.4 and 0.7. If one or both groups have a low threshold value, segregation is low. In the case of both having high threshold, the agents are never satisfied, and hence, segregation is also low, as in the homogeneous case (agents keep on moving). To approximate the posteriors, the methods are again given $N = 1000$ samples for training, and 100 values are sampled from the posteriors to calculate the RMSE with respect to the true values. This is repeated 10 times to arrive at an average RMSE per method and summary statistic, for each combination of the parameters.

For scalar summary statistics, the methods show similar performance (Fig. 5). However, adding extra information does benefit NPE and NRE more than ABC. These results are in line with the one-parameter case. Note that in the figure, μ_{blue} is varied along the x-axis and that these numbers are averaged over the various μ_{orange} values. Furthermore, NPE and NRE sometimes perform more than 50%–80% better than ABC (plots are available upon request), which appears to be mainly in the moderate-high segregation region and when both μ values are large (>0.7). Although ABC sometimes performs better than both NPE and NRE as well. In general, one could say the latter two are obtaining lower RMSEs in most of the parameter space, especially when increasing the amount of information provided by the summary statistics.

Furthermore, when the standard deviations of the tolerance thresholds are allowed to vary ($\sigma_{blue}, \sigma_{orange}$), the number of parameters goes from two to four. The standard deviations allow for control of the degree of heterogeneity within the group. Figure 6) shows three calibrations for different values of the four parameters. As four parameters must now be estimated, the sample size is set to 10,000. Although all methods perform better than the RMSE calculated on the prior values (except for two scalar summary statistic cases), NPE and NRE outperform ABC in almost all cases. Again, this may hint at the fact that these neural network methods are better able to learn in higher-dimensional

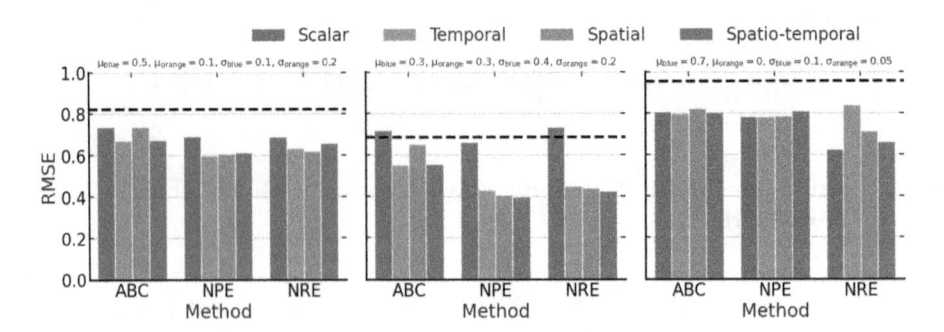

Fig. 6. RMSE of ABC, NPE and NRE when trained on a four parameter problem with 10,000 samples. The dashed line is the RMSE when using the prior values (i.e., training data) as estimates. Contrary to the other experiments, the RMSEs are for one training dataset only.

spaces and/or more complex relationships. In addition, information redundancy sometimes degrades performance. In the last case for NRE, the scalar summary statistic has a lower RMSE than the others and in the first case, both temporal and spatial information separately have lower RMSEs for NPE and NRE.

5 Conclusion

In most ABMs, it is often hard to perform traditional statistical inference, as the likelihood is analytically intractable or computationally expensive to sample from. Fortunately, methods from the field of SBI, do not rely on the likelihood function to conduct parameter inference. This study assessed the performance of two recently developed SBI techniques using neural networks, NPE [6] and NRE [11], and compared them to a more commonly used one: ABC. In addition, we proposed a general approach, suitable for many ABMs, to provide spatio-temporal information in the summary statistics, which is a necessary and crucial ingredient for calibration. Compared with previous studies [4,5], the methods were evaluated for a large part of the parameter space of the satisfaction threshold (i.e., tolerance parameter) in Schelling's model of residential segregation [13], instead of only one particular set of parameter values with relatively straightforward dynamics. This includes regions of high variance in output and non-equilibrium, as well as equilibrium dynamics. This could be (more) challenging dynamics to perform empirical calibration on. In addition, changes in the amount of spatial and/or temporal information in the summary statistics, altering training budgets, and increasing the number of parameters to be estimated were tested.

In general, NRE showed better performance in terms of RMSE than ABC and NPE in most regions of the parameter space, regardless of which summary statistics were used. However, especially when adding the most elaborate spatio-temporal information, a clear performance increase could be seen between NRE and NPE versus ABC, but even more so for NRE than NPE. These conclusions remain unchanged when the training budget was decreased or increased. Having to calibrate two or four parameters instead of one also led to NRE outperforming NPE, and both surpassed ABC, in most cases. This improved performance may be due to the difference in the principal ideas behind the methods. Firstly, calibration methods using neural networks might be better able to learn (complex) relationships in the data than the ABC method and being more efficient. However, neural networks take considerably longer to train and sample, which can be of importance when selecting a method. Additionally, NRE transforms the approximation of the posterior into a classification problem. This may be an easier (supervised) learning task compared to the (unsupervised) learning task of the direct posterior approximation used by NPE, which may explain the improved performance of NRE over NPE. Notable is the increase in RMSE ($\mu \in \{0.3, 0.4, 0.5\}$) when spatio-temporal information is used with NPE. In the same region, NRE obtains lower RMSEs for using only spatial or temporal information. Moreover, this difference does not disappear when the training budget is increased, suggesting that the problem is not due to an increase in dimensionality. The persistence of this result over multiple training budgets hints at a possible different, as yet unidentified, cause.

Most ABMs have more parameters to calibrate and slower runtimes than the Schelling model. In such cases, the performance of the calibration methods becomes even more crucial. Changing the hyper-parameters of the neural networks and using different architectures can improve performance compared to the default architectures used here. Assessing the performance when lowering the training budgets even more, as some models cannot be evaluated that many times, and increasing the number of parameters and/or summary statistics used (i.e., making the estimation problem more difficult) are interesting directions for future studies. One would expect a more rapid degradation in the performance of ABC, compared to NPE and NRE, because it throws away (many) samples. Additionally, in this specific setup, adding more information in the summary statistics is often beneficial, but for other problems, this might be different or not feasible to test various different summary statistics. Finding optimal high-level descriptions to capture the fundamental emergent patterns in the model through summary statistics and ensure the parameters are structurally identifiable could prove crucial in cases where systems are governed by more complex behaviour [17]. Moreover, all three methods have several extensions that can improve their performance. Sequential versions, which might improve efficiency but lose amortisation [5], could affect performance in different ways (i.e., some can be more efficient than others). In addition, learning summary statistics rather than handcrafting them can improve performance [4]. However, a general problem with neural networks is that it is not clear how these methods learn, learning summary statistics would only worsen this problem. A summary statistic that could have potentially improved performance here, discriminating between $\mu_{true} = 0.1$ and $\mu_{true} = 0.8$, is the total number of relocations. For the former, the number will be low as agents will be satisfied quickly, and for the latter, it will be high. Lastly, the actual true value is used for performance assessment (which is often unknown for empirical data), but there are also several other assessment metrics. One could compare with the true posterior distribution if it is available or use posterior predictive checks and simulation-based calibration [5].

Our general approach (suitable for many ABMs) of providing spatio-temporal information combined with methods from SBI, most notably NRE, makes it possible to accurately estimate the posterior distributions of ABMs. This is even true in difficult regimes, where the model output exhibits a high sensitivity to parameters and has multiple potential solutions. This approach can be applied to real-world empirical observations that contain longitudinal information and/or data that can be aggregated at different spatial scales, paving the way for more realistic and empirically calibrated ABMs.

Acknowledgments. This study is part of the Computational Modelling of Primary School Segregation (COMPASS) project which is funded by the Dutch Inspectorate of Education and the City of Amsterdam.

Disclosure of Interests. The authors declare that they have no competing interests.

A Appendix

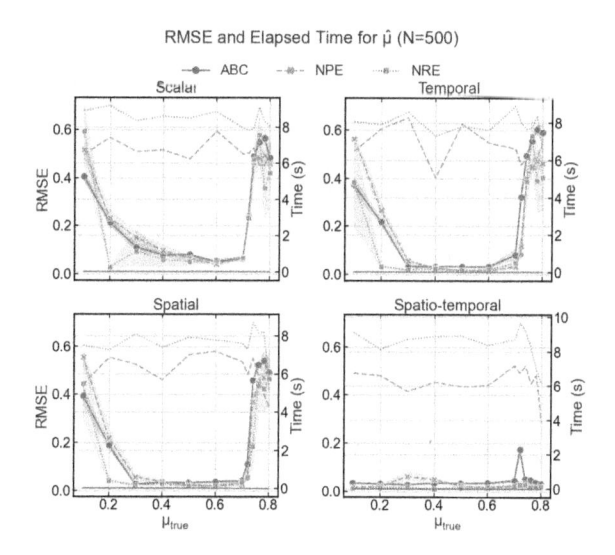

Fig. 7. RMSE for ABC, NPE, NRE and different summary statistics. For every μ value the methods are trained 10 times, with a different training set of 500 samples.

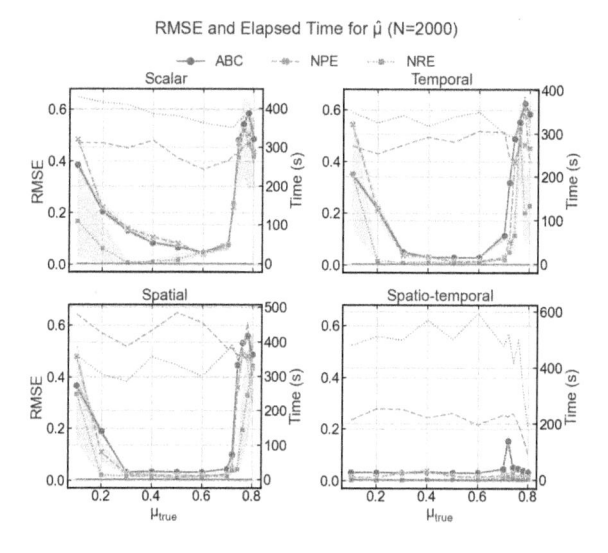

Fig. 8. RMSE for ABC, NPE, NRE and different summary statistics. For every μ value the methods are trained 10 times, with a different training set of 2000 samples.

References

1. Bonabeau, E.: Agent-based modeling: methods and techniques for simulating human systems. Proc. Natl. Acad. Sci. **99**(suppl 3), 7280–7287 (2002)
2. Cranmer, K., Brehmer, J., Louppe, G.: The frontier of simulation-based inference. Proc. Natl. Acad. Sci. **117**(48), 30055–30062 (2020)
3. De Marchi, S., Page, S.E.: Agent-based models. Annu. Rev. Polit. Sci. **17**, 1–20 (2014)
4. Dyer, J., Cannon, P., Farmer, J.D., Schmon, S.M.: Calibrating agent-based models to microdata with graph neural networks. arXiv preprint arXiv:2206.07570 (2022)
5. Dyer, J., Cannon, P., Farmer, J.D., Schmon, S.M.: Black-box Bayesian inference for agent-based models. J. Econ. Dyn. Control **161**, 104827 (2024)
6. Greenberg, D., Nonnenmacher, M., Macke, J.: Automatic posterior transformation for likelihood-free inference. In: International Conference on Machine Learning pp. 2404–2414. PMLR (2019)
7. Hermans, J., Begy, V., Louppe, G.: Likelihood-free MCMC with amortized approximate ratio estimators. In: International Conference on Machine Learning, pp. 4239–4248. PMLR (2020)
8. Lux, T.: Approximate Bayesian inference for agent-based models in economics: a case study. Stud. Nonlinear Dyn. Econ. **27**, 423–447 (2022)
9. Marin, J.M., Pudlo, P., Robert, C.P., Ryder, R.J.: Approximate Bayesian computational methods. Stat. Comput. **22**(6), 1167–1180 (2012)
10. Massey, D.S., Denton, N.A.: The dimensions of residential segregation. Soc. Forces **67**(2), 281–315 (1988)
11. Miller, B.K., Weniger, C., Forré, P.: Contrastive neural ratio estimation. Adv. Neural. Inf. Process. Syst. **35**, 3262–3278 (2022)
12. Papamakarios, G., Murray, I.: Fast ε-free inference of simulation models with Bayesian conditional density estimation. In: Advances in Neural Information Processing Systems, vol. 29 (2016)
13. Schelling, T.C.: Dynamic models of segregation. J. Math. Sociol. **1**(2), 143–186 (1971)
14. Sottoriva, A., Tavaré, S.: Integrating approximate Bayesian computation with complex agent-based models for cancer research. In: In: Lechevallier, Y., Saporta, G. (eds.) Proceedings of COMPSTAT 2010: 19th International Conference on Computational StatisticsParis France, 22-27 August 2010 Keynote, Invited and Contributed Papers, pp. 57–66. Springer (2010). https://doi.org/10.1007/978-3-7908-2604-3_5
15. Tejero-Cantero, A., et al.: SBI–A toolkit for simulation-based inference. arXiv preprint arXiv:2007.09114 (2020)
16. Turner, B.M., Van Zandt, T.: A tutorial on approximate Bayesian computation. J. Math. Psychol. **56**(2), 69–85 (2012)
17. Wolpert, D.H., Grochow, J.A., Libby, E., DeDeo, S.: Optimal high-level descriptions of dynamical systems. arXiv preprint arXiv:1409.7403 (2014)
18. Zbair, M., Qaffou, A., Hilal, K.: Approximate Bayesian estimation of parameters of an agent-based model in epidemiology. In: In: Melliani, S., Castillo, O. (eds.) International Conference on Partial Differential Equations and Applications, Modeling and Simulation, pp. 302–314. Springer, Cham (2021). https://doi.org/10.1007/978-3-031-12416-7_26

Emergent Communication in Merging Artificial Agent Populations

Piotr M. Kosela

Institute of Computer Science, Maria Curie-Skłodowska University, Lublin, Poland
piotr.kosela@mail.umcs.pl

Abstract. While emergent communication in artificial agents has been widely studied, interactions between previously separated populations remain underexplored, despite their real-world relevance. Our aim is to build a model of two pre-learned populations that meet and attempt to communicate. We develop an agent-based language evolution model, where agents are designed to resemble human internal development as closely as possible. These agents participate in 'language games'—atomic, scripted communication scenarios. When merging two pre-learned populations, we observe a significantly higher rate of successful communication compared to training all agents together from the beginning. This effect persists even after extended simulation of the merged population. Our findings suggest that merging pre-learned populations can enhance communication efficiency, offering practical insights for designing collaborative AI systems.

Keywords: Artificial Agents · Language Evolution · Language Games

1 Introduction

Human language evolves on various timescales and through diverse mechanisms, often too gradually to be observed within a single generation or even through historical records. This makes capturing the evolution of communication a challenging task. A significant breakthrough in studying language dynamics has been made possible with advances in computational methods, enabling empirical-like linguistic simulations. Various approaches in computational linguistics have yielded insights into these processes [1]. In this study, we examine differences between language evolution in a single, uniform population and communication developed by two distinct groups with pre-existing internal lexicons. We demonstrate that the latter scenario fosters more efficient communication.

Section 2 begins with an overview of agent-based modeling, followed by a review of computational studies examining the impact of social structures on language development. Section 3 details our model, while Sect. 4 presents the results obtained in the experiment. In Sect. 5, we analyze these findings, leading to the conclusions in Sect. 6. Finally, Sect. 7 outlines directions for future research.

2 Related Works

Performing experiments *in silico* became feasible with the rise of high-performance computing, allowing for the simulation of communication down to the level of individual utterances. Unlike mathematical formulations based on mean-field approximations, agent-based modeling enables the inspection of individual agents, making it a powerful tool for studying language development. This study adopts the *language games paradigm*, an approach pioneered by Steels in the well-known "talking heads" experiment [14,15], which has significantly shaped computational research on language evolution.

The "talking heads" experiment was originally designed to address the symbol grounding problem [2,14]. Its success led to a wave of research employing agent-based models to investigate linguistic dynamics. Spike et al. [13] provide an extensive, though not exhaustive, survey of such models, analyzing the minimal agent capabilities required for communication consensus. While such foundational models have theoretical and philosophical significance, their ability to explain large-scale linguistic phenomena is limited. Increasing agent complexity is necessary for studying population-wide linguistic behavior.

A crucial question in agent-based modeling is its alignment with real-world language emergence. While such cases are rare, empirical observations exist. Richie et al. [11] study homesigners—individuals who invent lexicons to communicate within deaf families—and compare their lexicon development to the slower progress observed in Nicaraguan Sign Language. Their computational model supports claims that social structure plays a crucial role in language evolution. In small home environments, interaction is centralized around a single deaf individual, whereas Nicaraguan Sign Language users engage in broader, diffused communication networks, mirroring findings from computational models.

Agent-based models have been instrumental in exploring the impact of social topology on language development. Labov's Harlem study [7] empirically demonstrated how social structure influences lexical evolution, inspiring computational models such as those by Fagyal et al. [4]. Their results corroborate Labov's findings, showing that communication frequency strongly affects linguistic consensus. Effective information processing within a group requires both leaders and loners—agents with high and low communication frequencies, respectively.

Further research investigates how network topology affects language evolution. Zubek [20] models interacting agent populations using language games, testing information flow efficiency under various social network configurations. Fully connected networks are optimal for information transfer, but star-like structures with designated leaders demonstrate greater adaptability to environmental change and higher overall communicative success. Gong et al. [5] identify key factors shaping linguistic development, including average vertex degree, shortcut connections between clusters, and network centrality. Their findings indicate that higher centrality reduces the number of linguistic categories required for communication.

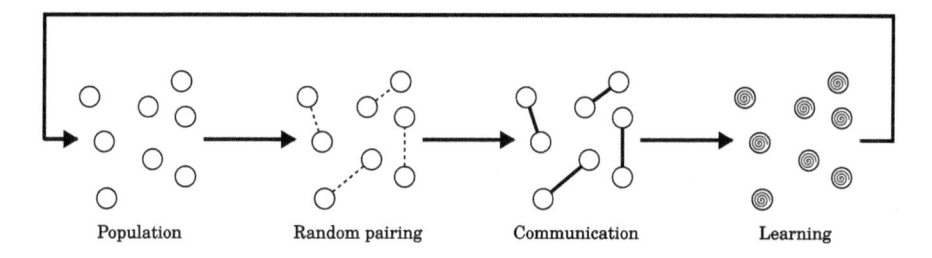

Fig. 1. Steps of simulation repeated in every epoch.

While prior studies focus on language evolution leading to eventual consensus, less attention has been given to how internal population diversity shapes linguistic equilibrium. Josserand et al. [6] employ Bayesian models to study the spread of sociolinguistic variants, demonstrating that even a small subset of biased agents can significantly influence the system's development and final linguistic landscape.

Existing research predominantly examines single populations reaching consensus from scratch. In contrast, real-world scenarios often involve the interaction of groups with pre-established lexicons [11]. In this study, we construct an agent-based computational model comparing two cases: one where a single population develops language from the ground up and another where two distinct populations interact. Our results suggest that merging populations leads to greater communicative success than learning from scratch.

3 Methods

The core objective of our model is to simulate the lexical dynamics of a population of agents engaged in simple linguistic interactions. This process is iterative, with each iteration referred to as an *epoch*. In every epoch, agents are paired uniformly at random[1], engage in scripted communication, and adjust their internal states based on the interaction outcome. Figure 1 illustrates this general process. The remainder of this section details the model's components, beginning with the internal architecture of agents, followed by communication mechanisms, and concluding with the parameter values and structures used in the experiments, distinguishing between control and research trials.

3.1 Agent Architecture

Embodied agents form the fundamental units of our model. Their capabilities are divided into three components [15]:

---

[1] Or *almost random*, as explained in Subsect. 3.3.

- Body
- Categorization system
- Lexicon

The body is the only publicly accessible component, facilitating communication. The categorization system and lexicon remain private to each agent.

Agents possess several specific abilities:

- Perceiving objects
- Perceiving (hearing) words
- Uttering words
- Pointing to objects
- Inferring pointing targets

While embodiment enables communication, further exploration of bodily aspects lies beyond this study's scope.

The core capability of agents is categorization, requiring an evolving internal category structure. Since categorization is dynamic, no predefined categories exist; they emerge through agent interactions. This is implemented using a modified SUSTAIN algorithm [9], which models human categorization. Agents categorize objects represented as d-dimensional vectors $v \in \mathbb{R}^d$, where features correspond to real-valued attributes.

A category abstracts similar objects. For instance, a blue pen and a green pen may form the *pen* category, while a yellow pencil belongs to *pencils*. Higher-level categories, such as *writing utilities*, may also emerge. The model accounts for prototypical examples, as observed in human cognition [12].

Each category consists of a set of prototype-object pairs, with weights denoting importance:

$$C = \{(p_0, w_0), (p_1, w_1), \ldots, (p_k, w_k)\},$$

where $p_i \in \mathbb{R}^d$ are prototypes, and $w_i \in \mathbb{R}+$ are their corresponding weights. Categorization of an object $x = (x_0, x_1, \ldots, x_d) \in \mathbb{R}^d$ is determined by computing the activation function for each category:

$$h_C(x) = \sum_{j=0}^{k} w_k \exp\left(-\frac{1}{2}\sum_{i=0}^{d}(x_i - p_{j,i})^2\right).$$

The category with the highest activation is selected. However, modifications to categorization occur based on language game outcomes (Subsect. 3.2), not categorization alone. Each category also maintains a *communicative success* score, representing the ratio of successful interactions.

To introduce system-wide dynamics, we define a *degeneration rate* δ. After each epoch, prototype weights decay:

$$w(t+1) = w(t)(1-\delta).$$

This mechanism weakens outdated categories unless actively maintained, preventing static structures and ensuring continual adaptation.

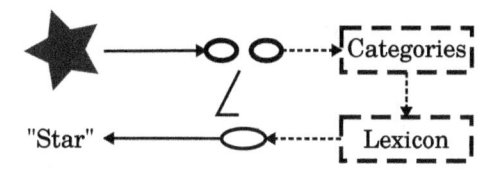

Fig. 2. Speaker perceiving an object and producing an utterance during a guessing game. Dashed lines denote internal agent processes.

Since categories are internal constructs, a lexicon is required to link them with words (Fig. 2). Each category may be associated with multiple words, with selection favoring previously successful terms. The lexicon is a many-to-many mapping between categories and words, with weights determining preference. Words are atomic and drawn from a global pool; no two agents independently invent the same word. Weights evolve dynamically, with lexicon links removed when weights fall below $\varepsilon = 10^{-7}$.

3.2 Language Games

Linguistic interactions are modeled using language games, commonly employed in similar simulations [16]. Specifically, we utilize the *discrimination game* and *guessing game*. Both require a small object set, termed the *environment*, containing five elements. Each object has three features drawn from a normal distribution:[2] One element is designated as the *topic*. Both games conclude with either SUCCESS or FAILURE, affecting agent adaptation.

The *discrimination game* ensures that an agent can uniquely categorize the topic within its environment. The process is as follows:

1. The agent perceives a random environment.
2. The agent assigns categories to objects.
3. If the topic shares a category with another object, the game results in FAILURE.
4. If the topic has a unique category, the game results in SUCCESS.

The *guessing game* involves two agents: a *speaker* and a *hearer*. Intuitively, the speaker perceives the topic, categorizes it, and utters a word, while the hearer attempts to identify the corresponding object. Communication proceeds as follows:

1. Participants are presented with a random, shared environment.
2. Speaker performs discrimination game.
3. If the discrimination game's result is FAILURE:
   - If the topic's category communicative success is greater than 0.95, the topic is added as a new prototype to this category in the speaker's internal memory. Otherwise, a new category centered on the topic is created.
   - Weights of all prototype in the speaker's categories structure are decreased.

---
[2] Means = (66.97, 18.65, 38.36), standard deviations = (20.73, 35.11, 39.77) [20].

– The guessing game ends with **FAILURE**.
4. If there is no word associated with the topic's category, the speaker invents one and immediately associates it with the category.
5. Speaker utters the word with the strongest association with the topic's category.
6. Hearer hears the spoken word.
7. If the hearer does not know the word:
   – Hearer performs the discrimination game. In case of **FAILURE**, if the topic's category communicative success is greater than 0.95, the topic is added as a new prototype to this category in the hearer's internal memory. Otherwise, a new category centered on the topic is created.
   – Topic's category is associated with the spoken word.
   – Weights of all prototypes in the hearer's categories structure are decreased.
   – The guessing game ends with **FAILURE**.
8. Hearer chooses the category with the strongest association with the spoken word.
9. Hearer points to the object, which suits the chosen category best.
10. If the hearer does not point to the topic meant by the speaker:
    – Speaker decreases the strength of association between the chosen category and spoken word.
    – Hearer decreases the strength of association between the chosen category and spoken word.
    – If the topic's category communicative success is greater than 0.95, the topic is added as a new prototype to this category in the hearer's internal memory. Otherwise, a new category centered on the topic is created.
    – Weights of all prototypes in the hearer's categories structure are decreased.
    – The guessing game ends with **FAILURE**.
11. The guessing game ends with **SUCCESS**.

Prototype weight reduction following an unsuccessful guessing game mirrors the degeneration process, but instead of the degeneration rate δ, we apply a distinct parameter, the *forgetting rate* ϕ, according to the equation:

$$w(t+1) = w(t)(1 - \phi).$$

This ensures that less effective prototypes gradually lose influence, facilitating adaptation to new linguistic patterns. The key difference is that degeneration follows every language game, whereas forgetting is applied only after failed games, serving as a penalty for incorrect classification.

3.3 Experimental Procedure

We detail the experimental procedure below. The control sample is generated using the method described at the beginning of Sect. 3, simulating 1,000,000 epochs with 200 agents.

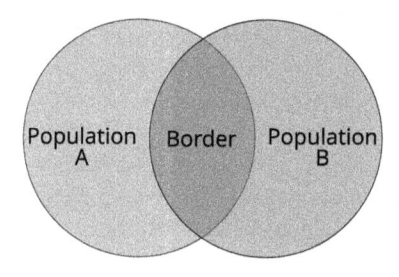

Fig. 3. Diagram of the merged populations, highlighting the border area.

The research trials consist of two stages. The first stage aims to develop basic categorization abilities and small lexicons in agents. Two separate populations of 100 agents each are simulated for 200,000 steps. This duration, chosen heuristically, ensures sufficient communicative success while maintaining agent flexibility to adapt to changes in population and environment.

The second stage combines the two populations into a single set of 200 agents, simulating an additional 1,000,000 epochs. The pairing rules are modified: 20 agents from each population are designated as *border agents*, with a probability $\beta = 0.8$ of interacting with border agents from the other population. One could think of them as interpreters or translators—individuals who engage with members of the other population, acquire their linguistic conventions, and transmit this knowledge back to their own group. Non-border agents interact only within their original populations. A schematic of this setup is shown in Fig. 3.

The research procedure was repeated independently 20 times to allow for result averaging. Key learning parameters were held constant: $\phi = 0.0005$ and $\delta = 0.0000625$. These values, chosen heuristically, balance communicative success, computational complexity, and population flexibility. Increasing ϕ and δ tends to destabilize lexicons, lowering the model's overall success, while decreasing them makes populations resistant to new variants introduced by border agents. Investigating these effects in large populations is challenging due to the substantial computational resources required by the simulation.

After each epoch, we compute several metrics:

- *Communicative Success (CS)*: The ratio of successful guessing games to total interactions.
- *Average Number of Categories (AVGC)*: The mean number of categories retained by agents.
- *Category Operations*: Total numbers of *created (C)*, *modified (M)*, and *deleted (D)* categories. Modifications refer to changes in category prototypes.

4 Results

In this section, we present the experimental results, comparing the development of the research sample (first and second stages) with the control sample. Each plot is derived by pointwise averaging data from 20 independent trials, followed by a moving average with a block size of 10,000. We begin with a comparison of the first $200k$ steps for both samples.

The control sample consists of 200 agents communicating uniformly. Figure 4 compares three populations: *A*, *B*, and *control*. Figure 4a shows communicative success, with the control sample stabilizing at 0.25, while the research sample achieves 0.32 due to its smaller, independent populations. This aligns with findings in [20].

Figure 4b reveals linear growth in the average number of categories per agent across all populations. Figure 4c shows minimal deletions and stable modifications (~ 0.2) after $15k$ epochs. The total number of created categories remains constant, except for initial generations where agents compensate for the lack of categories. Notably, the slowing of category creation does not trigger deletions.

After $200k$ steps, populations *A* and *B* are connected via border agents. Figure 5 presents post-merge results, with epoch 0 marking the merge.

Communicative success in the research sample drops to 0.3 post-merge (Fig. 5a), consistent with border agents. The control sample shows no change. Figure 5b shows the research sample stabilizing at $11k$ categories per agent by $300k$ epochs, while the control sample reaches $12k$ by $575k$ epochs.

Figure 5c shows higher category creation in the research sample post-merge (4 to 4.5). Modifications increase as category growth stabilizes, with deletions starting concurrently. The research sample exhibits two peaks in deletions, while the control sample stabilizes quickly. Figure 5d shows border agents with fewer creations and delayed deletions.

Finally, Fig. 6 presents heatmaps of word usage. Figure 6c shows the control sample, with early words reused successfully. Figure 6a depicts post-merge word usage, with new signs dominating early words. Figure 6b shows border agents following population trends, with successful interactions around generation $500k$.

5 Discussion

In this section, we analyze the findings of our model, beginning with an interpretation and explanation of the results. We then take a broader perspective on language evolution models, relating our observations to real-world situations.

Let us begin by focusing on the behavior of communicative success. It drops immediately after the merge. This decrease is likely caused by the onset of transpopulation interactions, while the groups do not share any words. However, the metric does not recover[3]. Thus, we conjecture that the drop in the research sample is an artifact resulting from model scaling issues [20]. CS at the border does not change over time. It is not significantly smaller, so it cannot affect the measure for the entire population. Nevertheless, the final population achieves better results than the control sample, even though it contains the same number of agents. The loss of CS from scaling is much smaller than in the case of agents learning together from the very beginning.

---
[3] We performed many more simulation steps to confirm this.

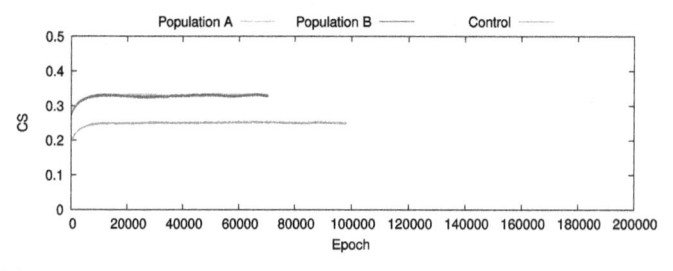

(a) Communicative success across research and control samples.

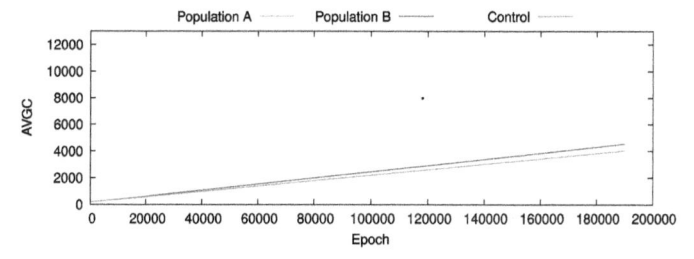

(b) Average number of categories per agent in research and control populations.

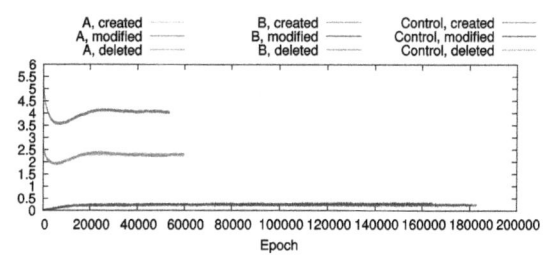

(c) Average total number of created, modified, and deleted categories in research and control samples.

Fig. 4. Comparison of communicative success, agent categories, and category operations between the control sample and research sample populations.

The CS of the research sample oscillated around 0.3 with fewer than $11k$ categories per agent. In contrast, the control sample had almost $12k$ categories on average, with CS at the level of 0.25. Overall, it appears that the population created from the two smaller ones performs better, surpassing the CS barrier established by the control sample. This suggests that building a population from smaller groups leads to easier adaptation and improved overall model performance. Despite having fewer categories, the research sample demonstrates better object recognition.

It was recently shown that, in the case of human categorization development, exposure to foreign cultures does not affect mature categorization networks [10]. We propose that an analogous effect may be present in our model; note that the weights in our categorization systems are not bounded, and the categories from

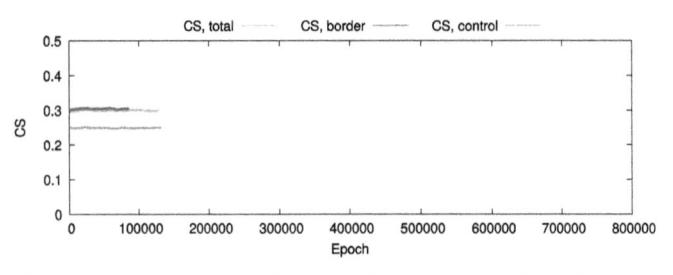

(a) Communicative success for control, research, and border agents.

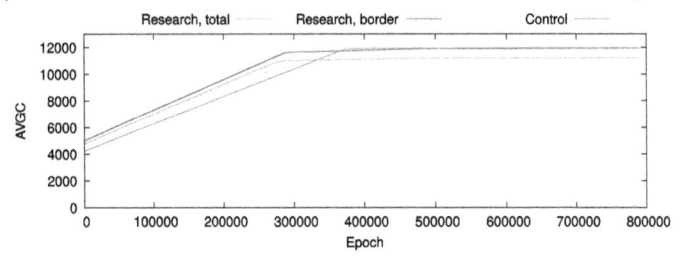

(b) Average number of categories per agent for control, research, and border agents.

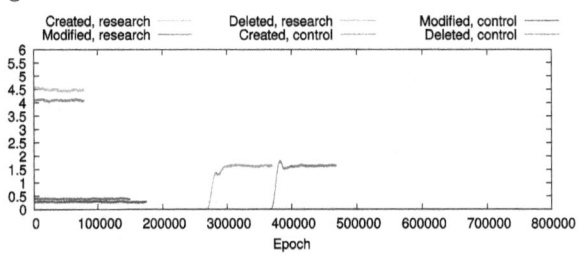

(c) Average total number of created, modified, and deleted categories for research and control samples.

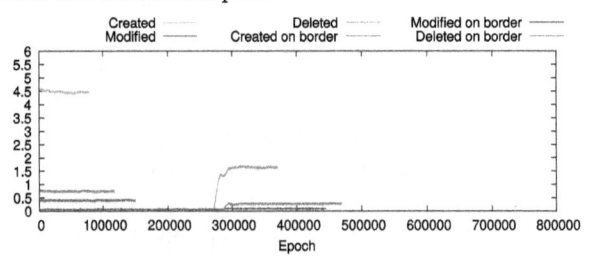

(d) Average total number of created, modified, and deleted categories for research sample and border agents.

Fig. 5. Communicative success and category dynamics in the second stage for control, research, and border agents.

(a) Word usage in merged populations.

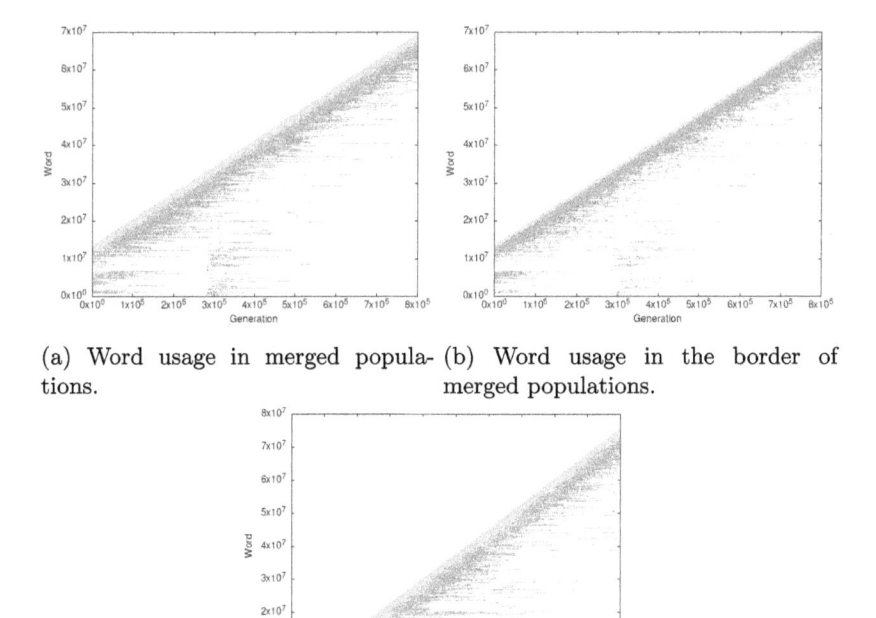

(b) Word usage in the border of merged populations.

(c) Word usage in control sample.

Fig. 6. Heatmaps showing word usage (Y-axis) across generations (X-axis). Colors indicate interaction results: green (success), red (failure), yellow (mixed). (Color figure online)

the preliminary stage of the simulation persist with border agents for an extended period. The average number of categories developed by the border agents is significantly greater than the population average. These rich categorization systems operate with more dynamic vocabularies. Notably, the horizontal lines, starting from generation $300k$ in Fig. 6a, are longer than those in Fig. 6b. Words do not survive long on the border but are exported to the rest of the population, where their usage is prolonged; this is consistent with [4]. These agents influence the main population and create trends. We believe this is the factor enabling the higher communicative success. The border does not, in fact, consist of leaders in the sense of [4]; Fig. 3 may be slightly misleading in this context. These agents interact with the rest of the population only 20% of the time. This subpopulation may function as a small, organized subsystem, exporting this well-designed innovation externally. We might better consider them a collective, more complex outsider, balancing the system.

6 Conclusion

The results of this study demonstrate that populations formed by merging smaller groups achieve higher communicative success compared to populations that learn from scratch. This enhanced performance can be attributed to the richer categorization systems developed by the border agents, who, despite interacting less frequently with the main population, export valuable innovations. These agents facilitate more efficient adaptation, resulting in a significant increase in overall model performance. In contrast, populations that evolve from scratch face greater challenges in reaching similar levels of communicative success, highlighting the advantage of merging pre-existing lexicons.

7 Further Research

While our study provides insights into language evolution in merged populations, several open questions remain. Here, we outline key areas for future research that could deepen our understanding of population interactions and model scalability.

One key difference between the control and research samples is the presence of additional peaks in category modifications and deletions in the latter. We believe this is not an inherent phenomenon in the control sample but rather a consequence of slower transition. While such peaks appear in both samples, their patterns differ. Investigating this variation could provide deeper insights into population interactions.

We find it worthwhile to incorporate spatial distribution into the model. While social network topology partially addresses this, no agent-based model we are aware of considers gradual environmental shifts based on agent locations. For instance, the need for specialized snow descriptions in the Arctic Circle compared to the Congo Valley is not well represented in existing models [19]. Social topology influences interaction probability alongside spatial factors. Spatial arrangement shapes populations into smaller societies, while uniform environments foster more sophisticated communication.

Simulating agents in a diverse world is feasible only if communities are large enough. Otherwise, the model would merely reflect a social topology framework with distinct environments for individual agents. At some scale, resource constraints hinder computational feasibility [17]. While [17] highlights model complexity, it underestimates categorization scaling, the most computationally demanding aspect. As [18] shows, similar tasks cannot be achieved in polynomial time. Efficient categorization remains crucial for advancing agent-based language models. Currently, no comprehensive mathematical frameworks exist for explanatory sociolinguistic models, aside from highly simplified cases [8]. While real-world language evolution models exist, their reliance on mathematical apparatus often limits explanatory power [3].

Our results show that merging two independent agent populations leads to higher communicative success, indicating that linguistic convergence benefits from increased diversity in lexical and categorical structures. This finding underscores the importance of cross-population dynamics in language evolution. Future research should extend this framework by incorporating spatially distributed environments, allowing for a more nuanced exploration of regional linguistic variation. Additionally, addressing the computational complexity of our models will be crucial for scaling simulations and capturing more intricate dynamics of category and lexicon formation. Exploring these aspects will deepen our understanding of emergent communication and its sensitivity to environmental constraints.

Acknowledgments. I would like to thank Jarosław Bylina and Piotr Giza for their valuable insights and thoughtful discussions, which greatly supported the development of this work.

References

1. Bailes, R., Cuskley, C.: The cultural evolution of language (2022). publisher: PsyArXiv
2. Bougsty-Marshall, S.: The dynamics of growing symbols: a ludics approach to language design by autonomous agents. In: Goertzel, B., Panov, A.I., Potapov, A., Yampolskiy, R. (eds.) Artificial General Intelligence, vol. 12177, pp. 44–53. Springer International Publishing, Cham (2020). https://doi.org/10.1007/978-3-030-52152-3_5, series Title: Lecture Notes in Computer Science
3. Dębowski, L.: Information Theory Meets Power Laws: Stochastic Processes and Language Models. Wiley, 1st edn. (2020). https://doi.org/10.1002/9781119625384, https://onlinelibrary.wiley.com/doi/book/10.1002/9781119625384
4. Fagyal, Z., Swarup, S., Escobar, A.M., Gasser, L., Lakkaraju, K.: Centers and peripheries: network roles in language change. Lingua **120**(8), 2061–2079 (2010). https://doi.org/10.1016/j.lingua.2010.02.001, https://linkinghub.elsevier.com/retrieve/pii/S0024384110000203
5. Gong, T., Baronchelli, A., Puglisi, A., Loreto, V.: Exploring the roles of complex networks in linguistic categorization. Artif. Life **18**(1), 107–121 (2011). https://direct.mit.edu/artl/article/18/1/107-121/2715
6. Josserand, M., Allassonnière-Tang, M., Pellegrino, F., Dediu, D.: Interindividual variation refuses to go away: a bayesian computer model of language change in communicative networks. Front. Psychol. **12**, 626118 (2021). https://doi.org/10.3389/fpsyg.2021.626118, https://www.frontiersin.org/articles/10.3389/fpsyg.2021.626118/full
7. Labov, W.: Language in The Inner City: Studies in the Black English Vernacular. University of Pennsylvania Press (1972). issue: 3
8. Loreto, V., Baronchelli, A., Puglisi, A.: Mathematical modeling of language games. In: Nolfi, S., Mirolli, M. (eds.) Evolution of Communication and Language in Embodied Agents, pp. 263–281. Springer, Heidelberg (2010). https://doi.org/10.1007/978-3-642-01250-1_15
9. Love, B.C., Medin, D.L., Gureckis, T.M.: SUSTAIN: a network model of category learning. Psychol. Rev. **111**(2), 309–332 (2004). https://doi.org/10.1037/0033-295X.111.2.309, http://doi.apa.org/getdoi.cfm?doi=10.1037/0033-295X.111.2.309

10. Naji, A.: Semantic representations of abstract and concrete categories in the mental lexicon of mono- and bilingual Jordanians: a prototype analysis. Alkalmazott nyelvtudomány **21**(2), eISSN 24984442 (2021). ISSN 1587-1061, https://doi.org/10.18460/ANY.2021.2.008, http://alkalmazottnyelvtudomany.hu/wordpress/wp-content/uploads/naji_tan.docx.pdf, publisher: Pannon Egyetem
11. Richie, R., Yang, C., Coppola, M.: Modeling the emergence of lexicons in homesign systems. Top. Cognit. Sci. **6**(1), 183–195 (2014). https://doi.org/10.1111/tops.12076, https://onlinelibrary.wiley.com/doi/10.1111/tops.12076
12. Rosch, E.: Cognitive representations of semantic categories. J. Exp. Psychol. General **104**(3), 192–233 (1975). https://doi.org/10.1037/0096-3445.104.3.192, http://doi.apa.org/getdoi.cfm?doi=10.1037/0096-3445.104.3.192
13. Spike, M., Stadler, K., Kirby, S., Smith, K.: Minimal requirements for the emergence of learned signaling. Cognit. Sci. **41**(3), 623–658 (2017). https://doi.org/10.1111/cogs.12351, https://onlinelibrary.wiley.com/doi/10.1111/cogs.12351
14. Steels, L.: The origins of ontologies and communication conventions in multi-agent systems. Auton. Agent. Multi-Agent Syst. **1**, 169–194 (1998)
15. Steels, L.: Evolving grounded communication for robots. Trends Cognit. Sci. **7**(7), 308–312 (2003). https://doi.org/10.1016/S1364-6613(03)00129-3, https://linkinghub.elsevier.com/retrieve/pii/S1364661303001293
16. Steels, L., Belpaeme, T.: Coordinating perceptually grounded categories through language: a case study for colour. Behav. Brain Sci. **28**(4), 469–489 (2005). https://doi.org/10.1017/S0140525X05000087, https://www.cambridge.org/core/product/identifier/S0140525X05000087/type/journal_article
17. Vogt, P., Divina, F.: Language evolution in large populations of autonomous agents: issues in scaling. In: Proceedings of AISB, pp. 80–87. Citeseer (2005)
18. Woensdregt, M., Cummins, C., Smith, K.: A computational model of the cultural co-evolution of language and mindreading. Synthese **199**(1-2), 1347–1385 (2021). https://doi.org/10.1007/s11229-020-02798-7, https://link.springer.com/10.1007/s11229-020-02798-7
19. Yule, G.: The Study of Language. Cambridge University Press, 7th edn. (2019). https://doi.org/10.1017/9781108582889
20. Zubek, J., Denkiewicz, M., Barański, J., Wróblewski, P., Rączaszek-Leonardi, J., Plewczynski, D.: Social adaptation in multi-agent model of linguistic categorization is affected by network information flow. PLOS ONE **12**(8), e0182490 (2017). https://doi.org/10.1371/journal.pone.0182490

MAVS: An Ensemble-Based Multi-agent Framework for Fake News Detection

Dhruv Tyagi[1(✉)], Anurag Singh[1], and Hocine Cherifi[2]

[1] National Institute of Technology Delhi, Delhi, India
{211220018,anuragsg}@nitdelhi.ac.in
[2] University de Bourgogne Europe, Dijon, France
hocine.cherifi@u-bourgogne.fr

Abstract. The Multi-Agent Verification System (MAVS) Framework aims to improve fake news detection by leveraging a multi-agent system that enhances decision-making through multidimensional evaluation, mitigating adversarial attack vulnerabilities.

MAVS utilizes four specialized agents (a GNN model and Generative AI models for fact-checking, stance-checking, and sentiment analysis) each operating independently and in parallel. The final classification is determined through a weighted aggregation of the agents' outputs, optimized using Stochastic Gradient Descent (SGD)-based Logistic Regression to ensure optimal weight distribution.

MAVS achieves an accuracy of 97.6% and an F1 score of 98%. Under a Multi-Agent Reinforcement Learning (MARL) attack, the system's accuracy drops to 74.19% and the F1 score to 71%, while maintaining a precision of 100%. This highlights the framework's resilience and ability to maintain high precision despite adversarial conditions.

The proposed framework strengthens fake news detection by combining multiple verification strategies, reducing susceptibility to adversarial attacks. Future work includes refining agent interactions and exploring real-time deployment for broader applicability.

Keywords: Multi-Agent Systems · Fake News Detection · Graph Neural Networks · Fact Checking · Sentiment Analysis · Stance Analysis

1 Introduction

The spread of misinformation undermines public trust, distorts democratic processes, and poses risks to public health and safety [13]. Fake news detection is essential to protect public discourse, ensuring informed debates and strengthening public dialogue. It is also crucial for safeguarding democracy, as misinformation can manipulate public opinion and influence elections [13]. Events like the COVID-19 pandemic have demonstrated the real-world harm of unchecked false information, necessitating robust detection mechanisms [2,14].

Traditional fact-checking methods are time-consuming and inefficient, prompting researchers to explore AI and machine learning-based solutions [3,15,16]. While NLP-based sentiment and stance analysis help analyze textual content, they fail to capture misinformation propagation patterns, making them susceptible to manipulation [2]. Deep learning approaches improved detection accuracy [11,12] but they require large labeled datasets and struggle with propagation analysis. This led to the adoption of Graph Neural Networks (GNNs), which model relationships and dependencies within news propagation networks [12]. By analyzing dissemination patterns, user interactions, and source credibility, GNNs provide a more comprehensive detection framework. However, they remain vulnerable to adversarial attacks, where fraudsters manipulate graph structures to spread misinformation [21].

To enhance robustness, we propose MAVS (Multi-Agent Verification System), an ensemble-based framework that integrates GNNs with generative AI models for fake news detection. MAVS combines multiple AI agents specializing in fact-checking, stance detection, and sentiment analysis to analyze both news content and its propagation context. The fact-checking agent retrieves external evidence using generative AI, while the stance and sentiment analysis agents assess contextual consistency and emotional bias [9]. The final classification is determined through a weighted fusion of agent outputs, ensuring a multidimensional evaluation [12].

To validate our approach, we conduct extensive experiments on the Politifact (UPFD) dataset [7], evaluating accuracy, precision, recall, and F1-score, where the F1-score is the harmonic mean of precision and recall. It provides a balanced measure and helps to evaluate the model's robustness in identifying both fake and real news accurately. We also analyze model interpretability by examining agent contributions and weight distributions in decision-making.

Contributions of this Research:

1. Evaluating adversarial vulnerabilities of the GNN.
2. Developing a multi-agent verification system integrating fact-checking, stance detection, and sentiment analysis.
3. Optimizing framework weights using machine learning.

By combining graph-based learning with AI-driven verification, MAVS offers a novel and effective approach to combating misinformation in the digital era.

2 Related Work

The rise of misinformation and advancements in AI have driven fake news detection techniques. Early approaches relied on rule-based systems and manual fact-checking, which were labor-intensive and lacked scalability. Machine learning models improved efficiency by classifying news based on writing style, sentiment, and credibility, but struggled with contextual understanding and adversarial attacks. Recent advancements incorporate NLP techniques, deep learning models, and GNNs to analyze both content and propagation patterns.

2.1 Fact-Checking Methods

Fact-checking plays a crucial role in fake news detection by verifying claims using external knowledge bases. These systems provide explainability but face challenges in dynamic scenarios due to their reliance on up-to-date knowledge [20].

2.2 NLP-Based Methods

Early detection methods used lexical features like BoW and TF-IDF to analyze textual content [2]. However, these models lacked contextual understanding and were vulnerable to adversarial manipulation [23].

2.3 Deep Learning for Fake News Detection

Deep learning models such as RNNs and CNNs improved contextual comprehension by capturing sequential dependencies [1,8]. However, they required extensive labeled datasets, lacked interpretability and often overlooked social interactions and propagation dynamics, making them susceptible to misinformation tactics [5].

2.4 Graph Neural Networks (GNNs)

GNNs effectively model social interactions and misinformation spread by analyzing propagation patterns [19]. However, GNNs face challenges such as high computational complexity, scalability issues, and adversarial vulnerabilities [4,6,22]. They remain vulnerable to adversarial attacks where malicious actors manipulate the graph structure to evade detection [21]. Real-world attackers add new nodes and edges to alter misinformation spread patterns just like in MARL attack shown in Fig. 1, making detection less effective.

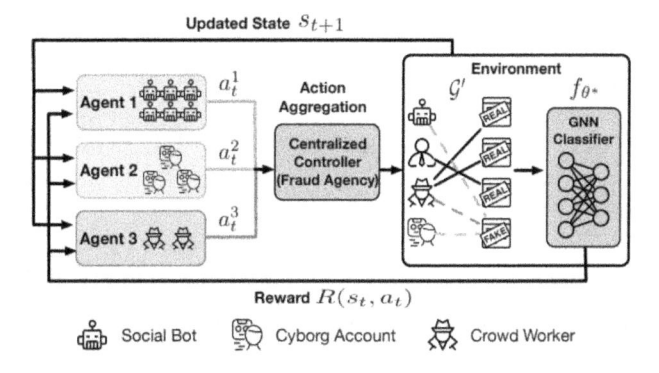

Fig. 1. MARL adversarial attack against GNN-based fake news detection, where manipulated edges and nodes degrade detection accuracy [21].

2.5 MAVS vs. Other Approaches

Table 1 presents a feature-wise comparison of detection methods. MAVS integrates GNNs with Fact, Stance, and Sentiment Checkers, offering a more robust and adaptable approach.

Table 1. Feature comparison across fake news detection models (✓: Yes, ✗: No, △: Partial).

Feature	GNN	Fact Checker	Stance Checker	Sentiment Checker	MAVS
Propagation Analysis	✓	✗	✗	✗	✓
Textual Analysis	✗	✓	✓	✓	✓
Credibility Checking	✓	✓	✗	✗	✓
Context Understanding	✗	✓	✓	✗	✓
Stance Detection	✗	✗	✓	✗	✓
Sentiment Analysis	✗	✗	✗	✓	✓
Fact Verification	✗	✓	✗	✗	✓
Adversarial Resilience	✗	✗	✗	✗	△

Key Takeaways:

- **Propagation Analysis:** GNNs excel at misinformation spread modeling, which traditional methods lack.
- **Context Understanding:** MAVS integrates multiple sources for more comprehensive evaluations.
- **Resilience to Adversarial Attacks:** MAVS reduces attack vulnerability by combining independent agents.

MAVS combines the best aspects of fact-checking, NLP, deep learning, and GNNs to create a robust, multi-perspective fake news detection system.

3 Methodology

The section discusses the methodology of the research in detail.

3.1 Proposed Framework

To effectively detect fake news, the proposed **MAVS (Multi-Agent Verification System)** framework leverages multiple AI agents for analyzing both content and propagation patterns in parallel and independently. The methodology is designed to enhance the reliability of fake news detection by integrating different perspectives, including **network propagation, factual consistency, stance alignment, and sentiment analysis**. This section outlines the dataset used, key components of the framework their roles, and how they contribute to the final classification decision.

3.2 Dataset

The dataset used in this study is the well-known **UPFD-Politifact** dataset [7], specifically curated for evaluating binary graph classification, graph anomaly detection, and fake news detection tasks. It is structured as a `PyTorch-Geometric` dataset object having field as shown in Table 2, enabling seamless integration with various GNN models.

The dataset consists of tree-structured graphs representing news propagation networks on Twitter. These graphs are constructed using fact-check information from *Politifact* and *Gossipcop*, originally extracted by FakeNewsNet [17]. The structure of each graph is as follows:

- **Root Node:** Represents the original news article.
- **Leaf Nodes:** Twitter users who retweeted the article.
- **Edges:** A directed edge exists from a user to the news node if they retweeted it. Two users are connected if one retweeted the news from the other.

The dataset includes three types of node features (used in our case):

- **SpaCy Features (300-dimensional):** Word2Vec embeddings generated using the `spaCy` library.
- **Profile Features (10-dimensional):** Extracted from Twitter user profiles, capturing metadata like follower count and verification status.
- **Content Features (310-dimensional):** Combines a 300-dimensional user comment word2vec embedding with the 10-dimensional profile feature.

Table 2. Dataset Columns and Descriptions

Column Name	Description	Example Value
id	Unique identifier	"politifact4190"
news_url	Source URL	http://www.c.gov/doc.pdf
title	Headline	"Budget and Economic Outlook"
tweet_ids	Related tweet IDs	"1102113056 1102113348 ..."
label	Real (0) or Fake (1)	1

3.3 Key Components

As shown in the architecture of proposed MAVS in Fig. 2 the key components are Graphical Neural Network (GNN), Fact-Checker, Stance-Checker and Sentiment-Checker. The detailed description of these components are as follows,

1. **Graphical Neural Network (GNN):-** The GNN Model operates on hierarchical tree-structured graphs as shown in Fig. 3

   The GNN captures the structure and influence patterns within this retweet graph, helping identify propagation trends [21]. The model leverages **Graph**

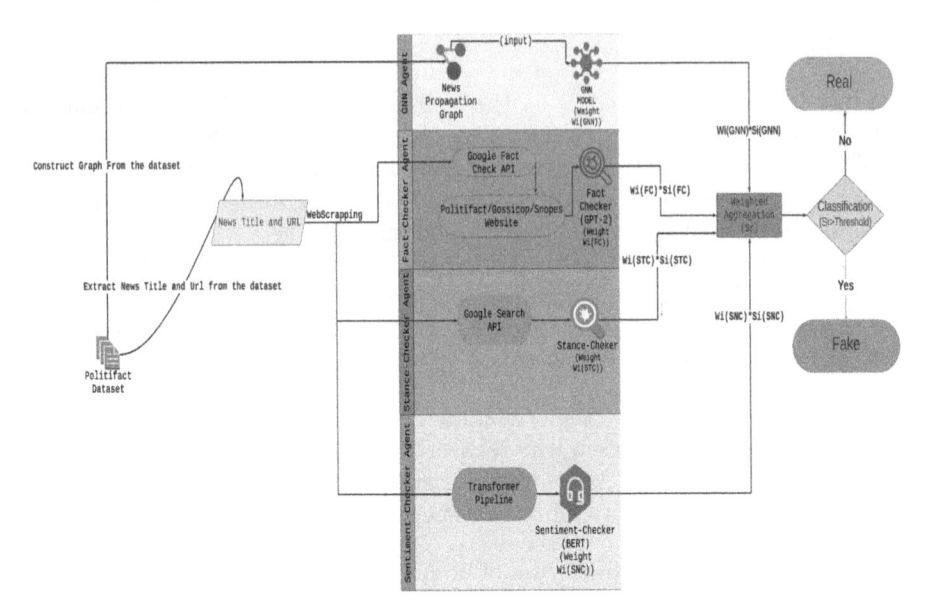

Fig. 2. The architecture of MAVS framework, integrating GNN and AI agents (Fact-Checker, Stance-Checker, Sentiment-Checker) for fake news detection.

Attention Networks (**GAT**) for learning propagation patterns. GNN consists of three stacked GATConv layers followed by a global max pooling operation, which aggregates graph-level embeddings. The final output is obtained through a fully connected readout layer with a sigmoid activation for binary classification.

2. **Fact-Checker**: The fact-checker agent uses Algorithm 1 which ensures that the information aligns with verified facts from trusted sources, helping to distinguish misinformation from truthful content [20]. The fact-checking process involves analyzing claims and assigning a weighted score based on their verdicts. Given a claim's verdict V_i, The final weighted score S_{weighted} is computed as the average of all S_i values. If S_{weighted} is negative, the statement is considered more likely to be true; otherwise, it is likely false.

3. **Stance-Checker**: The stance-checker agent determines how well your content is aligned from different perspectives using a variety of predefined stance detection algorithms. Whether the news concurs or would seem to refute identified affiliations and views, more easily detecting bias. Basically, using Algorithm 2 it evaluates the stance of the article relative to a claim or premise, classifying it as *supports*, *contradicts*, or *neutral* [9].

Algorithm 1. Fact-Checking Model Using GPT-2 and API

Input: S: Statement to verify, API_Key: API key for fact-checking.
Output: S_{weighted}: Final weighted score, Generated_Text: Explanation from GPT-2.
Step 1
Initialize tokenizer and text generator: Tokenizer $\leftarrow$ GPT2Tokenizer('gpt2'), Generator $\leftarrow$ pipeline('text-generation', model = 'gpt2')
Step 2
Fetch results: $F \leftarrow \text{API\_Request}(S, \text{API\_Key})$ **if** $status\_code \neq 200$ **then**
  return Error
Extract claims: $C \leftarrow F[\text{'claims'}]$
Step 3
for each claim $C_i \in C$ **do**
  Retrieve verdict V_i and compute score:
  $$S_i = \begin{cases} -1, & V_i \in \{\text{"true"}, \text{"mostly true"}, \text{"half true"}\} \\ 1, & V_i \in \{\text{"false"}, \text{"mostly false"}, \text{"pants on fire"}\} \\ 0, & \text{otherwise} \end{cases}$$ Accumulate: $S \leftarrow S + S_i$
Step 4
$S_{\text{weighted}} \leftarrow \frac{S}{|C|}$
Step 5
Construct prompt: Prompt $\leftarrow$ "Given the statement 'S' and the fact-check results:"
Generate explanation: Generated_Text $\leftarrow$ Generator(Prompt)
return S_{weighted}, Generated_Text

4. **Sentiment-Checker**: The sentiment-checker agent leverages generative AI models to assess the emotional tone of the content retweeted by users. Using Algorithm 3 it categorizes the sentiment as positive, negative, or neutral, helping in identifying emotional manipulation or polarizing content [23]. Each agent generates a score, which is aggregated to classify the news as real or fake.

Algorithm 2. Stance-Checking Model Using BART

Input: T: News title, URL: Article URL.
Output: L_{final}: Final stance label, S_{final}: Stance score.
Step 1
Initialize stance detection model: Model $\leftarrow$ pipeline("zero-shot-classification", model = "bart")
Step 2
Extract article content and summary: $C \leftarrow \text{ExtractText(URL)}, \quad S \leftarrow C[:300]$
Step 3
Compute stance of title w.r.t. content: $S_T \leftarrow w_i \cdot p(L_i \mid T, S)$
Step 4
Perform Google search for related URLs.
foreach related URL i **do**
  Extract content H_i (first 200 words), compute stance score: $S_{R_i} \leftarrow w_i \cdot p(L_i \mid T, H_i)$
  Append to stance list.
Compute average stance score: $S_R = \frac{1}{n}\sum_{i=1}^{n} S_{R_i}$
Step 5
Compute weighted final stance score: $S_{\text{final}} = 0.3 \cdot S_T + 0.7 \cdot S_R$
Step 6
if $0.7 \cdot S_C > 0.3 \cdot S_T$ **then**
  $L_{\text{final}} \leftarrow L_C$
else
  $L_{\text{final}} \leftarrow L_T$
Step 7
if L_{final} is "supports" **then**
  $S_{\text{adjusted}} \leftarrow -S_{\text{final}}$
else if L_{final} is "neutral" **then**
  $S_{\text{adjusted}} \leftarrow 0$
else
  $S_{\text{adjusted}} \leftarrow S_{\text{final}}$
return $L_{\text{final}}, S_{\text{final}}$

Algorithm 3. Sentiment Score Computation for News Articles

Input : T: News title, URL: Article URL.
Output: S_{final}: Sentiment score, Sentiment Label: Positive, Neutral, or Negative.
Step 1
Initialize model: Model ← pipeline("sentiment-analysis", model = "bert-multilingual")
Step 2
Compute sentiment score for title: (L_T, S_T) ← Model(T) **if** L_T is "4 stars" or "5 stars" **then**
  $S_T \leftarrow -S_T$
else if L_T is "3 stars" **then**
  $S_T \leftarrow 0$
Step 3
Extract and summarize article content: C ← ExtractText(URL), $S \leftarrow C[:200]$ Compute sentiment score: (L_C, S_C) ← Model(S) **if** L_C is "4 stars" or "5 stars" **then**
  $S_C \leftarrow -S_C$
else if L_C is "3 stars" **then**
  $S_C \leftarrow 0$
Step 4
Compute weighted sentiment score: $S_{\text{final}} = 0.3 \cdot S_T + 0.7 \cdot S_C$
Step 5
if $S_{\text{final}} \geq 0$ **then**
  Assign label as Positive.
else
  Assign label as Negative.
return S_{final}, Sentiment Label

3.4 Final Decision Making

The final classification of a news article is determined using a logistic regression model trained with Stochastic Gradient Descent (SGD) using Algorithm 4. Each agent provides a score, which is used as input features for classification. The feature vector for news item n_i is represented in Eq. (1) as:

$$X_i = [S_{i,\text{GNN}}, S_{i,\text{FC}}, S_{i,\text{STC}}, S_{i,\text{SNC}}] \quad (1)$$

where $S_{i,\text{GNN}}, S_{i,\text{FC}}, S_{i,\text{STC}}, S_{i,\text{SNC}}$ are the scores from the GNN, Fact-Checker, Stance-Checker, and Sentiment-Checker, respectively.

The final classification decision is obtained using a logistic regression model trained with Stochastic Gradient Descent (SGD). The model learns weight coefficients w_1, w_2, w_3, w_4, which were initially **randomly initialized** and subsequently optimized during training. These weights determine the contribution of each agent's output to the final classification.

The MAVS score, denoted as S_r, is computed as:

$$S_r = w_1 S_{i,\text{GNN}} + w_2 S_{i,\text{FC}} + w_3 S_{i,\text{STC}} + w_4 S_{i,\text{SNC}} \quad (2)$$

- $S_{i,\text{GNN}}$ - Score from the GNN model
- $S_{i,\text{FC}}$ - Score from the Fact-Checking Agent
- $S_{i,\text{STC}}$ - Score from the Stance-Checking Agent
- $S_{i,\text{SNC}}$ - Score from the Sentiment-Checking Agent

The final score S_r in Eq. (2) represents the **weighted fusion** of these agent outputs and is passed through a sigmoid activation function to determine the probability of the news being real or fake.

Applying the sigmoid activation function as shown in Eq. (3):

$$P(\text{Real News}) = \frac{1}{1 + e^{-S_r}} \quad (3)$$

The classification decision is made as follows:

$$\text{Classify } n_i = \begin{cases} \text{Fake}, & \text{if } P \geq 0.5, \\ \text{Real}, & \text{otherwise}. \end{cases} \quad (4)$$

The threshold $P \geq 0.5$ for classification used in Eq. (4) was chosen experimentally, aligning with the standard practice for sigmoid-based binary classification. Since the label encoding assigns 0 to real news and 1 to fake news, a lower sigmoid output (closer to 0) indicates a stronger belief in news being real, whereas a higher value indicates fake. Thus, the 0.5 cutoff serves as a balanced midpoint for binary decision-making.

Algorithm 4. MAVS-Based Fake News Classification Using SGD Logistic Regression

---
Input : Feature matrix X containing agent scores, Binary labels y (0 = Fake, 1 = Real).
Output: Trained SGD Logistic Regression Model, Classification result for new instances.
Step 1
Construct feature vectors: $X = [S_{i,\text{GNN}}, S_{i,\text{FC}}, S_{i,\text{STC}}, S_{i,\text{SNC}}]$ Assign labels and split dataset: $(X_{\text{train}}, y_{\text{train}}), (X_{\text{test}}, y_{\text{test}})$
Step 2
Initialize and train SGD Logistic Regression Model:
    model = SGDClassifier(loss = 'log_loss', max_iter = 1000, tol = $1e^{-3}$)
    model.fit($X_{\text{train}}, y_{\text{train}}$)
Step 3
Predict and compute accuracy: $y_{\text{pred}} = \text{model.predict}(X_{\text{test}})$,
    accuracy = $\frac{\text{Correct Predictions}}{\text{Total Predictions}}$
Extract feature weights: w_1, w_2, w_3, w_4, then compute raw MAVS score for the new instance using Eq. (2) and apply the sigmoid function as defined in Eq. (3)
Step 4
If $P \geq 0.5$, **then return** Real News (1),
else return Fake News (0).

The proposed MAVS framework presents a computationally efficient and scalable approach to fake news detection by integrating multi-agent verification strategies. In the following section, we detail the experimental setup, computational performance analysis, and evaluation metrics to assess the effectiveness of MAVS in real-world applications.

4 Experimental Setup

4.1 System Configuration and Data Processing

All experiments were conducted on an **Intel Core i5 system (16 GB RAM) running Ubuntu 20.04 LTS**. The software stack included **Python 3.8+, PyTorch 1.10+**, and **Torch Geometric 2.0+** for machine learning, **Hugging Face Transformers** for NLP, **NetworkX** for graph processing, **Selenium** and **BeautifulSoup** for web scraping, and **Stochastic Gradient Descent (SGD)** for optimization. This section details the data preparation, processing pipeline, evaluation metrics, and final classification algorithm (Algorithm 4) used in the MAVS framework for fake news detection.

UPFD Politifact dataset [7], containing **314 news propagation graphs** (157 fake), was used to train a **Graph Neural Network (GNN)** for fake news detection. The dataset was split into **70% training, 10% validation, and 20%**

testing, ensuring a balanced distribution of **115 fake** and **105 real** graphs as shown in Fig. 4.

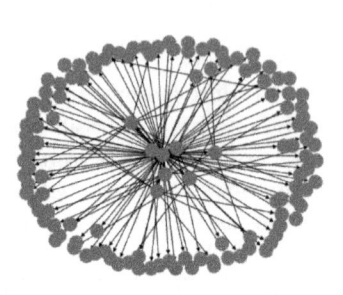

Fig. 3. Hierarchical Tree-Structured Graph Used in the GNN Model.

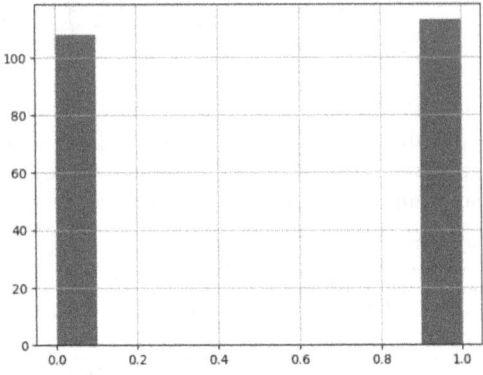

Fig. 4. Dataset labeling for MAVS.

The **MAVS framework** integrates GNNs with AI agents for fact-checking, stance detection, and sentiment analysis. News propagation was modeled using **NetworkX**, with **SpaCy Word2Vec embeddings** as input. A three-layer **GATConv** architecture (**310 input, 128 hidden, 128 output**) was optimized with **Adam (learning rate: 0.01, activation: LeakyReLU)**. To evaluate adversarial robustness, a **Multi-Agent Reinforcement Learning (MARL)** attack [21] was applied, modifying propagation graphs via **bot-driven adversarial edges**. Additionally, the **PolitiFact++ dataset** [18] was used to test the impact of **Human-written Fake news (HF)** and **LLM-generated Fake news (MF)**. Final classification was performed using **SGD**, where the outputs of all AI agents were weighted and fused, as outlined in Algorithm 4.

4.2 Evaluation Metrics

The performance of the MAVS framework was evaluated using standard classification metrics:

Accuracy

$$\text{Accuracy} = \frac{TP + TN}{TP + TN + FP + FN} \tag{5}$$

Precision

$$\text{Precision} = \frac{TP}{TP + FP} \tag{6}$$

Recall

$$\text{Recall} = \frac{TP}{TP + FN} \tag{7}$$

F1-Score

$$\text{F1-Score} = 2 \times \frac{\text{Precision} \times \text{Recall}}{\text{Precision} + \text{Recall}} \tag{8}$$

Figure 5 represents the structure of confusion matrix. The formulas for the evaluation metrics are given by Eqs. (5), (6), (7), and (8).

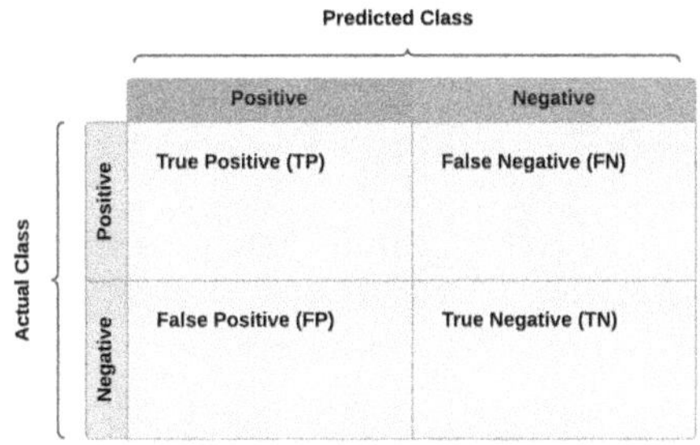

Fig. 5. Confusion Matrix.

The Accuracy and F1-score were compared across all baseline models: HGFND, UPFD-SAGE (GraphSAGE), UPFD-GAT, LSTM, BERT, HiSS, TextCNN, FactAgent with Expert Workflow, CNN, and RoBERTa with RoBERTa-base features, alongside our MAVS framework. Additionally, he impact of adversarial attacks was analyzed specifically for BERT, RoBERTa, GraphSAGE, MAVS by using a **4-fold cross-validation**, evaluating their performance under attack scenarios.

5 Results and Analysis

This section presents the evaluation of our MAVS framework, including the following mentioned topics

- Accuracy and F1-score comparisons with baseline models
- Post-attack performance analysis
- An ablation study regarding different AI Agents.

5.1 Accuracy and F1-Score Comparison of Baseline Models with MAVS

To evaluate the effectiveness of our MAVS framework, we compared its Accuracy and F1-score against several baseline models: HGFND, UPFD-SAGE (GraphSAGE), UPFD-GAT, LSTM, BERT, HiSS, TextCNN, FactAgent with Expert Workflow, CNN, and RoBERTa with RoBERTa-base features.

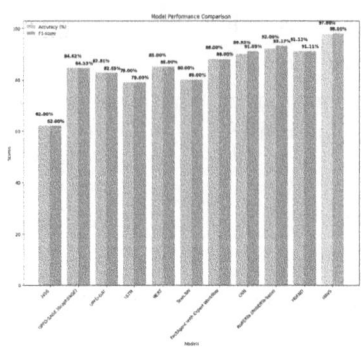

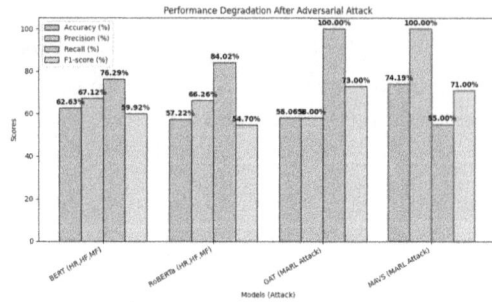

Fig. 6. Accuracy and F1-score comparison of baseline models and MAVS.

Fig. 7. Performance comparison after adversarial attack for BERT, RoBERTa, GraphSAGE, and MAVS.

The bar chart in Fig. 6 visualizes the performance comparison of different models. MAVS achieves the highest accuracy (97.60%) and F1-score (98.00), significantly outperforming other models. The closest competitor, RoBERTa (RoBERTa-base), attains an accuracy of 92.09% and an F1-score of 93.17%, indicating that MAVS leverages a more effective fusion of features to enhance predictive performance.

Models such as CNN [1], HGFND [10], and FactAgent with Expert Workflow [11] show relatively strong performance, achieving accuracy and F1-scores above 88%. However, traditional models like LSTM (79.00%) and TextCNN (80.00%) fall behind, demonstrating their limitations in capturing the complex relationships in fake news detection.HiSS was the worst performer with least accuracy (62.00%).

5.2 Performance Comparison After Adversarial Attack

Notably, GraphSAGE and MAVS were subjected to **MARL** attack as implemented by Wang et al. [21], where bot agents manipulate the news propagation graph by injecting misleading nodes and edges to mimic legitimate user behavior. This coordinated manipulation alters the graph structure to degrade detection performance. In contrast, BERT and RoBERTa were evaluated on adversarially perturbed versions of the **PolitiFact++** dataset.

The results in Fig. 7 reveal that adversarial attacks significantly degrade the performance of all models, but the extent of the degradation varies. BERT and RoBERTa experience a substantial drop, particularly in Recall and F1-score, indicating that they misclassify a significant portion of adversarially perturbed samples. GAT, despite maintaining a high Recall of 100%, suffers from a considerable Precision drop, suggesting that it incorrectly predicts a large number of false positives.

MAVS, however, demonstrates better robustness with an Accuracy of 74.19% and a balanced F1-score of 71%, outperforming the other models. The Precision score of 100% for MAVS suggests that it avoids false positives, but its lower Recall (55%) indicates some difficulty in capturing all real instances. This performance stability highlights MAVS's resilience to adversarial attacks, likely due to its fact-checking mechanisms.

Notably, one of the major findings is shown in the confusion matrices in Fig. 8 that reveals that after the adversarial attack, MAVS still maintains a balance between true positives (20) and false negatives (16) due to its fact-checking agent, indicating **partial resilience to adversarial interference**. Conversely, GNNs demonstrate a severe decline, completely failing to classify "Fake News" instances, as all predictions default to "Real." This highlights GNNs' heightened vulnerability to adversarial perturbations and the advantage of MAVS's architecture in handling such challenges.

5.3 Ablation Study

To analyze the contributions of individual components in the MAVS framework, we perform an ablation study on the GNN, Sentiment Checker, Stance Checker, and Fact Checker.

The results indicate that the **GNN** achieves the highest precision (94.29%), ensuring minimal false positives, while maintaining a balanced F1-score of 92.96%, demonstrating its ability to generalize well. The **Fact Checker** exhibits the highest recall (97.22%), making it the most effective at detecting fake news instances. The **Sentiment Checker** and **Stance Checker**, while weaker in precision, provide valuable complementary information. Their recall values indicate a tendency to detect more fake news instances, which is crucial in adversarial settings where attackers attempt to manipulate narratives.

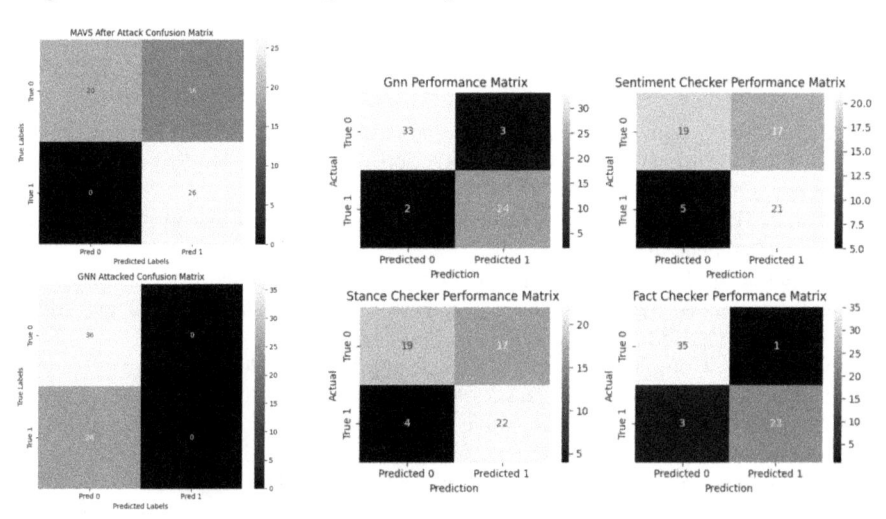

Fig. 8. Confusion Matrices for MAVS and GNN After Attacks.

Fig. 9. Confusion matrices for GNN, Sentiment Checker, Stance Checker, and Fact Checker.

The confusion matrices in Fig. 9 provide additional insights, illustrating that the Fact Checker exhibits strong classification ability, misclassifying only a few instances. In contrast, the Sentiment and Stance Checkers show higher false positives, underscoring the need for their weighted contribution in MAVS.

6 Conclusion and Future Work

This research presents **MAVS (Multi-Agent Verification System)**, an ensemble-based framework integrating **Graph Neural Networks (GNNs)** with **generative AI models** for fact-checking, stance detection, and sentiment analysis. By combining **news propagation modeling** with **text-based verification**, MAVS provides a robust and multidimensional approach to fake news detection.

Using a **weighted aggregation mechanism** optimized via **SGD-based logistic regression**, MAVS achieves **97.6% accuracy** and **98% F1-score**, outperforming **RoBERTa, CNNs, and GAT-based models**. The framework demonstrates **partial resilience to adversarial attacks**, with the fact-checking agent maintaining **74.19% accuracy** under adversarial conditions and ensuring **100% precision**. The fusion of **graph-based structural analysis, linguistic verification, and factual retrieval** enables MAVS to be a **scalable, interpretable, and resilient** misinformation detection system.

Future work will enhance MAVS by enabling real-time adaptive weighting using RL, strengthening multilingual and adversarial defenses, and integrating RAG-based APIs for instant fact-checking. Additionally, MAVS could be used in simulations to study misinformation's impact on public opinion dynamics.

References

1. Alghamdi, J., Lin, Y., Luo, S.: A comparative study of machine learning and deep learning techniques for fake news detection. Information **13**, 576 (2022). https://doi.org/10.3390/info13120576
2. Allcott, H., Gentzkow, M.: Social media and fake news in the 2016 election. J. Econ. Perspect. **31**, 211–236 (2017)
3. Arquam, M., Singh, A., Cherifi, H.: Impact of seasonal conditions on vector-borne epidemiological dynamics. IEEE Access **8**, 94510–94525 (2020). https://doi.org/10.1109/ACCESS.2020.2995650
4. Cherifi, H.: Complex networks and their applications (2014)
5. Conover, M., Gonçalves, B., Ratkiewicz, J., Flammini, A., Menczer, F.: Predicting the political alignment of twitter users. In: 2011 IEEE Third International Conference on Privacy, Security, Risk and Trust and 2011 IEEE Third International Conference on Social Computing, pp. 192–199 (2011). https://doi.org/10.1109/PASSAT/SocialCom.2011.34
6. Diop, I.M., Cherifi, C., Diallo, C., Cherifi, H.: Revealing the component structure of the world air transportation network. Appl. Netw. Sci. **6**(1), 1–50 (2021). https://doi.org/10.1007/s41109-021-00430-2

7. Dou, Y., Shu, K., Xia, C., Yu, P.S., Sun, L.: User preference-aware fake news detection. In: Proceedings of the 44th International ACM SIGIR Conference on Research and Development in Information Retrieval (2021)
8. Drif, A., Zerrad, H.E., Cherifi, H.: EnsVAE: ensemble variational autoencoders for recommendations. IEEE Access **8**, 188335–188351 (2020)
9. Hardalov, M., Hristov, T., Nakov, P., Koychev, I.: Few-shot stance detection for political claims. In: Proceedings of the 60th Annual Meeting of the Association for Computational Linguistics (ACL), pp. 7759–7771 (2022)
10. Jeong, U., Ding, K., Cheng, L., Guo, R., Shu, K., Liu, H.: Nothing stands alone: relational fake news detection with hypergraph neural networks (2022)
11. Li, X., Zhang, Y., Malthouse, E.C.: Large language model agent for fake news detection (2024)
12. Monti, F., Frasca, F., Eynard, D., Mannion, D., Bronstein, M.M.: Fake news detection on social media using geometric deep learning. In: Proceedings of the National Academy of Sciences, vol. 116, pp. 25738–25743 (2019)
13. Pennycook, G., Rand, D.G.: The psychology of fake news. Trends Cogn. Sci. **25**, 388–402 (2021). https://doi.org/10.1016/j.tics.2021.02.007
14. Qureshi, K., Malick, R., Sabih, M., Cherifi, H.: Complex network and source inspired Covid-19 fake news classification on twitter. IEEE Access **9**, 1–1 (2021). https://doi.org/10.1109/ACCESS.2021.3119404
15. Qureshi, K.A., Malick, R.A.S., Sabih, M., Cherifi, H.: Deception detection on social media: a source-based perspective. Knowl. Based Syst. **256**, 109649 (2022). https://doi.org/10.1016/j.knosys.2022.109649, https://www.sciencedirect.com/science/article/pii/S0950705122008346
16. Sharma, K., Qian, F., Jiang, H., Ruchansky, N., Zhang, M., Liu, Y.: Combating fake news: a survey on identification and mitigation techniques. ACM Trans. Intell. Syst. Technol. (TIST) **10**, 1–42 (2019)
17. Shu, K., Mahudeswaran, D., Wang, S., Lee, D., Liu, H.: FakeNewsNet: a data repository with news content, social context and dynamic information for studying fake news on social media. arXiv preprint arXiv:1809.01286 (2018)
18. Su, J., Zhuo, T.Y., Mansurov, J., Wang, D., Nakov, P.: Fake news detectors are biased against texts generated by large language models (2023)
19. Su, X., et al.: A comprehensive survey on community detection with deep learning. IEEE Trans. Neural Netw. Learn. Syst., 1–21 (2022). https://doi.org/10.1109/TNNLS.2021.3137396
20. Thorne, J., Chen, M., Myrianthous, G., Pu, J., Wang, X., Vlachos, A.: Fact extraction and verification (fever). In: Proceedings of the 2018 Conference of the North American Chapter of the Association for Computational Linguistics: Human Language Technologies, Volume 1 (Long Papers), pp. 809–819 (2018)
21. Wang, H., Dou, Y., Chen, C.H., Sun, L., Yu, P.S., Shu, K.: Attacking fake news detectors via manipulating news social engagement. In: Proceedings of the ACM Web Conference 2023 (WWW 2023), pp. 3978–3986 (2023)
22. Zhou, J., et al.: Graph neural networks: a review of methods and applications (2021)
23. Zhu, J., Peng, X., Wang, L.: Sentiment analysis meets fake news detection: a deep-learning approach. Expert Syst. Appl. **167**, 114171 (2021)

Evolutionary Game Selection Leads to Emergent Inequality

Isaak Mengesha and Debraj Roy

Computational Science Lab, University of Amsterdam, Amsterdam, The Netherlands
{i.mengesha,d.roy}@uva.nl

Abstract. The emergence of collective cooperation within competitive environments is well-documented in biology, economics, and social systems. Traditional evolutionary game models primarily investigate the evolution of strategies within fixed games, neglecting the simultaneous coevolution of strategies and the environment. Here, we introduce a game selection model where both the strategies employed by agents and the games themselves evolve dynamically through evolutionary processes. Our results demonstrate that these coevolutionary dynamics foster novel collective phenomena, including changed cooperative interactions. When applied to structured populations, the network's architecture, and agent properties such as risk-aversion and bounded rationality significantly influences outcomes. By exploring the interplay between these factors, our model provides novel insights into the persistent social dilemmas observable in real-world systems.

Keywords: Evolutionary Game Theory · Persistent Inequality · Social Networks · Multiple Equilibria · Agent-Based Modeling

1 Introduction

Wealth inequality has declined on a global scale over the past two centuries, but persistent and often worsening disparities remain a significant challenge. In 1800, around 80% of the global population was living below the international poverty line, whereas by 2022 that figure had fallen to approximately 8.5% [1]. This remarkable progress, documented by the World Bank, is tempered by persistent doubts about whether the international poverty line fully captures the reality of poverty and by the observed slowdown in poverty reduction over the last decade [2]. The COVID-19 pandemic, which adversely affected economies worldwide, has underscored the fragility of these gains [3].

Many researchers have sought to explain why low-income groups sometimes become trapped in cycles of poverty despite overall economic growth. They distinguish between "friction-driven" and "scarcity-driven" poverty traps. While the former requires multiple market failures (e.g., lack of credit and indivisibilities in production); the latter can exist without external frictions due deprivation leading to strong behavioural changes (e.g.low savings, underinvestment in human capital, and myopic decision-making) [4]. In the "poor but neoclassical" view, such adaptations are rational under the given constrains but as a consequence lead to a low-level equilibrium trap that individuals or communities struggle to escape from [5]. Contrary to this is the notion of classical macroeconomic theories like the Solow model of a single, universal equilibrium to which individuals and societies ultimately converge [6,7]. However, both perspectives face mixed evidence, with there being little evidence for the existence of strong traps [8], and simultaneously large and persistence difference in wealth accumulation across individuals and societies [9,10]. Identifying isolated causal mechanisms across heterogeneous populations is methodologically and ethically challenging. In response, modeling studies have become an attractive approach to probe how various micro-level factors can create and sustain wealth inequalities [11].

Risk aversion has been widely discussed as a catalyst for low-risk, low-reward behaviors, which may inhibit innovation and drive persistent wealth stratification [12]. Agents who are sensitive to risk in uncertain environments often prefer modest but reliable payoffs, a habit that can limit their capacity to accumulate resources in the long run. These mechanism are also studied from a modeling approach [13]. Bounded rationality introduces another layer to this dynamic [14]. Real-world decision-makers operate with finite computational abilities and incomplete information, often relying on heuristics rather than optimizing [15]. Furthermore we can see that the introduction of risk and uncertainty in the absence of comprehensive insurance can result in severe changes of the wealth distribution [16]. This condition can exacerbate poverty traps when agents fail to adopt better strategies even when they are locally observable.

Social capital, expressed through the structure of the network and the community's norms, represents a key dimension in explaining inequality [17,18]. Network connectivity allows for the exchange of knowledge, the adoption of profitable strategies, and the accumulation of social influence. Agents with more connections, or a higher network degree, can more easily access beneficial information and resources. In contrast, individuals who remain socially isolated have fewer opportunities to learn or innovate. For example, upward social mobility is greatly affected by the number of high SES connections [19]. These networks can co-evolve as agents establish or sever links based on shared characteristics or observed success, often reinforcing patterns of inequality. The notion that network degree functions as social capital suggests that the "rich-get-richer" effect may operate not only financially but also socially.

Computational models have further explored how simple local interaction dynamics can spontaneously generate persistent inequality, even absent explicit reinforcing mechanisms. For instance, the seminal Sugarscape model demonstrated that minimal behavioral rules alone can lead to substantial wealth disparities, despite agents sharing initially homogeneous conditions [20,21]. Network effects like preferential attachment and homophily further exacerbate these inequalities, allowing certain groups to accumulate disproportionate advantages in resources and information access [22]. Parallel to this, evolutionary game theory has investigated how cooperation can emerge and stabilize within competitive environments [23]. Axelrod's pioneering simulations of the iterated Prisoner's

Dilemma showed that repeated interactions enable reciprocity and foster persistent cooperative behaviors, even among self-interested agents [24]. Subsequent studies demonstrated that network structures and spatial clustering further promote cooperation by allowing cooperators to preferentially interact, thus protecting against exploitation by defectors [25–27]. These mechanisms highlight the significance of agent interaction structure in driving social outcomes, motivating our exploration of the role of social learning (game selection) on inequality.

In this paper, we develop a minimalistic agent-based model to examine how risk aversion, bounded rationality, and social capital can jointly lead to persistent inequality. By abstracting away many real-world complexities, the study aims to identify the core mechanisms of stratification that may persist in diverse contexts. We specifically avoided direct self-reinforcing mechanisms to define traps, or compounding returns to investment. The approach draws on concepts from evolutionary game theory and network science to capture how agents adapt both their strategies and type interactions they engage in (e.g. games played). The experimental design resembles Sadekar *et al.* work on the emergence of cooperative environments [28]. However, among other important distinctions, we focus on the resulting emergence of inequality and overall welfare. The results provide a theoretical account of how uneven wealth distributions may arise even when initial conditions are relatively homogeneous.

Our findings show that local decisions, strategy adaptation, and rewiring of social connections can compound even small initial differences into significant wealth disparities. Boundedly rational agents, who weigh risks conservatively, often settle on suboptimal behaviors, and those with slightly higher connectivity (social capital) gain preferential access to information (games), magnifying their wealth advantage over time. This outcome is consistent with real-world observations that social isolation hinders poverty alleviation and suggests that inequality can emerge and persist through minimal assumptions alone, without needing elaborate institutional or macroeconomic drivers. By illuminating how microlevel biases in learning and networking reinforce each other, this study underscores that small interventions or policy changes at the local level, such as reducing network fragmentation, could prove critical to preventing the entrenchment of persistent poverty.

2 Methods

To systematically investigate how minimal micro-level mechanisms generate and sustain inequality, we employ an agent-based evolutionary game-theoretic framework. This approach enables explicit modeling of individual heterogeneity, adaptive behavior, and dynamic interactions within structured populations. The model comprises four core components: heterogeneous agents characterized by distinct behavioral parameters, payoff matrices defining economic interactions, strategy selection driven by bounded rationality, and evolving network dynamics shaped by wealth-based homophily. Each component is selected to isolate fundamental drivers of inequality, abstracting complex real-world interactions into tractable, theoretically grounded processes. The following subsections formally detail each model component and its corresponding assumptions.

2.1 Individual Payoff Matrices

We use a general two-player, two-strategy symmetric game as our fundamental unit of analysis, due to its conceptual simplicity and analytical tractability. This formulation allows us to abstract essential strategic interactions into a generalizable framework, making it possible to clearly classify games based on payoff structures and systematically explore how different strategic conditions shape inequality dynamics. Formally, the payoff matrix for one agent (Alice) playing against another agent (Bob) is defined as follows:

$$\begin{pmatrix} R & S \\ T & P \end{pmatrix} \xrightarrow{(1)} \begin{pmatrix} R-P & S-P \\ T-P & 0 \end{pmatrix} \xrightarrow{(2)} \begin{pmatrix} 1 & \frac{S-P}{R-P} \\ \frac{T-P}{R-P} & 0 \end{pmatrix} \xrightarrow{(3)} \begin{pmatrix} 1 & U \\ V & 0 \end{pmatrix}. \tag{1}$$

Here, R is the payoff for Alice if both players choose action 1, S is the payoff for Alice if she chooses action 1 and Bob chooses action 2, and so on. Note that adding a constant to all payoffs does not change the strategic structure of the game. Similarly, subtracting P from each payoff does not alter the game's structure, which happens in step (1). By relabeling the strategies, we can assume that $R > P$. The payoffs lack a natural unit of measure, allowing us to rescale them by any positive number. In (2) we make a convenient choice is to rescale by $R - P$. Lastly we define $U = \frac{S-P}{R-P}$ and $V = \frac{T-P}{R-P}$. This representation allows us to analyze games by plotting them in two dimensions, as shown in the figure below (Fig. 1).

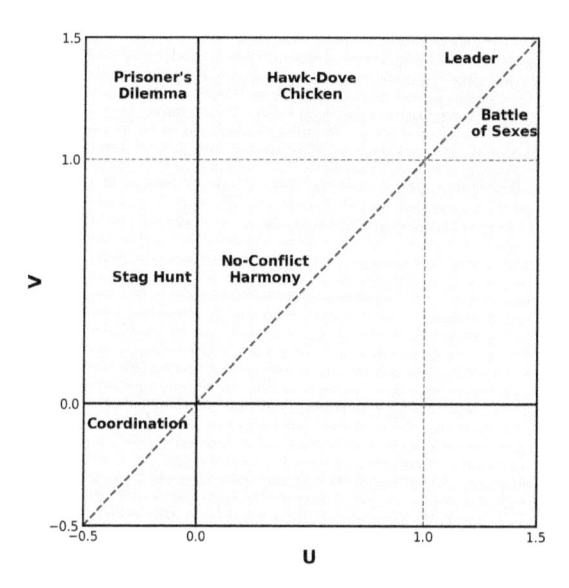

Fig. 1. Games classification - The U-V space of cooperation-defection games. Each agent is assigned one point in the U-V space. In this paper, we have constrained the space to $U > 0$ $V > 0$ space.

Note that there exist qualitative differences between the games depending on the possible orderings of U and V relative to 0 and 1. There are 12 possible orderings, corresponding to 12 different types of games, some of which are labeled in the figure. We limit ourself to the reduction to U and V as defined above.

2.2 Agent Attributes

The stochastic evolutionary game model is populated by heterogeneous agents, each with unique characteristics that influence their decision-making processes and economic outcomes. These agent-specific attributes form the foundation for the emergent inequality patterns observed in the simulation.

Each agent a_i in the model is characterized by a set of key attributes that evolve throughout the simulation. These attributes capture both accumulated resources and intrinsic behavioral tendencies:

W_i **Wealth**: The cumulative payoff accumulated over all previous time steps. Represents economic resources and serves as the primary measure of inequality. Compounds over time and influences both strategic decisions and network formation.

W_i^R **Recent Wealth**: The discounted sum of payoffs accumulated over the most recent five time steps. Reflects short-term performance and adaptability, providing a responsive measure of current economic trajectory.

η_i **Risk Aversion**: A fixed parameter indicating willingness to engage in risky decisions. Higher values correspond to higher risk aversion, leading to selection of strategies with lower potential returns despite uncertainty. Remains constant throughout the simulation.

λ_i **Bounded Rationality**: A fixed cognitive limitation parameter influencing decision-making precision. Higher values lead to more optimal strategy selection, while lower values result in more random choices. Captures heterogeneity in agents' ability to evaluate strategic opportunities.

The dynamic attributes—Wealth and Recent Wealth—are updated after each interaction based on the payoffs received. The Recent Wealth is calculated using a discounted sum of recent payoffs to reflect the greater relevance of more recent outcomes:

$$W_{i,t}^R = \sum_{j=0}^{4} (1-\delta_d)^j P_{i,t-j}, \tag{2}$$

where δ_d represents a constant discounting factor (set to 0.05 in our simulations), and $P_{i,t}$ is the payoff received by agent i at time t. For the initial time steps where $t < 5$, the recent wealth equals the total wealth: $W_{i,t}^R = W_{i,t}$. The total Wealth is simply the accumulated payoff over time. The risk aversion η changes the curvature in the isoelastic utility function:

$$U(\pi) = \begin{cases} \dfrac{\pi^{1-\eta}-1}{1-\eta}, & \eta \neq 1, \\ \ln(\pi), & \eta = 1, \end{cases}$$

where π represents consumption and η is the constant relative risk aversion parameter.

The fixed Risk Aversion (η) and (bounded) rationality (λ)—influences how agents select strategies and respond to opportunities. These parameters create heterogeneity in decision-making processes, allowing the model to explore how cognitive and behavioral differences contribute to economic stratification even in the absence of explicit institutional advantages or disadvantages. By incorporating these diverse agent attributes, the model captures essential aspects of economic inequality dynamics: the path-dependence of wealth accumulation, the role of risk preferences in economic outcomes, and the impact of decision-making limitations on long-term prosperity.

At $t = 0$, all agents are initialized with zero wealth ($W_i = 0$) and zero recent wealth ($W_i^R = 0$). The fixed personality traits are sampled from standard probability distributions:

$$\eta_i \sim \mathcal{N}(1, 0.5), \quad \lambda_i \sim \text{Log-}\mathcal{N}(0, 1). \tag{3}$$

The inverse risk aversion parameter η_i follows a normal distribution with mean 1 and standard deviation 0.5, creating a population where most agents have moderate risk preferences but with meaningful variation. Higher values of η_i indicate greater willingness to pursue high-risk, high-reward strategies. The bounded rationality parameter λ_i follows a lognormal distribution with location parameter 0 and scale parameter 1, resulting in a right-skewed distribution where most agents have relatively low cognitive precision while a small number have significantly higher decision-making capabilities. This distribution reflects empirical observations of skill heterogeneity in human populations [29, 30]. These fixed traits generate persistent heterogeneity in the agent population, as they remain constant throughout the simulation, influencing strategic choices and consequently wealth accumulation patterns over time.

2.3 Payoff Matrix and Updates

Each agent participates in a two-strategy asymmetric game, where their payoff matrix is given by:

$$M_i = \begin{bmatrix} 1 & U_i \\ V_i & 0 \end{bmatrix}. \tag{4}$$

Agents interact by playing against a randomly selected opponent, adjusting their payoffs dynamically based on their dependence parameter $\delta \in [0, 1]$. The dependence parameter determines the degree to which an agent's payoffs are influenced by their opponent's payoffs:

- $\delta = 0$: The agent retains its original payoffs (fully independent play).
- $\delta = 1$: The agent fully adopts the average payoffs between itself and its opponent (fully dependent play).

For an agent i interacting with opponent j, the adjusted payoffs are computed as:

$$U'_i = (1-\delta)U_i + \delta \frac{(U_i + U_j)}{2}, \tag{5}$$

$$V'_i = (1-\delta)V_i + \delta \frac{(V_i + V_j)}{2}. \tag{6}$$

Symmetrically, the opponent updates their payoffs. After updating their payoff matrices, agents independently select strategies based on their respective payoffs. The resulting payoffs are determined by their own updated matrices, rather than a shared game structure, preserving individual strategic diversity. This formulation ensures that when δ is low, agents retain distinct payoffs, leading to heterogeneous strategic behaviors. As δ increases, their payoff structures converge, promoting more homogeneous interactions while still allowing for dynamic adaptation over repeated games.

2.4 Strategy and Game Selection

Strategies are selected using the Logit Quantal Response Equilibrium (LQRE), an equilibrium concept that accounts for bounded rationality and decision-making errors by agents [31]. Unlike the classical Nash equilibrium, which assumes perfectly rational agents, the LQRE allows for stochastic behavior where the probability of choosing a particular strategy increases with its expected payoff but remains sensitive to payoff differences. Specifically, agents follow a logistic choice rule to probabilistically select strategies, given their individual rationality parameter λ_i. The probability of agent i selecting a strategy s_l is given by:

$$P_i(s_l) = \frac{e^{\lambda_i \pi_i(s_l)}}{\sum_k e^{\lambda_i \pi_i(s_k)}}, \tag{7}$$

where $\pi_i(s_j)$ denotes the expected payoff to agent i from playing strategy s_j. A higher rationality parameter λ_i indicates more precise optimization and thus less randomness in strategy choice.

In the specific context of two-strategy interactions, the probability that agent 1 selects strategy S_1 ($P^1_{S_1}$) in interaction with another agent who selects strategy S_1 with probability $P^2_{S_1}$ is explicitly represented as:

$$P^1_{S_1} = \frac{e^{\lambda_1(P^2_{S_1}\pi^1_{11} + (1-P^2_{S_1})\pi^1_{12})}}{e^{\lambda_1(P^2_{S_1}\pi^1_{11} + (1-P^2_{S_1})\pi^1_{12})} + e^{\lambda_1(P^2_{S_1}\pi^1_{21} + (1-P^2_{S_1})\pi^1_{22})}}, \tag{8}$$

where π^1_{11}, π^1_{12}, π^1_{21}, and π^1_{22} represent the payoffs to agent 1 corresponding to the strategy pairs (S_1, S_1), (S_1, S_2), (S_2, S_1), and (S_2, S_2) respectively. Together with the other agent this defines a system of two coupled equations with two unknowns that we solve numerically for each agent up to a certain error tolerance.

Additionally, agents need to make choices w.r.t. what games to play in the future. Taking into account the wealth difference between players δW_{ij} as well

λ_i and η_i. The utility function $U(\eta_i, \delta W_{ij})$ is the isoelastic Utility function. Specifically, the probability that agent i selects to interact with agent j in a future game is modeled using a softmax choice rule, closely resembling the QRE:

$$P_{i \to j} = \frac{\exp\left(\lambda_i \cdot U(\eta_i, \delta W_{ij})\right)}{\sum_k \exp\left(\lambda_i \cdot U(\eta_i, \delta W_{ik})\right)}, \tag{9}$$

This selection rule enables agents to strategically evaluate and probabilistically select opponents based on wealth-driven incentives, risk attitudes, and cognitive limitations, contributing to the dynamic evolution of the system's economic interactions.

2.5 Network Topology and Evolution

Homophily—the tendency for agents to associate with others who are similar to themselves—is a fundamental mechanism in social and economic networks. In wealth-driven settings, agents with similar wealth are more likely to interact, forming clusters that reflect their economic similarity. A formalization of such behavior must capture two essential aspects: first, the increased likelihood of forming connections when agents' wealth levels are similar; and second, the possibility of breaking connections when wealth disparities become significant. Agents interact on a network that updates adaptively based on wealth-driven homophily. The probability of forming a new link follows:

$$P_{\text{con}} = \frac{1}{1 + (1 - \beta \cdot \max(|W_i - W_k|, \epsilon))^{-\alpha}}, \tag{10}$$

where α and β modulate the strength of homophily. Here, $|W_i - W_k|$ represents the wealth difference between agents i and k, and ϵ is a small constant that ensures the expression is well-defined even when the wealth difference is very small. The parameter β scales the impact of the wealth difference, while α controls how sharply the connection probability decreases as this difference increases. Connections may be severed based on the inverse of P_con.

2.6 Simulation Process

At the start of each simulation, agents are placed onto an initial network topology generated using either a Random Regular, Watts-Strogatz, or Holme-Kim algorithm. In our simulations, results did not differ meaningfully across these topologies, suggesting robustness of the observed inequality patterns to the choice of initial network structure, as we demonstrate convergence to the same equilibrium network topology. Agents are initialized with zero wealth, and their fixed behavioral parameters (bounded rationality λ_i and inverse risk aversion η_i) are randomly drawn from the distributions described above. Optionally, entries of the payoff matrices may be normalized by the sum of all payoffs to facilitate comparison across games. Each simulation then proceeds for 100 discrete time steps. Within each time step, every agent sequentially executes the actions summarized in Table 1. Payoffs obtained during interactions are exclusively allocated to the focal agent, ensuring that the wealth effects are not dominated by network dynamics.

Table 1. Detailed simulation steps in the evolutionary game model.

1.	**Opponent Selection**	An agent selects an opponent randomly from its current network neighborhood.
2.	**Payoff Adjustment**	The agent adjusts its payoffs towards the opponent's payoffs based on the dependence parameter δ.
3.	**Strategy Selection**	The agent probabilistically selects a strategy using the Logit Quantal Response Equilibrium (LQRE), influenced by bounded rationality λ_i.
4.	**Play Game**	Agents play the selected game, receive payoffs, and update their Wealth W_i and Recent Wealth W_i^R.
5.	**Choose Game**	Agents choose future games based on past performance, wealth differences, and individual risk aversion η_i.
6.	**Network Update**	Network connections form or dissolve according to wealth-driven homophily rules.

Throughout the simulation, we specifically track two main outcomes: economic inequality—measured by agents' wealth distributions—and the learning rate, capturing how rapidly agents adapt their strategy and game choices in response to past experiences.

3 Results

In the standard experimental runs, we recover basic stylized facts about broader income distributions. We first investigate the complementary cumulative distribution of income to compare with empirical data [16]. The results qualitatively depend on whether the payoff matrices are normalized by the sum of their entries. If not normalized, the income distribution follows a classical power-law that slightly flattens over time and eventually saturates due to finite size effects. Similarly, in the normalized case, however, the income distribution approximates a heavy-tailed form, with visual resemblance to a power-law, although the absolute range is limited (recent wealth spans only a few discrete levels), preventing definitive tail characterization. This indicates the existence of multiple equilibria with varying productivity. Our inequality measures using the Gini coefficient confirm this, landing at 0.12 and 0.41 (normalized) respectively. We interpret the non-normalized case as reflecting technological innovation that increases baseline productivity. With normalization, there are no overall payoff advantages between games; instead, games differ primarily in their ease of coordination toward beneficial outcomes (Fig. 2).

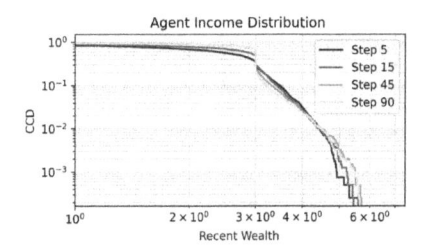

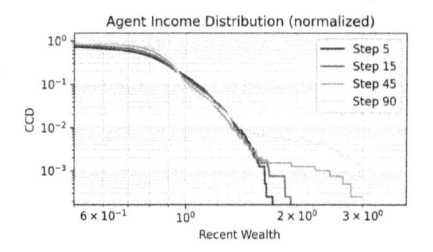

Fig. 2. The complementary cumulative distribution of income over a fixed time interval mimics a classical power-law for standard games and a log-normal shape for normalized games.

This brings us to learning, as illustrated in Fig. 3, where we observe a single equilibrium in the absence of normalization and multiple equilibria with normalization. Notice that in the normalized case, the real difference lies in the concentration level of payoffs, making it easier for players to coordinate toward them. These preferred states generally correspond to the corners of the UV state space. Also note that, due to compromises between payoff matrices, interacting with individuals playing a symmetric but different game effectively reduces one's payoff. This contributes to the overall lower wealth observed under normalization.

Lastly, we run experiments on system-level learning by varying the population-level rationality and risk-aversion parameters. We achieve this by adding a fixed constant to empirically justified distributions of these parameters. We approximate the learning speed of the system by measuring the rate at

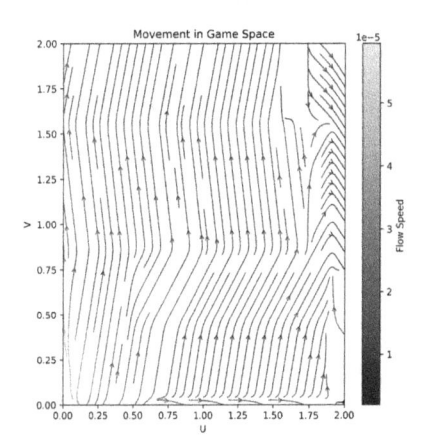

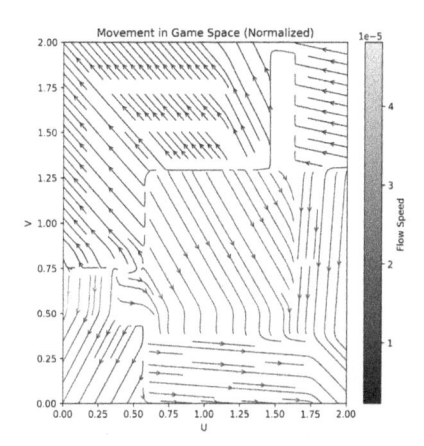

Fig. 3. Movement in the UV state space. Generally, attractors correspond to higher payoffs. In the non-normalized case, the upper right corner is the attractor state, preferred due to high payoffs for both players. In the normalized case, high singular values for both U and V are preferred as payoff concentrates value in one cell.

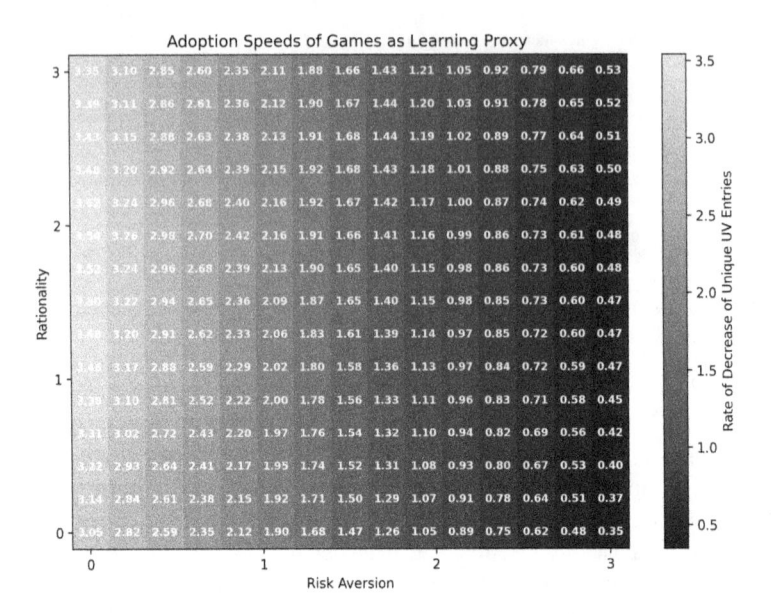

Fig. 4. While keeping the overall shapes of rationality and risk-aversion distributions fixed, we shift the distributions by adding a constant. This results in different system-level learning behaviors, approximated by the speed at which the system converges to the desired games.

which the number of unique games decreases within the population. As individuals converge toward equilibria, more individuals share similar games. We confirm that increased risk aversion decreases the learning speed, an effect stronger than the sensitivity observed with rationality. Interestingly, as rationality increases, adoption speed peaks before reaching maximum rationality values. We attribute this to the fact that when all players closely approach the Nash equilibrium, the normalized games become less differentiated. This mitigates coordination difficulty in a more rational population, thereby reducing payoff differences and slowing game adoption speed (Fig. 4).

4 Discussion

Our results indicate that the system can settle into multiple stable equilibria, leading to enduring wealth stratification. Even with identical initial conditions and rules, some agent communities gravitate toward high-payoff cooperative regimes, while others remain in low-payoff conventions. This finding reinforces the concept of poverty traps: stable low-level equilibria from which escape is difficult [4]. In line with the "poor but rational" perspective, we observe that purely adaptive behaviors under constraints (without external market failures) can lock populations into low-level wealth equilibria [5]. This suggests that persistent inequality arises endogenously through path-dependent dynamics, supporting theories that allow multiple long-run equilibria rather than a single universal outcome [7]. Thus, small differences can yield substantially divergent trajectories, explaining persistent wealth disparities amidst overall economic growth [8,9].

Normalization of payoffs, removing inherent technological advantages, underscores the importance of coordination among agents. Without exogenous innovation, agents prosper only through effective collective action, whereas miscoordination traps groups in suboptimal outcomes. Such dynamics mirror institutional traps described in growth theory, where societies fail to organize collective action necessary for development [4]. Practically, our findings emphasize social cohesion and norm alignment as critical to avoiding coordination failures, especially in innovation-poor contexts.

The interplay between rationality and risk aversion strongly influences equilibrium convergence. Highly rational, risk-averse agents swiftly settle into safe yet suboptimal equilibria, while moderate bounded rationality or reduced risk aversion facilitates exploration and potential discovery of superior outcomes. This outcome aligns with bounded rationality and risk-sensitive decision-making theories, where heuristic-driven choices and aversion to novel strategies prevent optimal outcomes [14–16]. Hence, neither extreme rationality nor excessive caution guarantees optimal collective outcomes—strategic exploration significantly improves long-term performance.

4.1 Limitations and Future Research

Our minimalist approach isolates essential inequality mechanisms but excludes many real-world complexities, potentially understating certain inequality drivers (e.g., inherited advantages or market institutions). Explicitly excluding compounding mechanisms such as capital investment returns further limits the realism [12]. Future work could also examine whether the emergent network structures exhibit real-world properties such as degree heterogeneity, clustering, or small-world topology. Additional extensions might include exploring network-based interventions, incorporating realistic agent behaviors (e.g., memory-driven learning), and validating models against empirical data. Such efforts would bridge theoretical findings with real-world implications, enhancing both the robustness and applicability of these insights.

Acknowledgment. We thank Karin Brinksma for her valuable contribution in developing the simulation code used in this study. We also acknowledge support and resources provided by University of Amsterdam.

Disclosure of Interests. The authors have no competing interests to declare that are relevant to the content of this article.

References

1. Roser M.: The history of global economic inequality. Our World in Data (2017). https://ourworldindata.org/the-history-of-global-economic-inequality

2. Alkire, S., Nogales, R., Quinn, N.N., Suppa, N.: On track or not? Projecting the global multidimensional poverty index. J. Dev. Econ. **165**, 103150 (2023)
3. Mahler, D.G., Yonzan, N., Lakner, C.: The impact of Covid-19 on global inequality and poverty (2022
4. Ghatak M.: Theories of poverty traps and anti-poverty policies. World Bank Econ. Rev. **29**(suppl_1), S77–S105 (2015)
5. Duflo, E.: Poor but rational. Underst. Poverty **24**, 367–379 (2006)
6. Acemoglu, D.: Introduction to Modern Economic Growth. Princeton university press (2008)
7. Banerjee, A.V., Duflo, E.: Growth theory through the lens of development economics. Handb. Econ. Growth **1**, 473–552 (2005)
8. Kraay, A., McKenzie, D.: Do poverty traps exist? Assessing the evidence. J. Econ. Perspect. **28**(3), 127–148 (2014)
9. Barrett, C.B., Carter, M.R.: The economics of poverty traps and persistent poverty: empirical and policy implications. J. Dev. Stud. **49**(7), 976–990 (2013)
10. Aiyar, S., Duval, R., Puy, D., Yiqun, W., Zhang, L.: Growth slowdowns and the middle-income trap. Jpn. World Econ. **48**, 22–37 (2018)
11. De Nardi, M.: Quantitative models of wealth inequality: a survey. National Bureau of Economic Research (2015). https://api.semanticscholar.org/CorpusID:35849844
12. Haushofer, J., Fehr, E.: On the psychology of poverty. Science **344**(6186), 862–867 (2014)
13. Iglesias, J.-R., Gonçalves, S., Abramson, G., Vega, J.-L.: Correlation between risk aversion and wealth distribution. XXPhys. A **342**(1–2), 186–192 (2004)
14. De Bruijn, E.-J., Antonides, G.: Poverty and economic decision making: a review of scarcity theory. Theor. Decis. **92**(1), 5–37 (2022)
15. Morton, J.M.: Reasoning under scarcity. Australas. J. Philos. **95**(3), 543–559 (2017)
16. Banerjee, A.V., Newman, A.F.: Risk-bearing and the theory of income distribution. Rev. Econ. Stud. **58**(2), 211–235 (1991)
17. Puttnam, R.: The role of social capital in development : an empirical assessment (2002). https://api.semanticscholar.org/CorpusID:197648850
18. Wooldridge, M., Jennings, N.R.: Pitfalls of agent-oriented development. In: Proceedings of the Second International Conference on Autonomous Agents, pp. 385–391 (1998)
19. Chetty, R., et al.: NBER working paper series social capital I: measurement and associations with economic mobility (2022). https://api.semanticscholar.org/CorpusID:251255471
20. Epstein, J.M., Axtell, R.: Growing Artificial Societies: Social Science from the Bottom Up. Brookings Institution Press, Washington, D.C. (1996). 9780262550253
21. Föllmer, H.: Random economies with many interacting agents. J. Math. Econ. **1**(1), 51–62 (1974)
22. DiMaggio, P., Garip, F.: Network effects and social inequality. Ann. Rev. Sociol. **38**, 93–118 (2012). https://doi.org/10.1146/annurev.soc.012809.102545
23. Sanchez Carrera, E.J.: Evolutionary dynamics of poverty traps. J. Evol. Econ. **29**(2), 611–630 (2019)
24. Axelrod, R., Hamilton, W.D.: The evolution of cooperation. Science **211**(4489), 1390–1396 (1981). https://doi.org/10.1126/science.7466396
25. Nowak, M.A.: Five rules for the evolution of cooperation. Science **314**(5805), 1560–1563 (2006). https://doi.org/10.1126/science.1133755
26. Santos, F.C., Pacheco, J.M.: Scale-free networks provide a unifying framework for the emergence of cooperation. Phys. Rev. Lett. **95**(9), 098104 (2005)

27. Kasthurirathna, D., Piraveenan, M.: Emergence of scale-free characteristics in socio-ecological systems with bounded rationality. Sci. Rep. **5**(1), 10448 (2015)
28. Sadekar, O., Civilini, A., Gómez-Gardeñes, J., Latora, V., Battiston, F.: Evolutionary game selection creates cooperative environments. Phys. Rev. E **110**(1), 014306 (2024)
29. Mincer, J.: Progress in human capital analyses of the distribution of earnings. In: The Personal Distribution of Incomes (Routledge Revivals), pp. 136–192. Routledge (2013)
30. Gualandi, S., Toscani, G.: Human behavior and lognormal distribution. A kinetic description. Math. Models Methods Appl. Sci. **29**(04), 717–753 (2019)
31. McKelvey, R.D., Palfrey, T.R.: Quantal response equilibria for normal form games. Games Econ. Behav. **10**(1), 6–38 (1995)

Computational Optimization, Modelling and Simulation

Enhancing Gaussian Mixture Model Fitting via Equiprobable Binning and Adaptive Differential Evolution

Wojciech Achtelik[✉] and Maciej Smołka

AGH University of Krakow, Kraków, Poland
{wachtelik,smolka}@agh.edu.pl

Abstract. Fitting Gaussian Mixture Models (GMMs) to one-dimensional data is a fundamental task in machine learning, traditionally addressed using the Expectation-Maximization (EM) algorithm. However, EM lacks inherent mechanisms to enforce separation between mixture components, a critical requirement in domains like medical research where distinct subgroups must be identified. Recently, the Distribution Optimization (DO) framework addressed this limitation by reformulating GMM estimation as a chi-squared goodness-of-fit minimization problem with an overlap penalty to enhance separation. However, its reliance on equiwidth binning and genetic algorithms can limit accuracy and scalability. In this paper, we refine the DO framework in two key ways: (1) replacing equiwidth binning with Mann–Wald's equiprobable cells to improve estimation accuracy, and (2) adopting advanced Differential Evolution (DE) for more robust optimization of the high-dimensional parameter space. Through extensive experiments on synthetic and real-world datasets, we demonstrate that our refined approach significantly enhances accuracy, stability, and scalability compared to the original DO method.

Keywords: Gaussian Mixture Model · Minimum Chi-squared Estimation · Expectation Maximization · Differential Evolution · Distribution Optimization

1 Introduction

Gaussian Mixture Models are a cornerstone of modern statistical modeling, offering a powerful framework for capturing complex, multimodal distributions. By representing a probability density as a weighted sum of Gaussian components, GMMs have become essential across a wide array of disciplines, including machine learning, signal processing, bioinformatics, and medicine. In medical research, GMMs are well-suited for uncovering latent structures in single-dimensional data, with applications such as clustering patient scores, analyzing pain levels in rheumatoid arthritis [10], assessing diabetes risk [12], and evaluating olfactory function in clinical settings [11]. This suitability stems from their theoretical foundation: many biological indicators arise from the cumulative effect of multiple underlying processes, aligning with the justification for normal distributions based on the central limit theorems.

Despite their widespread use, estimating GMM parameters accurately remains challenging, especially when overlap between components must be controlled or when reliable initial guesses are difficult to obtain. The classical approach is based on maximum likelihood estimation (MLE) and typically realized via the Expectation-Maximization algorithm. While EM offers computational efficiency and theoretical guarantees of convergence to a local optimum [23], it lacks built-in mechanisms for enforcing separation between mixture components, potentially complicating model interpretation in clinical applications where distinct, well-defined clusters are desired [7].

To overcome these limitations, alternative methods based on minimum chi-squared estimation (MCSE) have emerged. Notably, Lerch et al. [7] introduced the Distribution Optimization (DO) framework, which reimagines GMM parameter estimation as a constrained optimization problem. By minimizing a chi-squared goodness-of-fit statistic and incorporating an explicit penalty for component overlap, DO provides an intuitive mechanism to enforce mode separation, an advantage particularly relevant to medical applications where distinct subgroups must be identified. Yet, the original DO formulation has shortcomings: it employs equiwidth binning (Keating's formula), which can misalign with the data distribution and introduce bias, especially when components are well-separated or exhibit unequal variances, and it relies on genetic algorithms (GAs), which may lack the robustness required for higher-dimensional optimization landscapes.

This paper refines and extends the DO framework by addressing these critical challenges through two key innovations:

1. We replace equiwidth binning with Mann–Wald's equiprobable binning [15], which aligns bin boundaries with the underlying data distribution. This approach reduces bias and enhances estimation accuracy across diverse scenarios, from well-separated to overlapping components.
2. We substitute the original GA with advanced Differential Evolution (DE) variants, such as SHADE, L-SHADE, and iL-SHADE [2,19,20], known for their superior performance in continuous optimization. These adaptive algorithms improve convergence and scalability, particularly as the complexity of the mixture model grows.

Through rigorous experiments on synthetic and real-world datasets, we demonstrate that these refinements, equiprobable binning and DE-based optimization, significantly enhance the robustness, accuracy, and scalability of chi-squared-based GMM estimation.

2 Gaussian Mixture Model

The Gaussian Mixture Model is a parametric statistical model that represents a probability distribution as a weighted sum of multiple Gaussian components

[17]. In the one-dimensional case with M components, the probability density function is given by:

$$p(x \mid \lambda) = \sum_{i=1}^{M} \alpha_i \mathcal{N}(x \mid \mu_i, \sigma_i^2),$$

where $\mathcal{N}(x \mid \mu_i, \sigma_i^2)$ represents the i-th Gaussian component with mean μ_i and variance σ_i^2, and α_i is the corresponding mixing coefficient. The complete parameter set λ is defined as:

$$\lambda = \{\alpha_1, \ldots, \alpha_M, \ \sigma_1, \ldots, \sigma_M, \ \mu_1, \ldots, \mu_M\} \in \mathbb{R}^{3M},$$

subject to the constraints:

$$\alpha_i \geq 0, \quad \sum_{i=1}^{M} \alpha_i = 1, \quad \sigma_i > 0.$$

For notational convenience, we denote the parameters associated with the i-th component as:

$$\lambda_i = (\alpha_i, \sigma_i, \mu_i) \in \mathbb{R}^3.$$

2.1 Parameter Estimation

Typically, the parameters of a GMM are estimated through maximum likelihood estimation, which is commonly carried out using the Expectation-Maximization algorithm. This work builds upon and extends the chi-squared minimization framework for GMM parameter estimation introduced by Lerch et al. [7]. We first summarize their core approach.

Lerch et al. [7] redefined the task of parameter estimation in GMMs as an optimization problem, leveraging minimum chi-squared. Their approach was motivated by the need to directly address the challenge of mode separation within the GMM, a limitation often overlooked in conventional methods. The core of their method is an objective function that integrates two key metrics: the discrepancy between theoretical and empirical distributions quantified by χ^2 statistic, and a measure of overlap error between components. Conceptually, this can be viewed as a constrained optimization problem that penalizes mixtures exhibiting excessive overlap. In doing so, Lerch et al. introduce two important innovations. First, they minimize χ^2 instead of maximizing likelihood. Second, they incorporate an explicit overlap penalty to improve mode separation. While this penalty could theoretically be applied to a likelihood-based objective, the choice of χ^2 stands out for its computational simplicity. Specifically, χ^2 can be calculated efficiently using binned data, eliminating the need to evaluate likelihood function for every data point individually. As a result, the χ^2-based method offers a practical advantage, especially when working with large datasets.

To define the objective function we shall introduce some auxiliary notions. Let us denote the (one-dimensional) dataset on which the model is trained as X and

$$R_X = M_X - m_X, \quad M_X = \max(X), \quad m_X = \min(X), \quad n = \#X.$$

The χ^2 statistic assessing the goodness of fit is defined in the following way

$$\chi^2(\lambda) = \sum_{j=1}^{K} \frac{(\#(X \cap k_j) - P(x \in k_j | \lambda) \cdot n)^2}{P(x \in k_j | \lambda) \cdot n}, \tag{1}$$

where k_j denotes the j-th bin (further details are provided in the next section) and K is the total number of bins. The second component of the objective

$$\epsilon_{overlap}(\lambda) = \max_{m=1,\ldots,M} \sum_{i=1}^{n} \min\left(\frac{\max_{j=1,\ldots,M;\, j\neq m}(\alpha_j \cdot p(x_i|\lambda_j))}{\alpha_m \cdot p(x_i|\lambda_m)}, 1\right) \tag{2}$$

approximates the area overlapped between neighboring mixture components. The final objective function is calculated as

$$f(\lambda) = \begin{cases} +\infty & \text{if } \epsilon_{overlap}(\lambda) > \tau_{overlap}, \\ \chi^2(\lambda) & \text{otherwise.} \end{cases} \tag{3}$$

For a particular dataset, the value of the overlap threshold $\tau_{overlap}$ can be chosen in such a way that components are properly separated. To make the optimization problem tractable, Lerch et al. [7] further constrain the search domain to the following subset of a hypercube in $\mathbb{R}^{3M}$:

$$\mathcal{D} = \Big\{ \lambda \in \mathbb{R}^{3M} : \alpha_i \in [0,1],\ \sigma_i \in [0.001 \cdot R_X, 0.1 \cdot R_X],\ \mu_i \in [m_X, M_X], \\ \sum_{i=1}^{M} \alpha_i = 1 \Big\} \tag{4}$$

Thus, the final problem is to find $\lambda^* \in \mathcal{D}$ such that

$$f(\lambda^*) = \min_{\lambda \in \mathcal{D}} f(\lambda). \tag{5}$$

To solve this problem, Lerch et al. [7] propose the DO algorithm. It is a genetic algorithm (GA) that employs the uniform mutation, the simple arithmetic crossover and the tournament selection. The weights and the standard deviations of the initial population are sampled from the uniform distribution, the means are sampled from the following normal distribution

$$\mu_i \sim \mathcal{N}\left(m_X + i \cdot \frac{R_X}{M+1}, \frac{R_X}{5M}\right), \quad i = 1, \ldots, M. \tag{6}$$

3 Minimum Chi-Squared Estimation

Minimum chi-squared estimation is a classical technique for parameter estimation based on minimizing Pearson's chi-squared test statistic:

$$\chi^2 = \sum_{i=1}^{K} \frac{(O_i - E_i)^2}{E_i}. \tag{7}$$

Although it is applied less frequently than MLE, many statisticians have highlighted its theoretical and practical merits. In particular, Berkson [1] argued that MCSE represents a fundamental principle of statistical estimation.

A key issue in applying MCSE to continuous distributions is the necessity of discretizing data into bins, given that the chi-squared test is inherently designed for categorical data. This binning process can introduce bias and variability unless the bins are carefully selected. To address this, Mann and Wald [15] proposed using equiprobable bins and derived a formula for selecting the optimal number of such bins:

$$K = 4\left(\frac{2n^2}{c(\alpha)^2}\right)^{\frac{1}{5}}, \tag{8}$$

where $c(\alpha)$ is the $(1-\alpha)$ quantile of the standard normal distribution (commonly $\alpha = 0.05$). Subsequent studies [3,16] later proved that equiprobable binning yields an unbiased chi-squared test, whereas bins with unequal probabilities do not. Intuitively, equal-probability boundaries avoid having bins with extremely low counts, thus reducing the distortion in the chi-squared statistic. This approach, often referred to as the Mann and Wald formula, remains widely used in both minimum chi-squared estimation [6] and goodness-of-fit testing [9].

By contrast, Lerch et al. [7] used equiwidth bins rather than equiprobable bins. They selected the total number of bins, K, using a variant of Keating's formula [5], which first computes

$$\hat{\sigma} = \min\left(s, \frac{\text{IQR}}{1.349}\right), \tag{9}$$

where IQR is the interquartile range, s is the sample standard deviation. Then K is set by

$$K = \left\lceil \frac{R_X \, n^{\frac{1}{3}}}{3.49 \, \hat{\sigma}} \right\rceil. \tag{10}$$

Although these binning strategies might appear superficially similar, they can produce substantial differences in MCSE outcomes.

Figure 1 illustrates how these differing binning choices lead to noticeably different fitted GMMs. Specifically, using equiwidth bins, as in Lerch et al. [7], tends to shift the estimated mean values relative to those obtained via Mann and Wald's equiprobable bins.

Lerch's [7] equiwidth strategy is particularly problematic when some components of the distribution are well-separated or overlapping. In well-separated

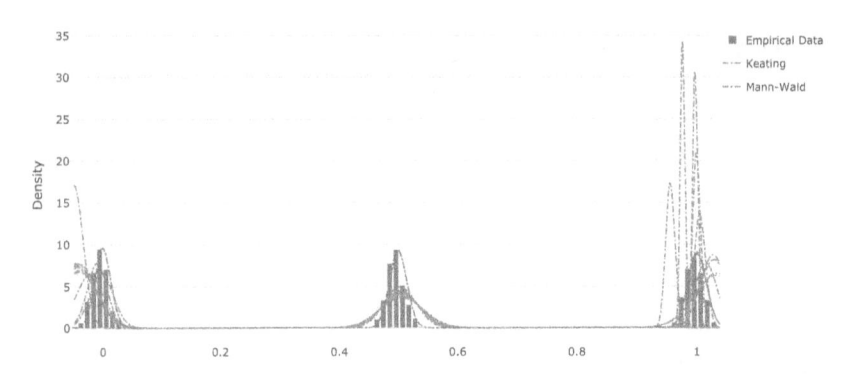

Fig. 1. Fitted GMMs under two different binning strategies, each run 10 times with different random seeds. The orange lines represent solutions using equal-width bins with Keating's formula, while green lines show solutions using equal-probability bins with the Mann-Wald method. The consistent pattern across multiple runs demonstrates that Keating's method systematically shifts the estimated mean values compared to the equal-probability approach (Color figure online).

scenarios, bins can end up near-empty in regions between modes. Such bins with expected counts near zero destabilize the chi-squared statistic, as χ^2 become more sensitive to small fluctuations when E_i is small. Conversely, if two overlapping components (especially those with small variance) end up in the same bin, MCSE struggles to resolve these components and yields parameter estimates with high variance. Furthermore, because equiwidth bin edges do not adapt to the observed data distribution, they frequently occur in low-probability regions. This placement systematically biases parameter estimates, pulling component means toward less accurate values, a phenomenon clearly illustrated in Fig. 1. In contrast, the equiprobable binning strategy proposed by Mann and Wald naturally aligns bin boundaries with high-density regions of the data, leading to more robust and accurate parameter estimates.

4 Differential Evolution

Differential Evolution (DE) is a population-based stochastic optimization algorithm for continuous search spaces introduced by Storn and Price [18]. DE evolves a population of candidate solutions through successive generations by applying mutation, crossover, and selection operators. In the widely adopted *DE/rand/1* scheme, a mutant vector is created by:

$$\mathbf{v}_i = \mathbf{x}_{r_1} + F\left(\mathbf{x}_{r_2} - \mathbf{x}_{r_3}\right), \tag{11}$$

where r_1, r_2, r_3 are distinct random indices different from i, and $F \in (0, 2]$ is the scale factor controlling the mutation step size. A trial vector is then generated through binomial crossover:

$$u_{i,j} = \begin{cases} v_{i,j}, & \text{if rand}(0,1) \leq CR \text{ or } j = j_{\text{rand}}, \\ x_{i,j}, & \text{otherwise}, \end{cases} \tag{12}$$

where $CR \in [0, 1]$ is the crossover rate and j_rand ensures at least one component from the mutant vector is inherited. In the selection step, the trial vector replaces the target vector if it yields a better objective function value. Standard parameter settings such as a population size of 100, $F = 0.5$, and $CR = 0.5$ often serve as a baseline.

4.1 SHADE Variants

To improve DE's robustness, several adaptive strategies have been proposed that dynamically adjust control parameters based on search performance. SHADE [19] implements *success-history based parameter adaptation*. It maintains a historical memory (of a fixed size) which stores promising F and CR parameter settings. After each generation, SHADE identifies the F and CR values that successfully produced offspring superior to their parents. These *successful* parameters are aggregated, using a weighted mean that gives greater importance to those yielding larger fitness improvements. This aggregated result then updates one entry in the historical memory. When generating new candidate solutions, SHADE draws upon values stored in this memory to guide the selection of F and CR for the next generation. This data-driven approach, learning from recent successful parameters, provides more reliable and adaptive control compared to static settings. L-SHADE [20] extends SHADE by incorporating *linear population size reduction*, starting with a large population for exploration and gradually reducing it to focus the search. iL-SHADE [2] further refines L-SHADE through several modifications aimed at enhancing convergence speed and solution quality. It incorporates mechanisms to promote higher CR values and employs a refined parameter memory update rule. Additionally, iL-SHADE restricts high F and low CR values during the initial search phase. A key feature is the linear adaptation of the control parameter p for the `current-to-pbest/1` mutation, adjusting it based on the search progression. These combined enhancements contribute to iL-SHADE's standing as a state-of-the-art DE variant [22].

4.2 Motivation

In the IEEE CEC 2006 and 2010 competitions on constrained real-parameter optimization [8,14], the majority of submitted algorithms were based on DE, and DE-variants ultimately claimed top positions in both events. The consistent success of DE in these competitions can be attributed to its robust mutation and recombination strategies, which balance exploration and exploitation effectively, even under complex constraints. Beyond constrained optimization, DE-based methods have exhibited state-of-the-art performance in single-objective box-constrained problems, as demonstrated in large-scale benchmarking studies such as those conducted on the BBOB function suite [22]. Algorithms like SHADE, L-SHADE, and il-SHADE have repeatedly shown competitive performance across a variety of problem classes, underscoring the versatility and effectiveness of the DE framework. These advances make DE and its variants well-suited for tackling constrained optimization problems, including those under death-penalty constraints.

In addition to DE-based methods, we also performed preliminary studies using CMA-ES [4], a covariance-matrix-adaptation evolution strategy widely recognized for its effectiveness in continuous optimization. However, our initial experiments revealed that CMA-ES struggled to handle the constraint effectively, yielding inferior performance compared to DE variants. Given the poor performance in our pilot tests we excluded CMA-ES from further analyses.

5 Experimental Procedure

To comprehensively evaluate binning methods and optimization algorithms for fitting one-dimensional GMMs we chose to work with synthetically generated datasets. Synthetic data allow us to control the underlying distribution and assess fitted models against the ground truth. We measure performance with the Jensen-Shannon Divergence (JSD), because it is independent from the optimization objectives of the methods compared (χ^2 for MCSE, likelihood for EM). JSD's symmetry, finiteness, and information-theoretic interpretation make it well-suited for comparing continuous distributions.

Moreover, because established benchmarks for one-dimensional GMMs are scarce, we opted to produce test datasets with diverse parameter configurations. Our generation procedure is based on an iterative rejection sampling strategy to ensure that each chosen parameter set lies within the feasible domain of our optimization problem (Eq. (4)). Specifically, we generate candidate mixtures by sampling (i) the mixture weights (enforcing a minimum weight to avoid degenerate components), (ii) the component means within a specified range, and (iii) the corresponding standard deviations within the bounds stipulated by our domain constraints. For each candidate, we then sample data by drawing random samples from the resulting mixture distribution. A dataset is accepted only if it meets the criteria that all component weights, means, and standard deviations remain valid and that the mixture components exhibit an overlap in a specified interval (which can be adjusted). This generation process guarantees that the ground truth distributions are feasible solutions – a crucial aspect that ensures both MCSE-based methods and EM can theoretically achieve the optimal solution, addressing a limitation in previous comparative studies [13] where the feasibility of the underlying distribution was often overlooked, potentially biasing results in favor of one method over the other.

By generating synthetic datasets, we can further investigate how varying the number of mixture components influences each method's robustness. To this end, we construct 100 datasets for each component count between 2 and 10, resulting in a total of 900 datasets. For each dataset, we run each method 10 times with different random seeds to assess initialization effects. The performance of each run is then quantified by computing the Jensen–Shannon Divergence between the fitted model and the ground truth. We aggregate these results by comparing methods through rank-based analyses and by reporting the average

Jensen–Shannon Divergence across all runs. Specifically, we benchmark the original Distribution Optimization approach (denoted as GA with Keating) against three alternatives: a Genetic Algorithm variant with Mann–Wald binning, an iL-SHADE algorithm with Mann–Wald binning, and the classical Expectation Maximization technique. This experimental design ensures a thorough assessment of how each method's stability and accuracy respond to changes in both the number of mixture components and the random seed.

Furthermore, our experimental design enables a detailed analysis of the individual contributions of binning strategies and optimization algorithms to the overall performance. By comparing the original GA with Keating and the GA with Mann–Wald binning, we can isolate and quantify the impact of the binning strategy, as both methods employ the same optimization algorithm. Any performance differences between these two methods can thus be attributed to the choice of binning. Similarly, the comparison between GA with Mann–Wald and iL-SHADE with Mann–Wald allows us to evaluate the effect of employing a more sophisticated optimization algorithm, iL-SHADE, while maintaining the same binning approach. In addition, by incorporating the classical EM algorithm into our benchmarking suite, we can assess the performance of the Distribution Optimization approach relative to a well-established standard in the field. Previous studies comparing the DO approach with EM have often relied on a much smaller number of datasets, which may not fully capture their behavior across diverse scenarios.

Finally, we compare optimization algorithms[1] (GA, DE, SHADE, L-SHADE, iL-SHADE) under fixed Mann-Wald binning. Each algorithm runs 10 times per dataset with different seeds, recording the best objective value f (Eq. (3)) after 10,000 evaluations – a standard budget sufficient for convergence across algorithms in our setting. We evaluate consistency, effectiveness, and robustness to the number of components, which scales the solution space dimensionality.

5.1 Real-World Datasets

To complement our synthetic experiments and evaluate the generalizability of the algorithms, we analyze three real-world datasets. These datasets were previously used to benchmark Gaussian Mixture Model implementations in R packages [13]. In the absence of ground truth distributions, we perform visual and qualitative assessments of the fitted GMMs.

1. **Truck Driving**: Sourced from the `AdaptGauss` R package [21], this dataset captures the time taken by trucks to reach seaports. We model it with a three-component GMM.
2. **Iris ICA**: This dataset comprises the first independent component obtained via Independent Component Analysis (ICA) applied to Fisher's Iris dataset. The one-dimensional projection corresponds to the three Iris species, justifying a three-component GMM to model the species-specific distributions.

---

[1] All implementation details are available in the accompanying GitHub repository at https://github.com/agh-a2s/distribution-optimization.

3. **Chromatogram Time**: Available in the `opGMMassessment` R package [13], this dataset contains chromatograms of five distinct lysophosphatidic acids. A five-component GMM is fitted to represent the separate peaks associated with each acid.

In our real-world study, we narrow our focus to two algorithms: the original DO approach, which uses a GA with Keating binning, and our enhanced version, which employs Mann-Wald binning and the iL-SHADE. This deliberate restriction allows us to directly evaluate the practical impact of our proposed modifications. By comparing these two methods, we can assess their effects on convergence stability and the variability of fitted mixture models, determining whether the improvements seen in synthetic experiments translate to real-world scenarios.

6 Results

Figure 2 evaluates the performance of four methods for GMM estimation: GA with Keating (original Distribution Optimization algorithm), GA with Mann-Wald, iL-SHADE with Mann-Wald, and EM across synthetic datasets with mixture components ranging from 2 to 10. The left panel displays the average rank based on JSD, where a lower rank indicates better performance, while the right panel shows the average JSD value, with lower values reflecting closer fits to the ground truth. For datasets with fewer components (2–4), methods using Mann-Wald binning (GA with Mann-Wald and iL-SHADE with Mann-Wald) achieve JSD values and ranks comparable to EM. However, as the number of components increases beyond 4, EM consistently outperforms the evolutionary methods, securing the lowest rank (near 1) and JSD values (0.01). This trend highlights EM's superior efficacy in optimizing complex mixtures. Among the evolutionary methods, iL-SHADE with Mann-Wald maintains the best performance, with JSD values rising modestly to approximately 0.02 by 10 components. In contrast, GA with Mann-Wald and GA with Keating show declining performance, with JSD values climbing to 0.06 and 0.08 respectively, as complexity grows. The transition from Keating to Mann-Wald binning with the GA delivers a consistent improvement, reducing JSD values by a nearly constant margin across all component counts. More striking, however, is the optimization algorithm's influence on performance scalability. While GA with Mann-Wald exhibits a rapid increase in JSD values with growing mixture complexity, iL-SHADE with Mann-Wald demonstrates a much slower growth rate, indicating that iL-SHADE's adaptive differential evolution enhances search efficiency in higher-dimensional spaces compared to genetic algorithms, enabling better handling of increasingly complex mixture models.

Figure 3 compares five optimization algorithms – DE, GA, SHADE, LSHADE, and iL-SHADE – using the final objective function value (Eq. 3) under fixed Mann-Wald binning. The left panel shows the average rank (lower is better), and the right panel presents the average function value (lower indicates

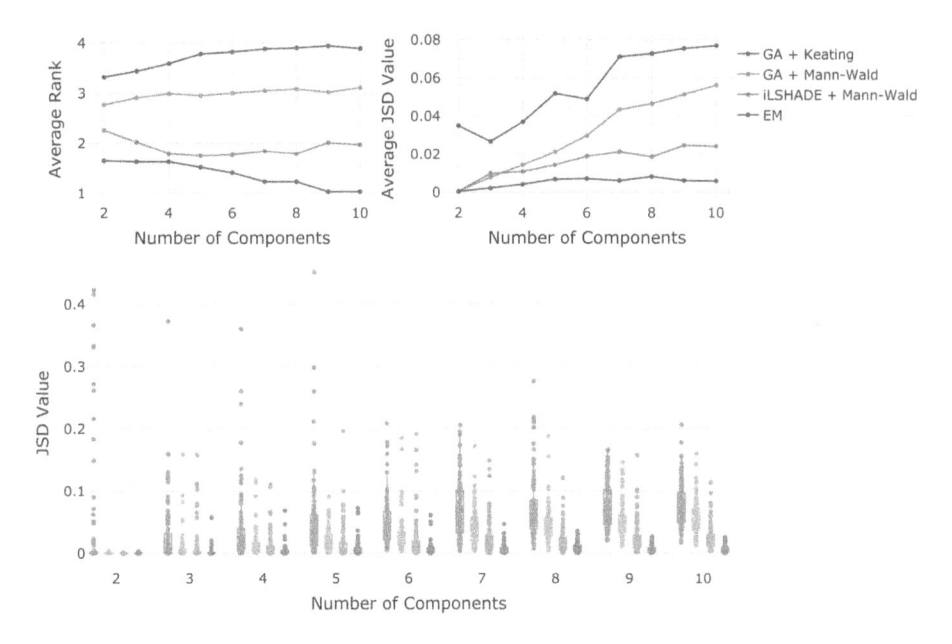

Fig. 2. The top left panel reports the average rank (based on JSD), while the top right panel presents the corresponding average JSD value. Both metrics are aggregated over 100 synthetic datasets per component count, with 10 runs per dataset. The plot in the second row presents distributions of JSD values.

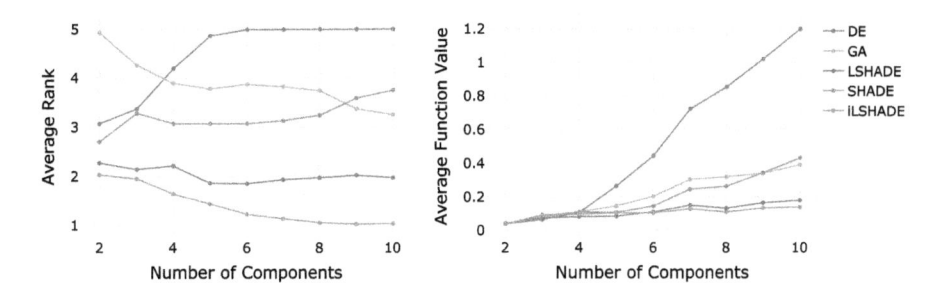

Fig. 3. The left panel reports the average rank (based on objective function value), while the right panel presents the corresponding average objective function value. Both metrics are aggregated over the same datasets as in Fig. 2.

better solutions) across component counts from 2 to 10. iL-SHADE consistently achieves the lowest rank and function values, demonstrating robust and effective optimization. LSHADE follows closely, maintaining strong performance, while SHADE and GA rank progressively higher (worse) and yield poorer function values, especially as the number of components increases. These findings underscore iL-SHADE's superior ability to handle multidimensional optimization problems.

6.1 Real-World Datasets

Fig. 4 presents three real-world datasets, each overlaid with the Gaussian Mixture Model fits from the GA using Keating and from iL-SHADE employing Mann-Wald, repeated ten times with distinct random seeds. Several clear patterns emerge:

- **Truck Driving (top).** The iL-SHADE solutions demonstrate consistently lower variability across runs and exhibit a closer alignment with the peaks. However, both models effectively capture the dataset's key structural features While subtle differences are present, it remains unclear which method offers a superior fit.
- **Iris ICA (middle).** Both methods capture the main mode near 1.0, but GA with Keating often broadens this peak excessively or shifts it (to the right). Meanwhile, iL-SHADE with Mann-Wald consistently converges on a sharper peak that more accurately matches the empirical histogram. Across repeated runs, GA-based fits fluctuate considerably and produce both under-estimated

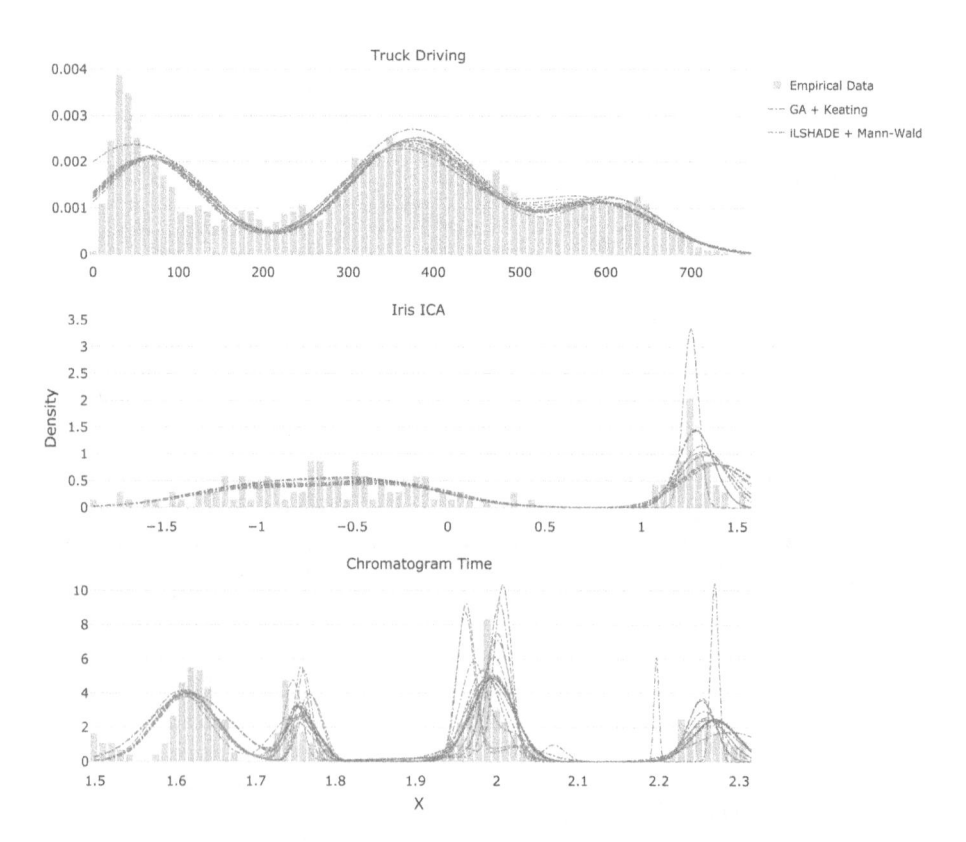

Fig. 4. Three real-world datasets (histograms) alongside the ten GMM fits of two methods: GA with Keating and iL-SHADE with Mann-Wald.

and over-estimated standard deviations, whereas iL-SHADE remains stable and yields nearly identical fits each time.
- **Chromatogram Time (bottom).** Here, the data contain five well-separated peaks. iL-SHADE with Mann-Wald again provides cohesive mixtures with tightly estimated component means and variances, resulting in visually robust fits for each distinct mode. In contrast, GA with Keating occasionally merges or splits peaks unnecessarily and shows a tendency to decrease the component widths.

Overall, the GA with Keating approach often struggles to estimate the correct mean locations and realistic standard deviations for the mixture components, often generating overly narrow or broadened peaks that fail to capture the true distributions. By comparison, iL-SHADE with Mann-Wald not only yields more stable solutions over multiple runs but also offers more precise alignment with the empirical data, indicating stronger convergence properties and more reliable mixture fits.

7 Conclusions

This paper examined strategies for estimating one-dimensional GMMs with a death-penalty constraint by focusing on two main aspects: the choice of binning scheme and the selection of optimization algorithm. Our study can be viewed as both an in-depth investigation and a refinement of the Distribution Optimization approach [7]. A key contribution of our work is the adoption of a binning method better suited to chi-squared-based estimation. Whereas Keating's equiwidth bins frequently misalign with the data, Mann–Wald's equiprobable bins adapt naturally to the underlying distribution. This refinement addresses previous difficulties in fitting GMMs by reducing bias. Among the evolutionary algorithms evaluated, iL-SHADE consistently outperformed the Genetic Algorithm and other DE variants, achieving lower objective function and JSD values. Furthermore, both iL-SHADE and LSHADE exhibit superior scalability, maintaining effective performance as the number of mixture components increases. While evolutionary methods were competitive for low-dimensional mixtures (fewer than four components), the EM algorithm achieved notably better fits for complex mixtures, consistently obtaining lower JSD values. This highlights EM's strengths particularly when the number of mixture components grows and the search space becomes more challenging. Nevertheless, evolutionary algorithms remain appealing when constraints or custom objective functions are required, scenarios in which EM might not be directly applicable or would require major modifications.

Building upon these findings, several avenues for future research emerge. A primary direction involves extending the proposed methodology to estimate multidimensional GMMs. Furthermore, the current work relied exclusively on the death-penalty approach for handling constraints. Future investigations could benefit from analyzing other constraint handling techniques.

Acknowledgments. The research presented in this paper was partially supported by the Polish National Science Center under grant No. 2023/49/B/ST6/01404, by the funds of the Polish Ministry of Science and Education assigned to the AGH University of Krakow, and by the program "Excellence initiative - research university" for the AGH University of Krakow.

Disclosure of Interests. The authors have no competing interests to declare that are relevant to the content of this article.

References

1. Berkson, J.: Minimum chi-square, not maximum likelihood! Ann. Stat. **8**(3), 457–487 (1980)
2. Brest, J., Maučec, M.S., Bošković, B.: iL-SHADE: improved L-SHADE algorithm for single objective real-parameter optimization. In: 2016 IEEE Congress on Evolutionary Computation (CEC), pp. 1188–1195 (2016). https://doi.org/10.1109/CEC.2016.7743922
3. Cohen, A., Sackrowitz, H.: Unbiasedness of the chi-square, likelihood ratio, and other goodness of fit tests for the equal cell case. Ann. Statist. **3**(4) 959–964 (1975)
4. Hansen, N.: The CMA evolution strategy: a tutorial. arXiv (2023). https://doi.org/10.48550/arXiv.1604.00772
5. Keating, J., Scott, D.: A primer on density estimation for the great home run race of 98. Stats **25**, 16–22 (1999)
6. Kominek, Z.: Minimum chi-squared estimation of stable distributions parameters: an application to the Warsaw stock exchange. J. Appl. Stat. **29**(5), 729–744 (2002)
7. Lerch, F., Ultsch, A., Lötsch, J.: Distribution optimization: an evolutionary algorithm to separate gaussian mixtures. Sci. Rep. **10**(1), 648 (2020). https://doi.org/10.1038/s41598-020-57432-w
8. Liang, J.J., et al.: Problem definitions and evaluation criteria for the CEC 2006 special session on constrained real-parameter optimization. J. Appl. Mech. **41**(8), 8–31 (2006)
9. Liu, T.C., Kalugin, P.N., Wilding, J.L., Bodmer, W.F.: GMMchi: gene expression clustering using [gaussian] mixture modeling. BMC Bioinform. **23**(1), 457 (2022)
10. Loetsch, J., Alfredsson, L., Lampa, J.: Machine-learning-based knowledge discovery in rheumatoid arthritis-related registry data to identify predictors of persistent pain. Pain **161**(1), 114–126 (2020)
11. Lötsch, J., Hummel, T.: Data science-based analysis of patient subgroup structures suggest effects of rhinitis on all chemosensory perceptions in the upper airways. Chem. Senses **46** (2021).https://doi.org/10.1093/chemse/bjab001
12. Lötsch, J., Hähner, A., Schwarz, P.E.H., Tselmin, S., Hummel, T.: Machine learning refutes loss of smell as a risk indicator of diabetes mellitus. J. Clin. Med. **10**(21) (2021). https://doi.org/10.3390/jcm10214971
13. Lötsch, J., Malkusch, S., Ultsch, A.: Comparative assessment of automated algorithms for the separation of one-dimensional Gaussian mixtures. Inform. Med. Unlocked **34**, 101113 (2022). https://doi.org/10.1016/j.imu.2022.101113, https://www.sciencedirect.com/science/article/pii/S2352914822002507
14. Mallipeddi, R., Suganthan, P.N.: Problem definitions and evaluation criteria for the CEC 2010 competition on constrained real-parameter optimization. Nanyang Technol. Univ. Singapore **24**, 910 (2010)

15. Mann, H., Wald, A.: On the choice of the number of class intervals in the application of the chi square test. Ann. Math. Stat. **13**(3), 306–317 (1942)
16. Moore, D.S.: Tests of chi-squared type. In: Goodness-of-Fit-Techniques, pp. 63–96. Routledge (2017)
17. Reynolds, D.: Gaussian mixture models. In: Li, S.Z., Jain, A. (eds.) Encyclopedia of Biometrics, pp. 659–663. Springer, Boston, MA (2009).https://doi.org/10.1007/978-0-387-73003-5_196
18. Storn, R., Price, K.: Differential evolution-a simple and efficient heuristic for global optimization over continuous spaces. J. Global Optim. **11**, 341–359 (1997)
19. Tanabe, R., Fukunaga, A.: Evaluating the performance of SHADE on CEC 2013 benchmark problems. In: 2013 IEEE Congress on Evolutionary Computation, pp. 1952–1959 (2013). https://doi.org/10.1109/CEC.2013.6557798
20. Tanabe, R., Fukunaga, A.S.: Improving the search performance of SHADE using linear population size reduction. In: 2014 IEEE Congress on Evolutionary Computation (CEC), pp. 1658–1665 (2014). https://doi.org/10.1109/CEC.2014.6900380
21. Ultsch, A., Thrun, M.C., Hansen-Goos, O., Lötsch, J.: Identification of molecular fingerprints in human heat pain thresholds by use of an interactive mixture model R toolbox (AdaptGauss). Int. J. Mol. Sci. **16**(10), 25897–25911 (2015)
22. Vermetten, D., Doerr, C., Wang, H., Kononova, A.V., Bäck, T.: Large-scale benchmarking of metaphor-based optimization heuristics. In: Proceedings of the Genetic and Evolutionary Computation Conference, pp. 41–49. GECCO 2024, Association for Computing Machinery, New York, NY, USA (2024). https://doi.org/10.1145/3638529.3654122
23. Wu, C.F.J.: On the convergence properties of the EM algorithm. The Annals of Statistics **11**(1), 95–103 (1983). http://www.jstor.org/stable/2240463

Asymptotics in Curve Estimation by Modified Cubic Spline and Exponential Parameterization

Ryszard Kozera[1,3](✉), Lyle Noakes[3], and Magdalena Wilkołazka[2]

[1] Warsaw University of Life Sciences - SGGW Institute of Information Technology, Ul. Nowoursynowska 159, 02-776 Warsaw, Poland
ryszard_kozera@sggw.pl

[2] Faculty of Natural and Health Sciences, The John Paul II Catholic University of Lublin, Ul. Konstantynów 1H, 20-708 Lublin, Poland
magda8310@kul.lublin.pl

[3] School of Physics, Mathematics and Computing, The University of Western Australia, 35 Stirling Highway, Crawley, Perth, WA 6009, Australia
ryszard.kozera@gmail.com, {ryszard.kozera,lyle.noakes}@uwa.edu.au

Abstract. The problem of fitting reduced data Q_m is discussed here. Reduced data form the ordered sequence of interpolation points $q_i = \gamma(t_i)$ in arbitrary Euclidean space. Here the corresponding unknown knots $\mathcal{T}$ are replaced with $\hat{\mathcal{T}}$ compensated by the so-called exponential parameterization determined by reduced data Q_m and a single parameter $\lambda \in [0,1]$. In sequel, a modified complete spline is used to interpolate Q_m with the aid of exponential parameterization. The main theoretical contribution of this work is to prove a linear convergence order in γ estimation by fitting Q_m (getting denser) with modified complete spline based on exponential parameterization for $\lambda \in [0,1)$. The latter holds for sufficiently smooth, regular curves sampled more-or-less uniformly. The asymptotics established here is subsequently verified numerically in affirmative as sharp. The respective tests are conducted on 2D and 3D curves.

Keywords: Spline Interpolation · Fitting Reduced Data · Curve Modeling · Convergence Orders and Approximation · Computer Graphics

1 Introduction

Let $\gamma : [0,T] \to \mathbb{E}^n$ be a smooth regular curve (i.e. $\dot{\gamma}(t) \neq \mathbf{0}$) over $t \in [0,T]$, for $0 < T < \infty$ (see e.g. [5]). The sequence of $m+1$ interpolation points $Q_m = \{q_i\}_{i=0}^m$ in arbitrary Euclidean space $\mathbb{E}^n$ is coined as reduced data. Additionally, it is assumed that for $q_i = \gamma(t_i)$ we have $q_{i+1} \neq q_i$. The interpolation knots $\mathcal{T} = \{t_i\}_{i=0}^m$ (with $t_i < t_{i+1}$) are here not supplied. In order to derive an interpolant $\hat{\gamma}$, first the knot estimates $\hat{\mathcal{T}} = \{\hat{t}_i\}_{i=0}^m \approx \mathcal{T}$ should be somehow guessed subject to $\hat{\gamma}(\hat{t}_i) = q_i$. Upon selecting specific interpolation scheme $\hat{\gamma} : [0, \hat{T}] \to \mathbb{E}^n$ and the substitutes $\hat{T}$ of T a natural question arises referring to the convergence order α (if any) while estimating γ with $\hat{\gamma}$ in norm infinity. The latter stipulates $m \to \infty$ which equivalently assumes Q_m as getting dense. The desirable choice of $\{\hat{t}_i\}_{i=0}^{m}$ should ensure convergence of $\hat{\gamma}$ to γ with possibly a fast order α.

A background information (see e.g. [17]) is now introduced.

Definition 1. *The interpolation knots $\{t_i\}_{i=0}^{m}$ are called admissible if:*

$$\lim_{m \to \infty} \delta_m \to 0^+, \quad \text{where} \quad \delta_m = \max_{1 \leq i \leq m} \{t_i - t_{i-1} : i = 1, 2, \ldots, m\}. \tag{1}$$

Recall now a special subfamily of admissible samplings i.e. the so-called *more-or-less uniform samplings* (see [35]):

Definition 2. *The sampling $\{t_i\}_{i=0}^{m}$ is more-or-less uniform if for some constants $0 < K_l \leq K_u$ and sufficiently large m the following holds:*

$$\frac{K_l}{m} \leq t_i - t_{i-1} \leq \frac{K_u}{m}, \tag{2}$$

for all $i = 1, 2, \ldots, m$. Note that, condition (2) can be substituted by the inequality $\beta \delta_m \leq t_{i+1} - t_i \leq \delta_m$ holding for some $0 < \beta \leq 1$ asymptotically (i.e. for sufficiently large m).

A good performance of any $\hat{\gamma}$ relies on appropriate guesses $\hat{T}$ of T. At this point recall a definition of *exponential parameterization* (see [28]):

$$\hat{t}_0^\lambda = 0 \quad \text{and} \quad \hat{t}_i^\lambda = \hat{t}_{i-1}^\lambda + \|q_i - q_{i-1}\|^\lambda, \tag{3}$$

for $i = 1, 2, \ldots, m$ and $\lambda \in [0, 1]$. If $\lambda = 0$, a *uniform* distribution of knots $\hat{t}_i^0 = i$ eventuates. The opposite case of $\lambda = 1$ is called a *cumulative chord parameterization* which yields $\hat{t}_i^1 = \hat{t}_{i-1}^1 + \|q_i - q_{i-1}\|$ (see [28] or [29]). Visibly the latter accounts for the geometrical dispersion of Q_m which is not reflected for $\lambda = 0$ in (3). Thus, it is expected that the latter should have an impact on $\alpha(\lambda)$. Indeed, recall first:

Definition 3. *Consider a family $\{f_{\delta_m}, \delta_m > 0\}$ of functions $f_{\delta_m} : I \to \mathbb{E}$. We say that f_{δ_m} is of order $O(\delta_m^\alpha)$ (denoted as $f_{\delta_m} = O(\delta_m^\alpha)$), if there is a constant $K > 0$ such that, for some $\bar{\delta} > 0$ the inequality $|f_{\delta_m}(t)| < K\delta_m^\alpha$ holds for all $\delta_m \in (0, \bar{\delta})$, uniformly over I. For the family of vector-value functions $F_{\delta_m} : I \to \mathbb{E}^n$ by $F_{\delta_m} = O(\delta_m^\alpha)$ it is understood that $\|F_{\delta_m}\| = O(\delta_m^\alpha)$.*

Definition 4. *For a given scheme $\hat{\gamma}$ interpolating Q_m with some knots' estimates $\hat{T} \approx T$ (and some chosen mapping $\phi : I \to \hat{I}$) the asymptotics $\gamma - \hat{\gamma} \circ \phi = O(\delta_m^\alpha)$ is sharp over I within the prescribed families of curves $\mathcal{J}$ and samplings $\mathcal{K}$, if there exist $\gamma \in \mathcal{J}$ and $T \in \mathcal{K}$ such that for some $t^* \in I$ and some $K > 0$ we have $\|\gamma(t^*) - (\hat{\gamma} \circ \phi)(t^*)\| = K\delta_m^\alpha + O(\delta_m^\theta)$, where $\theta > \alpha$.*

From now on we omit the superscript λ in $\hat{t}_i^\lambda = \hat{t}_i$ (see (3)).

2 Previous Results Versus This Work Contribution

It is proved [26] that $\hat{\gamma}_3$ (i.e. a Lagrange piecewise-cubic) based on (3) and Q_m results in $\alpha(1) = 4$ (for (1)) and in $\alpha(\lambda) = 1$ for $\lambda \in [0,1)$ (with (2)). This yields a left-hand side discontinuity of $\alpha(\lambda)$ at $\lambda = 1$ (see e.g. [20]). In fact we have:

Theorem 1. *Let γ be a regular $C^4([0,T])$ curve in $\mathbb{E}^n$ sampled more-or-less uniformly (2). Assume that $\{\hat{t}_i^\lambda\}_{i=0}^m$ are defined according to (3). Then there exists a piecewise-cubic C^∞ mapping $\psi : [0,T] \to [0,\hat{T}]$, such that over $[0,T]$:*

$$\begin{aligned}(\hat{\gamma}_3 \circ \psi)(t) - \gamma(t) &= O(\delta_m), \quad \text{for } \lambda \in [0,1) \\ (\hat{\gamma}_3 \circ \psi)(t) - \gamma(t) &= O(\delta_m^4), \quad \text{for } \lambda = 1.\end{aligned} \quad (4)$$

Here the mapping $\psi = \psi_3$ is defined as a piecewise-cubic Lagrange polynomial satisfying the conditions $\psi_3(t_i) = \hat{t}_i$ (for $i = 0, 1, \ldots, m$) to comply with the interpolation constraints $q_i = \hat{\gamma}(\hat{t}_i) = \hat{\gamma}(\psi(t_i))$.

Noticeably, the continuous interpolant $\hat{\gamma}_3$ is generically non-smooth at junction points $\{q_k\}_{k=1}^{3k}$, i.e. where two consecutive local piecewise-cubics are glued together.

One option is to consider any C^1 interpolation scheme based on extra provision of the unknown velocities $\{\boldsymbol{v}_i\}_{i=0}^m$ at Q_m. A possible remedy is proposed in [18,24], where a modified Hermite C^1 fitting scheme $\hat{\gamma}_H$ is introduced and analyzed in conjunction with (3). In particular, the asymptotics established for $\hat{\gamma}_H$ based on (3) coincides with (4) from Theorem 1.

The solution guaranteeing C^2 smoothness at Q_m resorts to various hybrids of C^2 cubic spline interpolants $\hat{\gamma}_3^S$ (see [4]) based on Q_m and (3).

First of them called *a complete cubic spline* $\hat{\gamma}_3^C$ (see [4]) requires an initial $\boldsymbol{v}_0 = \gamma'(0)$ and terminal $\boldsymbol{v}_n = \gamma'(T)$ velocities, generically not accompanying reduced data Q_m. This special case is discussed in [10] (though limited exclusively to $\lambda = 1$), where quartic order $\alpha(1) = 4$ for trajectory estimation by $\hat{\gamma}_3^C$ is established.

A possible alternative of C^2 class fitting scheme based merely on Q_m and (3) (with no reference to $\boldsymbol{v}_0$ and $\boldsymbol{v}_m$) is e.g. given in [25]. The latter introduces a *modified complete spline* $\hat{\gamma}_3^{MC}$, where both initial and terminal velocities are estimated by the derivatives of the first and the last components of the Lagrange piecewise-cubic $\hat{\gamma}_3$ (for which $\hat{\gamma}'_{3,0}(0) \approx \boldsymbol{v}_0$ and $\hat{\gamma}'_{3,m-3}(T) \approx \boldsymbol{v}_m$). More details concerning the construction of $\hat{\gamma}_3^{MC}$ are given next in Sect. 3.

The *main contribution of this paper* is to establish *a linear convergence order* in γ estimation with *a modified complete spline* $\hat{\gamma}_3^{MC}$ based on exponential parameterization (3), for $\lambda \in [0,1)$ - see Theorem 2. The latter holds for any regular curve $\gamma \in C^4$ sampled more-or-less uniformly (2) *on dense reduced data* Q_m. Here the analysis assumes $m \to \infty$ assuring a good approximation property in $\hat{\gamma} \approx \gamma$ for m getting large. *Numerical tests* are also performed on 2D and 3D regular curves to confirm the *asymptotics together with its sharpness* determined by (13) and (14).

Related work on *fitting sparse or dense reduced data* with optimal knots selection criterion based on using C^2 class splines (see Sect. 3) can also be found e.g. in [21–23] or [27]. Other curve interpolation schemes combined with various parameterizations (with some specific applications given) are also studied e.g. in [2,6,7,11,12,14,31,38] or [39].

3 Spline Construction

The construction of a *modified complete spline interpolant* $\hat{\gamma}_3^{MC}$ based on reduced data Q_m (see also [4]) and exponential parameterization (3) falls into the following steps (a similar procedure renders γ_3^{MC} for non-reduced data $(\mathcal{T}, Q_m)$):

1. Calculate the estimates $\{\hat{t}_i\}_{i=0}^{m}$ of the missing knots $\{t_i\}_{i=0}^{m}$ according to the exponential parameterization (3) (with $\lambda \in [0,1]$).
2. The so-called general C^2 piecewise-cubic spline $\hat{\gamma}_3^S$ interpolant (a sum-track of cubics $\{\hat{\gamma}_{3,i}^S\}_{i=0}^{m-1}$ - see [4]) fulfills the following condition (over each sub-segment $\hat{I}_i = [\hat{t}_i, \hat{t}_{i+1}]$):

$$\hat{\gamma}_{3,i}^S(\hat{t}_i) = q_i, \qquad \hat{\gamma}_{3,i}^S(\hat{t}_{i+1}) = q_{i+1},$$

$$\hat{\gamma}_{3,i}^{S'}(\hat{t}_i) = v_i, \qquad \hat{\gamma}_{3,i}^{S'}(\hat{t}_{i+1}) = v_{i+1}, \tag{5}$$

where $v_0, \cdots, v_m$ represent the unknown slopes (i.e. velocities) $v_i \in \mathbb{R}^n$. The internal velocities $\{v_1, v_2, \ldots, v_{m-1}\}$ must satisfy C^2 class $m-1$ constraints imposed on $\hat{\gamma}_3^S$ at junction points $\{q_1, \ldots, q_{m-1}\}$ i.e. by enforcing:

$$\hat{\gamma}_{3,i-1}^{S''}(\hat{t}_i) = \hat{\gamma}_{3,i}^{S''}(\hat{t}_i). \tag{6}$$

They can be uniquely computed (see [4] or (9) and Sect. 4) provided both v_0 and v_m are somehow known (or a priori given).
3. Assuming temporarily the provision of all velocities $\{v_i\}_{i=0}^{m}$, each cubic $\hat{\gamma}_{3,i}^S$ over $\hat{t} \in [\hat{t}_i, \hat{t}_{i+1}]$ reads as:

$$\hat{\gamma}_{3,i}^S(\hat{t}) = c_{1,i} + c_{2,i}(\hat{t} - \hat{t}_i) + c_{3,i}(\hat{t} - \hat{t}_i)^2 + c_{4,i}(\hat{t} - \hat{t}_i)^3, \tag{7}$$

where its respective coefficients (with $\Delta \hat{t}_i = \hat{t}_{i+1} - \hat{t}_i$) are equal to:

$$c_{1,i} = q_i, \quad c_{2,i} = v_i,$$

$$c_{3,i} = \frac{\frac{q_{i+1} - q_i}{\Delta \hat{t}_i} - v_i}{\Delta \hat{t}_i} - c_{4,i}\Delta \hat{t}_i, \quad c_{4,i} = \frac{v_i + v_{i+1} - 2\frac{q_{i+1} - q_i}{\Delta \hat{t}_i}}{(\Delta \hat{t}_i)^2}. \tag{8}$$

If also $v_i = \gamma'(t_i)$ are given then formulas (7) and (8) yield a well-known C^1 class Hermite spline. However, the required velocities $\{v_0, v_1, \ldots, v_m\}$ are not usually supplemented to Q_m. A scheme for computing the corresponding missing internal velocities $\{v_1, v_1, \ldots, v_{m-1}\}$ is recalled next (see [4]). Extending the latter a method of estimating $\{v_0, v_m\}$ is given in [18] - see below for more details.

4. Formulas (7) and (8) render $\hat{\gamma}_{3,i}^{S''}(\hat{t}_i) = 2c_{3,i}$ and $\hat{\gamma}_{3,i-1}^{S''}(\hat{t}_i) = 2c_{3,i-1} + 6c_{4,i-1}(\hat{t}_i - \hat{t}_{i-1})$ which combined with (6) leads to the linear system (for $i = 1, 2 \ldots m - 1$):

$$v_{i-1}\Delta\hat{t}_i + 2v_i(\Delta\hat{t}_{i-1} + \Delta\hat{t}_i) + v_{i+1}\Delta\hat{t}_{i-1} = b_i, \qquad (9)$$

where

$$b_i = 3\left(\Delta\hat{t}_i \frac{q_i - q_{i-1}}{\Delta\hat{t}_{i-1}} + \Delta\hat{t}_{i-1}\frac{q_{i+1} - q_i}{\Delta\hat{t}_i}\right). \qquad (10)$$

Assuming that the end-slopes v_0 and v_m are somehow given the tridiagonal system (9) solves uniquely in $\{v_i\}_{i=1}^{m-1}$ - see [4]. The latter yields a C^2 spline $\hat{\gamma}_3^S$ (which fits reduced data Q_m) defined as a track-sum of $\{\hat{\gamma}_{3,i}^S\}_{i=0}^{m-1}$ introduced in (7). If extra conditions hold, i.e. $\gamma'(t_0) = v_0$ and $\gamma'(T) = v_m$ then $\hat{\gamma}_3^S$ is called *a complete cubic spline* (denoted here as $\hat{\gamma}_3^{CS}$).

5. Since Q_m are usually deprived from both initial and terminal velocities $\{\gamma'(t_0) = v_0, \gamma'(T) = v_m\}$ a good estimate $\{v_0^a, v_m^a\}$ is therefore required. Of course, any choice of $\{v_0^a, v_m^a\}$ renders a unique explicit formula for modification of $\hat{\gamma}_3^{CS}$. This however is insufficient for our consideration. Indeed to preserve a good approximation property of $\hat{\gamma}$, still a good estimate of these two velocities is required so that (13) and (14) hold. In doing so, we apply Lagrange cubic $\hat{\gamma}_{3,0}^L : [0, \hat{t}_3] \to \mathbb{E}^n$ (and $\hat{\gamma}_{3,m-3}^L : [\hat{t}_{m-3}, \hat{T}] \to \mathbb{E}^n$), satisfying $\hat{\gamma}_{3,0}^L(\hat{t}_i) = q_i$ (and $\hat{\gamma}_{3,m-3}^L(\hat{t}_{m-3+i}) = q_{m-3+i}$), with $i = 0, 1, 2, 3$ - here the same $\lambda \in [0, 1]$ is applied in the derivation of $\hat{\gamma}_{3,0}^L, \hat{\gamma}_{3,m-3}^L$. With such velocities the resulting complete spline $\hat{\gamma}_3^C$ is called *a modified complete spline* (denoted as $\hat{\gamma}_3^{MC}$) for which $v_0^a = \hat{\gamma}_{3,0}^{L'}(0)$ and $v_m^a = \hat{\gamma}_{3,m-3}^{L'}(\hat{T})$.

6. However, to verify the asymptotics from (13) and (14) a candidate for a mapping $\psi : [0, T] \to [0, \hat{T}]$ is still required. In doing so, consider a C^2 complete spline $\psi = \psi_3^C : [0, T] \to [0, \hat{T}]$ satisfying the knots' interpolation constraints $\psi_3^C(t_i) = \hat{t}_i$, where $\{\hat{t}_i\}_{i=0}^m$ are defined according to (3) (in principle this procedure extends to any $\tilde{T}$). In addition, the initial and terminal velocities of $s_0 = \psi_3^{C'}(0)$ and $s_m = \psi_3^{C'}(T)$ are set similarly to the construction from above. The internal velocities $\{s_i\}_{i=1}^{m-1}$ (defined by $s_i = \dot\psi_3^C(t_i)$) satisfy the analogous constraints to those from (9) and (10) (for $i = 1, 2, \ldots m - 1$):

$$s_{i-1}\Delta t_i + 2(\Delta t_{i-1} + \Delta t_i)s_i + s_{i+1}\Delta t_{i-1} = a_i, \qquad (11)$$

where

$$a_i = 3\left(\Delta t_i \frac{\hat{t}_i - \hat{t}_{i-1}}{\Delta t_{i-1}} + \Delta t_{i-1}\frac{\hat{t}_{i+1} - \hat{t}_i}{\Delta t_i}\right). \qquad (12)$$

To generate both estimate of s_0 and s_m, define two Lagrange cubics $\psi_{3,0} : [0, t_3] \to [0, \hat{t}_3]$ and $\psi_{3,m-3} : [t_{m-3}, T] \to [\hat{t}_{m-3}, \hat{T}]$ satisfying interpolation conditions $\psi_{3,0}(t_i) = \hat{t}_i$ and $\psi_{3,m-3}(t_{m-3+i}) = \hat{t}_{m-3+i}$ (with $i = 0, 1, 2, 3$ and the same $\lambda \in [0,1]$ as for the construction of $\hat{\gamma}_3^C$), respectively. In sequel, one approximates here $s_0 = \psi_3^{C'}(0)$ with $\psi_{3,0}'(0)$ and $s_m = \psi_3^{C'}(T)$ with $\psi_{3,m-3}'(T)$. Such spline ψ_3^C is also called a modified complete spline and is analogously denoted here by ψ_3^{MC}.

This completes *a construction of a modified C^2 complete spline* $\hat{\gamma}_3^{MC}$ (and of $\hat{\psi}_3^{MC}$) based on reduced data Q_m and exponential parameterization (3). Noticeably, with m increasing the terminal velocities for $\hat{\gamma}_3^{MC}$ and $\hat{\psi}_3^{MC}$ must be re-estimated for each m in accordance with the procedure specified above.

Note that if $[\psi_i^{MC}(t_i) = \hat{t}_i, \psi_i^{MC}(t_{i+1}) = \hat{t}_{i+1}] \subsetneq \psi_{3,i}^{MC}([t_i, t_{i+1}])$ then one has to extend the domain of $\hat{\gamma}_{3,i}^{MC}$ from $[\hat{t}_i, \hat{t}_{i+1}]$ to $\mathbb{R}$ to enable calculation $\hat{\gamma}_{3,i}^{MC} \circ \psi_{3,i}^{MC}$. Such $\hat{\gamma}_{3,i}^{MC}$ is denoted by $\check{\gamma}_{3,i}^{MC}$ which obviously satisfies $\check{\gamma}_{3,i}^{MC}|_{[\hat{t}_i,\hat{t}_{i+1}]} = \hat{\gamma}_{3,i}^{MC}$. In fact the asymptotics established in Theorem 2 applies to the "extended version" $\check{\gamma}_{3,i}^{MC} \circ \psi_{3,i}^{MC}$ of $\hat{\gamma}_{3,i}^{MC} \circ \psi_{3,i}^{MC}$ over each I_i.

4 Main Result

We establish now the main contribution of this work. The following holds:

Theorem 2. *Let γ be a regular $C^4([0,T])$ curve in $\mathbb{E}^n$ sampled more-or-less-uniformly (3). Let $\mathbf{v}_0^a = \hat{\gamma}_3^{L'}(0)$ and $\mathbf{v}_m^a = \hat{\gamma}_3^{L'}(\hat{T})$, where $\hat{\gamma}_3^L$ defines a piecewise-cubic Lagrange based on Q_m and (3) with $\lambda \in [0,1]$. Assume also that $\hat{\gamma}_3^{MC}: [0,\hat{T}] \to \mathbb{E}^n$ define a modified complete spline based on Q_m, $(\mathbf{v}_0^a, \mathbf{v}_m^a)$ and (3). Then there is a piecewise-C^∞ mapping $\psi = \psi_3^{MC}: [0,T] \to [0,\hat{T}]$ such that over $[0,T]$ we either have for all $\lambda \in [0,1)$:*

$$\check{\gamma}_3^{MC} \circ \psi - \gamma = O(\delta_m) \tag{13}$$

or for $\lambda = 1$:

$$\check{\gamma}_3^{MC} \circ \psi - \gamma = O(\delta_m^4). \tag{14}$$

Proof. Taking into account that velocities $\mathbf{v}_0$, $\mathbf{v}_m$, s_0 and s_m are estimated (see Sect. 3) both (9) (with (10)) and (11) (with (12)) represent two quadratic tridiagonal linear systems of $m-2$ equations (each in $m-2$ unknowns) which are strictly row diagonally dominant. Thus each system has exactly one solution which can be found e.g. by Gauss elimination without pivoting. The following inequalities hold (see [4], Chap. 4, Problem 7):

$$\max_{0 \le i \le m} \|\mathbf{v}_i\| \le \max\{\|\mathbf{v}_0\|, \max_{1 \le j \le m-1} \frac{\|b_j\|}{\Delta \hat{t}_{j-1} + \Delta \hat{t}_j}, \|\mathbf{v}_m\|\} \tag{15}$$

and

$$\max_{0 \le i \le m} |s_i| \le \max\{|s_0|, \max_{1 \le j \le m-1} \frac{|a_j|}{\Delta \hat{t}_{j-1} + \Delta \hat{t}_j}, |s_m|\}. \tag{16}$$

The proof of Theorem 2 is performed here only for $\lambda \in [0,1)$ in (3). The case of $\lambda = 1$ exceeds the scope of this paper. By [26] (one assumes here $\gamma \in C^4$ sampled more-or-less uniformly along (2)), each pair of initial and terminal velocities satisfies (for $k = 0, m$):

$$v_k = O(\delta_m^{1-\lambda}) \quad \text{and} \quad s_k = O(\delta_m^{\lambda-1}),$$

thus yielding the following asymptotics (for $k = 0, m$):

$$\|v_k\| = O(\delta_m^{1-\lambda}) \quad \text{and} \quad |s_k| = O(\delta_m^{\lambda-1}). \tag{17}$$

In order to determine the asymptotics from the right-hand side of (15) (and of (16)) the remaining middle terms are now examined. Substituting (10) into (15) (and (12) into (16)) renders two expressions (for $j = 1, \ldots, m-1$):

$$I_v = 3 \left\| \frac{\Delta \hat{t}_j \frac{q_j - q_{j-1}}{\Delta \hat{t}_{j-1}} + \Delta \hat{t}_{j-1} \frac{q_{j+1} - q_j}{\Delta \hat{t}_j}}{\Delta \hat{t}_{j-1} + \Delta \hat{t}_j} \right\|, \quad I_s = 3 \left| \frac{\Delta t_j \frac{\hat{t}_j - \hat{t}_{j-1}}{\Delta t_{j-1}} + \Delta t_{j-1} \frac{\hat{t}_{j+1} - \hat{t}_j}{\Delta t_j}}{\Delta t_{j-1} + \Delta t_j} \right|$$

(18)

which asymptotics needs further analysis. In doing so, recall that the curve γ as a regular curve can be parameterized by arc-length parameterization (see e.g. [5] or [15]) yielding $\|\dot{\gamma}\| = 1$. Hence upon differentiating both sides $\|\dot{\gamma}(t)\|^2 = \langle \dot{\gamma}(t) | \dot{\gamma}(t) \rangle = 1$ the orthogonality condition $\langle \dot{\gamma}(t) | \ddot{\gamma}(t) \rangle = 0$ follows. Consequently Taylor expansion applied to γ renders the following (as $\|w\|^2 = \langle w|w \rangle$):

$$\begin{aligned}
\hat{t}_{j+1} - \hat{t}_j &= \|\gamma(t_{j+1}) - \gamma(t_j)\|^\lambda \\
&= (t_{j+1} - t_j)^\lambda \| \dot{\gamma}(t_j) + \frac{(t_{j+1} - t_j)}{2} \ddot{\gamma}(t_j) + O((t_{j+1} - t_j)^2)\|^\lambda \\
&= (t_{j+1} - t_j)^\lambda \left(\|\dot{\gamma}(t_j) + \frac{(t_{j+1} - t_j)}{2} \ddot{\gamma}(t_j) + O((t_{j+1} - t_j)^2)\|^2 \right)^{\frac{\lambda}{2}} \\
&= (t_{j+1} - t_j)^\lambda [1 + O((t_{j+1} - t_j)^2)]^{\frac{\lambda}{2}}.
\end{aligned} \tag{19}$$

Again Taylor expansion of $f(y) = (1+y)^{\frac{\lambda}{2}}$ yields $f(y) = 1 + \frac{\lambda}{2}(1+\xi)^{\frac{\lambda}{2}-1} y$ for some $\xi \in [0, y]$ or $\xi \in [0, y]$. Thus for such ξ (if y is bounded) the expression $\frac{\lambda}{2}(1+\xi)^{\frac{\lambda}{2}-1} = O(1)$ and therefore $f(y) = 1 + O(y)$. Substituting for $y = O((t_{j+1}-t_j)^2)$ in the latter together with (19) results in:

$$\hat{t}_{j+1} - \hat{t}_j = (t_{j+1}-t_j)^\lambda (1+O((t_{j+1}-t_j)^2)) = (t_{j+1}-t_j)^\lambda + O((t_{j+1}-t_j)^{2+\lambda}). \tag{20}$$

Combining the latter with Taylor expansion of γ leads to: $(q_{j+1} - q_j)/\Delta \hat{t}_j$

$$= \frac{\gamma(t_{j+1}) - \gamma(t_j)}{\hat{t}_{j+1} - \hat{t}_j} = \frac{\gamma(t_{j+1}) - \gamma(t_j)}{\|\gamma(t_{j+1}) - \gamma(t_j)\|^\lambda}$$

$$= \frac{(t_{j+1} - t_j)[\dot{\gamma}(t_j) + \frac{(t_{j+1}-t_j)}{2}\ddot{\gamma}(t_j) + O((t_{j+1} - t_j)^2)]}{(t_{j+1} - t_j)^\lambda [1 + O((t_{j+1} - t_j)^2)]}$$

$$= (t_{j+1} - t_j)^{1-\lambda}[\dot{\gamma}(t_j) + \frac{(t_{j+1} - t_j)}{2}\ddot{\gamma}(t_j) + O((t_{j+1} - t_j)^2)][1 + O((t_{j+1} - t_j)^2)]$$

$$= (t_{j+1} - t_j)^{1-\lambda}(O(1) + O((t_{j+1} - t_j)^2) + O((t_{j+1} - t_j)^4)$$

$$= O(\delta_m^{1-\lambda}) + O(\delta_m^{3-\lambda}) = O(\delta_m^{1-\lambda}). \tag{21}$$

Analogously one arrives at:

$$\frac{q_j - q_{j-1}}{\Delta \hat{t}_{j-1}} = O(\delta_m^{1-\lambda}). \tag{22}$$

Coupling $\Delta \hat{t}_{j-k}/(\Delta \hat{t}_{j-1} + \Delta \hat{t}_j) = O(1)$ (for $k = 0, 1$) with (21) and (22) renders the asymptotics of the first formula from (18) as $I_v = O(\delta_m^{1-\lambda})$. The latter together with (15) and (17) yields (for all $i = 0, 1, 2, \ldots, m$):

$$\|v_i\| = O(\delta_m^{1-\lambda}). \tag{23}$$

Similarly, by (20) the following holds (by more-or-less uniformity of $\mathcal{T}$):

$$\frac{\hat{t}_{j+1} - \hat{t}_j}{\Delta t_j} = (t_{j+1} - t_j)^{\lambda-1}(1 + O((t_{j+1} - t_j)^2)) = O(\delta_m^{\lambda-1}). \tag{24}$$

Analogously one obtains (again by more-or-less uniformity of $\mathcal{T}$):

$$\frac{\hat{t}_j - \hat{t}_{j-1}}{\Delta t_{j-1}} = (t_j - t_{j-1})^{\lambda-1}(1 + O((t_j - t_{j-1})^2)) = O(\delta_m^{\lambda-1}). \tag{25}$$

As previously, coupling $\Delta t_{j-k}/(\Delta t_{j-1} + \Delta t_j) = O(1)$ (for $k = 0, 1$) together with (24) and (25) guarantees the second formula in (18) as $I_s = O(\delta_m^{\lambda-1})$. Hence by (16) and (17) the following holds (for all $i = 0, 1, 2, \ldots, m$):

$$|s_i| = O(\delta_m^{\lambda-1}). \tag{26}$$

We are ready now to determine the asymptotics of the expression $f(t) = (\tilde{\gamma}_3^{MC} \circ \psi_3^{MC})(t) - \gamma(t)$ over $[0, T]$, which permits to establish the order in γ estimation by $\tilde{\gamma}_3^{MC} \circ \psi_3^{MC}$. Evidently, in doing so, it suffices to examine the latter over each sub-segment $I_i = [t_i, t_{i+1}]$ i.e. for each $f_i(t) = (\tilde{\gamma}_{3,i}^{MC} \circ \psi_{3,i}^{MC})(t) - \gamma(t)$. From now on, to abbreviate the notation shorter symbols $\tilde{\gamma}_{3,i} = \tilde{\gamma}_{3,i}^{MC}$ and $\psi_{3,i} = \psi_{3,i}^{MC}$ are used. Since $f_i(t_{i+k}) = \mathbf{0}$ (for $k = 0, 1$) by Hadamard's Lemma [33] and chain rule one arrives at (for $t \in I_i$):

$$f_i(t) = (t-t_i)(t-t_{i+1})O(\ddot{f}_i) = (t-t_1)(t-t_{i+1})O\left(\tilde{\gamma}_{3,i}''\dot{\psi}_{3,i}^2 + \tilde{\gamma}_{3,i}'\ddot{\psi}_{3,i} - \ddot{\gamma}\right). \tag{27}$$

Newton Interpolation formula [4] leads to:

$$\dot{\psi}_{3,i}(t) = \psi_{3,i}[t_i, t_i] + 2\psi_{3,i}[t_i, t_i, t_{i+1}](t - t_i) \\ + \psi_{3,i}[t_i, t_i, t_{i+1}, t_{i+1}](2(t - t_i)(t - t_{i+1}) + (t - t_i)^2). \quad (28)$$

Upon combining (2), (24), (25) with (26) one obtains:

$$\psi_{3,i}[t_i, t_i] = s_i = O(\delta_m^{\lambda-1}),$$
$$\psi_{3,i}[t_i, t_i, t_{i+1}](t - t_i) = (\psi_{3,i}[t_i, t_{i+1}] - s_i)\frac{t - t_i}{t_{i+1} - t_i} = (O(\delta_m^{\lambda-1}) + O(\delta_m^{\lambda-1}))O(1)$$
$$= O(\delta_m^{\lambda-1}),$$
$$\psi_{3,i}[t_i, t_i, t_{i+1}, t_{i+1}] = \frac{s_{i+1} - \psi_{i3}[t_i, t_{i+1}]}{(t_{i+1} - t_i)^2} - \frac{\psi_{3,i}[t_i, t_i, t_{i+1}](t_{i+1} - t_i)}{(t_{i+1} - t_i)^2} = O(\delta_m^{\lambda-3}). \quad (29)$$

Substituting now (29) into (28) renders:

$$\dot{\psi}_{3,i}(t) = O(\delta_m^{\lambda-1}) \quad \text{and} \quad \dot{\psi}_{3,i}^2(t) = O(\delta_m^{2\lambda-2}). \quad (30)$$

A simple inspection combined with (29) leads to:

$$\ddot{\psi}_{3,i}(t) = 2\psi_{3,i}[t_i, t_i, t_{i+1}] + 2\psi_{3,i}[t_i, t_i, t_{i+1}, t_{i+1}](2(t - t_i) + (t - t_{i+1})) \\ = O(\delta_m^{\lambda-2}) + O(\delta_m^{\lambda-2}) = O(\delta_m^{\lambda-2}). \quad (31)$$

In the next step, the asymptotics $\check{\gamma}'_{3,i}$ and $\check{\gamma}''_{3,i}$ is investigated. In doing so, Newton interpolation formula [4] applied to $\check{\gamma}_{3,i}$ yields:

$$\check{\gamma}'_{3,i}(\hat{t}) = \check{\gamma}_{3,i}[\hat{t}_i, \hat{t}_i] + 2\check{\gamma}_{3,i}[\hat{t}_i, \hat{t}_i, \hat{t}_{i+1}](\hat{t} - \hat{t}_i) \\ + \check{\gamma}_{3,i}[\hat{t}_i, \hat{t}_i, \hat{t}_{i+1}, \hat{t}_{i+1}](2(\hat{t} - \hat{t}_i)(\hat{t} - \hat{t}_{i+1}) + (\hat{t} - \hat{t}_i)^2). \quad (32)$$

Coupling together (20), (21) and (23) renders:

$$\check{\gamma}_{3,i}[\hat{t}_i, \hat{t}_i] = v_i = O(\delta_m^{1-\lambda}),$$
$$\check{\gamma}_{3,i}[\hat{t}_i, \hat{t}_i, \hat{t}_{i+1}](\hat{t} - \hat{t}_i) = \frac{\check{\gamma}_{3,i}[\hat{t}_i, \hat{t}_{i+1}] - v_i}{\hat{t}_{i+1} - \hat{t}_i}(\hat{t} - \hat{t}_i)$$
$$= \frac{O(\delta_m^{1-2\lambda})}{1 + O((t_{i+1} - t_i)^2)}(\hat{t} - \hat{t}_i) = O(\delta_m^{1-\lambda}). \quad (33)$$

To justify the last step in (33), note that Mean Value Theorem with (30) (for all $t \in I_i$) yield $\hat{t} - \hat{t}_i = \frac{\psi_{3,i}(t) - \psi_{3,i}(t_i)}{t - t_i}(t - t_i) = \dot{\psi}_{3,i}(\xi)(t - t_i) = O(\delta_m^{\lambda-1}) \cdot O(\delta_m) = O(\delta_m^{\lambda})$. Additionally, Taylor expansion applied to $f(x) = (1 + x^2)^{-1}$ renders $(1 + O(\delta_m^2))^{-1} = 1 + O(\delta_m^2)$. A similar argument as used in (33) (see also (29)) assures the following (for $t \in I_i$):

$$\check{\gamma}_{3,i}[\hat{t}_i, \hat{t}_i, \hat{t}_{i+1}, \hat{t}_{i+1}](2(\hat{t} - \hat{t}_i)(\hat{t} - \hat{t}_{i+1}) + (\hat{t} - \hat{t}_i)^2) = O(\delta_m^{1-3\lambda})O(\delta_m^{2\lambda}) = O(\delta_m^{1-\lambda}). \quad (34)$$

Consequently, both (33) with (34) result in the asymptotics:
$$\check{\gamma}'_{3,i}(\hat{t}) = O(\delta_m^{1-\lambda}) \quad \text{and} \quad \check{\gamma}''_{3,i}(\hat{t}) = O(\delta_m^{1-2\lambda}). \tag{35}$$

Finally, substituting (30), (31) and (35) into (27) renders the following asymptotics (over each I_i and $\lambda \in [0,1)$):

$$\begin{aligned} f_i(t) &= O(\delta_m^2)(O(\delta_m^{1-2\lambda})O(\delta_m^{2\lambda-2}) + O(\delta_m^{1-\lambda})O(\delta_m^{\lambda-2}) + O(1)) \\ &= O(\delta_m) + O(\delta_m^2) = O(\delta_m). \end{aligned} \tag{36}$$

This completes the proof of (13) in Theorem 2. The case of $\lambda = 1$ rendering a quartic convergence order (14) is here omitted. Note that in (36) the term $(t - t_i)(t - t_{i+1})\ddot{\gamma}(t) = O(\delta_m^2)$ forms the intrinsic quadratic barrier annihilating any improvement of the asymptotics in $O(\check{\gamma}''_{3,i}\dot{\psi}_{3,i}^2 + \check{\gamma}'_{3,i}\ddot{\psi}_{3,i})$ beyond $O(\delta_m^2)$. Thus the argument used herein prevails only for $\lambda \in [0,1)$ and as such needs modification for $\lambda = 1$. Alternatively, recall that $\lambda = 1$ in (3) is analyzed in [10] for complete spline $\hat{\gamma}_3^{CS}$ only, i.e. with v_0 and v_m a priori given. The adaptation of the latter to $\hat{\gamma}_3^{MC}$ forms an alternative tool to justify (14). □

The next section reports on numerical testing confirming the asymptotics together with its sharpness established in Theorem 2.

5 Experiments

In this section, a numerical verification of the asymptotics $\alpha(\lambda)$ (and its sharpness) from Theorem 2 is conducted. Recall that, given fixed $\lambda \in [0,1]$, by sharpness (see Definition 4) we understand the existence of at least one curve $\gamma \in C^4(0,T])$ and one special family $\mathcal{T}$ of more-or-less uniform sampling (2) such that the asymptotics $O(\delta_m^{\alpha(\lambda)})$ in difference $\check{\gamma}_3^{MC} \circ \psi_3^{MC} - \gamma$ (over $[0,T]$) is not faster than $\alpha(\lambda)$. A confirmation of (13) and (14) indicates again on an unexpected left-hand side discontinuity in $\alpha(\lambda)$ at $\lambda = 1$.

All tests are performed in *Mathematica 12.0* and use to two types of skew-symmetric more-or-less uniform samplings. The first one (for $t_i \in [0,1]$) is defined as follows:

$$t_i = \begin{cases} \frac{i}{m} + \frac{1}{2m}, & \text{for } i = 4k+1; \\ \frac{i}{m} - \frac{1}{2m}, & \text{for } i = 4k+3; \\ \frac{i}{m}, & \text{for } i \text{ even}; \end{cases} \tag{37}$$

with $K_l = (1/2)$ and $K_u = (3/2)$ introduced in (2). The second sampling reads as:
$$t_i = \frac{i}{m} + \frac{(-1)^{i+1}}{3m}, \tag{38}$$

with constants $K_l = (1/2)$ and $K_u = (5/3)$ from (2). For a given m, the error E_m, between γ and reparameterized spline $\check{\gamma}_3^{MC} \circ \psi_3^{MC}$ is determined by the formula:

$$E_m = \max_{t \in [0,1]} \|(\tilde{\gamma}_3^{MC} \circ \psi_3^{MC})(t) - \gamma(t)\|.$$

The latter is computed over each sub-interval $[t_i, t_{i+1}]$ (for $i = 0, \cdots, m-1$) by using *Mathematica* function - *FindMaximum* and then upon taking the maximal values from all segments' optima. In order to approximate $\alpha(\lambda)$ we calculate first E_m for $m_{min} \leq m \leq m_{max}$, where m_{min} and m_{max} are sufficiently large fixed constants. Then a linear regression yielding a function $y(x) = \bar{\alpha}(\lambda)x + b$ is applied to $\{(\log(m), -\log(E_m))\}_{m_{min}}^{m_{max}}$. Mathematica built-in function *LinearModelFit* extracts a coefficient $\bar{\alpha}(\lambda) \approx \alpha(\lambda)$. A full justification of this procedure to approximate $\alpha(\lambda)$ by $\bar{\alpha}(\lambda)$ is given in [17]. Note also that since both (13) and (14) have asymptotic character the constants $m_{min} < m_{max}$ should be taken as sufficiently large. On other hand, a potential negative impact of machine rounding-off errors stipulates these two constants not to exceed big values. In practice, the appropriate choices for $m_{min} < m_{max}$ are adjusted each time during the experimental phase. The tests conducted here employ three types of C^∞ regular curves: *an epitrochoid* γ_{ep} in $\mathbb{E}^2$ (i.e. planar a curve) and two curves *a conical spiral* γ_{cs} and *a quadratic helix* γ_{qh} both in $\mathbb{E}^3$ (i.e. 3D curves). All tested curves are sampled more-or-less uniformly (3) according to either (37) or (38).

Example 1. Consider *a regular planar epitrochoid* $\gamma_{ep} : [0,1] \to \mathbb{E}^2$,

$$\gamma_{ep}(t) = (4\cos(t) - 0.15\cos(4\pi t), 4\sin(t) - 0.15\sin(4\pi t)). \tag{39}$$

Figure 1(a) (or Fig. 1(b)) contains the plots of γ_{ep} sampled (here $m = 15$) according to either (37) (or (38)).

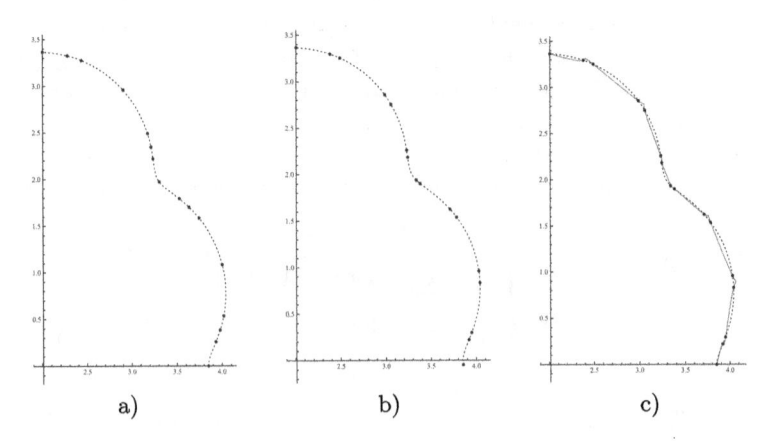

Fig. 1. *An epitrochoid* γ_{ep} (39) sampled along (dotted): a) (37) or b) (38) and c) fitted $\hat{\gamma}^{MC}$ with (38) & $\lambda = 0$ for $m = 15$.

The respective linear regression based estimates $\bar{\alpha}(\lambda) \approx \alpha(\lambda)$ (for various $\lambda \in [0,1]$) are computed here for $m_{min} = 60 \leq m \leq m_{max} = 120$. The numerical results contained in Table 1 confirm the sharpness of (13) and (14) for $\lambda \in \{0.0, 0.1, 0.3, 0.5, 0.7\}$ and yield marginally faster (though still consistent with asymptotics from Theorem (2)) $\alpha(\lambda)$ for $\lambda \approx 1$. Note that for $\lambda = 1$ we have $m_{min} = 240 \leq m \leq m_{max} = 270$.

Table 1. Computed $\bar{\alpha}(\lambda) \approx \alpha(\lambda)$ in (13) & (14) for γ_{ep} from (39) and various $\lambda \in [0, 1]$.

λ	0.0	0.1	0.3	0.5	0.7	0.9	1.0
$\bar{\alpha}(\lambda)$ for (37)	1.007	1.013	1.028	1.055	1.116	1.377	4.274
$\bar{\alpha}(\lambda)$ for (38)	1.037	1.036	1.042	1.066	1.143	1.483	4.259
$\alpha(\lambda)$ in Theorem 2	1.0	1.0	1.0	1.0	1.0	1.0	4.0

We pass now to the example with *a quadratic helix* in $\mathbb{E}^3$.

Example 2. Let *a quadratic helix* $\gamma_{qh} : [0, 1] \to \mathbb{E}^3$ be defined as:

$$\gamma_{qh}(t) = (1.5\cos(2\pi t), \frac{2\pi t}{4}\sin(2\pi t), t). \qquad (40)$$

Again Fig. 2(a) (or Fig. 2(b)) illustrates the trajectories of γ_{qh} sampled according to either (37) or (38), with $m = 15$.

As previously, a linear regression estimating $\bar{\alpha}(\lambda) \approx \alpha(\lambda)$ from Theorem 2 is used here, for m ranging between $100 \leq m \leq 160$ with various $\lambda \in [0, 1]$.

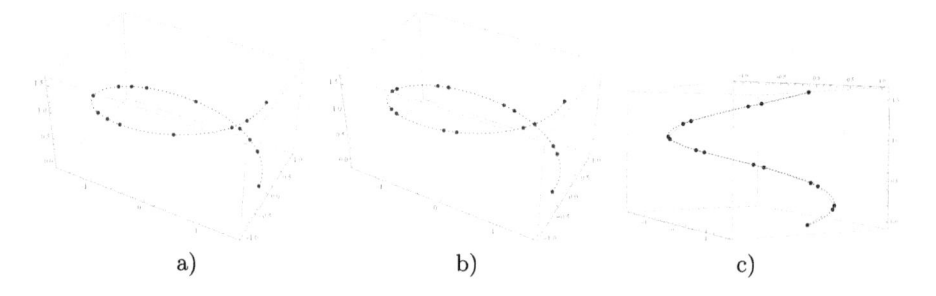

a)   b)   c)

Fig. 2. A *quadratic helix* γ_{qh} (40) sampled along (dotted): a) (37) or b) (38) and c) fitted $\hat{\gamma}^{MC}$ with (38) & $\lambda = 0.5$ for $m = 15$.

The coefficients $\bar{\alpha}(\lambda)$ (see Table 2) computed numerically sharply coincide with those specified in (13) and (14) (with marginally faster rates for $\lambda = 0.9$).

Finally, *a conical spiral* γ_{cs} in $\mathbb{E}^3$ is tested.

Table 2. Computed $\bar{\alpha}(\lambda) \approx \alpha(\lambda)$ in (13) & (14) for γ_{qh} from (40) and various $\lambda \in [0, 1]$.

λ	0.0	0.1	0.3	0.5	0.7	0.9	1.0
$\bar{\alpha}(\lambda)$ for (37)	1.001	1.002	1.007	1.016	1.038	1.187	3.916
$\bar{\alpha}(\lambda)$ for (38)	0.001	1.001	0.005	1.017	1.056	1.322	3.908
$\alpha(\lambda)$ in Theorem 2	1.0	1.0	1.0	1.0	1.0	1.0	4.0

Example 3. Let *a conical spiral* $\gamma_{cs} : [0, 1] \to \mathbb{E}^3$ be defined as follows:

$$\gamma_{cs}(t) = (2\sin(0.5\pi t)\cos(2\pi t), 2\sin(0.5\pi t)\sin(2\pi t), 2\cos(0.5\pi t)). \qquad (41)$$

Figure 3(a) (or Fig. 3(b)) contains the plots of γ_{cs} sampled more-or-less uniformly along either (37) or (38) (here $m = 15$).

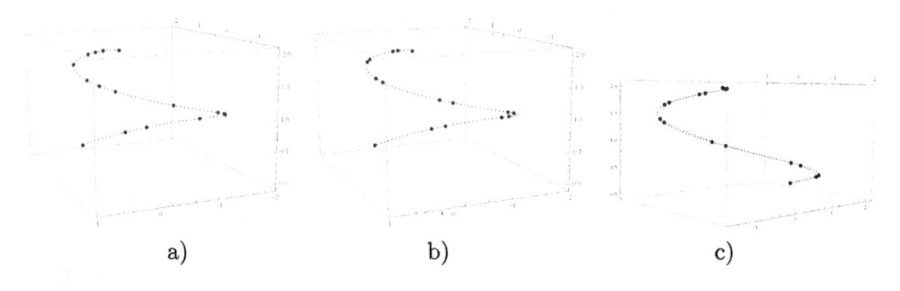

Fig. 3. A conical spiral γ_{cs} (41) sampled along (dotted): a) (37) or b) (38) and c) fitted $\hat{\gamma}^{MC}$ with (38) & $\lambda = 0.3$ for $m = 15$.

In order to compute $\bar{\alpha}(\lambda) \approx \alpha(\lambda)$ estimating the asymptotics from Theorem 2 again a linear regression is used (as explained at the beginning of this section) for $60 \le m \le 120$ and varying $\lambda \in [0, 1]$. Table 3 enlists numerically computed estimates $\bar{\alpha}(\lambda) \approx \alpha(\lambda)$ for various $\lambda \in [0, 1]$ and for samplings (37) and (38). Evidently these numerical results re-emphasize the sharpness of the asymptotics determined by (13) and (14), with marginally faster case for $\lambda = 0.9$.

Table 3. Computed $\bar{\alpha}(\lambda) \approx \alpha(\lambda)$ in (13) & (14) for γ_{cs} from (41) and various $\lambda \in [0, 1]$.

λ	0.0	0.1	0.3	0.5	0.7	0.9	1.0
$\bar{\alpha}(\lambda)$ for (37)	0.999	1.002	1.008	1.019	1.051	1.264	3.939
$\bar{\alpha}(\lambda)$ for (38)	0.991	0.992	0.999	1.018	1.078	1.448	3.955
$\alpha(\lambda)$ in Theorem 2	1.0	1.0	1.0	1.0	1.0	1.0	4.0

The experiments from this section confirm the asymptotics (and its sharpness) established in Theorem 2 - see (13) and (14).

6 Conclusion

This work examines the asymptotics in approximating a regular parametric curve γ in $\mathbb{E}^n$ by a modified complete spline $\hat{\gamma}_3^{MC}$ (see Sect. 3) based on reduced data Q_m (sampled more-or-less uniformly (2)). The unknown interpolation knots $\mathcal{T}$ are compensated by $\hat{\mathcal{T}}$ with the aid of exponential parameterization (3) depending on a single parameter $\lambda \in [0,1]$ and Q_m dispersion. The main theoretical contribution (see Theorem 2) proves a linear convergence order in γ estimation by $\hat{\gamma}_3^{MC}$ for any $\lambda \in [0,1)$. The numerical tests confirm *the sharpness* of both asymptotics from Theorem 2 including the case of $\lambda = 1$, where a quartic convergence order in (14) prevails. Noticeably, though the case of $\lambda \in [0,1)$ yields merely linear asymptotics (much slower than a quartic one for $\lambda = 1$) this case still provides one degree of freedom $\lambda \in [0,1)$ to model the interpolant, should extra constraints on fitting Q_m are imposed. In particular, one may select the knots within the family (3) (i.e. with the optimal parameter $\lambda_{opt} \in [0,1)$) to minimize the "acceleration mean" $\int_0^{\hat{T}} \|\hat{\gamma}''(\hat{t})\|^2 d\hat{t}$ (see e.g. [21–23]). In contrast, such flexibility representing additional curve controlling tool is not available anymore for arbitrary fixed λ including the case of $\lambda = 1$. Such degree of freedom can still be preserved (once $\lambda \in [0,1]$ is relaxed) at the cost of potentially decelerating the asymptotics (i.e. to a linear order) in trajectory estimation.

Related work and some applications (in *computer graphics and vision, image processing and engineering*) on fitting reduced data with various C^k (with $k = 0, 1, 2$) interpolation schemes $\hat{\gamma}$ based on alternative recipes for $\hat{\mathcal{T}}$ to compensate the unknown knots $\mathcal{T}$ can be found e.g. in [1,3,8–10,13,16,18,19,28,30,32,34,36,37] or [40].

References

1. Atkinson, K.: On the order of convergence of natural cubic spline interpolation. SIAM J. Numer. Anal. **5**(1), 89–101 (1968). https://doi.org/10.1137/0705007
2. Beccari, C.V., Casciola, G., Romani, L.: Non-uniform interpolatory curve subdivision with edge parameters built upon compactly supported fundamental splines. BIT Numer. Math. **51**, 781–808 (2011). https://doi.org/10.1007/s10543-011-0328-2
3. Bézier, P.E.: Numerical Control: Mathematics and Applications. John Wiley, New York (1972)
4. de Boor, C.: A Practical Guide to Spline. Springer, New York (1985)
5. do Carmo, M.P.: Differential Geometry of Curves and Surfaces. Prentice-Hall, Engelwoods Cliffs (1976)
6. Dubuc, S.: Interpolation through an iterative scheme. J. Math. Anal. Appl. **114**(1), 185–204 (1986). https://doi.org/10.1016/0022-247X(86)90077-6
7. Dyn, N., Floater, M.S., Harmann, K.: Four-point curve subdivision based on iterated chordal and centripetal parameterization. Comput. Aided Geom. Design **26**(3), 279–286 (2009). https://doi.org/10.1016/j.cagd.2008.09.006
8. Epstein, M.P.: On the influence of parameterization in parametric interpolation. SIAM J. Numer. Anal. **13**, 261–268 (1976). https://doi.org/10.1137/0713025

9. Farin, G.: Curves and Surfaces for Computer Aided Geometrical Design. Academic Press, San Diego (1993)
10. Floater, M.S.: Chordal cubic spline interpolation is fourth order accurate. IMA J. Numer. Anal. **26**, 25–33 (2006). https://doi.org/10.1093/imanum/dri022
11. Floater, M.S., Surazhsky, T.: Parameterization for curve interpolation. Stud. Comput. Math. **12**, 39–54 (2006). https://doi.org/10.1016/S1570-579X(06)80004-2
12. Haron, H., Rehman, A., Adi, D., Lim, S.P., Saba, T.: Parameterization method on B-spline curve. Math. Probl. Eng. **2012**(1), 1–22 (2012). https://doi.org/10.1155/2012/640472
13. Janik, M., Kozera, R., Kozioł, P.: Reduced data for curve modeling - applications in graphics, computer vision and physics. Adv. Sci. Tech. **7**(18), 28–35 (2013). https://doi.org/10.5604/20804075.1049599
14. Juhász, J., Hoffmann, M.: On parameterization of interpolating curves. J. Comput. Appl. Math. **216**(2), 413–424 (2008). https://doi.org/10.1016/j.cam.2007.05.019
15. Klingenberg, W.: A Course in Differential Geometry. Springer, Heidelberg (1978)
16. Kocić, Lj.M., Simoncelli, A.C., Della Vecchia, B.: Blending parameterization of polynomial and spline interpolants. Facta Univ. (NIS) Ser. Math. Inform. **5**, 95–107 (1990)
17. Kozera, R.: Curve modeling via interpolation based on multidimensional reduced data. Stud. Inform. **25**(4B-61), 1–140 (2004)
18. Kozera, R., Noakes, L.: C^1 interpolation with cumulative chord cubics. Fund. Inform. **61**(3–4), 285–301 (2004). https://doi.org/10.3233/FUN-2004-613-406
19. Kozera, R., Noakes, L.: Piecewise-quadratics and exponential parameterization for reduced data. Appl. Math. Comput. **221**, 620–638 (2013). https://doi.org/10.1016/j.amc.2013.06.060
20. Kozera, R., Noakes, L., Wilkołazka M.: Parameterizations and Lagrange cubics for fitting multidimensional data. In: Krzhizhanovskaya, V., et al. (eds.) Computational Science - ICCS 2020. LNCS, vol. 12138, pp. 124–140. Springer, Cham (2020). https://doi.org/10.1007/978-3-030-50417-5_10
21. Kozera R., Noakes, L., Wiliński, A.: Generic case of Leap-Frog Algorithm for optimal knots selection in fitting degenerate data. In: Paszyński, M., et al. (eds.) Computational Science - ICCS 2021. LNCS, vol. 12745, pp. 337–350. Springer, Cham (2021). https://doi.org/10.1007/978-3-030-77970-2_26
22. Kozera R., Noakes, L.: Non-generic case of Leap-Frog Algorithm for optimal knots selection in fitting degenerate data. In: Groen, J., et al. (eds.) Computational Science - ICCS 2022. LNCS, vol. 13352, pp. 341–454. Springer, Cham (2022). https://doi.org/10.1007/978-3-031-08757-8_29
23. Kozera, R., Noakes, L.: Optimal knots selection in fitting degenerate data. In: Mikyška, J., et al. (eds.) Computational Science - ICCS 2023. LNCS, vol. 10475, pp. 439–453. Springer, Cham (2023). https://doi.org/10.1007/978-3-031-36024_34
24. Kozera, R., Wilkołazka, M.: A note on modified Hermite interpolation. Math. Comput. Sci. **14**(2), 223–239 (2019). https://doi.org/10.1007/s11786-019-00434-3
25. Kozera, R., Noakes, L., Wilkołazka, M.: A modified complete spline interpolation and exponential parameterization. In: Computer information systems and industrial management - CISIM 2015. LNCS, vol. 9339, pp. 98–110. Springer (2015). https://doi.org/10.1007/978-3-319-24369-6_8
26. Kozera, R., Wilkołazka, M.: Convergence order in trajectory estimation by piecewise-cubics and exponential parameterization. Math. Model. Anal. **24**(1), 72–94 (2019). https://doi.org/10.3846/mma.2019.006

27. Kuznetsov, E.B., Yakimovich, A.Y.: The best parameterization for parametric interpolation. J. Comput. Appl. Math. **191**(2), 239–245 (2006). https://doi.org/10.1016/j.cam.2005.06.040
28. Kvasov, B.I.: Methods of Shape-Preserving Spline Approximation. World Scientific, Singapore (2000)
29. Lee, E.: Choosing nodes in parametric curve interpolation. Comput. Aided Design **21**(6), 363–370 (1989). https://doi.org/10.1016/0010-4485(89)90003-1
30. Lee, E.: Corners, cusps, and parameterization: variations on theorem of Epstein. SIAM J. Numer. Anal. **29**, 553–565 (1992). https://doi.org/10.1137/0729035
31. Lim, C.G.: Universal parameterization method in B-spline curve and surface interpolation. Comput. Aided Geom. Design **16**(5), 407–422 (1999). https://doi.org/10.1016/S0167-8396(99)00010-2
32. Marin, S.P.: An approach to data parameterization in parametric cubic spline interpolation problems. J. Approx. Theory **41**(6), 64–86 (1984). https://doi.org/10.1016/0021-9045(84)90121-7
33. Milnor, J.: Morse Theory, Annals of Mathematics Studies. Princeton University Press (1968)
34. Mørken, K., Scherer, K.: A general framework for high-accuracy parametric interpolation. Math. Comp. **66**(217), 237–260 (1997). https://doi.org/10.1090/S0025-5718-97-00796-5
35. Noakes, L., Kozera, R.: More-or-less uniform samplings and length of curves. Quart. Appl. Math. **61**(3), 475–484 (2003). https://doi.org/10.1090/QAM/1999832
36. Piegl, L., Tiller, W.: The NURBS Book. Springer, Heidelberg (1997)
37. Rababah, A.: High order approximation methods for curves. Comput. Aided Geom. Design **12**, 89–102 (1995). https://doi.org/10.1016/0167-8396(94)00004-C
38. Shalashilin, V.I., Kuznetsov, E.B.: Parametric Continuation and Optimal Parameterization in Applied Mathematics and Mechanics. Kluver Academic Publishers, Dordrecht (2003)
39. Shamsuddin, S.M.H, Ahmed, M.A.: A hybrid parameterization method for NURBS. In: Proceedings of the 2nd International Conference Computer Graphics, Imaging and Visualization, pp. 5–20. IEEE Computer Society, Penang Malaysia (2004). https://doi.org/10.1109/CGIV.2004.1323954
40. Williams, J.R., Thwaites, D.J.: Radiotherapy Physics in Practice. Oxford University Press, Oxford (2000)

Physics Informed Neural Networks for Non-stationary Material Science Problems

Paweł Maczuga, Tomasz Służalec, Łukasz Sztangret, Danuta Szeliga, Marcin Łoś, and Maciej Paszyński(✉)

AGH University of Krakow, Krakow, Poland
{pmaczuga,sluzalec,szt,szeliga,los,paszynsk}@agh.edu.pl

Abstract. Linear elasticity and Navier-Stokes equations are fundamental tools in material science, enabling the modeling of solid deformations and fluid flows under various conditions. These equations are widely used to simulate stresses, strains, and fluid interactions in processes like 3D printing, welding, casting, and extrusion. Physics-Informed Neural Networks (PINNs), introduced in 2019, have gained significant attention for solving complex physical problems, including fluid mechanics, wave propagation, and inverse problems. Despite their growing popularity, PINNs face challenges in training efficiency and accuracy. This paper investigates the applicability of modern PINN methodologies to material science problems involving Navier-Stokes and linear elasticity equations. For linear elasticity, a randomized selection of collocation points is employed to enhance training. For Navier-Stokes equations, hard constraints on initial and boundary conditions are implemented to avoid multi-objective optimization. These approaches aim to address training difficulties and improve PINN performance in simulating material science phenomena.

Keywords: Physics Informed Neural Networks · Transient linear elasticity · Navier-Stokes · Random selection of collocation points · Hard constraints

1 Introduction

Linear elasticity [21] and Navier-Stokes equations [35] are foundational in simulating various physical phenomena in material science. They are used to model the behavior of fluids and solids, respectively, and help researchers and engineers understand and design materials under different conditions. Linear elasticity equations describe the deformation of solid materials under applied loads. These equations are widely used in material science to model mechanical behavior. They are used to predict the distribution of stresses and strains within a solid material under mechanical load, to understand material strength, durability, and failure mechanisms. They are employed to simulate residual stresses and distortions that arise during manufacturing processes like welding or 3D printing. The Navier-Stokes equations describe the motion of fluids and are essential for studying fluid flow behavior. In material science, they are used to model how fluids flow through porous materials such as rocks, membranes, or composite materials. They can be combined with elasticity equations to simulate how a fluid interacts with a deformable solid boundary. They can also be employed to simulate fluid behavior in material science processes like casting, extrusion, and additive manufacturing, to optimize manufacturing processes, improve material properties, and reduce defects.

The family of Physics Informed Neural Network (PINN) solvers have been introduced by prof. George Karniadakis in 2019 [30]. The method gained exponential growth in the number of papers and citations, with several new papers and some modifications of the method introduced every year [13,29,39]. The method, however, has some difficulties of the training process, some of them discussed in [32,37]. PINNs have been successfully applied to solve a wide range of problems, from fluid mechanics [3,20], in particular Navier-Stokes equations [16,34,36], wave propagation [7,19,31], phase-filled modeling [9], biomechanics [2,15], quantum mechanics [12], electrical engineering [23], problems with point singularities [11], uncertainty qualification [38], dynamic systems [1,33], or inverse problems [4,18,22], among many others.

In this paper, we focus on the investigation of the applicability of the modern PINN methodology for modeling the Navier-Stokes and linear elasticity material science problems. In the linear elasticity method, we employ the randomized selection of the PINN colocation points to improve the training procedure. In the Navier-Stokes formulation, we employ the initial and boundary conditions using the hard constraints, avoiding a multi-objective optimization problem.

Our library is a simple alternative for other available PINN libraries, including the DeepXDE [17] and IDRLnet [28]. The first one is a very large library including ODEs and PDEs on complex geometries, with different initial and boundary conditions, enabling solving and forward and inverse problems. The DeepXDE supports TensorFlow, PyTorch, JAX, and PaddlePaddle. The second one allows for solving the wave equation, Allan-Cahn equations, Volterra integrodifferential equations, and variational minimization problems. The IDRLnet library uses pytorch, numpy, and Matplotlib.

2 Linear Elasticity

We start from the following time-dependent linear elasticity equations:

$$\begin{cases} \rho \frac{\partial^2 \boldsymbol{u}}{\partial t^2} = \nabla \cdot \boldsymbol{\sigma} + \mathbf{F} & \text{on } \Omega \times [0, T] \\ \boldsymbol{u}(x, 0) = u_0 & \text{for } x \in \Omega \\ \boldsymbol{\sigma} \cdot \hat{\boldsymbol{n}} = 0 & \text{on } \partial \Omega \end{cases} \quad (1)$$

where $\boldsymbol{u}$ is a displacement vector, $\sigma_{ij} = c_{ijkl}\epsilon_{lk} = \sum_{kl} c_{ijkl}\frac{1}{2}(\partial_j u_i + \partial_i u_j)$ is the stress tensor, $\epsilon_{lk} = \frac{1}{2}\left(\frac{\partial u_k}{\partial x_l} + \frac{\partial u_l}{\partial x_k}\right)$ is the strain tensor, $\boldsymbol{c}$ is a matrix describing the material properties. We assume that the computational domain is the cuboid $\Omega = [0, 1] \times [0, 1] \times [0, 5]$, namely $x \in [0, 1]$, $y \in [0, 1]$ and $t \in [0, 5]$.

For our simulation, we employ the 2D version of the problem: We seek the displacement vector field $\Omega \times [0,T] \ni (x,y;t) \to \mathbf{u}(x,y;t) = (u_1(x,y;t), u_2(x,y;t)) \in R^2$, where for $i,j \in \{0,1\}$ we have

$$\begin{bmatrix} \sigma_{11} & \sigma_{12} \\ \sigma_{21} & \sigma_{22} \end{bmatrix} = \begin{bmatrix} \lambda\left(\frac{\partial u_x}{\partial x} + \frac{\partial u_y}{\partial y}\right) & 0 \\ 0 & \lambda\left(\frac{\partial u_x}{\partial x} + \frac{\partial u_y}{\partial y}\right) \end{bmatrix} + 2\mu \begin{bmatrix} \frac{\partial u_x}{\partial x} & \frac{1}{2}\left(\frac{\partial u_x}{\partial y} + \frac{\partial u_y}{\partial x}\right) \\ \frac{1}{2}\left(\frac{\partial u_x}{\partial y} + \frac{\partial u_y}{\partial x}\right) & \frac{\partial u_y}{\partial y} \end{bmatrix} \quad (2)$$

and the two-dimensional linear elasticity equations read

$$\begin{cases} \rho \frac{\partial^2 u_x}{\partial t^2} = \frac{\partial \sigma_{11}}{\partial x} + \frac{\partial \sigma_{12}}{\partial y} + f_1 \\ \rho \frac{\partial^2 u_y}{\partial t^2} = \frac{\partial \sigma_{21}}{\partial x} + \frac{\partial \sigma_{22}}{\partial y} + f_2 \end{cases} \quad (3)$$

where we have assumed the forcing acting on the right-top corner

$$f_1(x,y,t) = -t^2(1-t)^2 r(x_1,x_2), \quad f_2(x,y,t) = -t^2(1-t)^2 r(x_1,x_2), \\ r(x,y) = 10\exp\left(-10\left[(x_1-1)^2 + (x_2-1)^2\right]\right). \quad (4)$$

We introduce the following boundary conditions

$$\begin{cases} \mathbf{u}(x_1,x_2,0) = u_0 = 0 \\ \sigma \cdot \hat{\mathbf{n}} = 0 \end{cases} \quad (5)$$

The Lame coefficients in our model problem are defined as $\lambda = \mu = 1$.

3 Physics Informed Neural Networks for Linear Elasticity

For the solution of the linear elasticity problem by using the Physics Informed Neural Networks, we assume 4 layers of the neural network with 200 neurons in each layer. The input to the neural network is (x,y,t), and the output from the neural network is a vector $(u_x, u_y, \frac{\partial u_x}{\partial x}, \frac{\partial u_x}{\partial y}, \frac{\partial u_x}{\partial t}, \frac{\partial u_y}{\partial x}, \frac{\partial u_y}{\partial y}, \frac{\partial u_y}{\partial t},)$. The initial configuration is set as a hard constraint, that is, it is enforced in the network architecture. The initial condition is simple: $\mathbf{u}(x_1,x_2,0) = u_0 = 0$ So we have:

$$\begin{cases} u_x = u_{x_{original}} \cdot t \\ u_y = u_{y_{original}} \cdot t \end{cases} \quad (6)$$

We can see that for $t = 0$, u will always be 0, and for $t > 0$, u will also be > 0, so it can be normally influenced by the neural network parameters. The loss function is divided into two parts, $LOSS_{PDE}$ and $LOSS_{boundary}$. $LOSS_{PDE}$ consists of 8 parts. The most important are the first two, the residuals of the system of equations (3) squared. The remaining 6 are derivatives of u_x and u_y with respect to all inputs minus the corresponding output of the network. It is possible to compute the second derivative in the residual directly, in which case the network would only have 2 outputs. However, we found that having extra outputs with first derivatives greatly improves PINN's convergence.

For training, we used the Adam optimizer [14] with the learning rate of 0.005. We run the training on 40 × 40 points grid with equally distributed collocation points for 60 time steps and 20 000 epochs. In order to train the neural network, we need to create a loss function such that minimizing it will push PINN towards the solution of the Eq. (3). The loss is divided into two parts: PDE residual loss of Eq. (3), and the boundary loss for the boundary conditions. Initial conditions are set using the hard constraints as mentioned earlier. The PDE residual loss is further divided into eight parts, where the first two are the most important and the rest is introduced as a way to improve the convergence.

$$\text{LOSS}_{\text{PDE}} = \sum_{i=1}^{8} \text{LOSS}_{\text{PDE}_i} \tag{7}$$

$$\text{LOSS}_{\text{PDE}_1} = \left(\mu \frac{\partial \text{PINN}_5}{\partial t} - \frac{\partial \sigma_{11}}{\partial x_1} - \frac{\partial \sigma_{12}}{\partial x_2} + f_1 \right)^2,$$

$$\text{LOSS}_{\text{PDE}_2} = \left(\mu \frac{\partial \text{PINN}_8}{\partial t} - \frac{\partial \sigma_{11}}{\partial x_1} - \frac{\partial \sigma_{12}}{\partial x_2} + f_2 \right)^2,$$

$$\text{LOSS}_{\text{PDE}_3} = \left(\frac{\partial \text{PINN}_1}{\partial x} - \text{PINN}_3 \right)^2, \quad \text{LOSS}_{\text{PDE}_4} = \left(\frac{\partial \text{PINN}_1}{\partial y} - \text{PINN}_4 \right)^2,$$

$$\text{LOSS}_{\text{PDE}_5} = \left(\frac{\partial \text{PINN}_1}{\partial t} - \text{PINN}_5 \right)^2, \quad \text{LOSS}_{\text{PDE}_6} = \left(\frac{\partial \text{PINN}_2}{\partial x} - \text{PINN}_6 \right)^2,$$

$$\text{LOSS}_{\text{PDE}_7} = \left(\frac{\partial \text{PINN}_2}{\partial y} - \text{PINN}_7 \right)^2, \quad \text{LOSS}_{\text{PDE}_8} = \left(\frac{\partial \text{PINN}_2}{\partial t} - \text{PINN}_8 \right)^2. \tag{8}$$

where

$$\begin{bmatrix} \sigma_{11} & \sigma_{12} \\ \sigma_{21} & \sigma_{22} \end{bmatrix} = \begin{bmatrix} \lambda \cdot (\text{PINN}_3 + \text{PINN}_7) + 2\mu \cdot \text{PINN}_3 & \mu \cdot (\text{PINN}_4 + \text{PINN}_6) \\ \mu \cdot (\text{PINN}_4 + \text{PINN}_6) & \lambda \cdot (\text{PINN}_3 + \text{PINN}_7) + 2\mu \cdot \text{PINN}_7 \end{bmatrix} \tag{9}$$

and f is the forcing term defined by defined by (4). Note that PINN_3 - PINN_8 are trained to be just derivatives of PINN_1 and PINN_2 with respect to all inputs. However, splitting it in this way proved to greatly improve the convergence. The loss responsible for enforcing boundary conditions is defined as follows:

$$\text{LOSS}_{\text{boundary}} = \text{LOSS}_{\text{boundary}_{down}} + \text{LOSS}_{\text{boundary}_{up}} \\ + \text{LOSS}_{\text{boundary}_{left}} + \text{LOSS}_{\text{boundary}_{right}} \tag{10}$$

$$\text{LOSS}_{\text{boundary}_{down}} = -\sigma_{12} - \sigma_{22}, \quad \text{LOSS}_{\text{boundary}_{up}} = \sigma_{12} + \sigma_{22},$$
$$\text{LOSS}_{\text{boundary}_{left}} = -\sigma_{11} - \sigma_{21}, \quad \text{LOSS}_{\text{boundary}_{right}} = \sigma_{11} + \sigma_{21}. \tag{11}$$

4 Numerical Results for Linear Elasticity

We have run the linear elasticity simulation for a time interval $[0,5]$ over a cube-shaped domain $[0,1]^2$. Notice that in the Physics Informed Neural Network simulations there are no time steps; the computational problem is solved over the space-time cuboid. In the initial configuration, the square-shaped body has fixed zero displacement. We hit the elastic body at the right top corner with the force defined by (4). The convergence of the training is presented in Fig. 1. For the linear elasticity computations, we minimize two loss functions. The first loss function is the residual of the PDE (called the Loss PDE on the second panel in Fig. 1). The second loss function is related with the minimization of the boundary condition (5) (called the Loss boundary on the third panel in Fig. 1). We do not know how to enforce this kind of tensorial boundary condition as the hard constraint of the neural network. Thus, we minimize the sum of the two losses, as it is presented in the first panel in Fig. 1.

The snapshots from the simulation are presented in Fig. 2. We have multiplied the displacements by a factor of 10 for a better visualization of the results. However, with the central point fixed in the body, it is possible that small displacements around the central point will be visualized as moving to the other side of the central point. This, however, does not happen in reality; it is just a drawback of our visualization method.

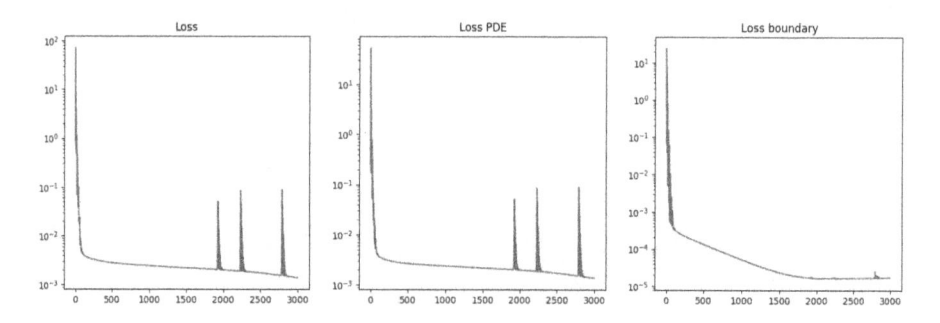

Fig. 1. The convergence of the training of the total loss function for the transient linear-elasticity problem. The total loss, the residual loss, and the boundary condition loss.

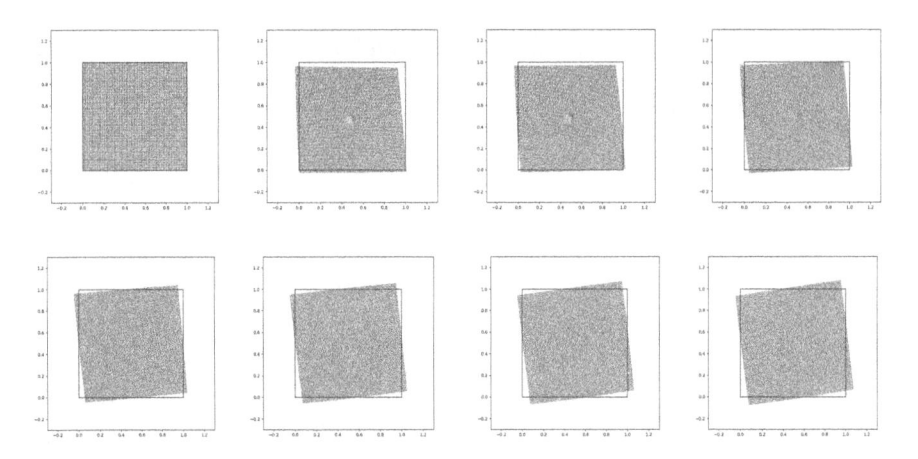

Fig. 2. Snapshots from the linear elasticity simulations. The body has fixed zero displacements in the central point.

5 Navier-Stokes Equations

Let us focus on the non-stationary cavity flow problem described with the Navier-Stokes equation for the incompressible fluid; see Fig. 3. The Dirichlet boundary condition drives the cavity flow for the velocity $u_x = 1$, $u_y = 0$ on the top boundary. On the remaining parts of the boundary, the velocity is equal to 0, and the ϵ thick transition zone in the left and right top corners ensures the possibility of a weak formulation. This problem exhibits pressure singularities at the two corners.

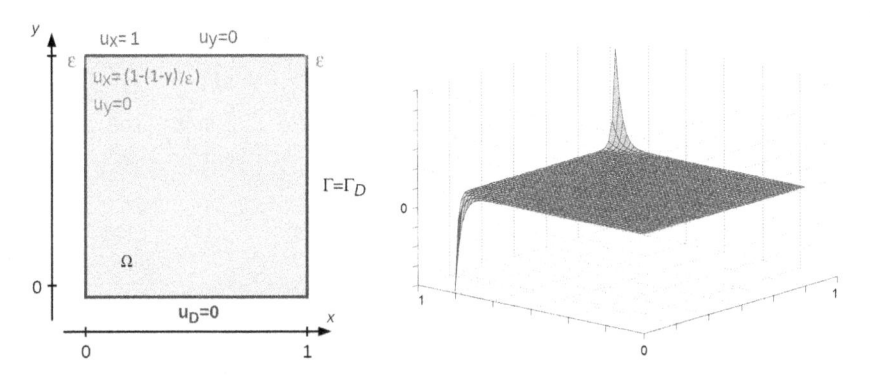

Fig. 3. Non-stationary cavity flow problem. Boundary conditions and pressure singularities.

Let $\Omega = (0,1)^2$ be the open boundary and $I = [0,T] \subset \mathcal{R}$ be the time interval. The problem reads: Find velocity u and pressure field p such that:

$$\begin{cases} \partial_t \mathbf{u} + (\mathbf{u} \cdot \nabla)\mathbf{u} - \Delta \mathbf{u} + \nabla p = 0 & \text{in } \Omega \times I, \\ \nabla \cdot \mathbf{u} = 0 & \text{in } \Omega \times I, \\ \mathbf{u} = h & \text{in } \Gamma \times I, \\ \mathbf{u}(0) = 0 & \text{in } \Omega, \end{cases} \quad (12)$$

where

$$h(x,y) = \begin{cases} 0 & x \in (0,1),\ y = 0 \\ 0 & x \in \{0,1\},\ y \in (0, 1-\epsilon) \\ 1 & x \in (0,1),\ y = 1 \\ \left(1 - \dfrac{(1-y)}{\epsilon}\right) & x \in \{0,1\},\ y \in (1-\epsilon, 1) \end{cases} \quad (13)$$

By incorporating a shift

$$\mathbf{u}_D = \left(\begin{cases} \left(1 - \dfrac{(1-y)}{\epsilon}\right)^2 & x \in (0,1),\ y \in (1-\epsilon, 1) \\ 0 & x \in (0,1),\ y \in (0, 1-\epsilon) \end{cases},\ 0\right) \quad (14)$$

this problem transforms into

$$\begin{cases} \partial_t \mathbf{u} + (\mathbf{u} \cdot \nabla)\mathbf{u} - \Delta \mathbf{u} + \nabla p = \mathbf{f} & \text{in } \Omega \times I, \\ \nabla \cdot \mathbf{u} = 0 & \text{in } \Omega \times I, \\ \mathbf{u} = 0 & \text{in } \Gamma \times I, \\ \mathbf{u}(0) = 0 & \text{in } \Omega, \end{cases} \quad (15)$$

where $\mathbf{f} = \left(0, \frac{2}{\epsilon}\left(1 - \frac{(1-y)}{\epsilon}\right)^3\right)$. Here $\mathbf{u} = (u_1, u_2)$ represents the velocity vector field, and p represents the scalar pressure field. The Γ denotes the boundary of the spatial domain Ω, and $\mathbf{f}$ is a given source resulting from the shift of the Dirichlet boundary conditions.

System (15) can be rewritten as

$$w_1(x_1,x_2,t) = \frac{\partial u_1(x_1,x_2,t)}{\partial x_1}, \quad w_2(x_1,x_2,t) = \frac{\partial u_1(x_1,x_2,t)}{\partial x_2},$$

$$z_1(x_1,x_2,t) = \frac{\partial u_2(x_1,x_2,t)}{\partial x_1}, \quad z_2(x_1,x_2,t) = \frac{\partial u_2(x_1,x_2,t)}{\partial x_2},$$

$$-\frac{\partial w_1(x_1,x_2,t)}{\partial t} - \frac{\partial w_1(x_1,x_2,t)}{\partial x_1} - \frac{\partial w_2(x_1,x_2,t)}{\partial x_2} + \frac{\partial p(x_1,x_2,t)}{\partial x_1} = f_1(x_1,x_2,t),$$

$$-\frac{\partial z_1(x_1,x_2,t)}{\partial t} - \frac{\partial z_1(x_1,x_2,t)}{\partial x_1} - \frac{\partial z_2(x_1,x_2,t)}{\partial x_2} + \frac{\partial p(x_1,x_2,t)}{\partial x_2} = f_2(x_1,x_2,t),$$

$$\frac{\partial u_1(x_1,x_2,t)}{\partial x_1} + \frac{\partial u_2(x_1,x_2,t)}{\partial x_2} = 0.$$

$$(16)$$

We define the following residual functions

$$RES_{6a}(u_\theta) = \frac{\partial u_1}{\partial x_2} - w_2, \ RES_{6b}(u_\theta) = \frac{\partial u_2}{\partial x_1} - z_1, \ RES_{6c}(u_\theta) = \frac{\partial u_2}{\partial x_2} - z_2,$$

$$RES_{6d}(u_\theta) = \frac{\partial w_1}{\partial t} - \frac{\partial w_1}{\partial x_1} - \frac{\partial w_2}{\partial x_2} + \frac{\partial p}{\partial x_1} - f_1,$$

$$RES_{6e}(u_\theta) = \frac{\partial z_1}{\partial t} - \frac{\partial z_1}{\partial x_1} - \frac{\partial z_2}{\partial x_2} + \frac{\partial p}{\partial x_2} - f_2,$$

$$RES_{6f}(u_\theta) = \frac{\partial u_1(x_1, x_2)}{\partial x_1} + \frac{\partial u_2(x_1, x_2)}{\partial x_2}, \ RES_{6g}(u_\theta) = \frac{\partial u_1}{\partial x_1} - w_1. \quad (17)$$

and the following total loss

$$RES(u_\theta) = RES_{6a}(u_\theta) + RES_{6b}(u_\theta) + RES_{6c}(u_\theta) + RES_{6d}(u_\theta)$$
$$+ RES_{6e}(u_\theta) + RES_{6f}(u_\theta) + RES_{6g}(u_\theta). \quad (18)$$

The Dirichlet boundary condition is obtained by multiplication of the output from the neural network by a summation of the four functions presented in Fig. 4 multiplied by the **g** function (definition of the Dirichlet b.c.). For the pressure approximation, we multiply the output from the neural network by the single value of the solution at the central point, see Fig. 4.

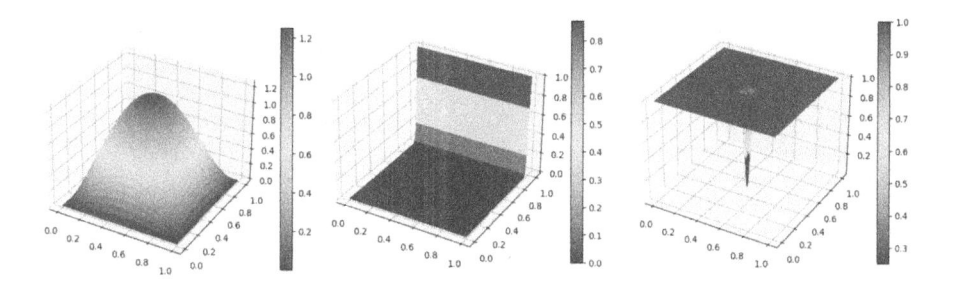

Fig. 4. The functions employed to enforce zero Dirichlet b.c., to enforce the boundary condition at the top of the domain, and the enforce zero pressure condition at the center of the domain.

6 Numerical Results for Transient Navier-Stokes

We have run the Navier-Stokes problem for the interval $[0, 1]$ over the cube-shaped cavity of dimensions $[0, 1]^2$. In the initial configuration, the velocity and the pressure inside the cavity are zero. The flow is driven by the boundary condition at the top edge of the computational domain, where the "river" flows from the left to the right. The convergence of the training is presented in Fig. 5.

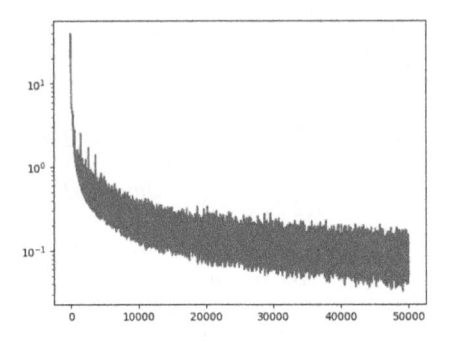

Fig. 5. The convergence of the training of the total loss function for the transient Navier-Stokes problem.

The boundary conditions are enforced by the hard constraint on the neural network, so there is no other loss function to minimize there.

The snapshots from the simulation are presented in Fig. 6.

We compare our PINN code with IGA-FEM code [41]. For the IGA-FEM we remove the non-linear term $u \cdot \nabla u$ since it requires special linearization treatment. For the finite element method formulation, following [10], we consider the singular perturbation of non-stationary Navier-Stokes problem

$$\begin{cases} \partial_t \mathbf{v}_\epsilon - \Delta \mathbf{v}_\epsilon + \nabla p_\epsilon = \mathbf{f} & \text{in } \Omega \times I, \\ \epsilon A \phi_\epsilon + \nabla \cdot \mathbf{v}_\epsilon = 0 & \text{in } \Omega \times I, \\ \epsilon \partial_t p_\epsilon = \phi_\epsilon - \chi \nabla \cdot \mathbf{v}_\epsilon & \text{in } \Omega \times I, \\ \mathbf{v}_\epsilon = 0 & \text{in } \Gamma \times I, \\ \mathbf{v}_\epsilon(0) = \mathbf{v}_0 & \text{in } \Omega, \\ p_\epsilon(0) = p_0 & \text{in } \Omega, \end{cases} \quad (19)$$

where A is an unbounded operator $A : D(A) \subset L_0^2(\Omega) \longrightarrow L_0^2(\Omega)$ and $\phi_\epsilon \in D(A)$. Here, ϵ is the perturbation parameter and $\chi \in [0, 1]$ is a user-defined parameter. We consider the alternating directions method presented in [10] with the Peaceman-Rachford scheme applied to the velocity update. We employ the residual minimization method with B-spline basis functions for discretization. The resulting IGA-FEM (isogeometric finite element) method is summarized in [40]. Figure 6 presents the visual comparison of results using 80 × 80 mesh for the time moment $t = 0.5$. The execution time for the IGA-FEM code, depending on the B-spline basis function used for discretization (see Table 1 in [40]), varies from 300 s (5 min) to 2100 s (35 min), using 1024 time steps. The training time for our PINN code is around 15 min on A100 card from Google Colab.

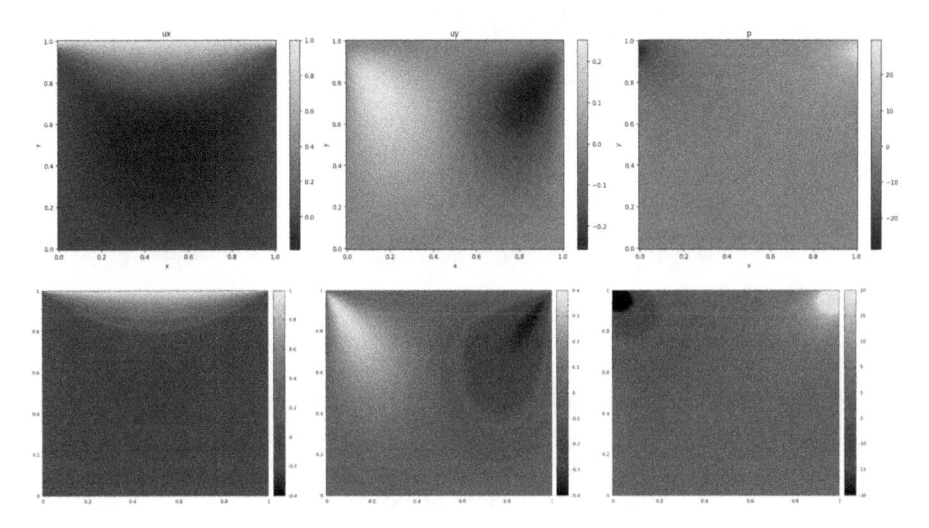

Fig. 6. The velocity component in the x direction, the velocity component in the y direction, and the pressure distribution p. **Top panel**: PINN solution extracted from the neural network solution at the time moment $t = 0.5$. **Bottom panel**: IGA-FEM solution.

7 Summary of the Code

The code for both simulations: linear elasticity and Navier-Stokes are independent Jupyter Notebooks run in a Google Colab environment:

https://colab.research.google.com/drive/
1CxCbbMfS1C2y-Q1w_mWA6YR1706N6M-C
https://colab.research.google.com/drive/
1lzq7qhlnIi5_Mz7b0PE0Fq_f958ANZlr

The code has the following structure. The code is tuned with neural network parameters such as LAYERS = 4, NEURONS_PER_LAYER = 200, training parameters such as LEARNING_RATE = 0.005, and EPOCHS = 20_000, as well as X_POINTS = 100 and Y_POINTS = 100 defining the grid of training points. We also define the plotting parameters X_PLOT = 100 and Y_PLOT = 100 the accuracy of the graphics. There are two parts of the code that require modification when implementing new simulation. The first one is the residual loss function. For example, the loss function for the Navier-Stokes equations, is defined as

```
def pde_loss(self, pinn: PINN):
    x, y, t = self.environment.get_interior_points()
    ux,uy,p,duxdx,duxdy,duydx,duydy = pinn(x, y, t)
    duxdt = df(ux, t); d2uxdx = df(duxdx, x)
    d2uxdy = df(duxdy, y); dpdx = df(p,x)
    duydt = df(uy, t); d2uydx = df(duydx, x)
    d2uydy = df(duydy, y); dpdy = df(p, y)
```

```
loss1 = duxdt −d2uxdx − d2uxdy + dpdx
loss2 = duydt −d2uydx − d2uydy + dpdy
loss3 = duxdx + duydy
loss_duxdx = duxdx − df(ux, x)
loss_duxdy = duxdy − df(ux, y)
loss_duydx = duydx − df(uy, x)
loss_duydy = duydy − df(uy, y)

return loss1.pow(2).mean() + \
       loss2.pow(2).mean() + \
       loss3.pow(2).mean() + \
       loss_duxdx.pow(2).mean() + \
       loss_duxdy.pow(2).mean() + \
       loss_duydx.pow(2).mean() + \
       loss_duydy.pow(2).mean()
```

The second modification defines the boundary conditions. They are defined by using the hard constraints. We force u_x equal to zero on the entire boundary except the top boundary where it is equal to 1, u_y equal to zero on the entire boundary, and pressure p equal to zero at the middle point. The hard constraints in PINNs look as follows:

```
def force_up_stream(x, y):
    return torch.exp(−1000*(y−1)**2)

def zero_at_middle(x, y):
    return −torch.exp(−1000*(y−0.5)**2) \
        *torch.exp(−1000*(x−0.5)**2) + 1.0

def ux_constraint(logits, x, y):
    return logits * zero_dirichlet(x, y) \
    + force_up_stream(x, y)

def uy_constraint(logits, x, y):
    return logits * zero_dirichlet(x, y)

def p_constraint(logits, x, y):
    return logits * zero_at_middle(x, y)
```

Conclusions. This paper explored the application of modern Physics-Informed Neural Networks (PINNs) to material science problems governed by linear elasticity and Navier-Stokes transient problems. For training of the linear elasticity non-stationary problem, we introduced randomized collocation point selection. For training of the Navier-Stokes time-dependent problem, we introduced the hard constraints for the initial and boundary conditions. The results demonstrated the potential of these methodologies to improve the performance and applicability of PINNs in modeling complex material science processes.

Main advantages of the PINN method compared to the classic simulations are the following. (1) Simplicity of the method. (2) Generic nature of neural networks - the approach is very similar for different problems and in many cases requires only a change in the loss function. There is no need to transform the equation to weak formulation and worry about choosing the right test and basis functions. (3) It can solve non-linear problems and inverse problems. (4) It can easily integrate measured data into training in addition to physics knowledge. (5) A single neural network can be trained for different sets of parameters.

However, there are certain disadvantages of PINNs. The solution is less accurate than the classical methods (like FEM); in more challenging problems to achieve proper convergence, certain "tricks" are required (like hard-constraint used in this paper). On top of that, there are issues related to all neural networks: choosing the right architecture and training parameters and difficulty in predicting what solution might help in certain problems. Overall, the topic of PINNs is definitely worth exploring, especially since the method is still relatively young. Neural networks are, after all, widely used in many various applications and perform extremely well. The main problem with PINNs, that is the accuracy of the solution, is improving fast, and it can even outperform classical simulators soon.

The future work following this paper may include: (1) Training generic PINN for different equation parameters. For example, linear elasticity equation has two so-called Lame coefficients (λ and μ), treated as constants in this work. It is possible to have them as additional input to the network. (2) Coupled multi-physics simulations including both elasticity and Navier-Stokes, or other challenging applications in material science. (3) Extension of the method to other classical problems solved by finite element method [5,6]. (4) Replacing PINNs by the Variational PINNs and including adaptive algorithms [8,24–27] for the test space.

Acknowledgements. Research project supported by the program "Excellence initiative - research university" for the AGH University of Krakow.

References

1. DPM: A novel training method for physics-informed neural networks in extrapolation. In: The Thirty-Fifth AAAI Conference on Artificial Intelligence, vol. 35. https://doi.org/10.1609/aaai.v35i9.16992. https://ojs.aaai.org/index.php/AAAI/article/view/16992
2. Alber, M., et al.: Integrating machine learning and multiscale modeling-perspectives, challenges, and opportunities in the biologica biomedical, and behavioral sciences. NPJ Digit. Med. **2** (2019). https://doi.org/10.1038/s41746-019-0193-y

3. Cai, S., Mao, Z., Wang, Z., Yin, M., Karniadakis, G.E.: Physics-informed neural networks (PINNs) for fluid mechanics: a review. Acta. Mech. Sin. **37**(12), 1727–1738 (2021)
4. Chen, Y., Lu, L., Karniadakis, G.E., Dal Negro, L.: Physics-informed neural networks for inverse problems in nano-optics and metamaterials. Opt. Express **28**(8), 11618–11633 (2020)
5. Demkowicz, L.: Computing with hp-Adaptive Finite Elements, vol. 1. Wiley (2006)
6. Demkowicz, L., Kurtz, J., Pardo, D., Paszynski, M., Rachowicz, W., Zdunek, A.: Computing with hp-ADAPTIVE FINITE ELEMENTS: Volume II Frontiers: Three Dimensional Elliptic and Maxwell Problems with Applications, 1st edn. Chapman and Hall/CRC (2007)
7. Geneva, N., Zabaras, N.: Modeling the dynamics of PDE systems with physics-constrained deep auto-regressive networks. J. Comput. Phys. **403** (2020). https://doi.org/10.1016/j.jcp.2019.109056
8. Goik, D., Jopek, K., Paszyński, M., Lenharth, A., Nguyen, D., Pingali, K.: Graph grammar based multi-thread multi-frontal direct solver with Galois scheduler. Procedia Comput. Sci. **29**, 960–969 (2014). https://doi.org/10.1016/j.procs.2014.05.086. https://www.sciencedirect.com/science/article/pii/S1877050914002634, 2014 International Conference on Computational Science
9. Goswami, S., Anitescu, C., Chakraborty, S., Rabczuk, T.: Transfer learning enhanced physics informed neural network for phase-field modeling of fracture. Theoret. Appl. Fract. Mach. **106** (2020). https://doi.org/10.1016/j.tafmec.2019.102447
10. Guermond, J., Minev, P.: A new class of massively parallel direction splitting for the incompressible Navier–Stokes equations. Comput. Methods Appl. Mech. Eng. **200**(23), 2083–2093 (2011). https://doi.org/10.1016/j.cma.2011.02.007. https://www.sciencedirect.com/science/article/pii/S0045782511000429
11. Huang, X., et al.: A universal PINNs method for solving Partial Differential Equations with a point source. In: Proceedings of the Fourteen International Joint Conference on Artificial Intelligence (IJCAI-2022), pp. 3839–3846 (2022)
12. Jin, H., Mattheakis, M., Protopapas, P.: Physics-informed neural networks for quantum eigenvalue problems. In: 2022 International Joint Conference on Neural Networks (IJCNN), pp. 1–8 (2022). https://doi.org/10.1109/IJCNN55064.2022.9891944
13. Kharazmi, E., Zhang, Z., Karniadakis, G.E.: hp-VPINNs: variational physics-informed neural networks with domain decomposition. Comput. Methods Appl. Mech. Eng. **374**, 113547 (2021). https://doi.org/10.1016/j.cma.2020.113547. https://www.sciencedirect.com/science/article/pii/S0045782520307325
14. Kingma, D.P., Ba, J.: Adam: a method for stochastic optimization. CoRR **abs/1412.6980** (2014). https://api.semanticscholar.org/CorpusID:6628106
15. Kissas, G., Yang, Y., Hwuang, E., Witschey, W.R., Detre, J.A., Perdikaris, P.: Machine learning in cardiovascular flows modeling: predicting arterial blood pressure from non-invasive 4d flow MRI data using physics-informed neural networks. Comput. Methods Appl. Mech. Eng. **358** (2020). https://doi.org/10.1016/j.cma.2019.112623
16. Ling, J., Kurzawski, A., Templeton, J.: Reynolds averaged turbulence modelling using deep neural networks with embedded invariance. J. Fuild Mech. **807**, 155–166 (2016). https://doi.org/10.1017/jfm.2016.615
17. Lu, L., Meng, X., Mao, Z., Karniadakis, G.E.: DeepXDE: a deep learning library for solving differential equations. SIAM Rev. **63**(1), 208–228 (2021). https://doi.org/10.1137/19M1274067

18. Lu, L., Pestourie, R., Yao, W., Wang, Z., Verdugo, F., Johnson, S.G.: Physics-informed neural networks with hard constraints for inverse design. SIAM J. Sci. Comput. **43**(6), B1105–B1132 (2021). https://doi.org/10.1137/21M1397908
19. Maczuga, P., Paszyński, M.: Influence of activation functions on the convergence of physics-informed neural networks for 1d wave equation. In: Mikyška, J., de Mulatier, C., Paszynski, M., Krzhizhanovskaya, V.V., Dongarra, J.J., Sloot, P.M. (eds.) Computational Science - ICCS 2023, pp. 74–88. Springer, Cham (2023)
20. Mao, Z., Jagtap, A.D., Karniadakis, G.E.: Physics-informed neural networks for high-speed flows. Comput. Methods Appl. Mech. Eng. **360**, 112789 (2020)
21. Marsden, J., Hughes, T.: Mathematical Foundations of Elasticity. Dover Publications, Inc. (1983)
22. Mishra, S., Molinaro, R.: Estimates on the generalization error of physics-informed neural networks for approximating a class of inverse problems for PDEs. IMA J. Numer. Anal. **42**(2), 981–1022 (2022)
23. Nellikkath, R., Chatzivasileiadis, S.: Physics-informed neural networks for minimising worst-case violations in DC optimal power flow. In: 2021 IEEE International Conference on Communications, Control, and Computing Technologies for Smart Grids (SmartGridComm), pp. 419–424 (2021). https://doi.org/10.1109/SmartGridComm51999.2021.9632308
24. Paszyńska, A., Paszyński, M., Grabska, E.: Graph transformations for modeling hp-adaptive finite element method with triangular elements. In: Bubak, M., van Albada, G.D., Dongarra, J., Sloot, P. (eds.) ICCS 2008. LNCS, vol. 5103, pp. 604–613. Springer, Heidelberg (2008). https://doi.org/10.1007/978-3-540-69389-5_68
25. Paszyńska, A., et al.: Quasi-optimal elimination trees for 2d grids with singularities. Sci. Program. (1), 303024 (2015)
26. Paszyński, M., Grzeszczuk, R., Pardo, D., Demkowicz, L.: Deep learning driven self-adaptive Hp finite element method. In: Paszynski, M., Kranzlmüller, D., Krzhizhanovskaya, V.V., Dongarra, J.J., Sloot, P. (eds.) ICCS 2021. LNCS, vol. 12742, pp. 114–121. Springer, Cham (2021). https://doi.org/10.1007/978-3-030-77961-0_11
27. Paszyński, M., Paszyńska, A.: Graph transformations for modeling parallel hp-adaptive finite element method. In: Wyrzykowski, R., Dongarra, J., Karczewski, K., Wasniewski, J. (eds.) Parallel Processing and Applied Mathematics, pp. 1313–1322. Springer, Heidelberg (2008)
28. Peng, W., Zhang, J., Zhou, W., Zhao, X., Yao, W., Chen, X.: IDRLNet: a physics-informed neural network library (2021). https://arxiv.org/abs/2107.04320
29. Qin, S., Li, M., Xu, T., Dong, S.: RAR-PINN algorithm for the data-driven vector-soliton solutions and parameter discovery of coupled nonlinear equations. ArXiv **abs/2205.10230** (2022). https://api.semanticscholar.org/CorpusID:248965018
30. Raissi, M., Perdikaris, P., Karniadakis, G.: Physics-informed neural networks: a deep learning framework for solving forward and inverse problems involving nonlinear partial differential equations. J. Comput. Phys. **378**, 686–707 (2019). https://doi.org/10.1016/j.jcp.2018.10.045. https://www.sciencedirect.com/science/article/pii/S0021999118307125
31. Rasht-Behesht, M., Huber, C., Shukla, K., Karniadakis, G.E.: Physics-informed neural networks (PINNs) for wave propagation and full waveform inversions. J. Geophys. Res. Solid Earth **127**(5), e2021JB023120 (2022)
32. Shin, Y., Darbon, J., Karniadakis, G.E.: On the convergence of physics informed neural networks for linear second-order elliptic and parabolic type PDEs. Commun. Comput. Phys. (2020). https://api.semanticscholar.org/CorpusID:225054225

33. Sun, F., Liu, Y., Sun, H.: Physics-informed spline learning for nonlinear dynamics discovery. In: Proceedings of the Thirtieth International Joint Conference on Artificial Intelligence (IJCAI 2021), pp. 2054–2061 (2021)
34. Sun, L., Gao, H., Pan, S., Wang, J.X.: Surrogate modeling for fluid flows based on physics-constrained deep learning without simulation data. Comput. Methods Appl. Mech. Eng. **361** (2020). https://doi.org/10.1016/j.cma.2019.112732
35. Taylor, C., Hughes, T.: Finite Element Programming of the Navier-Stokes Equations. Fluid Mechanics Its Applications. Pineridge Press (1981). https://books.google.pl/books?id=wo0eAQAAIAAJ
36. Wandel, N., Weinmann, M., Neidlin, M., Klein, R.: Spline-PINN: approaching PDEs without data using fast, physics-informed Hermite-spline CNNs. In: Proceedings of the AAAI Conference on Artificial Intelligence, vol. 36, no. 8, pp. 8529–8538 (2022). https://doi.org/10.1609/aaai.v36i8.20830. https://ojs.aaai.org/index.php/AAAI/article/view/20830
37. Wang, S., Yu, X., Perdikaris, P.: When and why PINNs fail to train: a neural tangent kernel perspective. J. Comput. Phys. **449**, 110768 (2022). https://doi.org/10.1016/j.jcp.2021.110768. https://www.sciencedirect.com/science/article/pii/S002199912100663X
38. Yang, Y., Perdikaris, P.: Adversarial uncertainty quantification in physics-informed neural networks. J. Comput. Phys. **394**, 136–152 (2019). https://doi.org/10.1016/j.jcp.2019.05.027
39. Yuan, L., Ni, Y.Q., Deng, X.Y., Hao, S.: A-PINN: auxiliary physics informed neural networks for forward and inverse problems of nonlinear integro-differential equations. J. Comput. Phys. **462**, 111260 (2022). https://doi.org/10.1016/j.jcp.2022.111260. https://www.sciencedirect.com/science/article/pii/S0021999122003229
40. Łoś, M., Muga, I., Muñoz-Matute, J., Paszyński, M.: Isogeometric residual minimization (IGRM) for non-stationary stokes and Navier–Stokes problems. Comput. Math. Appl. **95**, 200–214 (2021). https://doi.org/10.1016/j.camwa.2020.11.013. https://www.sciencedirect.com/science/article/pii/S0898122120304417, recent Advances in Least-Squares and Discontinuous Petrov–Galerkin Finite Element Methods
41. Łoś, M., Paszyński, M.: Parallel shared-memory open-source code for simulations of transient problems using isogeometric analysis, implicit direction splitting and residual minimization (IGA-ADS-RM). Adv. Eng. Softw. **196**, 103723 (2024). https://doi.org/10.1016/j.advengsoft.2024.103723. https://www.sciencedirect.com/science/article/pii/S0965997824001303

Adaptive Global Modeling Using Neural Networks with Deep Ensembles and Space-Filling Sequences

Pavankumar Koratikere[1(✉)], Leifur Leifsson[1], Slawomir Koziel[2,3], and Anna Pietrenko-Dabrowska[3]

[1] School of Aeronautics and Astronautics, Purdue University, West Lafayette, IN 47907, USA
{pkoratik,leifur}@purdue.edu
[2] Engineering Optimization and Modeling Center, Department of Engineering, Reykjavík University, Menntavegur 1, 102 Reykjavík, Iceland
koziel@ru.is
[3] Faculty of Electronics Telecommunications and Informatics, Gdansk University of Technology, Narutowicza 11/12, 80-233 Gdansk, Poland
anna.dabrowska@pg.edu.pl

Abstract. Global approximation models are often used in design and analysis activities in lieu of expensive simulations. Surrogate modeling with adaptive sampling is an efficient approach for creating these global models. There is a growing interest in global neural network (NN) modeling since it can approximate complex functions and can handle large datasets. Hence, this work proposes a novel method, called separate adaptive sampling (SAS), for creating global models using NNs. SAS performs exploration and exploitation using two criteria, unlike existing methods which use a single criterion. An exploration point is obtained from a space-filling sampling algorithm, while an exploitation point is obtained by maximizing the uncertainty in the NN prediction. Three existing global modeling algorithms are used for comparison. These algorithms are demonstrated on three test cases. The first two test cases are analytical functions with 3 and 8 dimensions, while the third test case is a physics-based airfoil modeling problem consisting of 16 dimensions. SAS performs best for the analytical cases while achieving comparable performance for the third case. Moreover, the NN training time for each method is nearly constant as the number of samples increase.

Keywords: global modeling · neural networks · adaptive sampling · space-filling sampling

1 Introduction

Computer simulations have become an essential tool for analyzing and designing systems. These simulations, such as computational fluid dynamics, solve a set of mathematical equations that represent a complex phenomenon. Hence, the computational resources required for these simulations is often considerable, limiting their direct application in design activities. This issue can be alleviated by using a surrogate model [2], which approximates the simulation output while offering faster evaluation times. Consequently, surrogate models have been widely adopted in various design and analysis activities [14].

Some of the popular models are kriging and neural networks (NNs). Kriging is an interpolating model that approximates function responses based on spatial correlation between samples [2]. However, the computational cost scales cubically with increasing number of samples [17]. Meanwhile, NNs can model complex functions and also scales better with larger datasets [4]. Hence, this work focuses on global modeling using NNs.

A global model approximates the underlying system across the entire design space [10]. There are two primary approaches for creating a global model: one-shot sampling and adaptive sampling [10]. In one-shot sampling, a design of experiment (DOE) is created using a sampling technique such as Latin hypercube sampling (LHS) [12]. The corresponding system output is then evaluated. Next, a surrogate model is created using the dataset and if the model's accuracy is insufficient, then the entire dataset is discarded, and the process is repeated with a larger DOE. This approach is inefficient for global modeling as it discards all the generated samples [1].

In adaptive sampling, a surrogate model is iteratively refined by adding new sample points [3]. Specifically, an initial DOE is created and the system is evaluated to obtain the corresponding observations. Next, a surrogate model is built using the dataset and if the model's accuracy is insufficient, then new samples are added to the DOE using an infill criterion. This iterative process continues until a stopping criterion is met. In this approach, the infill criterion plays a key role by balancing exploration and exploitation. Liu et al. [10] and Fuhg et al. [3] provide a detailed review of global modeling algorithms.

Many approaches have been proposed for global NN modeling. For instance, Gupta et al. [5] proposed an adaptive sampling strategy that uses information matrix and maximin distance criterion to select infill points. However, the computational cost for this method grows nonlinearly with increasing number of samples. Eason and Cremaschi [1] introduced the mixed adaptive sampling algorithm (MASA) that uses uncertainty in the NN prediction and the distance between samples for identifying the next point. Here, k-fold cross-validation was used to compute the uncertainty. However, recently more effective methods have been proposed for estimating this uncertainty such as deep ensembles [9].

In this work, a novel algorithm is proposed for global NN modeling, called separate adaptive sampling (SAS), that performs exploration and exploitation separately. Instead of using a single infill criterion, SAS consists of two separate criteria, one for exploration and the other for exploitation, and hence, adds two points in each iteration. The exploitation point is obtained by maximizing the uncertainty in the NN prediction. The exploration point is selected using a space-filling sampling sequence. SAS is demonstrated on three problems consisting of

3, 8, and 16 dimensions. For comparison, MASA is implemented, along with pure exploration- and pure exploitation-based methods.

The remainder of this paper is organized as follows. Section 2 describes the proposed adaptive global modeling technique along with the other three methods. Section 3 outlines the three test problems and discusses the results. Section 4 summarizes this work and provides some recommendations for future work.

2 Methods

This section first describes a general adaptive sampling framework used in this work for creating global NN models. Then, the space-filling sampling method is presented, which is used as a part of the infill criteria. Lastly, it describes four different methods investigated in this work.

2.1 Global Surrogate Modeling Using Adaptive Approach

A surrogate model is a simple, cheaper to evaluate approximation of a computationally expensive function $f: \mathbb{R}^n \to \mathbb{R}$. These models are used in design and analysis tasks that require repeated evaluations of the underlying function. Hence, a global model is often preferred, as it approximates a system over the entire design space. One of the ways to efficiently create a global model is to use an adaptive approach, where the model is gradually refined based on a criterion. Figure 1 illustrates an adaptive global modeling framework used in this work.

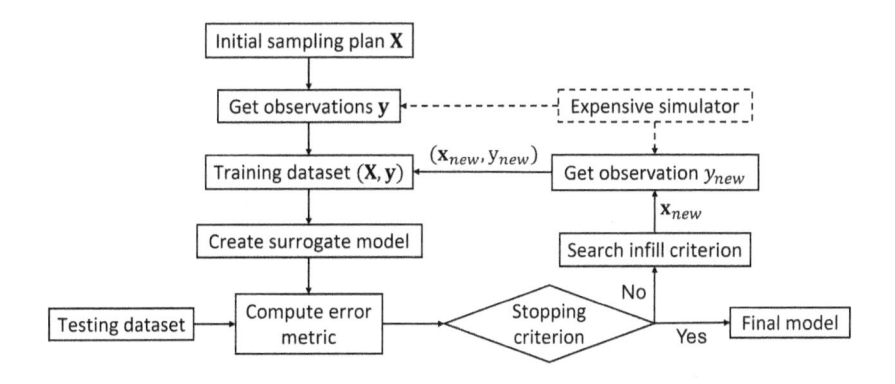

Fig. 1. The adaptive global modeling framework used in this work.

The process starts with an initial sampling plan $\mathbf{X}$ created using a DOE technique. The function f is then evaluated to obtain the output $\mathbf{y}$. Next, the iterative phase begins in which a surrogate model is created using the dataset $(\mathbf{X}, \mathbf{y})$. Then, an error metric is calculated using a test dataset to measure the model's accuracy. If the stopping criterion is not met, then a new sample $\mathbf{x}_{new}$ is selected using an infill criterion, and its output y_{new} is evaluated. The new data point ($\mathbf{x}_{new}, y_{new}$) is added to the training data and the surrogate model is retrained. This process repeats until the stopping criterion is met. In this work, the error metric is the normalized root mean squared error (NRMSE), which is written as

$$\text{NRMSE} = \frac{1}{y_{max} - y_{min}} \sqrt{\frac{1}{N} \sum_{i=1}^{N} \left[y^{(i)} - \hat{y}^{(i)} \right]^2}. \quad (1)$$

The y and $\hat{y}$ is the true and predicted value, respectively, while N is the number of samples. The y_{max} and y_{min} represent the maximum and minimum y values in the dataset, respectively. The maximum number of infill points is used as the stopping criterion.

Neural network (NN) is a machine learning model inspired by the human brain [4]. It consists of layers of interconnected nodes (neurons), where each node processes the input data using an activation function and passes the output to the next layer. The sigmoid linear unit is used as the activation function in this work. The number of layers and nodes isset aat thethe start aisis not changed during the iterations. Since the number of infill points is much larger than the initial DOE, NN is designed to overfit initially, ensuring enough flexibility at the start. As the training data increases, the model improves and the overfitting decreases.

The NN model is parameterized using weights and biases, which determine the strength of connections between nodes. These parameters are obtained by minimizing the mean squared error between the predicted and true values. In this work, the ADAM optimizer [7] is used for NN training with a learning rate of 10^{-2}. The optimization process is terminated when the NRMSE of the NN model for the training data is less than 10^{-4}. This ensures that the NN interpolates through the data without significant overfitting. Since only a few points are added in each iteration, the NN is not retrained from scratch. Instead, training resumes from the last iteration but with the updated dataset. PyTorch is used to build and train the NN model.

Another important aspect of the framework is the search infill criterion, which selects the next sample point by balancing exploration and exploitation. In this work, four different infill criteria are explored for global surrogate modeling. All of these criteria are described in the subsequent sections.

2.2 Space-Filling Sequential Sampling

In this work, the fully sequential space-filling (FSSF) [16] sampling technique is used within different methods, so it is introduced first. FSSF is a recently proposed two-phase sampling method. In the first phase, a large set of points (known as the candidate set) is generated using a space-filling technique, such as the Sobol' sequence. In the second phase, the desired sampling plan is created by sequentially selecting points from the candidate set using a distance criterion.

The pseudo-code for FSSF is outlined in Algorithm 1 as a class with three functions. The initialization function handles the first phase by defining variables

Algorithm 1. Fully sequential space-filling sampling class [16]

```
 1: procedure INITIALIZATION(N_max, n)
 2:     M ← 1000n + 2N_max
 3:     Generate candidate set C of size M using Sobol' sequence
 4:     Initialize D of size M
 5:     for i = 1, ..., M do
 6:         D_i ← distance of x_i from the closest design space boundary
 7:     end for
 8: end procedure
 9: procedure UPDATESAMPLING(x_new)
10:     for j = 1, ..., M do
11:         D_j ← min(D_j, d(x_j, x_new))       ▷ update distance criterion based on x_new
12:     end for
13: end procedure
14: procedure GENERATESAMPLES(N)
15:     for i = 1, ..., N do
16:         x_i ← find the point having maximum value in D
17:         UpdateSampling(x_i)
18:     end for
19:     return {x_1, ..., x_N}
20: end procedure
```

and generating the candidate set C. It takes N_{max} (total required points) and n (problem dimension) as inputs. The size of candidate set (M) is computed, and then Sobol' sequence is used to generate M points. Then, the distance criterion is computed for each point in the set.

The other two functions handle the second phase of FSSF. The update sampling function takes a newly added sample point x_{new} as input and adjusts the distance criterion for each point in C. This ensures that points near x_{new} will have a lower criterion value and are less likely to be selected in later iterations. This function is called whenever a point is added to the DOE.

The generate samples function takes N, the number of points to be generated, as input and uses an iterative process to generate them. The first step in the loop is to identify the next point x_i in the sequence , and then call the update sampling function with x_i as input. This function is used to create the initial DOE and then iteratively add new points without disrupting the space-filling properties.

2.3 Exploration-Based Global Neural Network Modeling

The exploration-based global NN modeling uses an infill criterion that performs pure exploration. The objective of this infill criterion is to add points in the unexplored regions. This can be achieved by using space-filling sampling techniques. In this work, the FSSF method [16] is used which is shown to have better space-filling properties than Sobol' sequences, Sect. 2.2 describes the FSSF method in detail.

Algorithm 2. Exploration-based global NN modeling (adapted from [2])

1: set the value of n, N_{init}, Q
2: initialization($N_{init} + Q, n$)   ▷ initialize FSSF, c.f. Algorithm 1
3: $\mathbf{X} \leftarrow$ generateSamples(N_{init})   ▷ initial DOE generated using FSSF
4: $\mathbf{y} = f(\mathbf{X})$ ▷ compute output for $\mathbf{X}$
5: $q = 0$ ▷ initialize number of infills
6: **while** $q \leq Q$ **do**   ▷ Q is total infill budget
7: $\quad \hat{y}(\mathbf{x}) \leftarrow$ fit a NN model to $(\mathbf{X}, \mathbf{y})$ dataset
8: $\quad$ Compute the NRMSE of the $\hat{y}$ model using testing data
9: $\quad \mathbf{x}_{new} \leftarrow$ generateSamples(1)   ▷ next sample in the sequence $\mathcal{S}$
10: $\quad \mathbf{X}, \mathbf{y} \leftarrow \mathbf{X} \cup \mathbf{x}_{new}, \mathbf{y} \cup f(\mathbf{x}_{new})$ ▷ append dataset
11: $\quad q = q + 1$ ▷ update number of infills
12: **end while**
13: **return** $\hat{y}$ model

Algorithm 2 presents a pseudo-code for the exploration-based global NN modeling. The problem dimension n, the initial DOE size N_{init}, and the maximum number of infill points Q are defined at start. The initialization function from Algorithm 1 is then executed with $N_{init} + Q$ and n as inputs. The initial DOE $\mathbf{X}$ is constructed using the FSSF method by running the generate samples function from Algorithm 1. The underlying system is then evaluated to obtain $\mathbf{y}$. The variable q is also initialized to zero, which denotes the number of infill points added to the dataset.

Next, the iterative phase begins, in which the first step is to create a $\hat{y}$ NN model using the dataset $(\mathbf{X}, \mathbf{y})$. Then, the NRMSE of the $\hat{y}$ model is computed using the test dataset. The next infill point $\mathbf{x}_{new}$ is obtained by running the generate samples function with $N = 1$ as input. This new sample and its corresponding output are added to the dataset, and q is incremented by 1. This loop continues until the stopping criterion of maximum number of infill points is met.

2.4 Exploitation-Based Global Neural Network Modeling

The exploitation-based global NN modeling aims to add infill points where the surrogate model's behavior is uncertain. Algorithm 3 provides a pseudo-code for this method, which is similar to Algorithm 2, but with a few differences. Firstly, the initial DOE can be generated using any sampling technique, but the FSSF method is used here to ensure consistency across all the methods. Secondly, the infill criterion is based on finding the point with the highest uncertainty in the NN prediction.

In this work, the deep ensembles (DE) [9] method is used to estimate the uncertainty due to its simplicity and ease of implementation. The DE approach consists of training an ensemble of NN models on the same dataset. The starting point for the NN training is randomly determined, which results in different predictions for a given input. The final prediction $\hat{y}(\mathbf{x})$ and the uncertainty $\hat{s}(\mathbf{x})$ is the mean and standard deviation of all predictions, respectively. The DE method

Algorithm 3. Exploitation-based global NN modeling (adapted from [2])

1: set the value of n, N_{init}, Q
2: initialization($N_{init} + Q, n$) ▷ initialize FSSF, c.f. Algorithm 1
3: $\mathbf{X} \leftarrow$ generateSamples(N_{init}) ▷ initial DOE generated using FSSF
4: $\mathbf{y} = f(\mathbf{X})$ ▷ compute output for $\mathbf{X}$
5: $q = 0$ ▷ initialize number of infills
6: **while** $q \leq Q$ **do** ▷ Q is total infill budget
7:     $\hat{y}(\mathbf{x}), \hat{s}(\mathbf{x}) \leftarrow$ fit NN models to $(\mathbf{X}, \mathbf{y})$ dataset using DE method
8:     Compute the NRMSE of the $\hat{y}$ model using testing data
9:     $\mathbf{x}_{new} \leftarrow \arg\max \hat{s}(\mathbf{x})$
10:    $\mathbf{X}, \mathbf{y} \leftarrow \mathbf{X} \cup \mathbf{x}_{new}, \mathbf{y} \cup f(\mathbf{x}_{new})$ ▷ append dataset
11:    $q = q + 1$ ▷ update number of infills
12: **end while**
13: **return** $\hat{y}$ model

---

is proposed for quantifying both aleatoric and epistemic uncertainty. However, it is assumed that the function to be modeled is deterministic, and hence, there is no aleatoric uncertainty. This assumption simplifies the DE method to a standard ensembles approach. The ensemble consists of five NN models, refer [9] for more details. In this work, differential evolution [18] is used to maximize the uncertainty. The remaining steps are similar to those in Algorithm 2.

2.5 Mixed Adaptive Sampling Algorithm

The mixed adaptive sampling algorithm (MASA) [1] uses an infill criterion that performs exploration and exploitation using a single criterion. The pseudo-code for MASA is similar to Algorithm 3, with the only difference being the infill criterion. Mathematically, the infill point is obtained as

$$\mathbf{x}_{new} = \arg\max_{\mathbf{x}} \hat{s}(\mathbf{x})/s_{max} + d(\mathbf{x}, \mathbf{X})/d_{max}. \qquad (2)$$

The exploitation part, $\hat{s}(\mathbf{x})/s_{max}$, accounts for the uncertainty in the NN prediction, while the exploration part, $d(\mathbf{x}, \mathbf{X})/d_{max}$, ensures new point is added in the unexplored regions. The $\hat{s}(\mathbf{x})$ denotes the uncertainty in the NN prediction which is computed using the DE approach discussed in Sect. 2.4. The $d(\mathbf{x}, \mathbf{X})$ denotes the euclidean distance of $\mathbf{x}$ to the closest point in the dataset $\mathbf{X}$. The s_{max} and d_{max} are used to normalize the $\hat{s}(\mathbf{x})$ and $d(\mathbf{x}, \mathbf{X})$, respectively. In this work, s_{max} is obtained by maximizing the $\hat{s}(\mathbf{x})$, and d_{max} corresponds to the euclidean distance between the farthest points in the dataset $\mathbf{X}$. In this work, differential evolution [18] is used to numerically solve (2).

2.6 Separate Adaptive Sampling Algorithm

The proposed separate adaptive sampling (SAS) algorithm performs exploitation and exploration using two different criteria, unlike MASA which uses a single

Algorithm 4. Separate adaptive sampling algorithm (this work)

1: set the value of n, N_{init}, Q
2: initialization($N_{init} + Q, n$) ▷ initialize FSSF, c.f. Algorithm 1
3: $\mathbf{X} \leftarrow$ generateSamples(N_{init}) ▷ initial DOE generated using FSSF
4: $\mathbf{y} = f(\mathbf{X})$ ▷ compute output for $\mathbf{X}$
5: **while** $q \leq Q$ **do** ▷ Q is total infill budget
6:     $\hat{y}(\mathbf{x}), \hat{s}(\mathbf{x}) \leftarrow$ fit NN models to $(\mathbf{X}, \mathbf{y})$ dataset using DE method
7:     Compute the NRMSE of the $\hat{y}$ model using testing data
8:     $\mathbf{x}_{exploit} \leftarrow \arg\max \hat{\sigma}(\mathbf{x})$ ▷ exploitation
9:     $\mathbf{X}, \mathbf{y} \leftarrow \mathbf{X} \cup \mathbf{x}_{exploit}, \mathbf{y} \cup f(\mathbf{x}_{exploit})$
10:    updateSampling($\mathbf{x}_{exploit}$) ▷ update FSSF about exploitation point
11:    $\mathbf{x}_{explore} \leftarrow$ generateSamples(1) ▷ exploration
12:    $\mathbf{X}, \mathbf{y} \leftarrow \mathbf{X} \cup \mathbf{x}_{explore}, \mathbf{y} \cup f(\mathbf{x}_{explore})$
13:    $q = q + 2$ ▷ update number of infills
14: **end while**
15: **return** $\hat{y}$ model

---

criterion. Algorithm 4 provides a pseudo-code for SAS which is similar to the previous algorithms. The $\mathbf{x}_{exploit}$ is obtained by maximizing the uncertainty in the NN prediction, similar to the exploitation-based method in Sect. 2.4. Before generating the exploration point, the update sampling function from Algorithm 1 is run with $\mathbf{x}_{exploit}$ as the input. This ensures that exploration point is not added around the exploitation points. Then, the $\mathbf{x}_{explore}$ is generated by running the generate samples function with $N = 1$ as the input. The variable q is incremented by 2 to account for both infill points. All other steps are similar to the previously discussed algorithms.

3 Numerical Experiments

This section presents three test cases used to demonstrate and compare the results of the global NN modeling algorithms described in Sect. 2. Each method is run 10 times to account for the randomness in NN training. The initial DOE is same across all methods to ensure a consistent starting point. The convergence history for all 10 runs is shown as a convergence band. The center-line represents the median, while the upper and lower limits represent 10th and 90th percentile, respectively. For all the problems, testing dataset consists of 50 LHS [12] points.

3.1 Ishigami Function

This section presents the Ishigami function and discusses the results of applying different global modeling methods. The Ishigami function [6] is written as:

$$f(\mathbf{x}) = \sin(x_1) + a\sin^2(x_2) + bx_3^4 \sin(x_1), \qquad (3)$$

where $a = 7$, $b = 0.1$, and $x_i \in [-\pi, \pi]$, $\forall\ i = 1, 2, 3$. The initial DOE consists of 25 samples generated using the FSSF method, and the maximum number of infill points is set to 100. The NN model has three layers, each with 20 neurons.

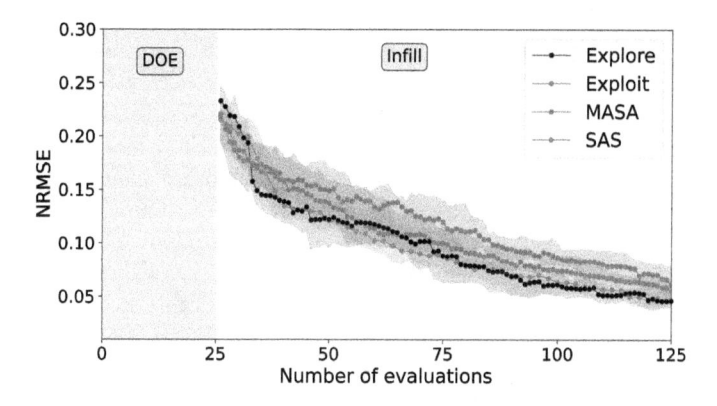

Fig. 2. Convergence history of the NRMSE for the Ishigami function.

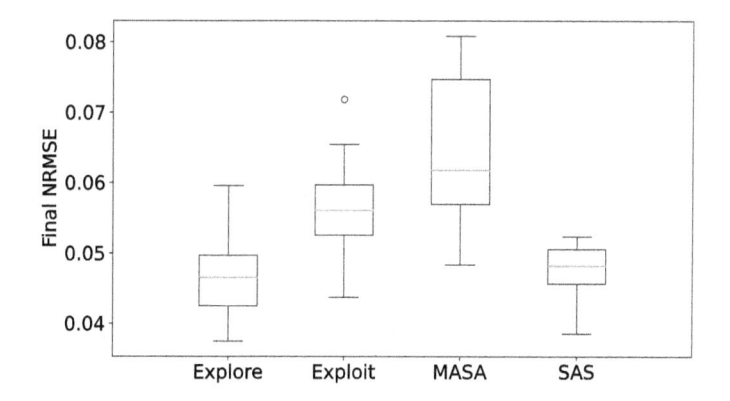

Fig. 3. Final NRMSE of the NN model for the Ishigami function.

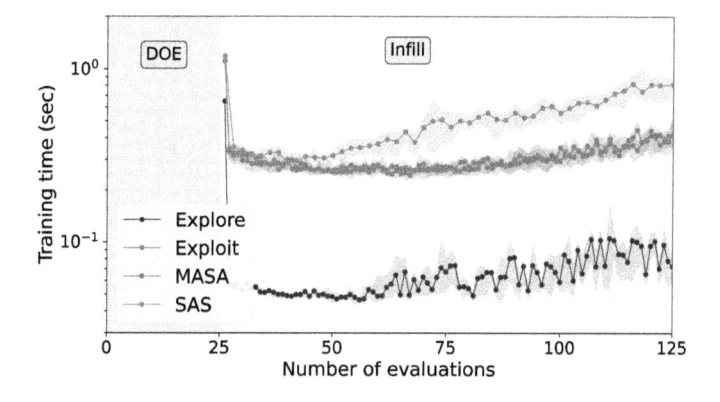

Fig. 4. Training time of the NN model for the Ishigami function.

Figure 2 shows the convergence history of the NRMSE as the number of samples increase. The exploration-based method and SAS have slightly better convergence rate, while MASA has the slowest convergence. However, all the methods perform comparably. Figure 3 compares the final NRMSE of the NN model obtained by each method. The exploration-based method has the best median NRMSE value but exhibits a large variation. On the other hand, SAS achieves a comparable median NRMSE value with least variation. The exploitation-based method and MASA do not perform as well for this problem. This indicates that pure exploration might be enough for this problem. Figure 4 shows the NN model training time as the number of samples increases. The exploration-based method has the shortest training time since it uses only one NN model, while the other methods use five. All methods scale much better than kriging, which scales cubically with the number of samples [17]. This makes global surrogate modeling with NN attractive for large datasets.

3.2 Borehole Function

The borehole function [13] models the flow of water through a borehole and is written as:

$$f(\mathbf{x}) = \frac{2\pi T_u (H_u - H_l)}{\ln(r/r_w) \cdot \left(1 + \frac{2LT_u}{\ln(r/r_w) r_w^2 K_w} + \frac{T_u}{T_l}\right)}, \quad (4)$$

where $\mathbf{x} = \begin{bmatrix} r_w & r & T_u & H_u & T_l & H_l & L & K_w \end{bmatrix}^T$. The upper and lower bound for the variables are provided in Table 1. The function f represents the water flow rate in m^3/yr. The initial DOE consists of 25 samples and the maximum number of infill points is set to 100. The NN model consists of two layers, each having 20 neurons.

Table 1. Bounds for the variables of the borehole function

Variable	Lower bound	Upper bound
r_w (m)	0.05	0.15
r (m)	100	50000
T_u (m^2/yr)	63070	115600
H_u (m)	990	1110
T_l (m^2/yr)	63.1	116
H_l (m)	700	820
L (m)	1120	1680
K_w (m/yr)	9855	12045

Figure 5 shows the evolution of NRMSE as the number of samples are increased. For this test case, SAS outperforms all the other methods, especially

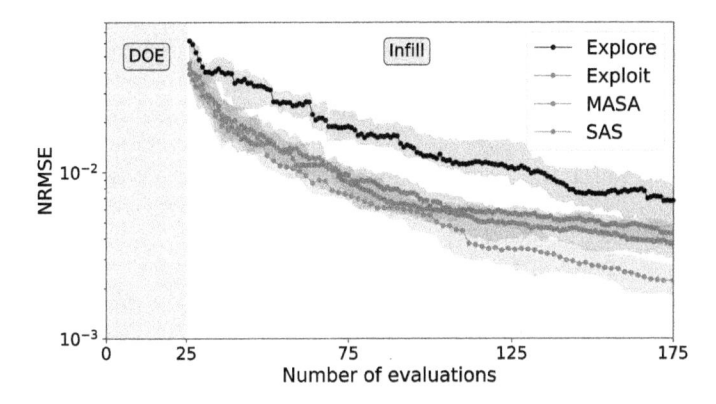

Fig. 5. Convergence history of the NRMSE for the borehole function.

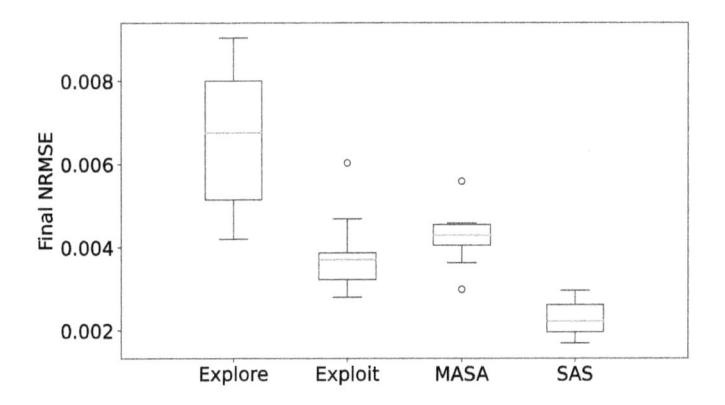

Fig. 6. Final NRMSE of the NN model for the borehole function.

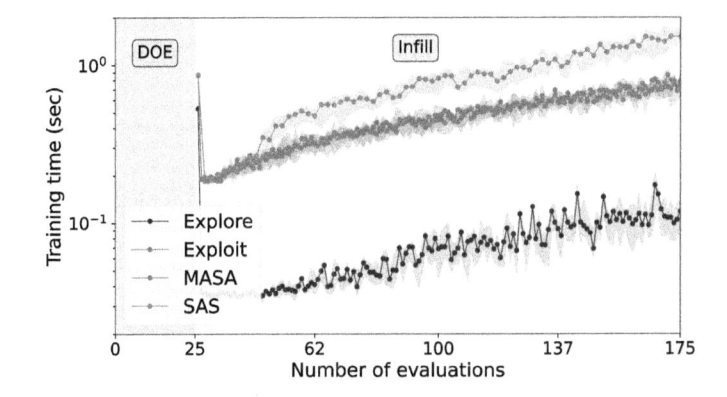

Fig. 7. Training time of the NN model for the borehole function.

after 100 function evaluations. The exploitation-based method and MASA perform comparably to SAS up to 100 evaluations, but their convergence rate slows down beyond that point. The exploration-based method has the slowest rate of convergence among all methods. This suggests that performing pure exploration is not enough for this test case.

Figure 6 compares the final NRMSE values obtained by each method. SAS achieves the best final NRMSE as compared to other methods, with the least variation. The exploration-based method has the highest NRMSE value and the most variation. Both exploitation-based method and MASA yield comparable final NRMSE value. This suggests that the distance-based exploration term used in MASA does not improve the performance for this problem. Figure 7 shows how NN training time changes as the number of samples increase. The exploration-based method has the shortest training time since it uses only one NN model, while the other methods use five NN models. Since two infill points are added in each iteration, the training time for SAS is slightly longer.

3.3 Airfoil Drag Coefficient Modeling

The airfoil modeling problem involves predicting the drag coefficient (C_d) of a given airfoil in viscous transonic flow conditions. The variables in the problem include the shape of the airfoil, the angle of attack (α), and the free-stream mach number (M_∞). The airfoil shape is parameterized using the class-shape transformation (CST) [8] technique. In this work, 14 CST coefficients are used to represent the airfoil shape, 7 for the upper surface and 7 for the lower surface. In total, the problem consists of 16 variables.

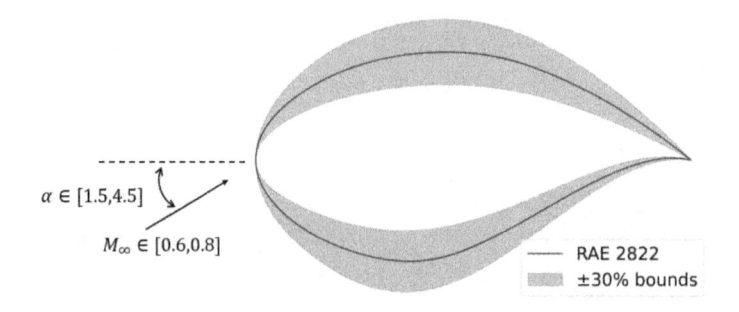

Fig. 8. Schematic describing the design space of the airfoil modeling problem.

The upper and lower bounds for the CST coefficients are determined by perturbing the CST coefficients for the RAE 2822 airfoil by ±30%. The $\alpha \in [1.5, 4.5]$ and $M_\infty \in [0.6, 0.8]$, while the Reynolds number (Re_∞) is fixed at 6 million. Figure 8 illustrates these variables and their respective bounds. The flow around the airfoil is computed using ADflow [11], an open-source finite volume solver. The computational grid uses an o-mesh generated using pyHyp [15]. The initial DOE consists of 50 samples generated using the FSSF method and the maximum number of infill points is set to 100. The NN model consists of two hidden layers with 20 neurons in each layer.

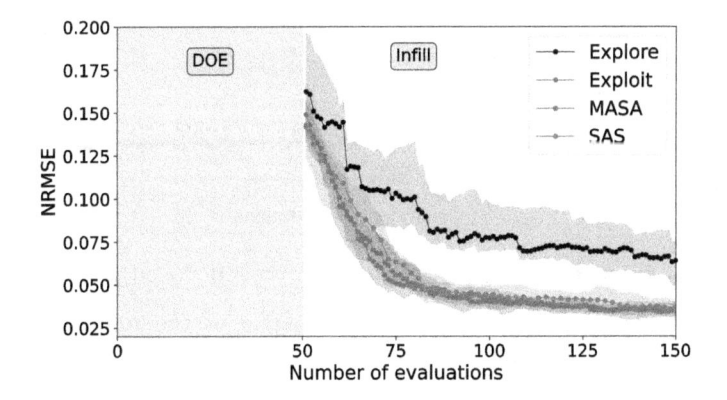

Fig. 9. Convergence history of the NRMSE for the airfoil modeling problem.

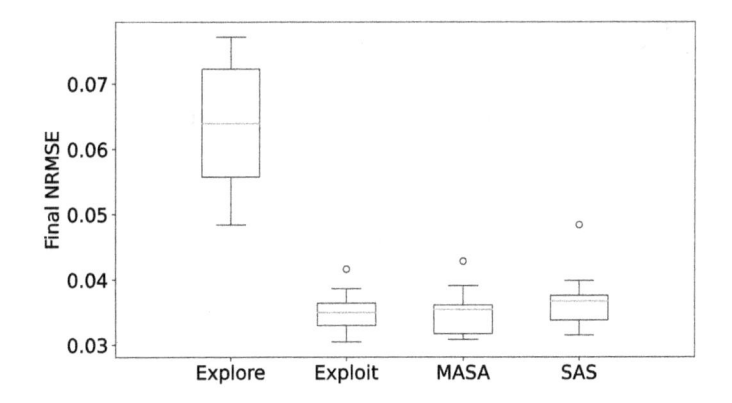

Fig. 10. Final NRMSE of the NN model for the airfoil modeling problem.

Figure 9 shows the evolution of NRMSE as the number of samples increase. The exploitation-based method, MASA, and SAS perform comparably, with similar convergence rates. However, the exploration-based method fails to reduce the NRMSE after 100 function evaluations. Since the C_d is highly nonlinear in the transonic regime, the pure exploration-based strategy does not perform well.

Figure 10 shows the final NRMSE of the NN model. The SAS, MASA, and exploitation-based method have comparable median final NRMSE value, with SAS exhibiting the least variation. While some final NRMSE values from MASA and the exploitation-based method are better than SAS, both methods show larger variation. The exploration-based method has the highest final NRMSE and the largest variation for the reasons described earlier.

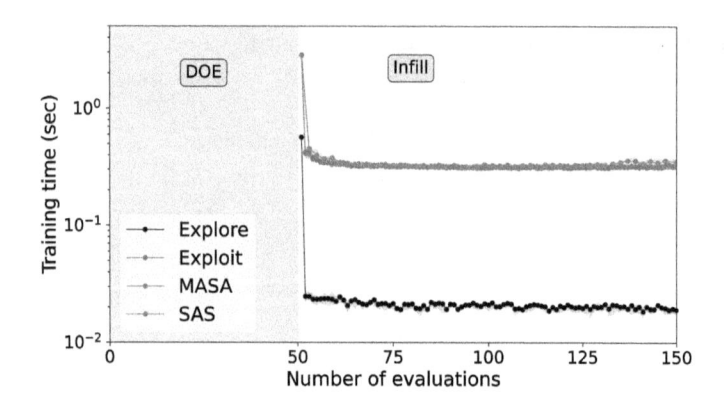

Fig. 11. Training time of the NN model for the airfoil modeling problem.

Figure 11 shows the NN model training time as the number of samples increase. As in previous test problems, the exploration-based method has the shortest training time. Additionally, the training time for all the methods remains nearly constant, even with 16 variables in the problem. This demonstrates the effectiveness of the NN for the global NN modeling.

4 Conclusion

This work proposes a novel infill criterion for global neural network (NN) modeling. The proposed method, called separate adaptive sampling (SAS), performs exploration and exploitation using different criterion, and hence, adds two infill points in each iteration. The first point is obtained by maximizing the uncertainty in the NN prediction, while the second point is obtained from a space-filling sequential sampling method. Uncertainty in the NN prediction is estimated using the deep ensembles approach. The study also explores the mixed adaptive sampling algorithm, which balances exploration and exploitation with a single criterion, as well as pure exploitation- and pure exploration-based methods.

The algorithms described in this work are demonstrated on three test cases with varying levels of difficulty. SAS performed the best for the first two test cases and performed comparably for the third problem. The pure exploration-based method did not yield good results for highly nonlinear test cases. However, for all problems, NN training time scales well as the dataset size increased, highlighting its advantage over the kriging model.

Future work will focus on testing the proposed method on high-dimensional global modeling problems requiring many infill points. The uncertainty in the NN prediction can be estimated using other methods to improve the exploitation. Moreover, a non-uncertainty based criterion can be explored for exploitation, as SAS decouples exploration and exploitation.

Acknowledgments. This work was supported in part by the U.S. National Science Foundation (NSF) award number 2223732 and by the Icelandic Centre for Research (RANNIS) award number 239858.

References

1. Eason, J., Cremaschi, S.: Adaptive sequential sampling for surrogate model generation with artificial neural networks. Comput. Chem. Eng. **68**, 220–232 (2014)
2. Forrester, A.I.J., Sóbester, A., Keane, A.J.: Engineering Design via Surrogate Modelling: A Practical Guide. Wiley (2008)
3. Fuhg, J.N., Fau, A., Nackenhorst, U.: State-of-the-art and comparative review of adaptive sampling methods for kriging. Arch. Comput. Methods Eng. **28**(4), 2689–2747 (2020)
4. Goodfellow, I., Bengio, Y., Courville, A.: Deep Learning. MIT Press (2016)
5. Gupta, S., Paudel, A., Thapa, M., Mulani, S.B., Walters, R.: Adaptive sampling-based artificial neural network for surrogate modeling. In: AIAA SCITECH 2022 Forum. American Institute of Aeronautics and Astronautics (2022)
6. Ishigami, T., Homma, T.: An importance quantification technique in uncertainty analysis for computer models. In: Proceedings. First International Symposium on Uncertainty Modeling and Analysis, pp. 398–403. IEEE Computer Society Press (1990)
7. Kingma, D.P., Ba, J.: Adam: a method for stochastic optimization (2014)
8. Kulfan, B.M.: Universal parametric geometry representation method. J. Aircr. **45**(1), 142–158 (2008)
9. Lakshminarayanan, B., Pritzel, A., Blundell, C.: Simple and scalable predictive uncertainty estimation using deep ensembles. Adv. Neural Inf. Process. Syst. **30** (2017)
10. Liu, H., Ong, Y.S., Cai, J.: A survey of adaptive sampling for global metamodeling in support of simulation-based complex engineering design. Struct. Multidiscip. Optim. **57**(1), 393–416 (2017)
11. Mader, C.A., Kenway, G., Yildirim, A., Martins, J.: ADflow: an open-source computational fluid dynamics solver for aerodynamic and multidisciplinary optimization. J. Aerosp. Inf. Syst. **17**(9), 508–527 (2020)
12. Mckay, M.D., Beckman, R.J., Conover, W.J.: A comparison of three methods for selecting values of input variables in the analysis of output from a computer code. Technometrics **42**(1), 55–61 (2000)
13. Morris, M.D., Mitchell, T.J., Ylvisaker, D.: Bayesian design and analysis of computer experiments: use of derivatives in surface prediction. Technometrics **35**(3), 243–255 (1993)
14. Queipo, N.V., Haftka, R.T., Shyy, W., Goel, T., Vaidyanathan, R., Kevin Tucker, P.: Surrogate-based analysis and optimization. Prog. Aerosp. Sci. **41**(1), 1–28 (2005)
15. Secco, N.R., Kenway, G., He, P., Mader, C., Martins, J.: Efficient mesh generation and deformation for aerodynamic shape optimization. AIAA J. **59**(4), 1151–1168 (2021)
16. Shang, B., Apley, D.W.: Fully-sequential space-filling design algorithms for computer experiments. J. Qual. Technol. **53**(2), 173–196 (2021)
17. Snoek, J., et al.: Scalable bayesian optimization using deep neural networks. In: International Conference on Machine Learning, pp. 2171–2180. PMLR (2015)
18. Storn, R., Price, K.: Differential evolution-a simple and efficient heuristic for global optimization over continuous spaces. J. Global Optim. **11**(4), 341–359 (1997)

Automated Antenna Design Using Computational Intelligence and Numerical Optimization

Slawomir Koziel[1,2]($\boxtimes$), Anna Pietrenko-Dabrowska[2], and Leifur Leifsson[3]

[1] Engineering Optimization and Modeling Center, Department of Engineering, Reykjavík University, Menntavegur 1, 102 Reykjavík, Iceland
`koziel@ru.is`
[2] Faculty of Electronics Telecommunications and Informatics, Gdansk University of Technology, Narutowicza 11/12, 80-233 Gdansk, Poland
`anna.dabrowska@pg.edu.pl`
[3] School of Aeronautics and Astronautics, Purdue University, West Lafayette, IN 47907, USA
`leifur@purdue.edu`

Abstract. Modern antenna design is a daunting task aimed at fulfilling diverse performance requirements and constraints imposed by specific application areas. Traditional techniques are heavily based on engineering experience. This limits the options considered for conventional architectures and leads to sub-optimal results. This study introduces an innovative approach to automated development of antennas. The introduced methodology incorporates computational intelligence techniques and numerical optimization procedures to carry out unsupervised antenna topology generation and fine-tuning its geometry parameters. The crucial component of the suggested method is a versatile parameterization involving elliptical-shaped patches and gaps, which can realize a massive variety of different shapes. Computational intelligence is used to execute a purely specification-driven antenna evolution process. The decision variables include a mixture of discrete and continuous parameters handled by a customized evolutionary algorithm (Stage I) and local optimizer (Stage II – fine tuning). The procedure is fully specification-based and requires no human-expert interaction whatsoever. The proposed framework has been comprehensively demonstrated by designing several devices of distinct characteristics (broadband, multi-band, compact). The findings underscore the versatility of the technique and its suitability to produce nonconventional structures with acceptable computational costs.

Keywords: Design automation · antenna systems · computational intelligence · optimization · unsupervised design

1 Introduction

Antennas are vital building blocks of wireless communication systems, including mobile phones, satellite communications, radio-frequency identification, medical imaging, etc. Traditional design methods are typically based on the existing antenna topologies (e.g., available in the literature), which are modified to achieve the required functionality and performance. The process is iterative and involves trying out different architectural variations interleaved with parametric studies [1], or, recently, rigorous optimization, [2]. Optimization can be carried out in the local [3], global [4, 5], or multi-objective sense [6, 7]. The fundamental underlying tool is electromagnetic (EM) analysis. However, simulation-driven design is CPU intensive, which becomes a severe bottleneck when repetitive simulations are involved. Addressing this issue fostered the development of expedited methods, which include surrogate-assisted techniques [8, 9], machine learning (ML) routines [10, 11], response feature algorithms [12], or variable-resolution approaches [13]. A comprehensive review of metamodel-driven antenna design procedures is available in [14].

Boosting performance and realization of extra functionality is typically achieved by altering fundamental structures (e.g., patches, dipoles, etc.). The final product usually resembles the initial topology [15], while reaching it is laborious. This is highly restrictive regarding the number of alternative antenna geometries that may be investigated. Alternative methods include topology optimization (TO). It often involves spatial discretization and optimization-based assignment of individual cells (filled with metal or empty) [16, 17]. This approach is associated with the necessity of solving complex combinatorial tasks, even if only part of the antenna is discretized [18]. Another option is pixel antennas with a predefined arrangement of metal patches whose interconnections are decided upon through optimization [19, 20]. Free form TO is yet another technique [21, 22]. It offers significant flexibility but typically requires fast custom EM solvers to maintain acceptable computational expenses [23, 24]. Consequently, it cannot be integrated with commercial engineering design automation (EDA) tools. Also, it relies on gradient-based optimizers, which necessitates reasonable initial starting points to ensure the quality of the final structure.

This study outlines an innovative approach to the unsupervised design of antenna structures, which introduces several original contributions. The main focus is on ensuring flexibility, which is understood as a broad range of distinct geometries that can be generated and the capability of realizing diverse functionality (multi-band, wide-band operation, etc.). Other prerequisites are integrality with commercial EM solvers and reasonable computational efficiency, which are critical from the practical engineering standpoint. Our methodology leverages versatile parameterization consisting of elliptical-shaped patches and gaps that can relocate within the substrate area and adjust their size to assemble antenna geometry through Boolean operations. Varying the number of building blocks and relocating them allows for the implementation of many topologies based on a restricted number of independent parameters. The design is entirely specification-based and utilizes computational intelligence to realize a global search stage (geometry evolution) and local optimization tools for final dimension tuning. The design problem is formulated to realize the assumed functionality and boost the antenna performance regarding the target operating frequency ranges and impedance matching level. The proposed approach has been demonstrated by designing several broadband and multi-band antennas of diverse characteristics, some experimentally validated for supplementary illustration. The obtained results corroborate the capability of the developed technique to produce high-performance unconventional antenna structures while ensuring computational efficiency. At the same time, the design process does not require any human-expert interaction. These findings underscore the framework's suitability for developing high-performance antennas for demanding applications in both academic and industrial settings.

2 Antenna Design Using Computational Intelligence

This section explains the proposed unsupervised design framework. Section begins with discussing parameterization (Sect. 2.1), followed by a description of the computational model (Sect. 2.2) and the algorithms employed for antenna development in Sect. 2.3. The entire procedure is elucidated in Sect. 2.4.

2.1 Antenna Building Blocks

Geometry parameterization of the critical component of the presented framework. The prerequisites are as follows: (i) simplicity to ensure straightforward handling, (ii) flexibility to enable the construction of diverse topologies, (iii) a restricted number of decision variables to keep the underlying optimization task computationally tractable. The antenna parameters should include continuous ones, allowing for local tuning and discrete (to control the architecture's complexity). The components of the assumed parameterization are presented in Fig. 1. For illustration, a rectangular substrate is taken along with a rectangular ground plane. The antenna is excited through a relocatable discrete port. The front-side metallization is composed using N_P elliptical patches and N_G of gaps. These numbers may be fixed or treated as design variables. The positions and dimensions of all building blocks are relative to the substrate width W and length L. All parameters are aggregated into a vector x. Figure 2 shows some randomly generated geometries demonstrating the flexibility of the discussed parameterization even when using a limited number of patches and gaps ($N_P = 5$ and $N_G = 3$). The antenna size can be optimized or fixed depending on the intended application.

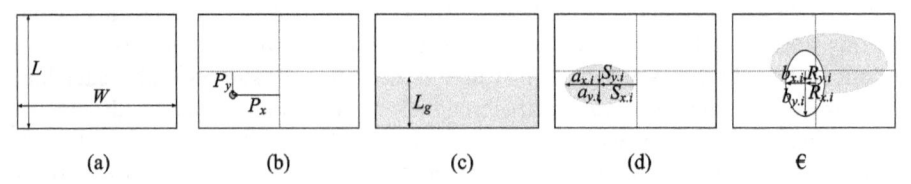

Fig. 1. Antenna parameterization building blocks: (a) substrate, (b) discrete port, (c) ground plane, (d) ith elliptical patch ($i = 1, ..., N_P$), (e) ith elliptical gap ($i = 1, ..., N_G$).

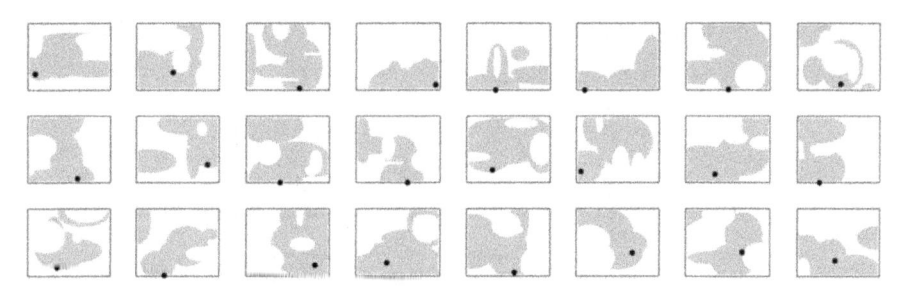

Fig. 2. Parameterization flexibility demonstrated through randomly generated architectures for $N_P = 5$ and $N_G = 3$. Front-side metallization marked gray; discrete port marked as a black dot.

2.2 Computational Model

In this research, CST Microwave Studio [25] is utilized to implement and simulate the antenna's computational model. The template CST project incorporates ten and six patches and gaps, respectively, which is sufficient for real-world applications. When evaluating the device, the parameter vector x is recalculated into absolute dimensions and the excessive metal parts are trimmed to the substrate. The EM analysis is executed in a batch mode. Using a Visual Basic script, the template project is updated with the current antenna dimensions. The underlying programming environment for the presented framework is Matlab. Communication with CST is arranged using a custom-developed Matlab-CST interface, which also performs post-processing of the results exported upon accomplishing EM simulation. The operating flow of the process has been shown in Fig. 3.

2.3 Antenna Development

The design task considered here is a realization of multi-band antennas to ensure acceptable impedance matching, i.e., maintaining $|S_{11}| \leq -10$ dB over the frequencies. Let $F = [f_1 - B_1/2, f_1 + B_1/2] \cup [f_2 - B_2/2, f_2 + B_2/2] \cup \ldots \cup [f_K - B_K/2, f_K + B_K/2]$, denote target operating bands, where f_k and B_k are the center frequencies and the respective bandwidths. In rigorous terms, the objective is to solve the

$$x^* = \arg\min_{x \in X} U(x) \qquad (1)$$

where the cost function is given as

$$U(x) = \max_{f \in F}\{|S_{11}(x,f)|\} \qquad (2)$$

in which $|S_{11}(x,f)|$ is the modulus of the reflection coefficient at design x and frequency f. The design space is a box-constrained domain with lower/upper bounds imposed on antenna parameters (cf. Fig. 1). Note that mixing discrete and continuous parameters allows us to efficiently control the antenna's architecture and its specific dimensions. It is also possible to impose additional conditions, e.g., concerning the maximum antenna size, requirement for minimum gain, etc.). These scenarios will be considered elsewhere.

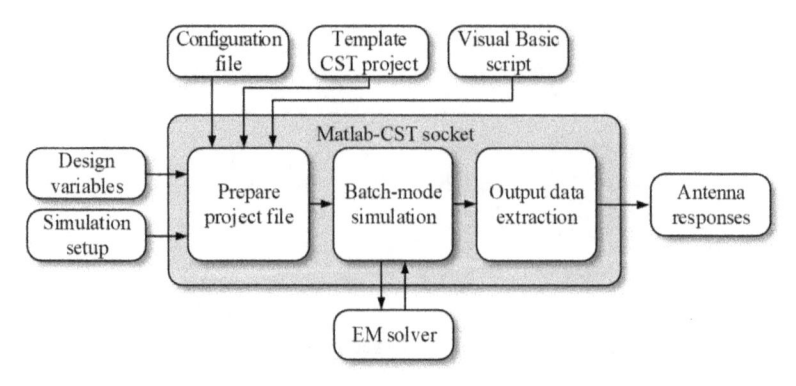

Fig. 3. Antenna's computational model. The EM model template uses decision variables (vector x) and simulation setup to render the project file. Following the analysis, antenna responses are extracted from the EM data.

In algorithmic terms, problem (1) is solved in two stages. The first is a global search process realized using a floating-point evolutionary algorithm incorporating elitism and adaptive mutation rate, which improves the exploitation capability of the procedure when close to convergence, see Fig. 4. At this stage, the antenna topology is decided upon along with a rough adjustment of its dimensions. The second stage is local tuning, which is executed with the trust-region (TR) gradient-based algorithm [27] with antenna response sensitivity computed using finite differentiation (FD) [28]. The algorithm's outline is shown in Fig. 5. At this stage, the antenna's topology has already been established, and only its dimensions are adjusted to boost the device's performance regarding the merit function (2). The TR algorithm produces approximate solutions using predictions from a linear model of antenna outputs (cf. (6)). The problem (6) is solved using the Sequential Quadratic Approximation (SQP) routine [29] available in Matlab [30]. Using FD entails the cost of $n + 1$ EM analyzes per iteration, where n is the overall number of decision variables. These expenses are reduced by eliminating the variables that have minor effects on antenna characteristics upon initial pre-screening. Also, the global search stage employs coarse-discretization EM analysis, which further improves computational efficiency. The accurate (high-fidelity) model is only used for final tuning.

2.4 Design Framework

The operation of the suggested unsupervised antenna design methodology is illustrated in Fig. 6. The process is purely specification-driven and does not involve any human-expert interaction. The input data consists of the intended substrate parameters (thickness, relative permittivity), the size (if not optimized), and, most importantly, the target operating frequency ranges. The number of unit cells, NP and NG, can be decided upon beforehand. The antenna topology evolves during the global search stage. It is further refined through final tuning.

The algorithm's structure is similar to evolutionary procedures (e.g., [26]). Main features:
- Generational model (a new population entirely replaces the previous one);
- Population size $N = 20$;
- Selection scheme: binary tournament;
- Elitism: a single best individual is stored and inserted to the next population;
- Recombination: a mixture of intermediate and arithmetic crossover:
  - Parent individuals: $\mathbf{x} = [x_1 \ldots x_n]^T$ and $\mathbf{y} = [y_1 \ldots y_n]^T$; offspring: $\mathbf{z} = [z_1 \ldots z_n]^T$;
  - Intermediate crossover: $z_i = ax_i + (1-a)y_i$ with $0 \le a \le 1$ (a selected randomly);
  - Arithmetic crossover: $\mathbf{z} = a\mathbf{x} + (1-a)\mathbf{y}$ with $0 \le a \le 1$ (a selected randomly).
  - Crossover probability is set to $p_m = 0.8$;
- Mutation: Random mutation with nonuniform probability distribution. It is applied individually to each parameter vector component so that $x_i \to x_i' + \Delta x_i$, where Δx_i is a random deviation defined as

$$\Delta x_i = \begin{cases} (x_{i\,max} - x_i) \cdot (2(r - 0.5))^\beta & \text{if } r > 0.5 \\ (x_{i\,min} - x_i) \cdot (2(0.5 - r))^\beta & \text{otherwise} \end{cases} \qquad (3)$$

where $r \in [0,1]$ is a random number and $\beta = 3$;
- Algorithm termination: maximum number of $N_i = 100$;
- Adaptive adjustment of mutation rate p_m. Let P_D be a population diversity defined as

$$P_D = \frac{1}{n}\sum_{k=1}^{n} std\left([x_{1,k}\ x_{2,k}\ \ldots\ x_{N,k}]\right) \qquad (4)$$

where $\mathbf{x}^j = [x_{j,1} \ldots x_{j,n}]^T$ is jth member of the population, and $x_{j,k}$ is its kth entry. The mutation rate $p_m^{(i+1)}$ for iteration i of the algorithm is set as follows (initial value $p_m^{(0)} = 0.2$): if $i < N_i/2$ then adjust p_m based on P_D, else $p_m^{(i+1)} = p_m^{(N_i/2)}[2(N_i-1)/N_i]^2$. Adaptation for P_D: if $P_D < P_{Dmin}$ then $p_m^{(i+1)} = p_m^{(i)}m_{incr}$, elseif $P_D > P_{Dmax}$ then $p_m^{(i+1)} = p_m^{(i)}/m_{decr}$.

Remark: Setup: $P_{Dmin} = 0.05$, $P_{Dmax} = 0.1$, and $m_{incr} = 1.3$, $m_{decr} = 1.2$. P_{Dmin} and P_{Dmin} are set considering antenna parameters are relative (i.e., between zero and unity).

Fig. 4. The global search stage of antenna development: evolutionary algorithm [26].

3 Results

This part of the work showcases the performance of the unsupervised design procedure. It is used to develop several antenna structures based on different design specifications concerning the target operating frequency bands. The results are encapsulated in Sects. 3.1 through 3.4, whereas Sect. 3.5 discusses the findings. Furthermore, Sect. 3.6 provides experimental validation of the selected designs.

All designs are realized on FR-4 dielectric substrate ($\varepsilon_r = 4.4$, h = 1.0 mm) or the size $W = 30$ mm and $L = 20$ mm (except Example II, which uses $W = 25$ mm and $L = 15$ mm). The model complexity is $N_P = 5$, $N_G = 3$. The evolutionary algorithm runs for 100 iterations with a population size of 20. The global search stage uses a low-fidelity EM model (simulation time ~20 s). The final tuning is based on the high-fidelity model (simulation time ~60 s).

3.1 Example I

The first case is a single-band antenna operating from 5 GHz to 6 GHz. The structure generated by the suggested framework, the impedance matching characteristic, and the history of the development process can be found in Fig. 7. Note that global search has

Final tuning task:

$$x^* = \arg\min_{x \in X} U(x) \quad (5)$$

TR algorithm: the procedure generates a sequence $x^{(i)}$, $i = 0, 1, 2, \ldots$ as

$$x^{(i+1)} = \arg\min_{x, \|x - x^{(i)}\| \leq d^{(i)}} U_L(x) \quad (6)$$

The merit function U_L is the same as U but with S_{11} evaluated using a linear model:

$$S_{11L}^{(i)}(x, f) = S_{11L}^{(i)}(x^{(i)}, f) + G_{11}(x^{(i)}, f) \cdot (x - x^{(i)}) \quad (7)$$

Here, $G_{11}(x, f)$ is the gradient of $S_{11}(x, f)$ at x and frequency f, computed using FD [28].

Other operating details:
- $x^{(i+1)}$ is accepted if the gain ratio $r = [U(x^{(i+1)}) - U(x^{(i)})]/[U_L(x^{(i+1)}) - U_L(x^{(i)})] > 0$ (i.e., EM-simulated cost function is improved); otherwise, the iteration is repeated with reduced TR size;
- The trust region size $d^{(i)}$ is adaptively adjusted based on r; $d^{(i+1)} = d^{(i)} m_{incr}$ if $r > r_{incr}$, and $d^{(i+1)} = d^{(i)}/m_{decr}$ if $r < r_{decr}$; the control parameters are $r_{incr} = 0.75$, $r_{decr} = 0.25$, $m_{incr} = 1.5$, $m_{decr} = 2$ [27];
- Termination criteria: convergence in argument ($\|x^{(i+1)} - x^{(i)}\| < \varepsilon$) or sufficient reduction of the TR size ($d^{(i)} \leq \varepsilon$); the termination threshold is set to $\varepsilon = 10^{-3}$.

Fig. 5. Final tuning using the TR algorithm.

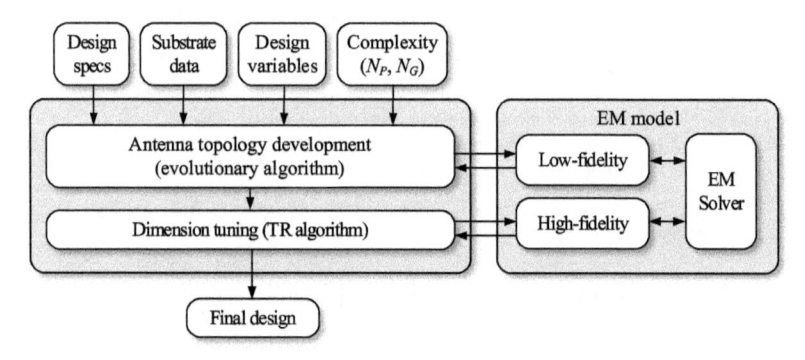

Fig. 6. Proposed design framework: the flow diagram.

produced a satisfactory outcome. It is further improved through local tuning. It should be stressed that the entire procedure is purely specification-based. No human expert is necessary whatsoever.

3.2 Example II

The next validation case is a compact ultra-wideband (UWB) antenna. The target frequency range is from 3.1 GHz to 10.6 GHz. The antenna substrate is diminished to only 15 mm × 25 mm compared to the remaining test cases. The obtained architecture and the history of the development process can be found in Fig. 8.

3.3 Example III

The third case is a dual-band antenna. The target ranges are from 2.4 GHz to 2.5 GHz (lower band) and from 5.2 GHz to 5.4 GHz (upper band). As indicated in Fig. 9, the proposed framework yields a design fulfilling the specifications already at the geometry evolution stage. Local tuning only slightly improves the impedance matching.

3.4 Example IV

The last test case is a triple-band antenna. The target frequency ranges are from 2.4 GHz to 2.5 GHz, 5.2 GHz to 5.4 GHz, 7.5 GHz to 8.0 GHz. Figure 10 shows the antenna structure found by the presented framework and the development history. As observed, the specifications for this challenging example are met for the lower and middle bands. Only a slight violation is observed in the upper band.

3.5 Discussion

The data showcased in Sects. 3.1 through 3.4 underscores the capability of the presented framework to successfully develop antenna structures for diverse performance specifications. The design process is unsupervised, with the only input data being the intended number of operating bands and the target frequency ranges. The topologies produced by the proposed algorithm are highly unconventional yet evolved to adequately utilize all antenna components, which is illustrated using surface current distributions shown in Fig. 11.

It is also important to emphasize that all considered antennas were obtained using identical algorithm setups and at an acceptable computational cost from a practical engineering perspective (about fifteen hours of CPU time).

3.6 Experimental Validation

The antennas developed in Sects. 3.2 and 3.4 (Examples II and IV) were manufactured and experimentally validated for additional demonstration. The results are shown in Figs. 12 and 13. As observed, the alignment between EM simulations and measurements is satisfactory. Minor discrepancies are due to fabrication tolerances and assembly inaccuracies.

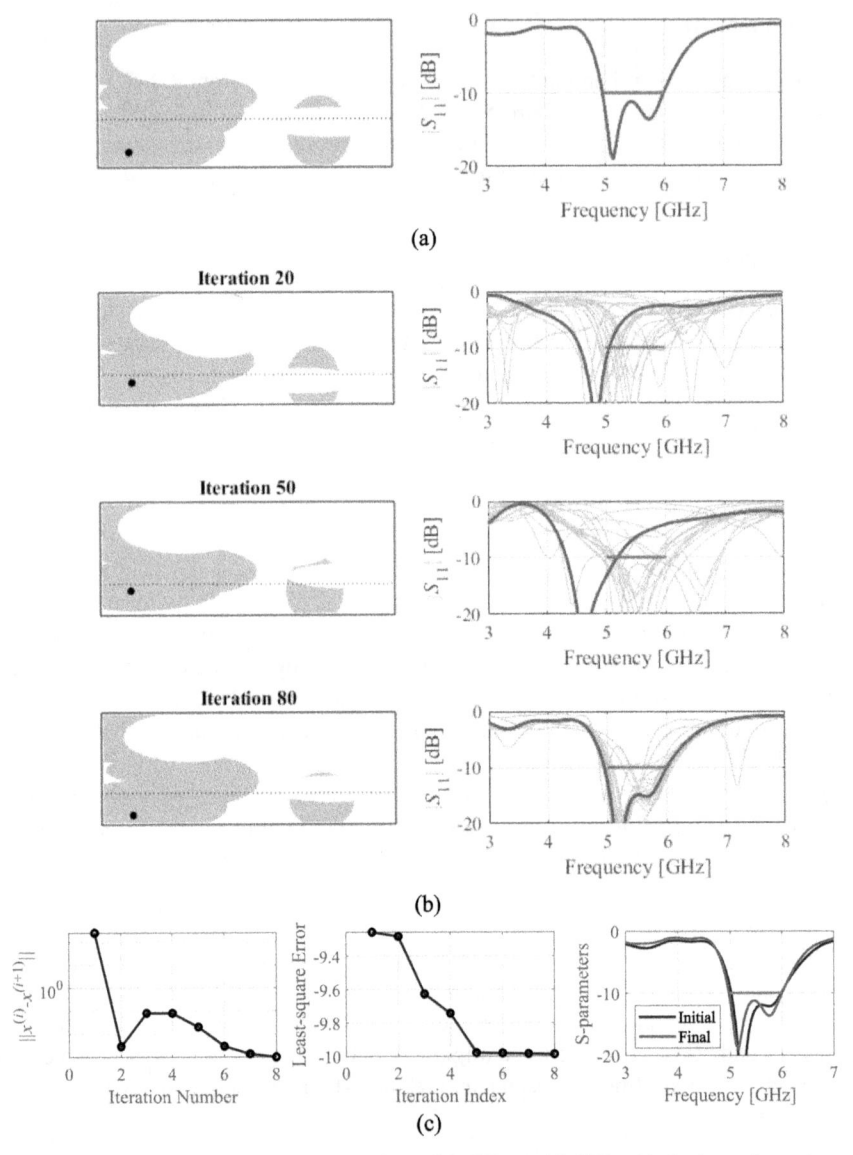

Fig. 7. Example I: target frequency range from 5.0 GHz to 6.0 GHz: (a) final topology (ground plane ⋯⋯) and $|S_{11}|$ characteristic; (b) topology evolution (the best architecture marked using the blue line; antenna outputs at the current population marked using gray lines); (c) local tuning (convergence plot, merit function vs. iteration index, and initial/final reflection response). (Color figure online)

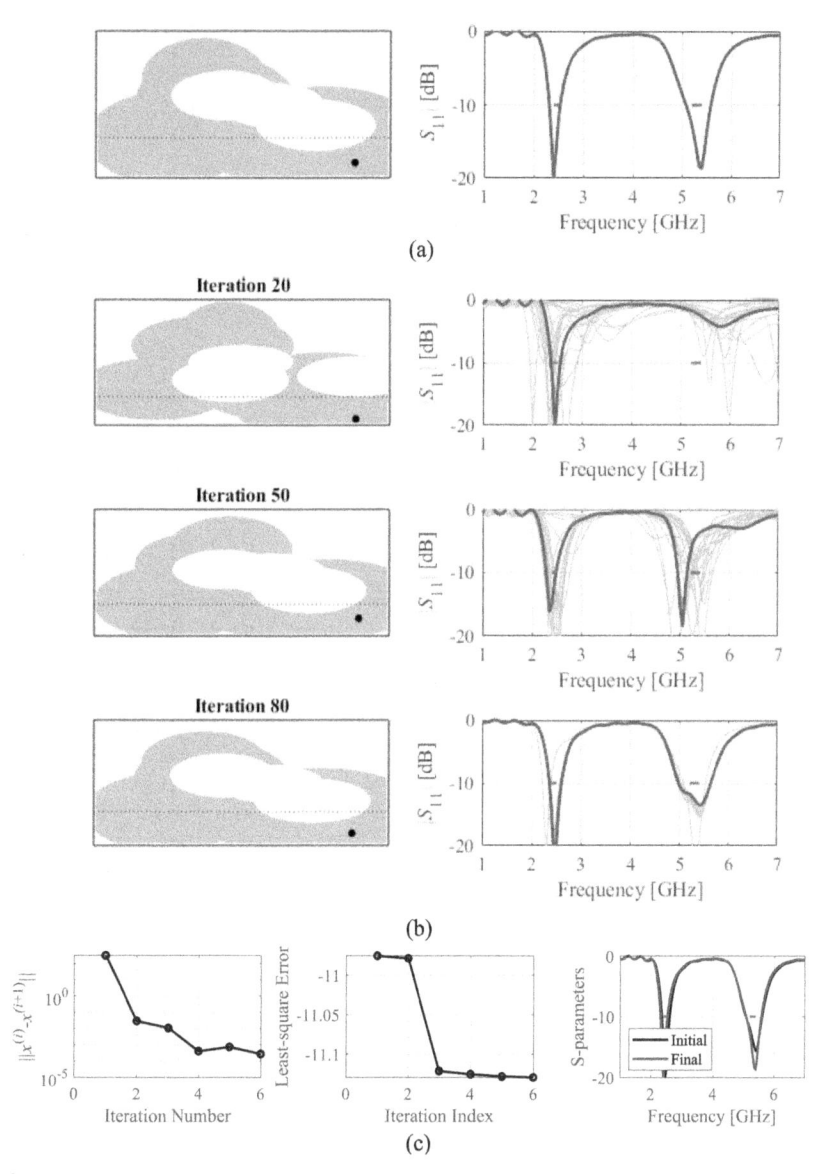

Fig. 8. Example II: an ultra-wideband antenna; target frequency range from 3.1 GHz to 10.6 GHz: (a) final topology (ground plane ····) and $|S_{11}|$ characteristic; (b) topology evolution (the best architecture marked using the blue line; antenna outputs at the current population marked using gray lines); (c) local tuning (convergence plot, merit function vs. iteration index, and initial/final reflection response). (Color figure online)

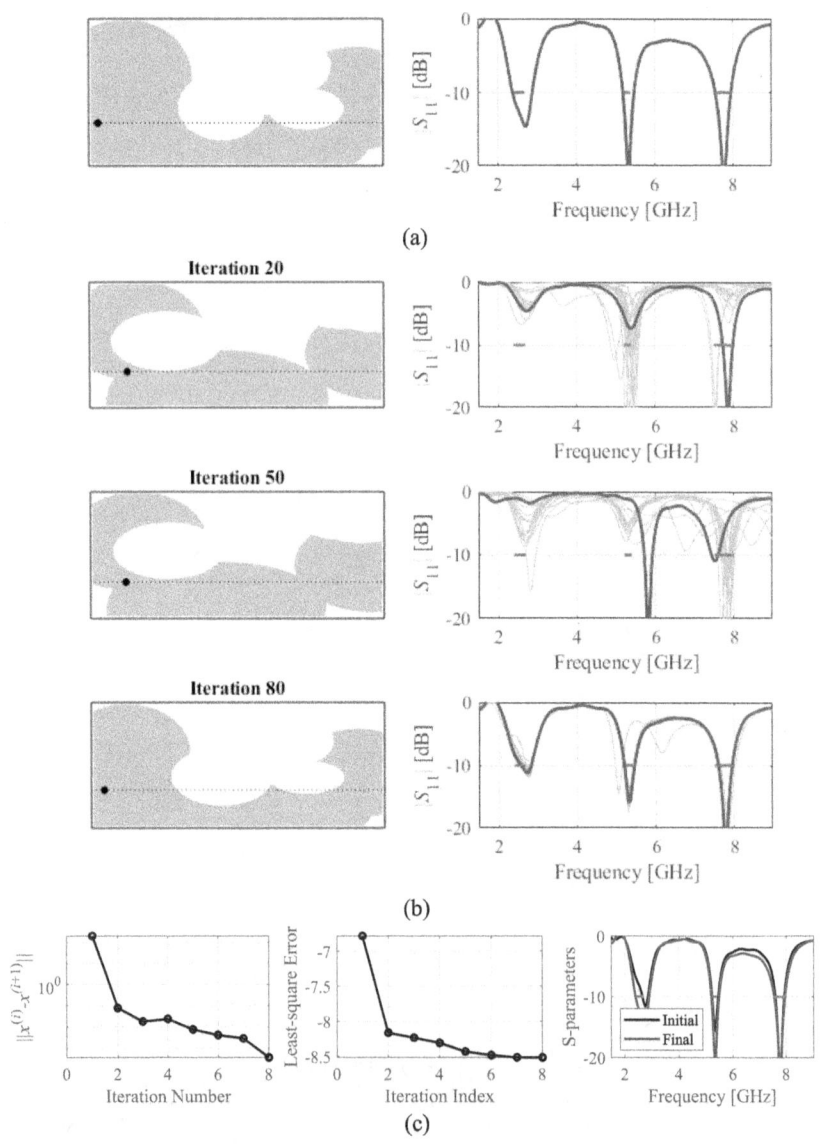

Fig. 9. Example III: target frequency ranges from 2.4 GHz to 2.5 GHz and from 5.2 GHz to 5.4 GHz: (a) final topology (ground plane ····) and $|S_{11}|$ characteristic; (b) topology evolution (the best architecture marked using the blue line; antenna outputs at the current population marked using gray lines); (c) local tuning (convergence plot, merit function vs. iteration index, and initial/final reflection response). (Color figure online)

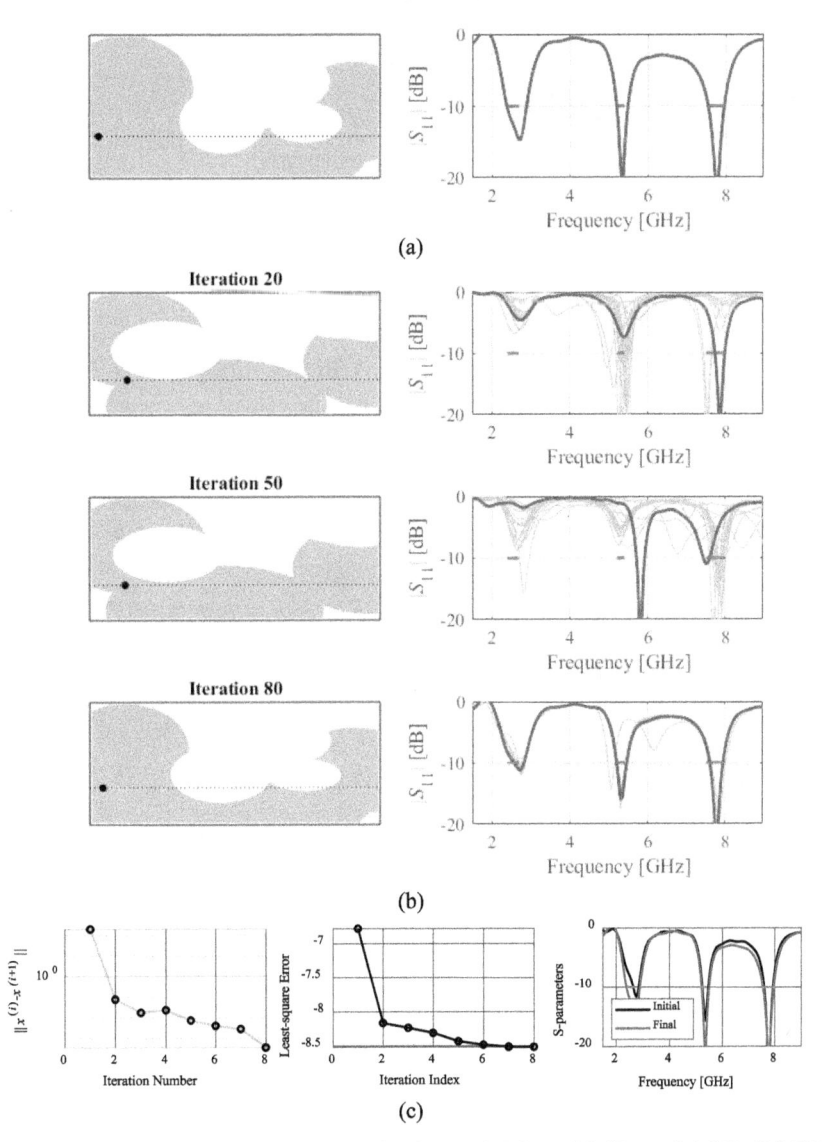

Fig. 10. Example IV: target frequencies from 2.4 GHz to 2.5 GHz, 5.2 GHz to 5.4 GHz, 7.5 GHz to 8.0 GHz: (a) final topology (ground plane ····) and $|S_{11}|$ characteristic; (b) topology evolution (the best architecture marked using the blue line; antenna outputs at the current population marked using gray lines); (c) local tuning (convergence plot, merit function vs. iteration index, and initial/final reflection response). (Color figure online)

4.0 GHz   6.0 GHz   8.0 GHz   10.0 GHz

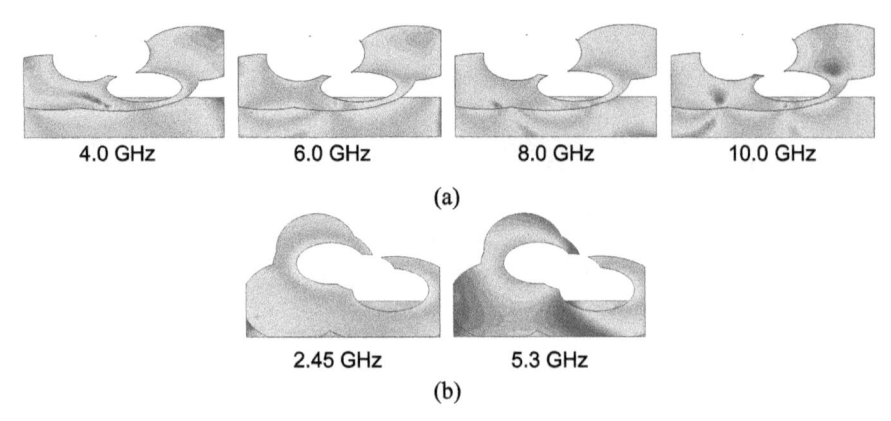

(a)

2.45 GHz   5.3 GHz

(b)

Fig. 11. Surface currents for (a) Example II, (b) Example III. Observe the use of the diverse antenna components at diverse frequencies (increasing current density corresponds to the transition from blue to red through green and yellow color). This demonstrates the importance of each building block employed to assemble the antenna structures. (Color figure online)

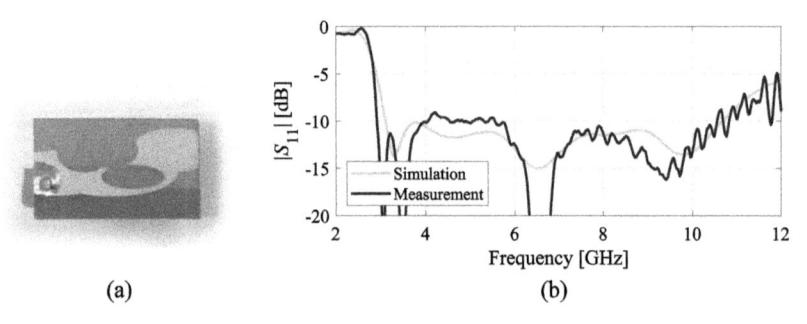

(a)   (b)

Fig. 12. Example II: (a) prototype, (b) simulated and measured $|S_{11}|$.

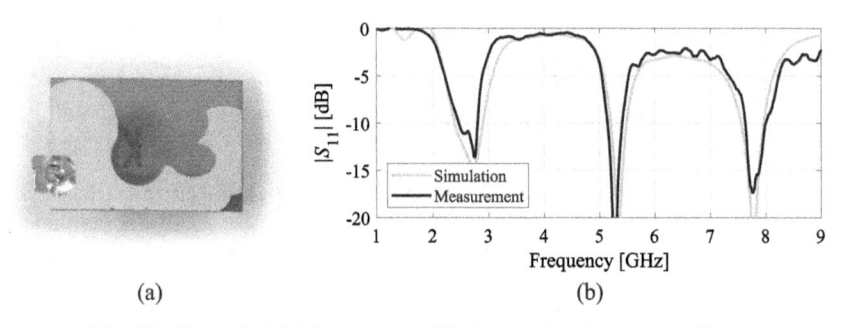

(a)   (b)

Fig. 13. Example IV: (a) prototype, (b) simulated and measured $|S_{11}|$.

4 Conclusion

This work introduces an innovative methodology for an automated design of antennas, which employs computational intelligence and numerical optimization methods. Capitalizing on flexible parameterization and simultaneous adjustment of discrete and continuous parameters determining the antenna architecture and its dimensions, the proposed technique can produce high-quality designs in a purely specification-driven manner. This has been extensively demonstrated using several examples of single-, dual-, triple-, and broadband devices and further corroborated through experimental validation of selected designs. The developed method can be viewed as an attractive approach to the automated development of unconventional antenna structures for demanding applications using reasonable computational resources.

Acknowledgments. The authors thank Dassault Systemes, France, for making CST Microwave Studio available. This work is partially supported by the Icelandic Centre for Research (RANNIS) Grant 239858 and by the National Science Centre of Poland Grant 2020/37/B/ST7/01448.

References

1. Ullah, U., Al-Hasan, M., Koziel, S., Ben, M.I.: Circular polarization diversity implementation for correlation reduction in wideband low-cost multiple-input-multiple-output antenna. IEEE Access **8**, 95585–95593 (2020)
2. Zeng, Y., Qing, X., Chia, M.Y.W.: A wideband circularly polarized antenna with a nonuniform metasurface designed via multiobjective Bayesian optimization. IEEE Ant. Wireless Propag. Lett. **23**, 1739–1743 (2024)
3. Lucchini, F., Torchio, R., Bettini, P., Dughiero, F.: TopIE: an integral equation tool for topology optimization in electromagnetics. IEEE Trans. Ant. Propag. **72**, 683–692 (2024)
4. Koziel, S., Pietrenko-Dabrowska, A.: Global EM-driven optimization of multi-band antennas using knowledge-based inverse response-feature surrogates. Knowl. Based Syst. **227**, 107189 (2021)
5. Tan, J., Shao, Y., Zhang, J., Zhang, J.: Efficient antenna modeling and optimization using multifidelity stacked neural network. IEEE Trans. Ant. Propag. **72**, 4658–4663 (2024)
6. Koziel, S., Pietrenko-Dabrowska, A.: Recent advances in accelerated multi-objective design of high-frequency structures using knowledge-based constrained modeling approach. Knowl. Based Syst. **214**, 106726 (2021)
7. Li, J., Sun, G., Duan, L., Wu, Q.: Multi-objective optimization for UAV swarm-assisted IoT with virtual antenna arrays. IEEE Trans. Mobile Comput. **23**, 4890–4907 (2024)
8. Zhang, Z., Chen, H.C., Cheng, Q.S.: Surrogate-assisted quasi-newton enhanced global optimization of antennas based on a heuristic hypersphere sampling. IEEE Trans. Ant. Propag. **69**, 2993–2998 (2021)
9. He, Y., Huang, J., Li, W., Zhang, L., Wong, S.W., Chen, Z.N.: Hybrid method of artificial neural network and simulated annealing algorithm for optimizing wideband patch antennas. IEEE Trans. Ant. Propag. **72**, 944–949 (2024)
10. Liu, Y., Chen, P., Tian, J., Xiao, J., Noghanian, S., Ye Q.: Hybrid ANN-GA optimization method for minimizing the coupling in MIMO antennas. AEU – Int. J. Electron. Commun. **175**, 155068 (2024)
11. Yasmeen, K., Mishra, K.V., Subramanyam, A.V., Ram, S.S.: Circularly polarized Fabry-Pérot cavity sensing antenna design using generative model. IEEE Sens. Lett. **7**, 1–4 (2023)

12. Pietrenko-Dabrowska, A., Koziel, S.: Response feature technology for high-frequency electronics. Optimization, Modeling, and Design Automation. Springer, New York (2023)
13. Koziel, S., Pietrenko-Dabrowska, A.: Fast EM-driven nature-inspired optimization of antenna input characteristics using response features and variable-resolution simulation models. Sci. Rep. **14**, 10081 (2024)
14. Sarker, N., Podder, P., Mondal, M.R.H., Shafin, S.S., Kamruzzaman, J.: Applications of machine learning and deep learning in antenna design, optimization, and selection: a review. IEEE Access **11**, 103890–103915 (2023)
15. Zhong, Y., Renner, P., Dou, W., Ye, G., Zhu, J., Liu Q.H.: A machine learning generative method for automating antenna design. IEEE Trans. Ant. Propag. **68**, 6858–6866 (2020)
16. Liu, P., Chen, L., Chen, Z.N.: Prior-knowledge-guided deep-learning-enabled synthesis for broadband and large phase shift range metacells in metalens antenna. IEEE Trans. Ant. Propag. **70**, 5024–5034 (2022)
17. Jiang, F., Chiu, C.-Y., Shen, S., Cheng, Q.S., Murch, R.: Pixel antenna optimization using N-port characteristic mode analysis. IEEE Trans. Ant. Propag. **68**, 3336–3347 (2020)
18. Chen, Y.-S., Chiu, Y.-H.: Application of multiobjective topology optimization to miniature ultrawideband antennas with enhanced pulse preservation. IEEE Ant. Wireless Propag. Lett. **15**, 842–845 (2016)
19. Jiang, F., et al.: Pixel antenna optimization based on perturbation sensitivity analysis. IEEE Trans. Ant. Propag. **70**(1), 472–486 (2022)
20. Zheng, W., Li, H.: Designing antennas with quasi-isotropic radiation patterns using pixel structures. IEEE Trans. Ant. Propag. **71**, 7813–7823 (2023)
21. Zhu, S.-H., Yang, X.-S., Wang, J., Wang, B.-Z.: Design of MIMO antenna isolation structure based on a hybrid topology optimization method. IEEE Trans. Ant. Propag. **67**, 6298–6307 (2019)
22. Wang, J., Yang, X.-S., Ding, X., Wang, B.-Z.: Topology optimization of conical-beam antennas exploiting rotational symmetry. IEEE Trans. Ant. Propag. **66**, 2254–2261 (2018)
23. Tucek, J., Capek, M., Jelinek, L., Sigmund, O.: Density-based topology optimization in method of moments: Q-factor minimization. IEEE Trans. Ant. Propag. **71**, 9738–9751 (2023)
24. Wang, L.L., Yang, X.S., Ma, C.J.: An efficient gradient-based hybrid parameter-topology optimization for antenna design. IEEE Trans. Ant. Propag. **71**, 9477–9486 (2023)
25. CST Microwave Studio, ver. 2023, Dassault Systemes, France (2023)
26. Michalewicz, Z.: Genetic Algorithms + Data Structures = Evolution Programs. Springer, New York (1996)
27. Conn, A.R., Gould, N.I.M., Toint, P.L.: Trust Region Methods, MPS-SIAM Series on Optimization (2000)
28. Levy, H., Lessman, F.: Finite Difference Equations. Dover Publications Inc., New York (1992)
29. Nocedal, J., Wright, S.J.: Numerical Optimization, 2nd edn. Springer, New York (2006)
30. Matlab, ver. 2023a, MathWorks Inc., Natick, MA, USA (2023)

Near-Optimal Mixed Partial Replications Versus Uniform Replication

Ralf Vamosi[✉]

Faculty of Computer Science, University of Vienna, Vienna, Austria
ralf.vamosi@cs.univie.ac.at
https://informatik.univie.ac.at

Abstract. Replication involves creating and managing multiple duplicates of objects on various storage devices, workers, or computers. Replication is widely used in RAID (Redundant Array of Independent Disks) systems, database systems, storage clusters, cloud storage networks, and compute clusters, among others. The primary goal of duplicating or replicating data is to enhance its availability, thereby improving fault tolerance and performance. The replication factor (degree of duplication, multiplicity) determines the number of copies or replicas of objects, indicating the level of redundancy and thus giving a guarantee. In many cases, such as multi-disk storage, distributed file systems, and database systems, policies fix the replication factor to a global value such as 3, for example, to balance storage expenses and performance. This research presents a global search using an evolutionary optimization to find a set of partial replications, combining them to fit the objects into the system and further reduce costs.

Keywords: replication · placement · optimization · evolution

1 Introduction

Replication involves generating duplicates or replicas of objects (e.g., files) and managing them in terms of access, updates, and lifetime. Multiple locations or nodes guarantee redundancy and accessibility. Replication and placement are part of data management that improve metrics such as latency, transfer rates, and local load and ensure fault tolerance. *Data tiering* is a placement technique that aims to classify objects into tiers. Hot objects (e.g., data) used or read more frequently are placed in the high tier. High-tier locations or sites provide high bandwidth or faster storage, increasing performance. The different tiers depend on the definition of performance or goodness and so on a metric or measure of some dimension.

Replication defines how often objects are replicated/duplicated/copied (the term depends on the field and is sometimes arbitrary). The *replication factor*/multiplicity/degree indicates the cardinality of objects as replicas/copies/duplicates among various locations such as nodes, servers, caches, nodes etc. A replication factor (Rf) of 1 implies only one primary copy of a particular object(s). This scenario is often described as a single-replica or non-replicated configuration. A replication factor of 2 provides a single-point failsafe, implying that a single node can fail and that the full extension or operation can still be provided. Higher-fold replications are chosen to improve this guarantee or expectation on dimensions such as average latency, bandwidth, throughput, etc. The number of replications required depends on the desired guarantee. It is based on cost-benefit analysis (trade-off) by architects, administrators, or users. Replication always incurs certain costs for storage, communication, and management.

The *replication problem* covers three partial problems:

1. The replication factor or multiplicity - how many copies shall be produced
2. Which subset of objects are subject to replication
3. Where should the objects (subset) be placed or located

This work shows a replication method for 1) and 2) cases in which 'optimal' placement does not play a role due to location independence or lack of information. There are different use cases for that:

– One case may be an entire storage provided as one tier in one device. One tier means equal performance for the entire placement. This can be a logical device such as a file server or a network-attached storage (NAS).
– In another case, no information is available on the nodes (servers, storages, caches) or tiers. This would include, for example, a case with an external, managed storage or with alternating use patterns of objects that hinder the application in improving the placement for cost decrease.

Contribution: This work presents an algorithm that, in principle, generates a near-optimal replication with a superposition of partial replications. This covers tasks 1) and 2). The novel result indicates that optimal replication consists of several replication factors applied to subsets. The optimization model can update the allocation by running on a new set of objects and, hence, combine the previous and current allocation outputs.

Assumptions: Objects to be replicated are i.i.d. Observed values for the cost function (network) are accurate (sufficient). This means that the regime does not change.

2 State-of-the-Art

Replication has been a fundamental aspect since the advent of distributed databases and parallel computing. It enhances availability by distributing resources across multiple endpoints to mitigate the effects of individual endpoints becoming overloaded, unresponsive, or offline. The term 'running hot' refers to the condition where an endpoint is subjected to excessive requests, overwhelming its processing capacity. Replication also improves runtime performance, evidenced by reductions in latency. Many systems use process and data replication

[1,2]. A form of replication is data replication, which manages copies/replicas for distribution to machines or locations. Data management offers techniques like tiering, caching, or replication [1–3]. Publications either focus on practical applications and best operational practices or on theoretical work showcasing optimization benefits. Applied methods include simplifications, local and greedy optimization, and various models such as linear, deterministic, and stochastic.

Numerous studies demonstrate the NP-hard nature of optimally resolving data placement and replication problems. It is evident from the combinatorial growth that replication falls within NP. [4] illustrates this by converting a decision/satisfaction problem in replication to an NP-complete partitioning issue. Additionally, they contend that certain instances of the replication problem are non-approximable, necessitating the use of heuristics or simplifications.

To manage the complexity of replica placement, algorithms often simplify elements of the problem or its solution. Similar to other complex issues, local optimization is examined and applied to replication challenges. Some approaches integrate properties like caches, bandwidth, or input data. In the research by [3], a greedy cooperative cache management strategy is introduced, dividing the input set into N caches, each of which is optimized locally. A different greedy approach focused on cost metrics per object and network costs is introduced in [5]. This method uses a greedy replication algorithm to handle caching. Objects are ranked on the basis of their cost contributions, determining their placement in specific locations.

In [6], a segment of the cache servers' storage is dedicated to holding the targeted content data. The authors suggest enhancing these cache portions to strike a balance between improved network performance and computational expense.

A number of studies incorporate the concept of data popularity as a weighting factor. This popularity is often referred to as (local) probability, weight, or hotness. As noted in [3], popularity can be influenced by the user community. Additionally, a temporal aspect can affect popularity. According to [7], the probability of data utilization is examined over time and location, accounting for varying resources and user requests. This research also investigates an uncertain job dispatching mechanism that impacts the cost function.

For high-cardinality problems, the machine learning method of clustering may be used. Instead of dealing with single objects, similar objects are grouped into clusters that can be handled as input objects [6,7].

Multiple requests to websites, especially to those with high demand, can cause the site to become overwhelmed. Strategically placing duplicates in advance helps to spread the potential load during peak times [1]. The endpoints in consideration can be geographically dispersed, even on a global scale, such as in cloud or grid computing, where replication techniques are frequently used. Duplicating processes allows for parallel processing and enhances fault tolerance, while data replication is essential for ensuring fault resilience and boosting performance [2].

Database systems implement protection and replication strategies for failover and load balancing, ensuring high availability (HA) and fast data access. Data parallelization requires intricate coding within the database management system

(DBMS) because the DBMS must write, read, and update data using multiple table replicas concurrently. [8] Managing and updating replicas necessitates a flexible approach.

Data replication plays a vital role in RAID systems (Redundant Array of Independent Disks), which are utilized in personal computers, servers, and network-attached storage (NAS). The hardware involved may consist of various technologies like disk or solid-state storage. RAID configurations employ data stripping and replication to distribute data segments across multiple persistent storage units. The specific numbers vary based on the RAID level and system, guaranteeing data accessibility and integrity even during storage failures. Upon disk replacement, duplicate data is restored to the new disk to maintain secure and regular operation. [9] RAID continues to see active development, such as integration into Unix kernel modules [10].

Storage clusters must balance network load across nodes to ensure reliable, fast access through redundancy [11]. The Hadoop Distributed File System (HDFS) is intended for large volumes of files and stores files on commodity hardware, offering a large capacity for computations or video content. It is highly fault-tolerant and provides high data throughput due to replication. Files are partitioned into blocks, each replicated multiple times by default with a factor of 3, adjustable at the application level [12].

The vast landscape of grid computing, where data- and compute-intensive applications typically comprise multiple data centers, can be considered from a storage cost and energy cost perspective [13,14].

Some approaches select a fixed number of replicas (replication factor) based on constraints or best practices. These approaches miss the opportunity to vary the number of replications to expand the search space and eventually find a better solution.

3 Network and Cost

A network has several costs depending on the resources and user, such as CPU wall time, energy per request, communication cost, transfer time, etc. A specific communication cost is chosen to showcase the optimization algorithm, which is the latency of files that affects the delays. Such delays are central to many applications and perceived by users, especially for data-intensive computing or content networks.

Several network architectures exist depending on their use and extension, such as local, wide, and global networks. The network of content providers or grids may be hierarchical. Content providers have autonomous systems (AS) connected to internet service providers (ISPs) and add several caching servers to provide the (replicated) content. In grids, there may be some high-tier sites with connected low-tier nodes (nucleus) each.

An example network is depicted in Fig. 1. The network may be different in shape, which must be considered in the cost function.

This distributed network can be modeled as a graph in Fig. 2. $client_x$ and $node_x$ represent nodes nearby at location x. This locality implies the shortest

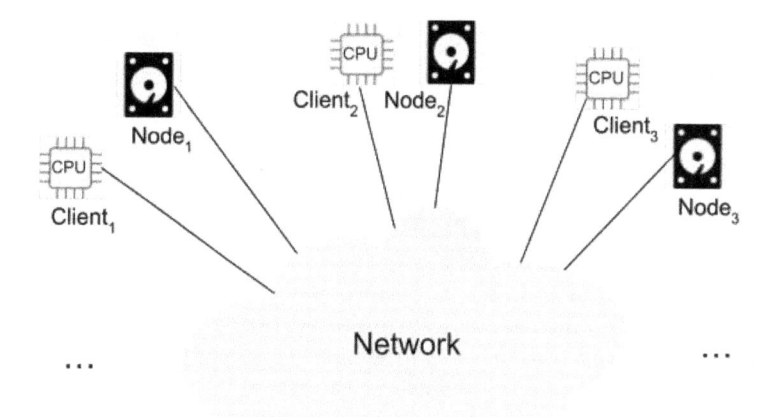

Fig. 1. Storage network connected to a network. Clients request objects from the storage nodes.

path between them modeled as one hop. Clients trigger actions on objects. For all M = 8 nodes, storage is defined by capacity. The capacity values for those logical bins $bin_1, ..., bin_M$, $S_1, ..., S_M$, are randomly chosen in the interval $[0.2, 1]$ each. The capacity values are then normalized to provide the necessary capacity for all replicated objects. Only relative values and not absolute values are interesting.

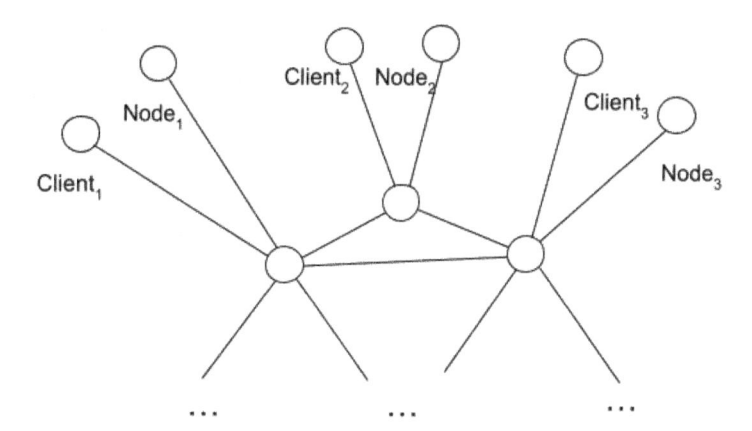

Fig. 2. Storage network connected to a dense network. Clients request objects from the storage nodes.

The core network is densely connected, so some inner nodes are interconnected, providing the communication network as would be for a *wide area network (WAN)*. Clouds, grids, communication networks, and content networks are

WANs that are part of the Internet. The experiment shall impose a variation of architectures on the optimization model. A link between two inner nodes is established with probability 0.5 (50%), which decides whether a link exists. The inner nodes are routers/ core switches. The transfer rate depends on the transfer speeds of the involved nodes, so the hops the communication path extends.

The **use patterns** of objects affect the cost. Since read operations (retrieving a page, playing a video file, reading measurement data, etc.) are much more likely than write operations, the average ratio is set to 3:1 between read and writes, reflected as a factor of 3 in the probabilities. Read operations request object copies (pull), which have been improved due to replication. Let $\boldsymbol{P}^r = [P_1^r, P_2^r, \ldots, P_N^r]^T$ be the read probabilities of the object. T stands for transposed. Write operations update all the replicas of the related object, which results in a higher effort (cost). $\boldsymbol{P}^w = [P_1^w, P_2^w, \ldots, P_N^w]^T$ the write probabilities for objects with the same index. The probabilities express the demands for objects to be fetched or updated. The values are randomly sampled from a Zipf distribution. Zipf means that the probability of observed frequency is inversely proportional to a top-to-bottom order. The distribution is observed in many applications. Some items are frequently in demand, while the majority are rarely used. Since there is no location-dependent information on the clients, the cost function incorporates only a random client selection (uninformed).

A *replicated placement* of objects is defined by the parameter $\boldsymbol{A} = [a_{mn}]$. $a_{mn} = 1$ if and only if a replica of the object n is stored at m, 0 otherwise. The index n always points to the object (prototype) from which replicas are placed. The index m points to the location (endpoint, node, storage, etc.) that is virtually considered bin as depicted in Fig. 3 and Fig. 4. A cost is generally defined based on the system and the intended dimension on which to act and depends on $\boldsymbol{A}$. The total cost related to the selected replication, $c = c(\boldsymbol{A})$, factorizes into N terms, c_n, per object (prototype) n:

$$
\begin{aligned}
c(\boldsymbol{A}) &= \frac{1}{N} \sum_{n=1,\ldots,N} c_n(\boldsymbol{A}) \\
&= \frac{1}{N} \sum_{n=1,\ldots,N} (P_n^r t_{mn}^r(\boldsymbol{A}) + P_n^w t_{mn}^w(\boldsymbol{A}))|_{m=client()}
\end{aligned}
\tag{1}
$$

$$
\boldsymbol{A} = \begin{bmatrix} \boldsymbol{a}_1 \\ \boldsymbol{a}_2 \\ \vdots \\ \boldsymbol{a}_M \end{bmatrix} = \begin{bmatrix} a_{11} & a_{12} & \cdots & a_{1N} \\ a_{21} & a_{22} & \cdots & a_{2N} \\ \vdots & \vdots & \ddots & \vdots \\ a_{M1} & a_{M2} & \cdots & a_{MN} \end{bmatrix} \quad a_{ij} \in \{0,1\},
$$

in which $t_{mn}^r(\boldsymbol{A})$ is a composed cost function for read operations from client m to object n, and $t_{mn}^w(\boldsymbol{A})$ is a composed cost function for write operations analog as before, and $client : \Omega \to \{1, 2, \ldots, M\}$, with $\mathbb{P}(client = m) = \frac{1}{M}$ $\forall m \in \{1, 2, \ldots, M\}$ selects a random client. Before evaluation, M^2 minimum routes between M clients and M endpoints for t_{mn}^r and t_{mn}^w must be stored (number of hops as a predictor distance and thereby delay time).

$t_{mn}^r(\boldsymbol{A})$ seeks the replicas of the object n and outputs an approximate read latency. Since the bandwidth can be shared, the replication factor/multiplicity changes the weighing with the inverse of the replication number $1/|n|$. For each replica, the shortest route is selected, equal to a route with the minimal number of hops, and the transfer t is calculated by $t = s_n \cdot c_n^r \cdot |hops|/|n|$ where s_n is the size of the object n transferred, $|n|$ is the replication factor/multiplicity of n, c_n^r is a category-dependent cost multiplier, and $|hops|$ is number of hops between the source and the destination. Each category is given a random cost multiplier in $[.25, 1]$ at the beginning of each evaluation. Such categories may be based on the use profiles of user groups (personas) or resource dependency of the object. The calculated t's for all replicas of n are then averaged, obtaining the cost $t_{mn}^r(\boldsymbol{A})$.

$t_{mn}^w(\boldsymbol{A})$ seeks the longest of the shortest paths from the client m to all replicas of n and outputs an estimated update time. This is an approximated worst-case for updating replicas of n in a protocol with multi-peer communication. With the one selected path, the final update time is calculated as $t_{mn}^w = s_n \cdot c_n^w \cdot |hops|$ with the variables as before, but with c_n^w as a category-dependent cost multiplier for updates of n. This is also randomly selected for each category in $[.25, 1]$.

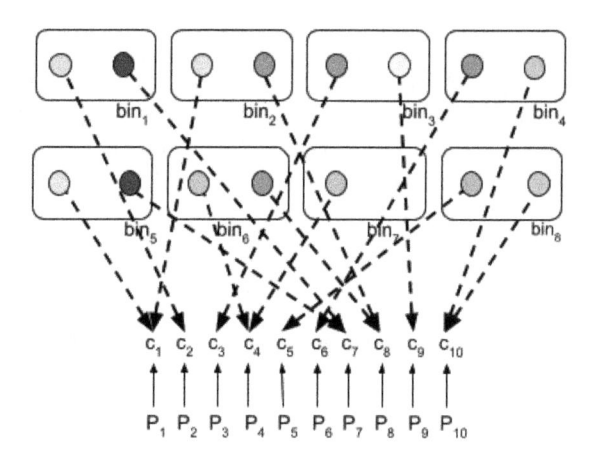

Fig. 3. Single cost values c_n dependent on topology, replication, and object properties.

4 Optimization

Replication is a selection problem with particular boundaries. Depending on the use case, an algorithm must generate sets of replicas and place them at locations such as sites or machines. If M is the number of bins (subsets), there are maximal M-fold replicated objects. For the replication factors $\boldsymbol{Rf} = [Rf_1, Rf_2+2, \ldots, M]$, the same number of replication subsets correspond. If all possible replication factors are computationally feasible, then $\boldsymbol{Rf} = [1, 2, \ldots, M]$ for M bins. A case with eight bins and ten different objects is shown in Fig. 4. With these variables, the optimization task is the following:

$$\underset{\hat{\boldsymbol{A}}}{\text{argmin }} c(\boldsymbol{A}_{t=1})$$

$$\boldsymbol{Rf} = [Rf_1, Rf_1+1, \ldots, Rf_{max}] \quad \mathbb{N} \ni Rf_{max} \leq M > 1$$
$$\mathbb{N} \ni Rf_1 \geq 1$$

$$\boldsymbol{A}_{t=1} = place(\hat{\boldsymbol{A}}) \quad \hat{\boldsymbol{A}} = \begin{bmatrix} Rf_1 \cdot \boldsymbol{a}_1 \\ (Rf_1+1) \cdot \boldsymbol{a}_2 \\ \vdots \\ Rf_{max} \cdot \boldsymbol{a}_{Rf_{max}} \end{bmatrix} \quad (2)$$

s.t.

$$\langle \boldsymbol{A}_{t=1}, \boldsymbol{s} \rangle \leq \boldsymbol{S}$$

where $\hat{\boldsymbol{A}}$ are the partial replication sets $\boldsymbol{a}_1, \boldsymbol{a}_2, \ldots, \boldsymbol{a}_M$ of the objects denoted as a matrix, $\boldsymbol{A}_{t=1}$ the replicated objects placed to the bins, and the cost $c(\cdot)$ has been defined before. $\hat{\boldsymbol{A}}$ is not an allocation matrix as was introduced with Boolean entries since it also indicates the multiplicity/replication number. $\boldsymbol{s}$ is the sizes of the objects in vector form, $\boldsymbol{S}$ is the size vector of the bins.

In the dynamic case, objects are added (altered) to the system (others are removed). The number of nodes is constant, and the optimization considers a steady state in which the values of the system (properties of the system) do not change. Otherwise, the mapping of the bins and the cost function must be changed.

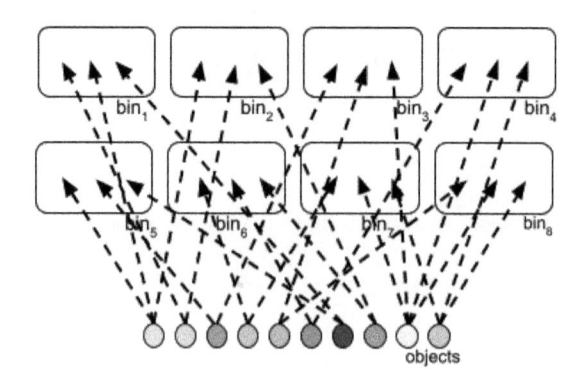

Fig. 4. Replication of objects into bins.

The optimization method presented separates the replication and placement steps for computational feasibility. The optimization model needs to perform the following steps:

1. The model selects feasible replication factors, $\boldsymbol{Rf} = [Rf_1, Rf_1 + 1, \ldots, Rf_{max}]$. The first replication factor, Rf_1, depends on the guarantee and may be one for persistence without fail safety or 2 with single-point fail safety. The maximum in $\boldsymbol{Rf}$ depends on the total capacity relative to the object sizes and the number of bins (subsets). An object in a bin has no cardinality other than 0 or 1. The search for subsets must occur within a sufficient but feasible range $\boldsymbol{Rf}$ to approach the global optimum.
2. The model selects partial subsets $[\boldsymbol{a}_1, \boldsymbol{a}_2, \ldots, \boldsymbol{a}_M]$, corresponding to the numbers in $\boldsymbol{Rf}$. Each $\boldsymbol{a}_m$ has length N (objects) and indicates a selected object n with 1 at position n, otherwise 0. These vectors are Boolean and are disjoint from a set perspective. If an object has been selected for a replication factor, it cannot be selected again. A subset can be varied in size as long as the capacity allows, as was denoted in Eq. 2. The subsets $\boldsymbol{a}_1 + \boldsymbol{a}_2 + \cdots + \boldsymbol{a}_M$ are further related by the size condition imposed on them. The capacity occupied by a $\boldsymbol{Rf}$ is the weighted sum $Rf_1 \langle \boldsymbol{s}, \boldsymbol{a}_1 \rangle + \cdots + (Rf_1 + M) \langle \boldsymbol{s}, \boldsymbol{a}_M \rangle$ where $\langle \cdot, \cdot \rangle$ is the inner product.
3. After sets of replicas/copies are generated, placement is necessary in distributed systems. $place(\cdot)$ is defined as hash binning to be applied. Objects are assigned to bins (storage locations) to obtain the $\boldsymbol{A}_{t=1}$. Hash binning is a weighted placement: Objects are hashed into an extensive range of virtual bins mapped into the M bins in proportional numbers. If the capacities of all bins were equal, the same number of virtual bins would be assigned per bin. If there is no prediction on the client side reading and updating the objects, then this process cannot be improved. In a uniform system, where transactions are uniformly distributed (load balancing); just the replication factor determines the performance/cost. With a feasible placement/final allocation $\boldsymbol{A}_{t=1}$ (capacities $\boldsymbol{S}$), a cost value can be calculated to measure the goodness (fitness in an evolutionary context).

5 Method

$\boldsymbol{Rf}$ determines the number of subsets in the optimization and, therefore, the complexity of the search. Simplifications may be necessary depending on the length of $\boldsymbol{Rf}$. The complexity of the search increases exponentially with an increase in length of $\boldsymbol{Rf}$. This work leaves the optimal range open to discussion. There must be a balance of computational cost and accuracy. For the search of subsets to replication factors, an Allocation Optimization Model has been developed. Within this model, solution candidates are in-memory objects that use clustering for subset generation. There are $\boldsymbol{a}_1, \boldsymbol{a}_2, \ldots, \boldsymbol{a}_M$ subsets for which cluster methods are used. $\boldsymbol{a}_m$ can be further decomposed into subclusters $\boldsymbol{a}'_m, \boldsymbol{a}''_m, \ldots, \boldsymbol{a}^{max_m}_m$ whereas it always starts with one basic cluster $\boldsymbol{a}'_m$. A dynamic max_m increases the computational effort to vary the output subset $\boldsymbol{a}_m$. So, the allocation model should increase max_k of a random k and trial and error to improve the outcome. In the same way, it can decrease this number to make the search simpler. This is a subordinate random walk in the parameter

max_k in the evolutionary process. The decomposed subsets are generated by clustering methods illustrated in Fig. 5. These clustering methods are part of the allocation model of the solution candidate. The allocation model has controlled clustering based on a nearest-neighbor search. The control in the clustering means that the clustering parameters can be modified in different trajectories in the evolutionary process:

- The center of the cluster
- The size of the cluster can be varied within the size constraint under Eq. 2 in which replication factors affect the output size of replicated subsets.

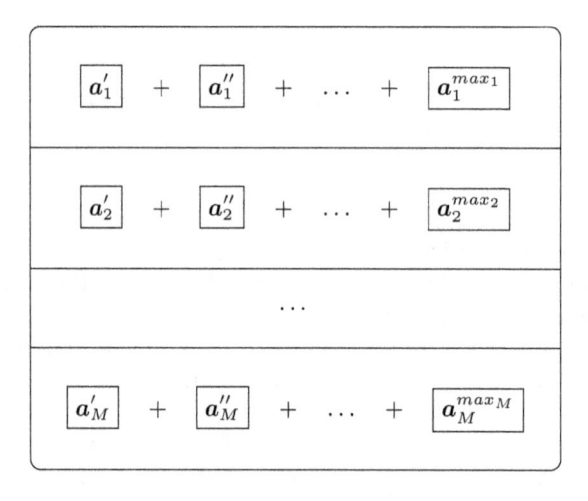

$$\boxed{a'_1} + \boxed{a''_1} + \ldots + \boxed{a_1^{max_1}}$$

$$\boxed{a'_2} + \boxed{a''_2} + \ldots + \boxed{a_2^{max_2}}$$

$$\ldots$$

$$\boxed{a'_M} + \boxed{a''_M} + \ldots + \boxed{a_M^{max_M}}$$

Fig. 5. The allocation model outputs subsets $a_1, a_2, \ldots, a_M$ that are composed of subsets from clusterings.

For improving the solution to the nonconvex problem, the Allocation Optimization Model is based on an evolutionary algorithm. In an evolutionary context, multiple solution candidates live in a pool of candidates that communicate and compete with each other. The evolution contains candidate solutions. Evolutionary genetic techniques change how variation evolves. Evolution accelerates the convergence to an optimal solution with different operators, such as selection and propagation. The parameters are modified iteratively and shared among solution candidates. Candidates for a better solution are more likely to persist (chance). The solution candidates run in parallel (processes), and any solution may take the lead. The solution candidates in the evolution pool/population are simplified in Fig. 6. The colored boxes represent the collected subsets.

Genetic/evolutionary operators must be adapted to the solution parameters to provide the full potential of the meta-optimization. The set of operators is at least the following:

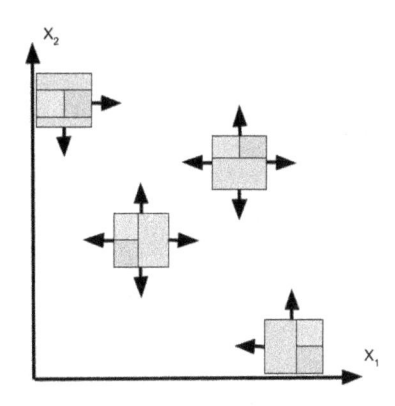

Fig. 6. Solution candidates 'walk' in their parameter space and probe the replication subsets (in colors).

- The selection operator selects candidate solutions from the population based on their fitness to pick winners and losers. Fitness is the inverse cost.
- Crossover or recombination combines parameters from two solutions to produce an offspring. A child z by single-point crossings x and y is represented as $z = (x_0, x_1, \ldots, x_{p-1}, y_p, y_{p+1}, \ldots, y_{n-1})$
  A child z from a 2-point crossover between x and y is represented as: $z = (x_0, x_1, \ldots, x_{p-1}, y_p, y_{p+1}, \ldots, y_q, x_{q+1}, x_{q+2}, \ldots, x_{n-1})$ The exchanged parameters are the subsets of the allocation $a_1, a_2, \ldots, a_M$ with their individual clusterings as illustrated in Fig. 5.
- Mutation or variation randomly changes the parameters to match the changing parameters to introduce diversity into the pool (random walk).
- Discard or annihilation selects candidate solutions to discard.

6 Justification

A validation has been carried out with the global optimization as discussed. All properties of the network model and objects are randomly sampled for each experiment run, as discussed in Sect. 3. The number of bins are fixed to 8. There are $N = 100'000$ objects that represent files in the considered showcase and, for which the following data are given, $\boldsymbol{X} = [\boldsymbol{s}, [\boldsymbol{k}]_{0/1}, \boldsymbol{P}^r, \boldsymbol{P}^w]$ with the columns size $\boldsymbol{s}$, columns category in Boolean encoding $[\boldsymbol{k}]_{0/1}$, read and write probabilities columns $\boldsymbol{P}^r, \boldsymbol{P}^w$ per object. Nominal features of the category are represented in 1-hot encoding.

The observed relative gains of the repetitions of the experiments are averaged. Each experiment consists of a run for each total capacity value as multipliers such as 1x, 2x, etc. ticked on the x-axis in Fig. 7.

- The *complexity of the calibration* of the model is very high due to the number of iterations needed to probe the points in the search space. For each opti-

mization run, one million iterations and more are executed, but this number by far does not exhaust the search.
- With the calibration of the model, the *static case* is done.
- *Dynamic* means that objects are passed to the model that have been unseen before. Usually, new objects are inbound from a stream of data from users, computations, or measurements. The Allocation Optimization Model is aware of a previous placement, $\boldsymbol{A}_{t=0}$. An update, $\boldsymbol{A}_{t=1}$, must be complementary to the previous subsets defined with a prior $\boldsymbol{A}_{t=0}$. With this in mind, the final placement is $\boldsymbol{A}_{t=0} + \boldsymbol{A}_{t=1}$. This term is plugged into Eq. 2. This leads to argmin of the cost for the prior and update placements

$$\underset{\hat{\boldsymbol{A}}}{\operatorname{argmin}}\ c(\boldsymbol{A}_{t=0} + \boldsymbol{A}_{t=1}).$$

such that $\langle \boldsymbol{A}_{t=0} + \boldsymbol{A}_{t=1}, \boldsymbol{s} \rangle \leq \boldsymbol{S}$ whereas $\boldsymbol{S}$ must be updated for the additional allocation. The control variable $\boldsymbol{Rf}$ affects only $\boldsymbol{A}_{t=1} = \boldsymbol{A}_{t=1}(\boldsymbol{Rf})$.

The first step of the optimization model is to read a batch of 5, 6, 7, 8, 9, and 10% of input objects (calibration set). Small batch inputs are desirable to achieve the number of optimization iterations or evolution steps. The entire data population is used for validation in the static replication case. The best optimization run among these batches is selected. Figure 7 shows the yield versus the benchmark case with a uniform global replication factor (1) measured by the defined cost function. There are three cases in total:

1. Uniform global replication is the benchmark with one replication factor at a time, $\boldsymbol{Rf} = [Rf']$ for $Rf' = 1, 2, \ldots, 8$ as depicted on the x-axis in Fig. 7. The replication factor Rf' defines the total capacity in terms of the order of magnitude of the size of the input objects for all cases (1.), (2.), and (3.). The cost values are normalized to 1 (0%).
2. The search includes all possible replication factors for the M = 8 bins, $\boldsymbol{Rf} = [1, 2, 3, 4, 5, 6, 7, 8]$. The **static case** covers all objects at once. The entire set is the validation set. The **dynamic case** means that the capacity is extended by the size of the added objects (50% of objects in the first static run and 50% of objects in the dynamic run). The proportion of bin capacities is conserved and could be changed due to the adaptability of $place(\cdot)$. The subsets are searched using the capacity available. Other parameters could be changed in the Allocation Optimization Model for update runs. It can be seen that the yield of the dynamic case drops a bit. This is likely because global optimization in a static case becomes two iterative local optimizations that combine the outputs. On the other hand, 50% implies that the convergence is faster, which can be investigated in more depth with feasibility and the definition of $\boldsymbol{Rf}$ in another study.

The minimum and maximum cases, 1x and 8x, cannot be improved in contrast to a uniform replication: 1x is the edge case where one and only one replica/copy exists per object (primary copy), and no additional capacity is left. In the introduced notation, the variation would be searched for in $\boldsymbol{Rf} = [1]$ and

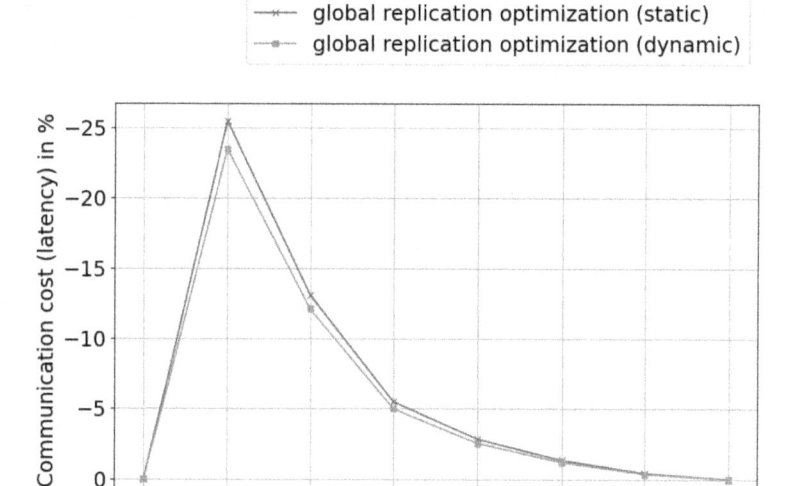

Fig. 7. Replication methods versus uniform, global replication.

nothing can be done. With 8x, all objects are maximally replicated; in this case, the search would occur in $\boldsymbol{Rf} = [1, 2, 3, 4, 5, 6, 7, 8]$ but no further improvement can be achieved. The global optimization method outperforms uniform replication, especially with replication factors RF $= 2, 3$. The relative improvement at low replication factors is higher. At high replication factors (with higher total capacity), the effect of an optimization is weakened asymptotically toward 0%.

7 Conclusion and Future Work

The findings imply that uniform global replication, for example, with replication factor 3, can be improved without including placement in optimization. This is valuable when placement dependency does not play a role or cannot be predicted. Spatial load may be balanced or change continuously.

Not only can a minimal replication factor Rf_1 be defined with the intended range of replication factors, but a search for replication factors in relation to storage or transfer cost can also be analyzed (not covered in this paper). The minimal replication factor of objects in the system should follow a policy according to a service-level agreement (SLA). The maximal replication factor can be varied to observe the gain over a cost function. This cost function can consider the storage cost/management cost, or may be balanced versus such cost (vector optimization).

The novel algorithm adapts to the provided capacity (multiple of the input size of the objects) and exhibits potential for broad application in distributed systems using static (one-time) and dynamic replication (objects are added or changed). Clustering and evolution work in a broad field of data cases. The one shown is just an example. Non-continuous cost functions are feasible for evolutionary algorithms as well.

The demonstrated optimization model can be plugged into a replication system with a fixed global replication, which lacks near-optimal replication to improve the system. The range of replication factors can be set. Prominent storage systems like Hadoop and Ceph have a default global replication factor (number of copies, e.g., 3), and some of them provide a configuration-based replication configuration per keyspace or bucket, or object-wise [15–19].

In the future, the heuristic may be investigated in more detail to introduce an improvement to guide the coordination of the partial replication sets. Furthermore, the problem has affine properties, and some parameters of the replication subsets may be approximated with an affine model (in a continuous space instead of a discrete one). From such a model, there may be insight for the overall optimization.

References

1. Baentsch, M., Baum, L., Molter, G., Rothkugel, S., Sturm, P.: Enhancing the Web's infrastructure: from caching to replication. IEEE Internet Comput. **1**(2), 18–27 (1997)
2. Dabrowski, C.: Reliability in grid computing systems. Concurr. Comput. Pract. Exp. **21**(8), 927–959 (2009)
3. Borst, S., Gupta, V., Walid, A.: Distributed caching algorithms for content distribution networks. In: 2010 Proceedings IEEE INFOCOM, San Diego, CA, USA, pp. 1–9. IEEE (2010)
4. Čibej, U., Slivnik, B., Robič, B.: The complexity of static data replication in data grids. Parallel Comput. **31**(8–9), 900–912 (2005)
5. Kangasharju, J., Roberts, J., Ross, K.W.: Object replication strategies in content distribution networks. Comput. Commun. **25**(4), 376–383 (2002)
6. Li, Y., Xie, H., Wen, Y., Zhang, Z.-L.: Coordinating in-network caching in content-centric networks: model and analysis. In: 2013 IEEE 33rd International Conference on Distributed Computing Systems, Philadelphia, PA, USA, July 2013, pp. 62–72. IEEE (2013)
7. Vamosi, R., Schikuta, E.: Dynamic data replication for short time-to-completion in a data grid. In: Mikyška, J., et al. (eds.) Computational Science – ICCS 2023. LNCS, vol. 10475, pp. 668–675. Springer, Cham (2023)
8. Singh, T., Sandhu, P.S., Singh Bhatti, H.: Replication of data in database systems for backup and failover – an overview. Int. J. Comput. Commun. Eng., 535–538 (2013)
9. Bhargava, B., Riedl, J.: Implementation of RAID. In: Proceedings [1988] Seventh Symposium on Reliable Distributed Systems, Columbus, OH, USA, pp. 157–166. IEEE Computer Press, Society (1988)
10. Li, J., Gerofi, B., Trahay, F., Cai, Z., Liao, J.: Rep-RAID: an integrated approach to optimizing data replication and garbage collection in RAID-enabled SSDs. In: Proceedings of the 24th ACM SIGPLAN/SIGBED International Conference on Languages, Compilers, and Tools for Embedded Systems, Orlando FL USA, June 2023, pp. 99–110. ACM (2023)

11. Leslie, M., Davies, J., Huffman, T.: A comparison of replication strategies for reliable decentralised storage. J. Netw. **1**(6) (2006)
12. Borthakur, D.: The Hadoop Distributed File System: Architecture and Design (2007)
13. Boru, D., Kliazovich, D., Granelli, F., Bouvry, P., Zomaya, A.: Energy-efficient data replication in cloud computing datacenters (2013)
14. Kim, J., Rotem, D.: Using Replication for Energy Conservation in RAID Systems
15. Shvachko, K., Kuang, H., Radia, S., Chansler, R.: The hadoop distributed file system. In: 2010 IEEE 26th Symposium on Mass Storage Systems and Technologies (MSST), pp. 1–10. IEEE (2010)
16. Ghemawat, S., Gobioff, H., Leung, S.-T.: The google file system. In: Proceedings of the Nineteenth ACM Symposium on Operating Systems Principles, pp. 29–43 (2003)
17. Weil, S., Brandt, S.A., Miller, E.L., Long, D.D.E., Maltzahn, C.: Ceph: a scalable, high-performance distributed file system. In: Proceedings of the 7th Conference on Operating Systems Design and Implementation (OSDI 2006), pp. 307–320 (2006)
18. DeCandia, G., et al.: Dynamo: Amazon's highly available key-value store. ACM SIGOPS Oper. Syst. Rev. **41**(6), 205–220 (2007)
19. Lakshman, A., Malik, P.: Cassandra: a decentralized structured storage system. ACM SIGOPS Oper. Syst. Rev. **44**(2), 35–40 (2010)

Hybrid Subgradient and Simulated Annealing Method for Hemivariational Inequalities

Piotr Bartman-Szwarc[1,2](✉), Adil M. Bagirov[3], and Anna Ochal[1]

[1] Chair of Optimization and Control, Jagiellonian University, Krakow, Poland
piotr.bartman@uj.edu.pl
[2] Doctoral School of Exact and Natural Sciences, Jagiellonian University, Krakow, Poland
[3] Centre for Smart Analytics, Institute of Innovation, Science and Sustainability, Federation University Australia, Ballarat, VIC, Australia

Abstract. In this paper, we employ a global aggregate subgradient method for the numerical solution of hemivariational inequality problems arising in contact mechanics. The method integrates a global search procedure to identify starting points for a local minimization algorithm. The algorithm consists of two types of steps: null steps and serious steps. In each null step, only two subgradients are utilized: the aggregate subgradient and the subgradient computed at the current iteration point, which together determine the search direction. Furthermore, we compare the performance of the proposed method with selected solvers using a representative contact mechanics problem as a case study.

Keywords: Hemivariational Inequalities · Nonsmooth Optimization · Subgradient Method

1 Introduction

Hemivariational inequalities are generalizations of variational inequalities. In most cases hemivariational inequalities can be reformulated as substationary point problems of the corresponding nonsmooth nonconvex energy functions. The theory and some applications of hemivariational inequalities can be found in [11,13]. To date, various methods have been developed for solving hemivariational inequalities in different applications including in contact mechanics [4,5,16]. A nonsmooth optimization approach to such problems is studied in [7,10].

The aim of this paper is to develop a nonsmooth nonconvex optimization method for the numerical solution of hemivariational inequalities. The proposed method is a combination of the local search subgradient method and a global search simulated annealing method. The use of the subgradient method allows to address nonsmoothness of the problem and the use of the simulated annealing method allows to deal with the nonconvexity of this problem. More specifically, we apply the subgradient method to find stationary points of the so-called energy function and then use the simulated annealing method to escape from these stationary points to find a better starting point for the local search method.

The paper is structured as follows. In Sect. 2, first, we introduce the subgradient method for solving hemivariational inequalities and study its convergence. Then we describe the hybrid method. The application of the hybrid method for solving the contact mechanics problems is discussed in Sect. 3. Numerical results are presented in Sect. 4. Section 5 provides some concluding remarks.

2 A Subgradient Method for Hemivariational Inequalities

We consider the following minimization problem

$$\begin{cases} \text{minimize} & \mathcal{L}(\mathbf{u}) \\ \text{subject to} & \mathbf{u} \in \mathbb{R}^n, \end{cases} \quad (1)$$

where

$$\mathcal{L}(\mathbf{u}) = \frac{1}{2}\langle A\mathbf{u}, \mathbf{u}\rangle + \langle \mathbf{b}, \mathbf{u}\rangle + J(\mathbf{u}). \quad (2)$$

Here $A \in \mathbb{R}^{n \times n}$ is a symmetric, positively defined matrix, $\mathbf{b} \in \mathbb{R}^n$ is a vector and $J \colon \mathbb{R}^n \to \mathbb{R}$ is a locally Lipschitz function, in general, nonsmooth nonconvex. In what follows we denote by $\mathbb{R}^n$ the n-dimensional Euclidean space, $\langle \mathbf{u}, \mathbf{w}\rangle = \sum_{i=1}^n u_i w_i$ is the inner product of vectors $\mathbf{u}, \mathbf{v} \in \mathbb{R}^n$ and $\|\mathbf{u}\| = \langle \mathbf{u}, \mathbf{u}\rangle^{1/2}$ is the associated norm. $B_\epsilon(\mathbf{u}) = \{\mathbf{w} \in \mathbb{R}^n : \|\mathbf{u} - \mathbf{w}\| < \epsilon\}$ is an open ball centered at $\mathbf{u}$ with the radius $\epsilon > 0$. S_1 is the sphere of the unit ball in $\mathbb{R}^n$.

In addition, we assume that for any $\mathbf{u} \in \mathbb{R}^n$ and $\mathbf{d} \in S_1$ function $\mathcal{L}$ satisfies

$$\mathcal{L}(\mathbf{u} + \tau\mathbf{d}) - \mathcal{L}(\mathbf{u}) \leq \tau\langle \mathbf{v}, \mathbf{d}\rangle, \quad \mathbf{v} \in \partial\mathcal{L}(\mathbf{u} + \tau\mathbf{d}), \quad \tau > 0. \quad (3)$$

Here and below $\partial\mathcal{L}(\mathbf{v})$ denotes the Clarke subdifferential of $\mathcal{L}$ at point $\mathbf{v}$ [11]. Moreover, let us remark that the above assumption (3) is satisfied when J is for instance a difference-of-convex function [3]. The objective function $\mathcal{L}$ in problem (1) is represented as the sum of three terms. The third term in this representation is a nonconvex and nonsmooth function. This makes the problem nonconvex, having many local minimizers. Therefore, we propose an algorithm consisting of two phases. In the first phase, using the current starting point, we apply the aggregate subgradient method to find a local minimizer of this problem, and then we apply a special procedure to escape from this local minimizer and find a better starting point for the aggregate subgradient method. Different versions of the subgradient and aggregate subgradient methods can be found in [1–3].

The method proceeds as follows. First, we choose tolerances $\varepsilon > 0$, $\delta > 0$ and three constants $\gamma \in (0, 1), c_1 \in (0, 1)$ and $c_2 \in (0, c_1)$.

Algorithm 1: Solving problem (1) for a given starting point **u**

1: [Initialization of outer iteration] Set $l \leftarrow 1$ and $\eta \leftarrow 1$.
2: [Initialization of inner iteration] Set $k \leftarrow 1$, select any direction $\mathbf{d}_k \in S_1$ and compute $\mathbf{v}_k \in \partial \mathcal{L}(\mathbf{u} + \eta \mathbf{d}_k)$. Set $\tilde{\mathbf{v}}_k \leftarrow \mathbf{v}_k$.
3: Solve the following problem

$$\text{minimize } \varphi_k(\lambda) \equiv \|\lambda \mathbf{v}_k + (1-\lambda)\tilde{\mathbf{v}}_k\|^2 \text{ subject to } \lambda \in [0,1].$$

Let λ_k be a solution to this problem. Set

$$\bar{\mathbf{v}}_k \leftarrow \lambda_k \mathbf{v}_k + (1-\lambda_k)\tilde{\mathbf{v}}_k.$$

4: [Switching phase] If

$$\|\bar{\mathbf{v}}_k\| \leq \delta \qquad (4)$$

then go to Step 9.
5: Compute the search direction by $\mathbf{d}_{k+1} \leftarrow -\|\bar{\mathbf{v}}_k\|^{-1}\bar{\mathbf{v}}_k$.
6: If

$$\mathcal{L}(\mathbf{u} + \eta \mathbf{d}_{k+1}) - \mathcal{L}(\mathbf{u}) \leq -c_1 \eta \|\bar{\mathbf{v}}_k\|, \qquad (5)$$

then go to Step 8,
7: Compute $\mathbf{v}_{k+1} \in \partial \mathcal{L}(\mathbf{u} + \eta \mathbf{d}_{k+1})$. Set $\tilde{\mathbf{v}}_{k+1} \leftarrow \bar{\mathbf{v}}_k$, $k \leftarrow k+1$ and go to Step 3.
8: Compute $\mathbf{u} \leftarrow \mathbf{u} + \sigma_l \mathbf{d}_{k+1}$, where σ_l is defined as follows

$$\sigma_l = \max\left\{\sigma \geq \eta : \mathcal{L}(\mathbf{u} + \sigma \mathbf{d}_{k+1}) - \mathcal{L}(\mathbf{u}) \leq -c_2 \sigma \|\bar{\mathbf{v}}_k\|\right\}.$$

Set $l \leftarrow l+1$ and go to Step 2.
9: Set $\eta \leftarrow \gamma \eta$. If $\eta < \varepsilon$ then STOP. Otherwise go to Step 2.

Algorithm 1 consists of two loops: inner and outer loops. For a given value of $\eta > 0$ the search directions are calculated in the inner loop (Steps 2–7). The new iteration is calculated and also the parameter η is updated in the outer loop. First, we prove that for any fixed $\eta > 0$ the inner loop terminates after finite number of iterations.

Proposition 1. *Suppose that $\mathcal{L}: \mathbb{R}^n \to \mathbb{R}$ is a locally Lipschitz function, $\mathbf{u} \in \mathbb{R}^n$, $\eta > 0$ and the constant $C_1 < +\infty$ is such that*

$$C_1 = \max\left\{\|\mathbf{v}\| : \mathbf{v} \in \partial \mathcal{L}(\mathbf{u} + \eta \mathbf{d}), \mathbf{d} \in S_1\right\}. \qquad (6)$$

If $c_1 \in (0,1)$ and $\delta \in (0, C_1)$, then the inner loop in Algorithm 1 terminates after finite many iterations $m > 0$, where

$$m \leq 2\log_2(\delta/C_1)/\log_2 C_2 + 1, \quad C_2 = 1 - [(1-c_1)(2C_1)^{-1}\delta]^2.$$

Proof. The inner loop in Algorithm 1 will terminate the search for a descent direction only when either condition (4) or (5) is satisfied. To prove that the search for the descent direction concludes in at most m steps, it suffices to establish an upper bound on the number of steps where condition (4) is satisfied. The proof is conducted in three stages: first, we demonstrate that each step identifies a new subgradient $\mathbf{v}_{k+1}$ that is not a convex combination of $\mathbf{v}_k$ and $\bar{\mathbf{v}}_k$; next, we show that this subgradient is the best fit in terms of minimizing the corresponding functional φ_k, i.e., $\varphi_{k+1}(\lambda_{k+1}) < \varphi_k(\lambda_k)$; and finally, we establish that the norm of the aggregated subgradient becomes smaller than any fixed δ within a finite number of steps. Let us assume, therefore, that none of the stopping conditions are met; thus, we have

$$\mathcal{L}(\mathbf{u} + \eta \mathbf{d}_{k+1}) - \mathcal{L}(\mathbf{u}) > -c_1 \eta \|\bar{\mathbf{v}}_k\|.$$

It follows from (3) that

$$\mathcal{L}(\mathbf{u} + \eta \mathbf{d}_{k+1}) - \mathcal{L}(\mathbf{u}) \leq \eta \langle \mathbf{v}_{k+1}, \mathbf{d}_{k+1} \rangle,$$

so we have

$$-c_1 \eta \|\bar{\mathbf{v}}_k\| < \eta \langle \mathbf{v}_{k+1}, \mathbf{d}_{k+1} \rangle,$$

which can be simplified by using definition of $\mathbf{d}_{k+1}$ to

$$\langle \mathbf{v}_{k+1}, \bar{\mathbf{v}}_k \rangle < c_1 \|\bar{\mathbf{v}}_k\|^2. \tag{7}$$

Moreover, since $\bar{\mathbf{v}}_k$ is the minimal value of $\varphi_k(\cdot)$ it can be shown that

$$\langle \lambda \mathbf{v}_k + (1-\lambda)\widetilde{\mathbf{v}}_k, \bar{\mathbf{v}}_k \rangle \geq c_1 \|\bar{\mathbf{v}}_k\|^2, \quad \text{for all } \lambda \in [0,1]. \tag{8}$$

Thus, based on equations (7) and (8), we can deduce that

there is no $\lambda \in [0,1]$ such that $\mathbf{v}_{k+1} = \lambda \mathbf{v}_k + (1-\lambda)\widetilde{\mathbf{v}}_k$,

which is what we aimed to demonstrate in the first step. Using the definition of $\bar{\mathbf{v}}_{k+1}$ and the fact that λ_{k+1} minimizes $\varphi_{k+1}(\cdot)$, we can write

$$\|\bar{\mathbf{v}}_{k+1}\|^2 = \|\lambda_{k+1}\mathbf{v}_{k+1} + (1-\lambda_{k+1})\widetilde{\mathbf{v}}_{k+1}\|^2$$
$$\leq \|\lambda \mathbf{v}_{k+1} + (1-\lambda)\widetilde{\mathbf{v}}_{k+1}\|^2 \quad \forall \lambda \in [0,1],$$

which is equivalent to

$$\|\bar{\mathbf{v}}_{k+1}\|^2 \leq \lambda^2 \|\mathbf{v}_{k+1} - \widetilde{\mathbf{v}}_{k+1}\|^2 + \|\widetilde{\mathbf{v}}_{k+1}\|^2 + 2\lambda \langle \mathbf{v}_{k+1} - \widetilde{\mathbf{v}}_{k+1}, \widetilde{\mathbf{v}}_{k+1} \rangle \quad \forall \lambda \in [0,1].$$

From the boundedness assumption (6), we know that $\|\mathbf{v}_{k+1} - \widetilde{\mathbf{v}}_{k+1}\| \leq 2C_1$, and furthermore, using (7) for $k > 1$, we obtain

$$\langle \mathbf{v}_{k+1} - \widetilde{\mathbf{v}}_{k+1}, \widetilde{\mathbf{v}}_{k+1} \rangle = \langle \mathbf{v}_{k+1}, \bar{\mathbf{v}}_k \rangle - \langle \bar{\mathbf{v}}_k, \bar{\mathbf{v}}_k \rangle \leq (c_1 - 1)\|\bar{\mathbf{v}}_k\|^2.$$

Thus, we derive

$$\|\bar{\mathbf{v}}_{k+1}\|^2 \leq (2\lambda C_1)^2 + (1 + 2\lambda(c_1 - 1)\|\bar{\mathbf{v}}_k\|^2)\|\bar{\mathbf{v}}_k\|^2 \quad \forall \lambda \in [0,1]. \tag{9}$$

Choosing arbitrary $\lambda_0 \in [0,1]$

$$\lambda_0 = \frac{(1-c_1)\|\bar{\mathbf{v}}_k\|^2}{(2C_1)^2},$$

and substituting it into inequality (9) we get

$$\|\bar{\mathbf{v}}_{k+1}\|^2 \leq \left(1 - \left(\frac{(1-c_1)\|\bar{\mathbf{v}}_k\|}{2C_1}\right)^2\right)\|\bar{\mathbf{v}}_k\|^2.$$

From assumption that (4) is not satisfied we know that $\delta < \|\bar{\mathbf{v}}_k\|$. Denoting $C_2 = 1 - ((1-c_1)\delta)^2(2C_1)^{-2}$ and using the boundedness assumption (6) for $\|\bar{\mathbf{v}}_k\|^2$ it follows

$$\|\bar{\mathbf{v}}_k\|^2 < C_1^2 C_2^k.$$

From above and the fact that $C_2 \in (0,1)$ we deduce that (4) is satisfied if

$$\delta^2 \geq C_1^2 C_2^k,$$

what is true if $k = m$. □

For a given $\eta > 0$ consider the following convex hull of the set of subgradients

$$W_\eta(\mathbf{u}) = \operatorname{conv}\left\{\mathbf{v} \in \mathbb{R}^n : \mathbf{v} \in \partial\mathcal{L}(\mathbf{u} + \eta\mathbf{d}), \mathbf{d} \in S_1\right\}.$$

Let $\delta > 0$ be given. A point $\bar{\mathbf{u}}$ is called an (η, δ)-stationary point of the function $\mathcal{L}$ iff

$$0 \in W_\eta(\bar{\mathbf{u}}) + B_\delta(\mathbf{0}).$$

Proposition 2. *Suppose that function $\mathcal{L}$ is bounded below*

$$\mathcal{L}^* = \inf\{\mathcal{L}(\mathbf{u}) : \mathbf{u} \in \mathbb{R}^n\} > -\infty. \tag{10}$$

Let $\mathbf{u}_0 \in \mathbb{R}^n$ be an initial point. Then Algorithm 1 terminates after finite many iterations $M > 0$ and produces (η, δ)-stationary point $\mathbf{u}_M$ where

$$M \leq M_0 \equiv \left\lfloor \frac{\mathcal{L}(\mathbf{u}_0) - \mathcal{L}^*}{c_2 \eta \delta} \right\rfloor + 1. \tag{11}$$

Proof. We conduct proof by contradiction. Let us assume that the sequence $\{\mathbf{u}_l\}$ generated by Algorithm 1 is infinite and for any $l \in \mathbb{N}^+$ the point $\mathbf{u}_l$ is not (η, δ)-stationary point, i.e.

$$\min\{\|\mathbf{v}\| : \mathbf{v} \in W_\eta(\mathbf{u}_l)\} > \delta, \quad \forall l \in \mathbb{N}^+.$$

Then we know that the descent direction d_{k+1} can be found so that

$$\mathcal{L}(\mathbf{u}_l + \eta\mathbf{d}_{k+1}) - \mathcal{L}(\mathbf{u}_l) < -c_1\eta\|\bar{\mathbf{v}}_k\| \leq -c_2\eta\|\bar{\mathbf{v}}_k\|$$

It follows from the definition that $\sigma_l \geq \eta$. Therefore, we get
$$\mathcal{L}(\mathbf{u}_{l+1}) - \mathcal{L}(\mathbf{u}_l) = \mathcal{L}(\mathbf{u}_l + \eta \mathbf{d}_{k+1}) - \mathcal{L}(\mathbf{u}_l) < -c_2 \sigma_l \|\bar{\mathbf{v}}_k\| \leq -c_2 \eta \|\bar{\mathbf{v}}_k\|,$$
in addition the condition $\|\bar{\mathbf{v}}_k\| > \delta$ is satisfied that implies
$$\mathcal{L}(\mathbf{u}_{l+1}) \leq \mathcal{L}(\mathbf{u}_0) - (l+1)c_2 \eta \delta.$$
Therefore, $\mathcal{L}(\mathbf{u}_l) \to -\infty$ as $l \to \infty$, which contradicts (10). Clearly, the upper bound for the number of iterations M necessary to find the (η, δ)-stationary point is M_0 given by (11). □

To describe the hybrid algorithm we will use the Metropolis function
$$R(\mathbf{w}, \mathbf{u}, T) = \min\left\{1, \exp((\mathcal{L}(\mathbf{u}) - \mathcal{L}(\mathbf{w}))/T)\right\}$$
where $\mathbf{u}, \mathbf{w} \in \mathbb{R}^n$ and $T > 0$. Let $\mathbf{e}_i \in \mathbb{R}^n$ be the i-th standard unit vector.

Algorithm 2: Hybrid subgradient and simulated annealing method for solving problem (1)

1: [Initialization] Select the initial point $\mathbf{u}_0 \in \mathbb{R}^n$, the initial temperature $T_0 \in (1, \infty)$, the minimum temperature $T_{min} < T_0$, the temperature reduction factor $\alpha \in (0, 1)$. Set $\mathbf{u}_{best} \leftarrow \mathbf{u}_0, \mathcal{L}_{best} \leftarrow \mathcal{L}(\mathbf{u}_0), \bar{\mathbf{u}} \leftarrow \mathbf{u}_0$ and $k \leftarrow 0$.
2: Apply Algorithm 1 starting from the point $\bar{\mathbf{u}}$ and find the stationary point of problem (1). Denote it by $\mathbf{u}_{k+1}$. If $\mathcal{L}(\mathbf{u}_{k+1}) < \mathcal{L}_{best}$ then update $\mathbf{u}_{best} \leftarrow \mathbf{u}_{k+1}, \mathcal{L}_{best} \leftarrow \mathcal{L}(\mathbf{u}_{k+1})$.
3: Generate a uniformly distributed random number μ from $[0, 1]$, randomly select $i \in \{1, \ldots, n\}$ and calculate a trial point $\bar{\mathbf{w}} \leftarrow \mathbf{u}_{k+1} + \mu \mathbf{e}_i$.
4: If $\mathcal{L}(\bar{\mathbf{w}}) < \mathcal{L}_{best}$, then update $\mathbf{u}_{best} \leftarrow \bar{\mathbf{w}}, \mathcal{L}_{best} \leftarrow \mathcal{L}(\bar{\mathbf{w}})$, set $\bar{\mathbf{u}} \leftarrow \bar{\mathbf{w}}, k \leftarrow k + 1$ and go to Step 2.
5: Sample a uniformly distributed random number β from $[0, 1]$. If $\beta \leq R(\bar{\mathbf{w}}, \mathbf{u}_{k+1}, T)$, then set $\bar{\mathbf{u}} \leftarrow \bar{\mathbf{w}}$ and go to Step 2.
6: Set $T \leftarrow \alpha T$. If $T < T_{min}$ then STOP. $\mathbf{u}_{best}$ is a solution. Otherwise go to Step 3.

Remark 1. Algorithm 2 is based on the combination of Algorithm 1 and the simulated annealing method. We apply Algorithm 1 to find the stationary point of the function $\mathcal{L}$ and then apply the simulated annealing method to escape from this point and find a new starting point for Algorithm 1. The convergence of the simulated annealing method is studied, for example, in [9]. In the proposed hybrid algorithm, the simulated annealing method is only used to find starting points for the local search algorithm. This means that the hybrid method developed in this paper converges to the global minimizer of the function $\mathcal{L}$ with probability one.

3 Contact Mechanics Problem

In this section, we present an example of contact mechanics problem with the relevant physical context and notations.

An elastic body is considered to occupy a domain $\Omega \subset \mathbb{R}^d$, where $d = 2, 3$ in applications. The boundary Γ of the domain is divided into three measurable parts: Γ_D, Γ_C, and Γ_N, with Γ_D having positive measure. The boundary Γ is Lipschitz continuous ensuring the existence of the outward normal vector $\boldsymbol{\nu}$ almost everywhere on Γ. Boundary conditions specify that the body is clamped on Γ_D meaning the displacement satisfies $\boldsymbol{u} = \boldsymbol{0}$ there. A surface force with density $\boldsymbol{f}_N$ acts on Γ_N, while a body force density $\boldsymbol{f}_0$ is applied throughout Ω. The contact interaction on Γ_C are govern by using general subdifferential inclusions. For the sake of simplicity we consider frictionless case. The objective is to determine the displacement of the body in static equilibrium.

The scalar product and Euclidean norm in $\mathbb{R}^d$ or $\mathbb{S}^d$ (the space of symmetric second-order tensors) are denoted by "$\cdot$" and $\|\cdot\|$, respectively. The normal and tangential components of displacement $\boldsymbol{u}$ and stress $\boldsymbol{\sigma}$ on Γ_C are represented by $u_\nu = \boldsymbol{u} \cdot \boldsymbol{\nu}$, $\boldsymbol{u}_\tau = \boldsymbol{u} - u_\nu \boldsymbol{\nu}$, $\sigma_\nu = \boldsymbol{\sigma}\boldsymbol{\nu} \cdot \boldsymbol{\nu}$, and $\boldsymbol{\sigma}_\tau = \boldsymbol{\sigma}\boldsymbol{\nu} - \sigma_\nu \boldsymbol{\nu}$, respectively. The small strain tensor is defined as $\boldsymbol{\varepsilon}(\boldsymbol{u}) = (\varepsilon_{ij}(\boldsymbol{u}))$, where: $\varepsilon_{ij}(\boldsymbol{u}) = \frac{1}{2}(\frac{\partial u_i}{\partial x_j} + \frac{\partial u_j}{\partial x_i})$.

Problem P: Find a displacement $\boldsymbol{u}\colon \Omega \to \mathbb{R}^d$ and a stress $\boldsymbol{\sigma}\colon \Omega \to \mathbb{S}^d$ satisfying

$$\boldsymbol{\sigma} = \mathcal{A}(\boldsymbol{\varepsilon}(\boldsymbol{u})) \quad \text{in } \Omega, \tag{12}$$
$$\operatorname{Div} \boldsymbol{\sigma} + \boldsymbol{f}_0 = \boldsymbol{0} \quad \text{in } \Omega, \tag{13}$$
$$\boldsymbol{u} = \boldsymbol{0} \quad \text{on } \Gamma_D, \tag{14}$$
$$\boldsymbol{\sigma}\boldsymbol{\nu} = \boldsymbol{f}_N \quad \text{on } \Gamma_N, \tag{15}$$
$$-\sigma_\nu \in \partial j(u_\nu) \quad \text{on } \Gamma_C, \tag{16}$$
$$\boldsymbol{\sigma}_\tau = \boldsymbol{0} \quad \text{on } \Gamma_C. \tag{17}$$

Here, Eq. (12) represents an elastic constitutive law and $\mathcal{A}$ is an elasticity operator. Equilibrium Eq. (13) reflects the fact that problem is static. Eq. (14) represents clamped boundary condition on Γ_D and (15) represents the action of the traction on Γ_N. Inclusion (16) describes the response of the foundation in normal direction, where j is a given potential. Eq. (17) means that contact is frictionless.

The Hilbert spaces for the problem are

$$\mathcal{H} = L^2(\Omega; \mathbb{S}^d), \quad V = \{\boldsymbol{v} \in H^1(\Omega)^d \mid \boldsymbol{v} = \boldsymbol{0} \text{ on } \Gamma_D\},$$

the latter with norm defined through strain tensors and Korn's inequality ensuring completeness. The trace operator $\gamma\colon V \to L^2(\Gamma_C)^d$ is continuous by the Sobolev trace theorem.

Using the Green formula and the definition of generalized subdifferential, a weak formulation of Problem P is derived as a hemivariational inequality.

Problem P_{hvi}: Find $\boldsymbol{u} \in V$ such that

$$(\mathcal{A}(\boldsymbol{\varepsilon}(\boldsymbol{u})), \boldsymbol{\varepsilon}(\boldsymbol{v}))_\mathcal{H} + \int_{\Gamma_C} j^0(u_\nu; v_\nu)\, da \geq \langle \boldsymbol{f}, \boldsymbol{v}\rangle_{V^* \times V} \quad \text{for all } \boldsymbol{v} \in V,$$

where for $\boldsymbol{f}_0 \in L^2(\Omega)^d$, $\boldsymbol{f}_N \in L^2(\Gamma_N)^d$

$$\langle \boldsymbol{f}, \boldsymbol{v}\rangle_{V^* \times V} = \int_\Omega \boldsymbol{f}_0 \cdot \boldsymbol{v}\, dx + \int_{\Gamma_N} \boldsymbol{f}_N \cdot \gamma \boldsymbol{v}\, da.$$

Below we present the rest of necessary assumptions for existence of the solution to Problem P_{hvi} (see more [7]).

The elasticity operator $\mathcal{A} \colon \Omega \times \mathbb{S}^d \to \mathbb{S}^d$ satisfies

(a) $\mathcal{A}(\boldsymbol{x}, \boldsymbol{\tau}) = (a_{ijkh}(\boldsymbol{x})\tau_{kh})$ for all $\boldsymbol{\tau} \in \mathbb{S}^d$, a.e. $\boldsymbol{x} \in \Omega$, $a_{ijkh} \in L^\infty(\Omega)$,
(b) $\mathcal{A}(\boldsymbol{x}, \boldsymbol{\tau}_1) \cdot \boldsymbol{\tau}_2 = \boldsymbol{\tau}_1 \cdot \mathcal{A}(\boldsymbol{x}, \boldsymbol{\tau}_2)$ for all $\boldsymbol{\tau}_1, \boldsymbol{\tau}_2 \in \mathbb{S}^d$, a.e. $\boldsymbol{x} \in \Omega$,
(c) $\exists\, m_\mathcal{A} > 0$ such that $\mathcal{A}(\boldsymbol{x}, \boldsymbol{\tau}) \cdot \boldsymbol{\tau} \geq m_\mathcal{A} \|\boldsymbol{\tau}\|^2$ for all $\boldsymbol{\tau} \in \mathbb{S}^d$, a.e. $\boldsymbol{x} \in \Omega$.

The potential $j \colon \Gamma_C \times \mathbb{R} \to \mathbb{R}$ satisfies

(a) $j(\cdot, \xi)$ is measurable on Γ_C for all $\xi \in \mathbb{R}$ and there exists $e \in L^2(\Gamma_C)$ such that $j(\cdot, e(\cdot)) \in L^1(\Gamma_C)$,
(b) $j(\boldsymbol{x}, \cdot)$ is locally Lipschitz continuous on $\mathbb{R}$ for a.e. $\boldsymbol{x} \in \Gamma_C$,
(c) there exist $c_0, c_1 \geq 0$ such that

$$|\partial j(\boldsymbol{x}, \xi)| \leq c_0 + c_1 |\xi| \quad \text{for all } \xi \in \mathbb{R},\ \text{a.e. } \boldsymbol{x} \in \Gamma_C,$$

(d) there exists $\alpha \geq 0$ such that

$$j^0(\boldsymbol{x}, \xi_1; \xi_2 - \xi_1) + j^0(\boldsymbol{x}, \xi_2; \xi_1 - \xi_2) \leq \alpha |\xi_1 - \xi_2|^2$$

for all $\xi_1, \xi_2 \in \mathbb{R}$, a.e. $\boldsymbol{x} \in \Gamma_C$.

We define the functional $J \colon L^2(\Gamma_C)^d \to \mathbb{R}$ by

$$J(\boldsymbol{v}) = \int_{\Gamma_C} j(v_\nu)\, da \quad \text{for all } \boldsymbol{v} \in L^2(\Gamma_C)^d.$$

To find a solution of the discrete version of hemivariational inequality (Problem P_{hvi}) we use an optimization-based method described in detail in [7]. Let $V^h \subset V$ be a finite dimensional subspace with a discretization parameter $h > 0$. The corresponding optimization problem in the case of Problem P_{hvi} is as follows.

Problem P_{opt}^h: Find $\boldsymbol{u}^h \in V^h$ that minimizes functional $\mathcal{L} \colon V \to \mathbb{R}$ defined by

$$\mathcal{L}(\boldsymbol{v}) = \frac{1}{2}(\mathcal{A}(\varepsilon(\boldsymbol{v})), \varepsilon(\boldsymbol{v}))_\mathcal{H} - \langle \boldsymbol{f}, \boldsymbol{v}\rangle_{V^* \times V} + J(\gamma \boldsymbol{v}) \quad \text{for all } \boldsymbol{v} \in V.$$

To numerically solve the above optimization problem, we use the finite element method. We discretize the elasticity operator, the forces $\boldsymbol{f}$, and the contact condition, which allows us to construct the functional in the form (2).

4 Numerical Results

To verify the accuracy and efficiency of the proposed method, we conducted a series of simulations. To keep this work concise, we focus on the simplified model introduced in the previous section. The schematic representation of the modeled 2D beam is shown in Fig. 1. The beam thickness is 10 mm, length 210 mm and it is clamped at both ends and rests directly on a soft obstacle made of a composite material. A force is applied to the beam from above:

$$f_N(x) = \left(0, -L\,S\left(1 - \frac{(x - 105\,\text{mm})^2}{(105\,\text{mm})^2}\right)\right),$$

where beam upper surface $S = 1000\,\text{mm}^2$ and L is applied loading.

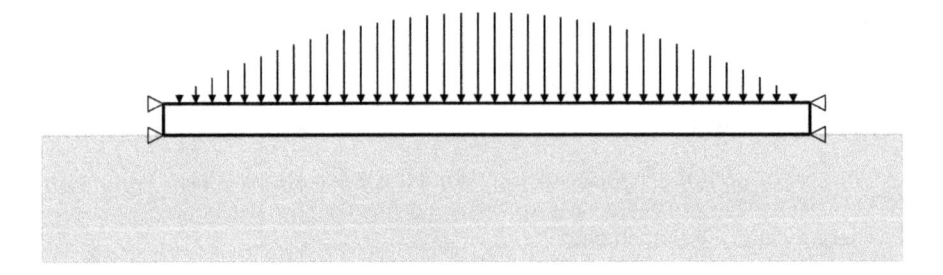

Fig. 1. Modeled beam under loading.

As the beam deflects under the applied force and penetrates the obstacle. The composite nature of the foundation is characterized by the subgradient of the functional j_n. Examples of the functional j_n for $n = 2$ (black) and $n = 7$ (gray) are illustrated in Fig. 2. For more examples of composite materials, see e.g. [14,15]. The presented contact law corresponds to a material whose reaction force increases with penetration until a critical point is reached, at which a composite layer cracks, causing a reduction in the foundation's reaction force.

The function j_2 represents two composite layers represents a soft base covered by a thin (3 mm) protective layer, which is responsible for the initial increase in reaction force with penetration. Once the critical penetration threshold is exceeded, the protective layer cracks, and the reaction force ceases to increase, stabilizing at a constant level. Similarly, in the function j_7 represents seven layers, where are six progressively thicker protective layers, each contributing to the force response until crack occurs. The total thickness of all protective layers is constant and equal to 3 mm. For any $n > 2$ function j_n is nondifferentiable and nonconvex.

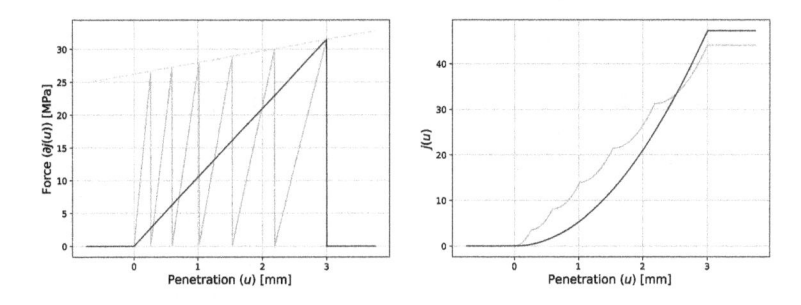

Fig. 2. A nonmonotone contact law and the corresponding functional $j(u)$.

For a sake of simplicity, we neglect the body internal forces, i.e., $\boldsymbol{f}_0 = \boldsymbol{0}$. Elasticity tensors are given by

$$(\mathcal{A}\boldsymbol{\tau})_{ij} = \frac{E\kappa}{(1+\kappa)(1-2\kappa)}(\tau_{11} + \tau_{22})\delta_{ij} + \frac{E}{1+\kappa}\tau_{ij}$$

$$\forall \boldsymbol{\tau} = (\tau_{ij}) \in \mathbb{S}^2, \ i,j = 1, 2.$$

Here and below δ_{ij} is the Kronecker delta, and E and κ are Young's modulus and Poisson's ratio of the body material, respectively. We chose $E = 9.646 \cdot 10^7$ MPa and $\kappa = 0.4$.

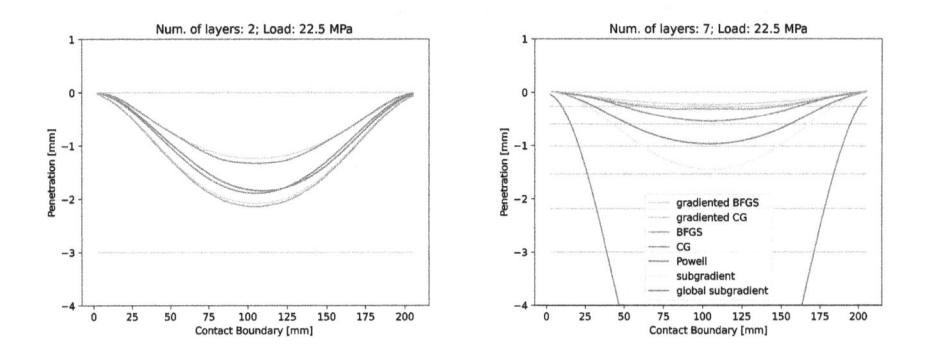

Fig. 3. Deflection of the beam under loading.

We present the results of numerical experiments conducted for the contact law based on the functions j_2, j_3, j_7, and j_{10}. For each case, we performed 10 simulations under varying load conditions. The optimization problem in each simulation was solved using six different methods which was: four optimization techniques from the numpy package [6] and two proposed in the following paper. A mesh with a finite element size not exceeding 1.75 mm was used.

Figure 3 highlights the increased differences between methods observed in multi-layered materials, further illustrating the complexities introduced by additional layers in the composite structure. In the left plot, we see the deformation of the body in the case of two layers (protective and base), while the right plot shows the deformation for seven layers.

The methods - BFGS, Conjugate Gradient Method (CG), "gradiented BFGS", and "gradiented Conjugate Gradient Method" ("gradiented CG") - exhibited short optimization times but failed to yield satisfactory results. The prefix "gradiented" indicates that the method explicitly utilized the computed subgradient of the cost function, as opposed to its non-prefixed counterparts, which estimated the gradient. These methods are marked in shades of gray in the presented plots.

A more notable approach is Powell's method, which demonstrated significantly better results. Although Powell's method generally exhibits high precision, it is computationally expensive. Moreover, as illustrated in the left column of Fig. 4, it does not always find the optimal solution. Notably, its performance fluctuates around intermediate force values, which correspond to the progressive rupture of individual material layers - an inherently nonlinear phenomenon.

Our proposed methods, denoted as "subgradient" and "global subgradient", perform notably better when dealing with a larger number of composite material layers. Due to the possibility of multiple local minima, the basic subgradient method may occasionally become trapped. To mitigate this issue, we developed a strategy for selecting promising initial points for the subgradient method. Specifically, we employed a search algorithm that identified five starting points for each optimization run.

Note that in Fig. 4 the y-axis of the plots employs a logarithmic scale, and the cost function values are presented with a negative sign to enhance readability - thus, higher values indicate better results. The right column of Fig. 4 presents the computational time required for optimization. The markers distinguish three categories of results: squares indicate the lowest achieved cost function value $\mathcal{L}_{\text{best}}$, circles represent values within section $(0.9\,\mathcal{L}_{\text{best}} + 0.01, \mathcal{L}_{\text{best}})$, and "x" denotes results that deviate more significantly from $\mathcal{L}_{\text{best}}$. These plots reveal that the "global subgradient" method outperforms the others in finding the global minimum while also being computationally more efficient than Powell's method.

It is worth emphasizing that subgradient and global subgradient methods were implemented in Python. We employ our original open-source package *conmech* [12], a user-friendly Pythonic tool designed for contact mechanical simulations. The package is a comprehensive simulation framework, allowing for straightforward definition of body geometry and material properties, automatic generation of computational meshes, and empirical error analysis. It supports simulations for static, quasistatic, and dynamic problems in both two and three dimensions. The software is designed with modularity in mind, enabling seamless extension of existing models with additional physical effects. To improve computational robustness in *conmech* we utilize the just-in-time compiler *Numba* [8]. Wall times for CPU depicted in right column of Fig. 4 are obtained on commodity hardware: MacBook Pro M1 16 GB.

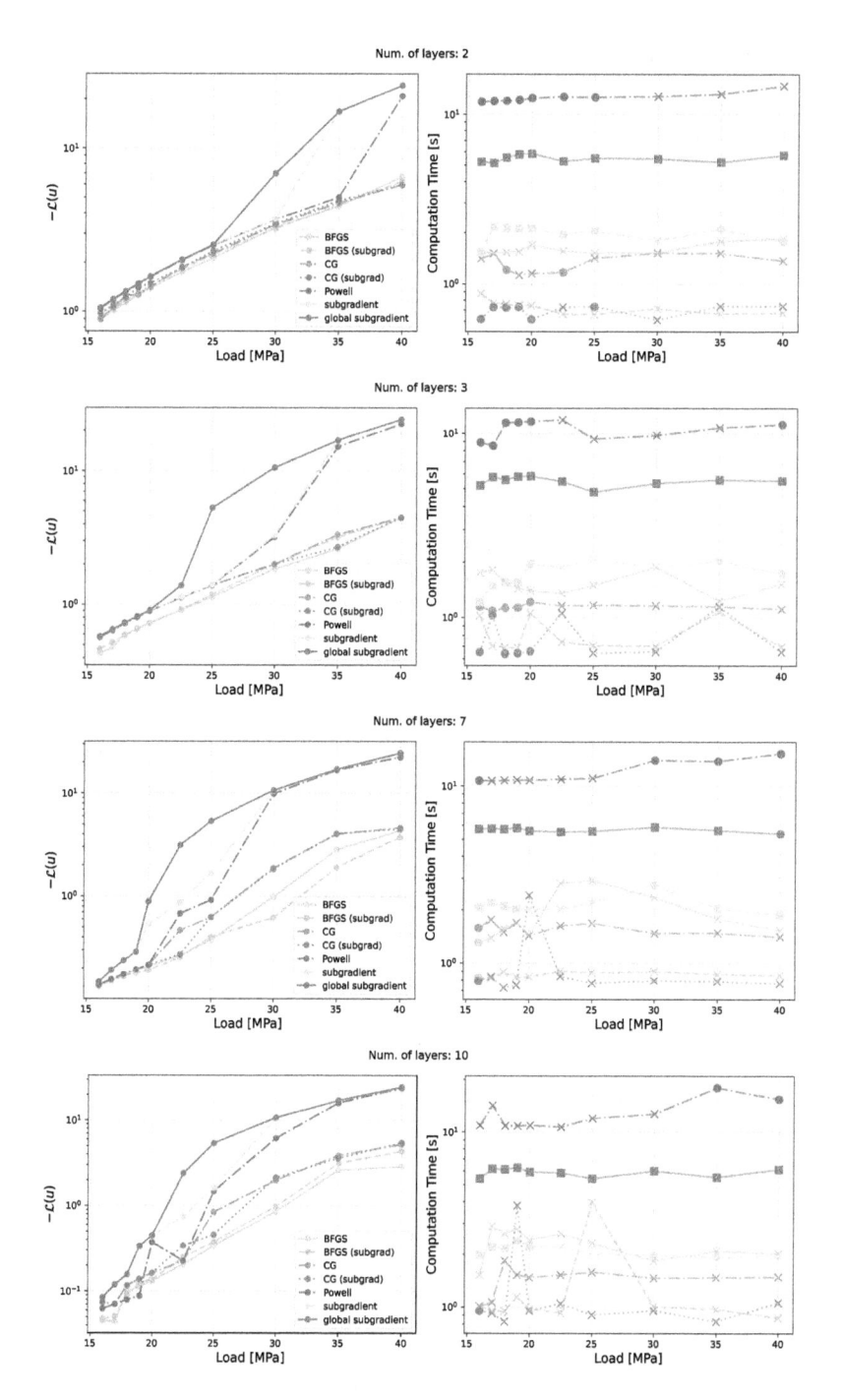

Fig. 4. Energy function values for selected simulations (left column) and the corresponding computation time (right column).

5 Conclusion

In this paper, we developed a method for the numerical solution of hemivariational inequalities. The hemivarional inequalities problem is reduced to the minimization of the so-called energy function. This function is both nonsmooth and nonconvex. The proposed method is a hybrid of the subgradient and the simulated annealing methods. The subgradient is applied to find stationary points of nonsmooth energy function and the simulated annealing is used to escape from these stationary points and find better starting points for the subgradient method. In this way the proposed hybrid method can efficiently deal with both nonsmoothness and nonconvexity of the problem under consideration.

Acknowledgement. This project has received funding from the European Union's Horizon 2020 Research and Innovation Programme under the Marie Sklodowska-Curie Grant Agreement No 823731 CONMECH. The first and third authors are supported by National Science Center, Poland, under project OPUS no. 2021/41/B/ST1/01636.

Disclosure of Interests. The authors have no competing interests to declare that are relevant to the content of this article.

References

1. Bagirov, A., Karmitsa, N., Mäkelä, M.: Introduction to Nonsmooth Optimization. Springer, Berlin/New York (2014). https://doi.org/10.1007/978-3-319-08114-4
2. Bagirov, A., Jin, L., Karmitsa, N., Al Nuaimat, A., Sultanova, N.: Subgradient method for nonconvex nonsmooth optimization. J. Optim. Theory Appl. **157**, 416–435 (2013). https://doi.org/10.1007/s10957-012-0167-6
3. Bagirov, A., Taheri, S., Joki, K., Karmitsa, N., Mäkelä, M.: Aggregate subgradient method for nonsmooth DC optimization. Optim. Lett. **15**, 83–96 (2021). https://doi.org/10.1007/s11590-020-01586-z
4. Han, W., Sofonea, M.: Numerical analysis of hemivariational inequalities in contact mechanics. Acta Numer. **28**, 175–286 (2019). https://doi.org/10.1017/S0962492919000023
5. Han, W., Sofonea, M.: Numerical analysis of a general elliptic variational-hemivariational inequality. J. Nonlinear Var. Anal. **6**, 517–534 (2022). https://doi.org/10.23952/jnva.6.2022.5.06
6. Harris, C.R., et al.: Array programming with NumPy. Nature **585**(7825), 357–362 (2020). https://doi.org/10.1038/s41586-020-2649-2
7. Jureczka, M., Ochal, A.: A nonsmooth optimization approach for hemivariational inequalities with applications to contact mechanics. Appl. Math. Optim. **83**, 1465–1485 (2021). https://doi.org/10.1007/s00245-019-09593-y
8. Lam, S., Pitrou, A., Seibert, S.: Numba: a LLVM-based Python JIT compiler. In: Proceedings of the Second Workshop on the LLVM Compiler Infrastructure in HPC, LLVM 2015. ACM (2015). https://doi.org/10.1145/2833157.2833162
9. Locatelli, M.: Simulated annealing algorithms for continuous global optimization: convergence conditions. J. Optim. Theory Appl. **104**, 121–133 (2000). https://doi.org/10.1023/A:1004680806815

10. Mäkelä, M., Miettinen, M., Lukšan, L., Vlček, J.: Comparing nonsmooth nonconvex bundle methods in solving hemivariational inequalities. J. Global Optim. **14**, 117–135 (1999). https://doi.org/10.1023/A:1008282922372
11. Naniewicz, Z., Panagiotopoulos, P.: Mathematical Theory of Hemivariational Inequalities and Applications. Marcel Dekker, Basel/New York (1995). https://doi.org/10.1002/zamm.19960760104
12. Ochal, A., Jureczka, M., Bartman, P.: A survey of numerical methods for hemivariational inequalities with applications to contact mechanics. Commun. Nonlinear Sci. Numer. Simul. **114**, 106563 (2022). https://doi.org/10.1016/j.cnsns.2022.106563
13. Panagiotopoulos, P.: Hemivariational Inequalities. Springer, Berlin/New York (1993)
14. Tairidis, G., Foutsitzi, G., Stavroulakis, G.E.: A Multi-layer Piezocomposite Model and Application on Controlled Smart Structures, pp. 365–385. Springer, Cham (2018). https://doi.org/10.1007/978-3-319-70563-7_17
15. Tairidis, G.K., Foutsitzi, G., Stavroulakis, G.E.: Optimal Design of Smart Composites, pp. 185–217. Springer, Cham (2019). https://doi.org/10.1007/978-3-030-12767-1_10
16. Wang, F., Wu, B., Han, W.: The virtual element method for general elliptic hemivariational inequalities. J. Comput. Appl. Math. **389**, 113330 (2021). https://doi.org/10.1016/j.cam.2020.113330

Reduced-Order Modeling of Compressible Flows Using Supervised Dimensionality Reduction

Abhijnan Dikshit[1]($\boxtimes$), Leifur Leifsson[1], Slawomir Koziel[2,3], and Anna Pietrenko-Dabrowska[3]

[1] School of Aeronautics and Astronautics, Purdue University, West Lafayette, IN 47907, USA
{adikshit,leifur}@purdue.edu
[2] Engineering Optimization and Modeling Center, Department of Engineering, Reykjavík University, Menntavegur 1, 102 Reykjavík, Iceland
koziel@ru.is
[3] Faculty of Electronics Telecommunications and Informatics, Gdansk University of Technology, Narutowicza 11/12, 80-233 Gdansk, Poland
anna.dabrowska@pg.edu.pl

Abstract. Data-driven reduced-order modeling (ROM) methods are widely used in aerodynamic flow modeling. They are mainly used to predict distributed quantities obtained from high-fidelity simulations such as surface pressure distributions and flow fields. However, such models only consider the distributed quantities and do not consider the data of the aerodynamic coefficients which are also available from high-fidelity simulations. This work proposes a novel supervised ROM architecture that learns from both distributed quantities and aerodynamic coefficients. The proposed model is characterized using a transonic airfoil modeling problem and compared with a standard ROM and neural network model. The results demonstrate that the supervised ROM architecture can outperform a standard neural network by 12% in predicting aerodynamic coefficients when only using 25 training samples. Even with 300 training samples, the supervised ROM can outperform the neural network by 1 or 2%. This can be achieved while maintaining the same level of accuracy as a standard ROM in predicting airfoil surface pressure distributions. This demonstrates that the proposed model can lead to sample-efficient aerodynamic modeling by reducing computational cost and enhancing model accuracy.

Keywords: Reduced-order modeling · Multi-task learning · Autoencoders · Deep learning · Compressible flows · Aerodynamics

1 Introduction

Aerodynamic modeling has been a cornerstone of the design process in the fields of aerospace and automotive engineering. Traditionally, high-fidelity numerical methods, such as computational fluid dynamics (CFD), have been used extensively to evaluate aerodynamic quantities of interest (QoI), such as flow fields and surface distributions. However, the usage of CFD methods for aerodynamic modeling and design can be prohibited by the high computational cost of these methods. To mitigate the issue of computational cost, surrogate modeling methods have been utilized to create cheap-to-evaluate models based on data obtained from high-fidelity numerical simulations [24].

Data-driven reduced order modeling (ROM) methods are a popular surrogate modeling strategy for aerodynamics [4]. ROMs provide a parametric mapping between the inputs and outputs of a high-fidelity simulation that is efficient and cheap to evaluate. These models are primarily used to predict high-dimensional aerodynamic data, such as surface pressure distributions and flow fields [12]. These methods are also non-intrusive and do not require the original simulation code to be modified for implementation. They can be implemented using opensource machine learning packages, further enhancing the appeal of such modeling methods. ROMs typically work by using a dimensionality reduction method to transform the high-dimensional aerodynamic data into a latent space. The latent space is then modeled using an interpolation or regression model, with the parameter space used to generate the data used as an input.

Proper orthogonal decomposition (POD) has been used extensively to create ROMs and the effectiveness of the method has been proven in the case of aerodynamic modeling [3,14]. However, previous studies have shown that POD is unable to capture the correct trends in compressible flow data with nonlinearities such as shock waves [24]. To increase the accuracy of ROMs in compressible flow, manifold learning and deep learning have been used. Decker et al. [4] and Iyengar et al. [13] developed ROMs using manifold learning methods and demonstrated them on flows in the compressible flow regime. However, the main drawback of manifold learning methods is that it is necessary to formulate and solve an optimization problem to reconstruct the high-dimensional solution from the latent space of the model [7]. As an alternative, deep learning methods were considered to create nonlinear models with an efficient means to reconstruct the high-dimensional solution. Wang et al. [27] used a variational autoencoder ROM to successfully model compressible flow fields around different airfoil shapes. Halder et al. [10] combined a convolutional autoencoder and a Gaussian process (GP) model to predict lid-driven cavity flows.

While the literature on ROMs has steadily grown in the past, such models ignore the integrated quantities, such as lift and drag coefficients, that are also available as part of the data generated from CFD simulations. Due to this, the use of ROMs has been limited in aerodynamic design work. This is because while the lift and drag coefficients can be computed from flow fields or surface quantities via numerical integration, the results may not be accurate enough to perform design work [19]. Including the data of the aerodynamic coefficients in the modeling architecture will enhance the applicability of such models to aerodynamic design and may also improve the generalizability and accuracy of the predictions.

To that end, this work explores a supervised dimensionality reduction method [18] that creates a latent space that accounts for the lift and drag coefficients. Dimensionality reduction methods used in previous ROM architectures, like most dimensionality reduction methods, are unsupervised. This means that the data used to train the dimensionality reduction method has no labels. On the other hand, a supervised method will utilize these labels in the model to enhance accuracy and generalization. In this way, this work proposes a novel nonlinear ROM architecture that learns from both aerodynamic coefficient data and surface pressure distribution data by a supervised encoding generated through an autoencoder and the corresponding loss function. The use of the autoencoder method is motivated by the ease of generating the latent space and reconstructing the high-dimensional data with a simple forward pass of an NN model. The latent space is modeled using a multi-task GP (MTGP) model [2] for enhanced prediction accuracy. The characteristics of the novel model are showcased on a transonic airfoil modeling problem. This demonstrates the capabilities of the new ROM architecture in the context of compressible flows.

These contributions produce a ROM that can combine multiple forms of data for enhanced surrogate modeling of aerodynamic flows. The remainder of the paper is organized as follows. Section 2 will describe the novel ROM architecture proposed in this work while Sect. 3 will describe benchmarking methods used as points of comparison. Section 4 provides a description of the transonic flow test problem and the results obtained for the problem. Finally, Sect. 5 highlights the main conclusions obtained in this work.

2 Supervised Reduced Order Modeling

This section introduces the novel reduced-order modeling architecture that uses a supervised embedding of high-dimensional aerodynamic data. In this work, high-dimensional aerodynamic data consists of the distributed surface pressure and skin friction values. The low-dimensional embedding is modeled using a multi-task GP model. The proposed architecture enables fast and accurate prediction of aerodynamic fields and coefficients. A graphical representation of the proposed modeling architecture is shown in Fig. 1.

2.1 Supervised Embedding Procedure

The novelty of the modeling procedure proposed in this work lies in the creation of a supervised embedding of the high-dimensional aerodynamic data. The supervised nature of the embedding will allow the model to use both aerodynamic coefficients and field data to increase the accuracy of the model. The embedding procedure is carried out using deep neural networks in the form of a supervised autoencoder [18]. Autoencoder neural networks [11] are a special type of a neural network model that create a nonlinear embedding of high-dimensional data.

The neural network architecture consists of two separate neural networks, an encoder and a decoder network. The encoder neural network, f_{enc}, compresses

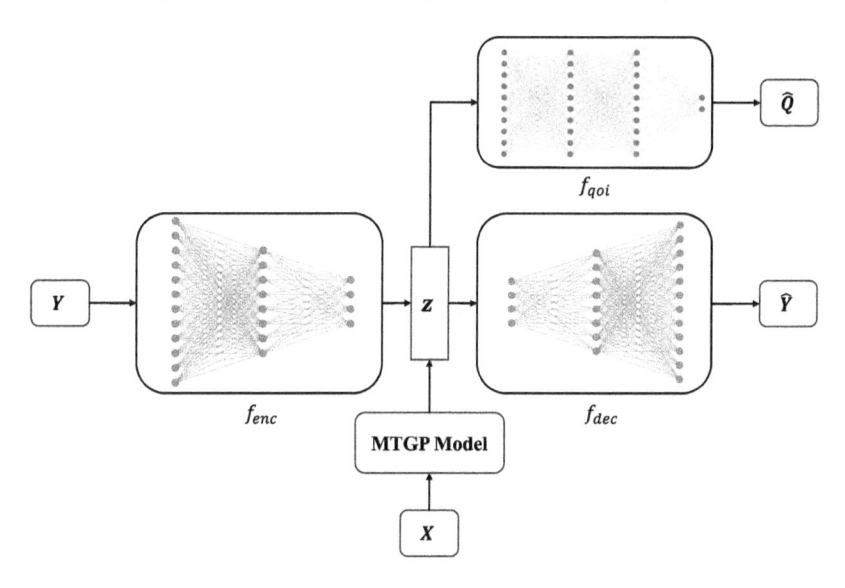

Fig. 1. Graphical representation of reduced order model with supervised embedding architecture proposed in this work.

the high-dimensional data, $\mathbf{Y} = [\mathbf{y}_1, \mathbf{y}_2, ..., \mathbf{y}_n] \in \mathbb{R}^{n \times m}$, to a low-dimensional representation, $\mathbf{z} \in \mathbb{R}^{n \times k}$, where k is much smaller than the original dimensionality m. The decoder network, f_{dec}, then projects the latent vector $\mathbf{z}$ to the original high-dimensional state to obtain a reconstruction of the high-dimensional data, $\hat{\mathbf{Y}}$. In this way, the autoencoder can create a low-dimensional representation of the original high-dimensional data. It can also regenerate the high-dimensional data from a given latent vector.

To enable a supervised embedding of the high-dimensional data, a subnetwork, f_{qoi}, is added to the autoencoder which takes the latent vector and transforms it into the supervised labels of the problem. In this case, the labels are the aerodynamic coefficients, i.e., lift and drag coefficients which are denoted by $\mathbf{Q} = [\mathbf{q}_1, \mathbf{q}_2, ..., \mathbf{q}_n] \in \mathbb{R}^{n \times 2}$. This is done as the aerodynamic coefficients are directly correlated with the aerodynamic surface variables, i.e., the surface pressure and skin friction. By adding the aerodynamic coefficients to the autoencoder model, a multi-task model is created that can obtain better generalization and accuracy for each task considered by predicting correlated sets of data [30].

The hyperparameters of the neural network used for the numerical experiments performed in this work are shown in Table 1. These can different for other flow modeling problems of interest. A fully connected symmetric architecture is used for the autoencoder. Each layer of the encoder uses the SiLU activation function [6] to introduce nonlinearity in the dimensionality reduction process. The neural network is implemented using Pytorch [21].

The loss function of the autoencoder can be written as the sum of two losses. The first loss is the mean squared error (MSE) between the original high-dimensional data and the reconstruction of the autoencoder. The second loss function is the MSE between the original and predicted aerodynamic coefficients. The combined loss function can then be written as

$$\mathcal{L} = w\frac{1}{N}\sum_{m=1}^{N}(\hat{\mathbf{Y}}^{(m)} - \mathbf{Y}^{(m)})^2 + (1-w)\frac{1}{N}\sum_{m=1}^{N}(\hat{\mathbf{Q}}^{(m)} - \mathbf{Q}^{(m)})^2, \quad (1)$$

where w determines the weighting of each term in the loss function. The loss function is optimized using the adaptive moments (ADAM) optimization algorithm [16], and the gradients are computed using backpropagation [23].

Once the autoencoder is trained, high-dimensional aerodynamic predictions can be made for new input parameters using the corresponding latent vector. However, this latent vector is not known for a new set of input parameters. It is, therefore, necessary to create a parametric mapping between the input space and the latent space to be able to obtain the latent vector at a new design point.

Table 1. Hyperparameter selection for the supervised autoencoder.

Hyperparameter	Value used
f_{enc} Hidden Layer Neurons	[512, 128, 64]
f_{dec} Hidden Layer Neurons	[64, 128, 512]
f_{qoi} Hidden Layer Neurons	[32, 16, 8]
Latent dimension	20
Activation function	SiLU
Learning Rate	1e−4
Number of Epochs	10000

2.2 Latent Space Interpolation

The parametric mapping between the parameter space and the latent space is created using a MTGP model. This model will use the parameter space $\mathbf{X} = [\mathbf{x}_1, \mathbf{x}_2, ..., \mathbf{x}_n] \in \mathbb{R}^{n \times p}$ as inputs and the latent space vector $\mathbf{z}$ as the outputs. MTGPs are multi-task models created in the context of Gaussian processes. MTGP models learn to predict multiple related outputs simultaneously. The model not only learns the outputs but also the correlation between the outputs. This allows the outputs to be modeled using a multivariate Gaussian distribution with the correlation between the outputs providing a boost in the prediction capability of the model.

In this work, the free-form parameterization formulation of MTGPs [2] is used. In this formulation, the correlation between the outputs is treated as an additional hyperparameter that is optimized during the training process of the MTGP. The models are implemented using GPyTorch [8] and BoTorch [1],

Python packages for creating GP models and training them using GPU acceleration. The radial basis kernel function was used to train the models and the optimization of the hyperparameters was carried out using the maximum log-likelihood criterion [22].

3 Benchmarking Methods

This section briefly describes the benchmarking modeling methods that are used as a point of comparison for the model architecture proposed in this work. The methods cover modeling approaches that consider surface quantities or aerodynamic coefficients alone.

3.1 Unsupervised ROM Architecture

The first benchmark modeling method that is used in this work is the conventional ROM architecture that utilizes an unsupervised autoencoder method, similar to a previously used deep learning ROM [5], to make aerodynamic predictions. For a fair comparison, the unsupervised embedding is created using a fully connected autoencoder, following much of the same procedure as described in Sect. 2.1. The only difference is that the autoencoder does not contain a subnetwork that predicts the aerodynamic coefficients, making it a completely unsupervised dimensionality reduction process. In this case, the loss function only contains the first term in (1). The optimization and implementation procedure is the same. The architecture and hyperparameters of the encoder and decoder network are also the same as shown in Table 1.

The latent space of the unsupervised ROM is also modeled using a MTGP model with the same implementation as the supervised ROM proposed in this work. Maintaining these similarities in the implementations will isolate the effect of the supervised embedding on the performance of the ROM. This will provide insight into the performance benefits of the supervised embedding.

3.2 Standard Neural Network Model

Another comparison that must be made in this study is the comparison to a standard NN model that has been previously used to predict lift and drag coefficients for a given set of inputs [28]. The standard NN model directly predicts the aerodynamic coefficients. This will be a multi-output NN model where the output layer contains two outputs, corresponding to the lift, and drag coefficients. The input layer of the NN model contains the parameter space of the problem given by **X**. The architecture of the NN model is the same as the f_{qoi} subnetwork that is described for the supervised ROM in Sect. 2.1. Each layer uses the SiLU activation function [6] and the NN model is trained for 10000 epochs. The ADAM optimization algorithm [16] is used along with backpropagation [23] to calculate the gradients. In this case, the loss function for the NN model contains the second term of (1) which represents the MSE between the original and predicted aerodynamic coefficients.

4 Numerical Experiments

This section describes the numerical experiments conducted to characterize the proposed supervised ROM. The test case used in this work is an airfoil modeling problem.

4.1 Problem Setup

In this test case, the surface pressure distribution (C_p) of airfoils is used as a distributed aerodynamic quantity with the lift (C_l) and drag coefficients (C_d) being the supervised labels of the autoencoder. Figure 2 shows the RAE 2822 airfoil, force coefficients, and a typical pressure distribution of the airfoil. The figure indicates the non-dimensionalized lift and drag force which act perpendicular to and along the direction of the incoming freestream flow, respectively. The aerodynamic coefficients, C_l and C_d, are set as the output $\mathbf{Q}$ of the ROM while the entire set of pressure distribution values, C_p, is set as the output $\mathbf{Y}$ of the ROM. The airfoil is assumed to be operating in a flow with a Mach number (M_∞) of 0.734 and a Reynolds number of 6.5×10^6. The parameter space of the problem contains the shape variables describing the airfoil shape as well as the angle of attack (α) of the airfoil shown in Fig. 2. The blue shaded region around the airfoil in Fig. 2 indicates the bounds of the airfoil shape variables with RAE 2822 as the baseline shape. Twelve class shape transformation (CST) variables [17] are used to describe the shape of the airfoil. The angle of attack varies between $-3°$ and $5°$. Figure 2 also indicates the shock location and shock strength on the surface pressure distribution. The shock location is a point of sudden rise in pressure on the airfoil surface because of local acceleration of flow. The rise in pressure that occurs at this point is called the shock strength.

The data for this test problem is generated through a design of experiments created using Latin Hypercube sampling (LHS) [15]. The number of samples in the training data varies between 25 and 300 samples. There is a separate testing data set of 75 samples that is also generated through LHS.

4.2 Computational Modeling

The high-fidelity aerodynamic data of various airfoil shapes at different angles of attack was generated using ADFlow [20]. The solver uses a combination of the Approximate Newton-Krylov method [29] and the full Newton-Krylov method to solve the steady compressible Reynolds-averaged Navier Stokes (RANS) equations with the Spalart-Allmaras turbulence model [26]. The grid topology used in this work is an O-grid topology shown in Fig. 3. The grid is created by extruding an airfoil surface mesh using the hyperbolic marching method implemented using pyHyp [25]. The geometry manipulation and airfoil parameterization are implemented using pyGeo [9], an open-source CAD-free geometry framework. For every new geometry, pyGeo is used to alter the design variables, and the mesh is regenerated once the new airfoil shape has been generated.

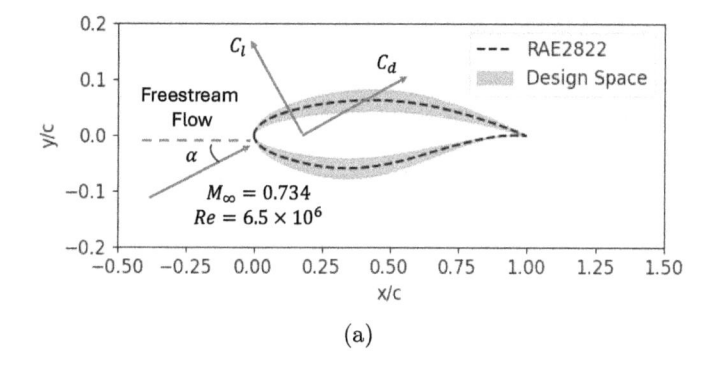

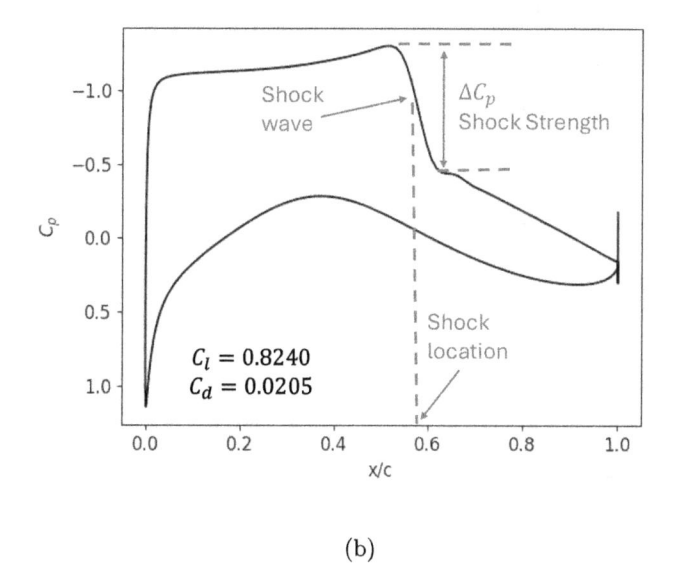

Fig. 2. The problem setup with (a) the RAE 2822 airfoil and the design space, and (b) its surface pressure distribution. (Color figure online)

To balance accuracy and computational cost, a mesh convergence study is performed as shown in Table 2. The mesh convergence study was conducted at the designated flow conditions and a fixed lift coefficient of 0.824. The computations were run in parallel on 64 processors using a high-performance computing cluster. For the airfoil computations performed in this work, the L1 mesh is chosen as refining the mesh further does not significantly change the value of the drag coefficient and the angle of attack for achieving a lift coefficient of 0.824. This mesh will balance accuracy and computational cost.

4.3 Model Evaluation Metrics

The performance of the proposed and benchmark modeling methods is evaluated using the metric defined in this section. To evaluate the prediction of the distributed airfoil surface quantities, the mean relative error will be used as a performance metric. The mean relative error is computed as

$$e_{rel} = \frac{1}{N} \sum_{i=1}^{N} \frac{||\hat{\mathbf{Y}} - \mathbf{Y}||_2}{||\mathbf{Y}||_2}, \qquad (2)$$

where $\hat{\mathbf{Y}}$ is the prediction, $\mathbf{Y}$ is the CFD solution and N is the number of testing samples.

The prediction of aerodynamic coefficients is assessed using the normalized root mean squared error (NRMSE) metric. Although, the models used in this work produce both the lift and drag coefficient simultaneously as two outputs, the NRMSE of each will be calculated independently. The NRMSE of the model prediction can be obtained as

$$\text{NRMSE} = \frac{\sqrt{\frac{1}{N} \sum_{i=1}^{N} (y_i - \hat{y}_i)^2}}{\max(y) - \min(y)}, \qquad (3)$$

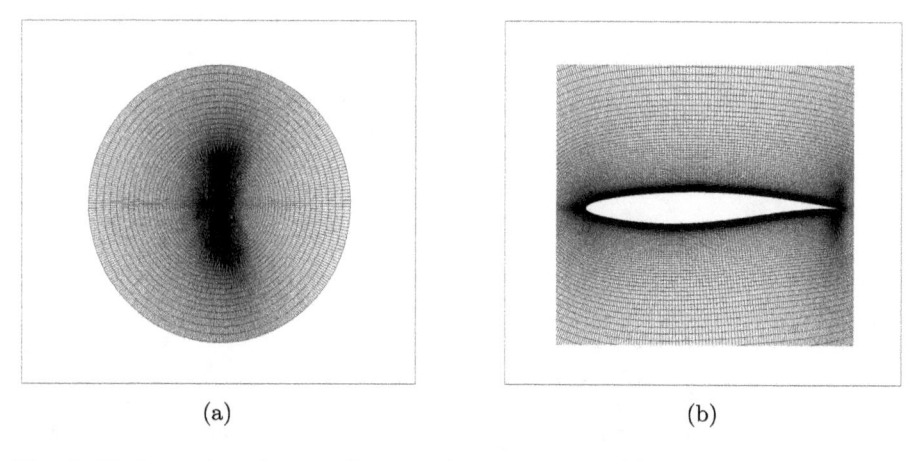

Fig. 3. Mesh topology for aerodynamic data generation (a) in the far-field, and (b) near the airfoil surface.

Table 2. Mesh Convergence study of RAE2822 airfoil.

Level of Mesh	Size of Mesh	C_d (10^{-4})	C_l	α (deg.)
L0	256,000	205.12	0.824	2.93
L1	128,000	205.15	0.824	2.95
L2	32,000	214.86	0.824	3.05

where y is the C_l or C_d and $\hat{y}$ is the prediction of the C_l or C_d. The RMSE is normalized using the range of values of the C_l or C_d in the testing dataset and this is calculated as $\max(y) - \min(y)$.

4.4 Results

Figure 4 shows the variation of the mean relative error of the testing dataset with the number of training samples. The results illustrate that the mean relative error of the supervised ROM architecture is dependent on the value of the weight in (1) and the value of the weight is varied to demonstrate this variation in performance. Increasing the value of the weight in (1) improves the performance of the supervised ROM architecture. Once the value of the weight in (1) is high enough, it can match the performance of an unsupervised ROM architecture. With the current setup and architecture, the supervised ROM can match the performance of an unsupervised ROM, however, the supervised ROM still provides the benefit of simultaneously predicting the aerodynamic coefficients.

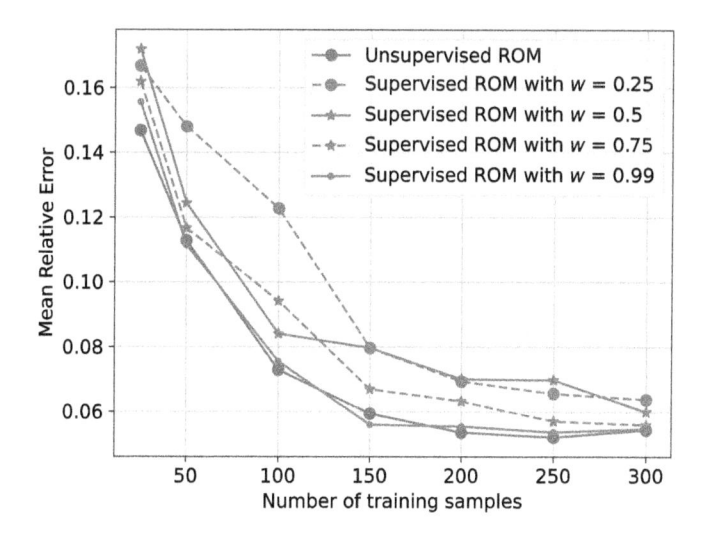

Fig. 4. Mean relative error versus number of training samples for the prediction of surface pressure distribution (C_p).

To assess the prediction of lift coefficients, Fig. 5 shows a plot of the variation of NRMSE values with the number of training samples. The supervised ROM architecture achieves a NRMSE of approximately 3.5% with 25 samples and approximately 1% with 300 samples. On the other hand, the standard neural network achieves a NRMSE of approximately 15.5% with 25 samples and approximately 2.1% with 300 samples. The results illustrate that the supervised ROM architecture achieves a significant improvement in the prediction of the lift coefficient over the standard neural network architecture. The supervised

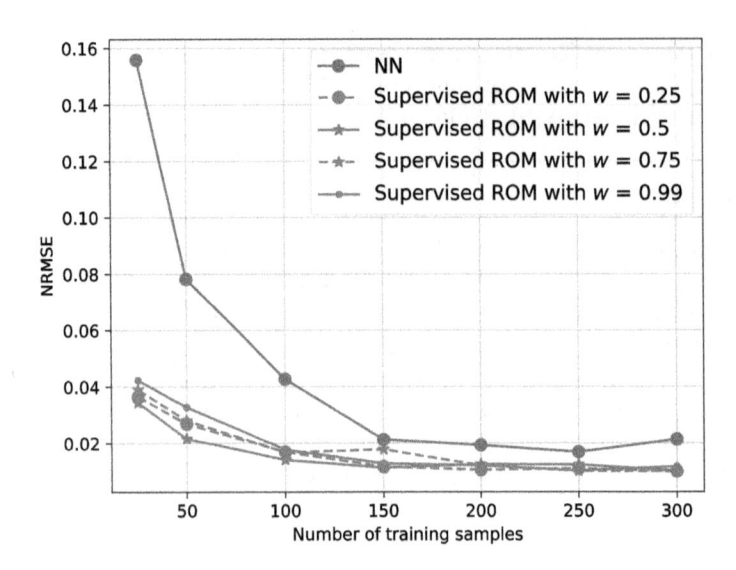

Fig. 5. NRMSE versus number of training samples for the prediction of lift coefficient (C_l).

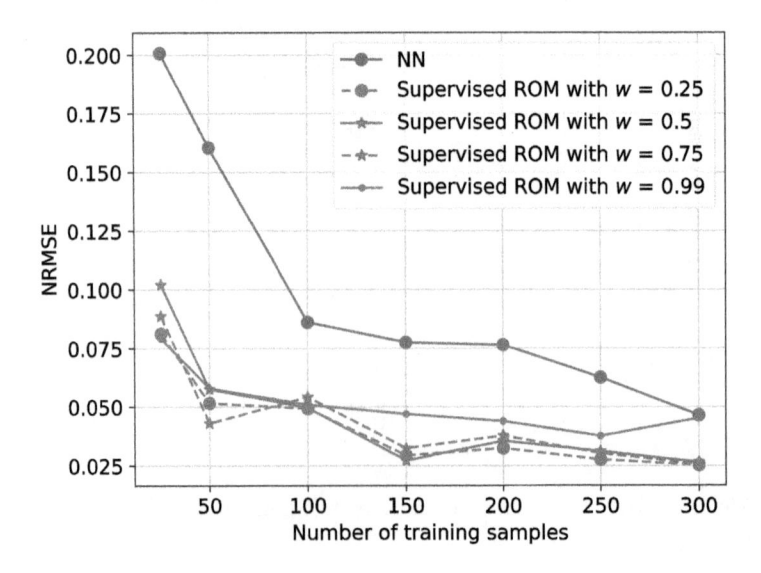

Fig. 6. NRMSE versus number of training samples for the prediction of drag coefficient (C_d).

ROM is also more sample efficient which means that it has a lower NRMSE value with lower number of training samples. The supervised ROM improves the NRMSE by almost 12% for a sample size of 25. Eventually, as the number of training samples increase the performance of the two models becomes similar, but the supervised ROM still has slightly better performance. A similar trend

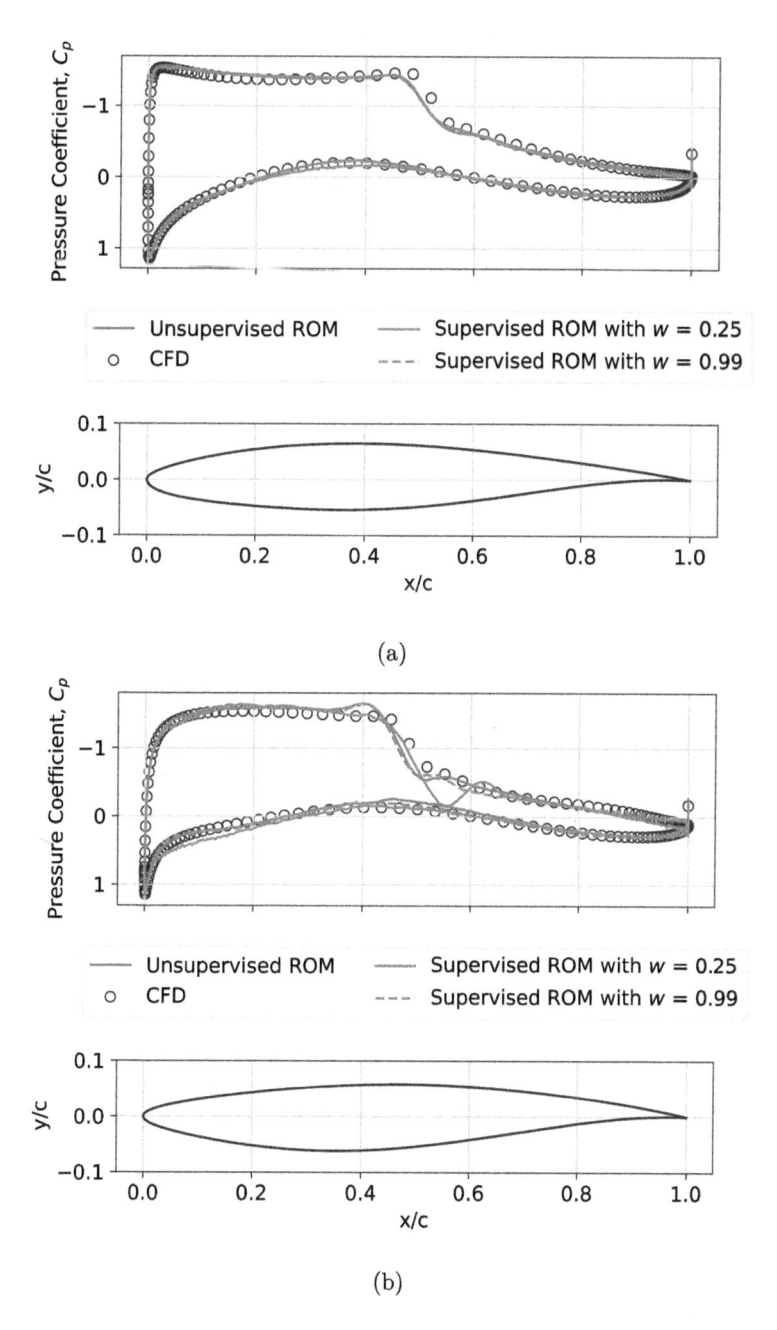

Fig. 7. Prediction of surface pressure distributions for (a) a test airfoil at $\alpha = 4.733°$ and (b) a test airfoil at $\alpha = 3.667°$ using an unsupervised ROM and the proposed supervised ROM.

can be observed for the drag coefficient in Fig. 6. The supervised ROM achieves an NRMSE of approximately 8% at 25 samples which is a 12% improvement over the standard neural network. At 300 samples, the best performance of the supervised ROM provides a 2% benefit over the standard neural network architecture. The prediction of the drag coefficient does seem to be affected by the value of the weighting of each term in (1). As the weighting of the pressure prediction is increased, the NRMSE of the drag coefficient prediction increases. Even accounting for the variations caused by the weighting, the supervised ROM architecture can outperform a standard neural network architecture in the prediction of the drag coefficient. This demonstrates the superiority of the multi-task learning paradigm that is introduced in the supervised ROM architecture.

To demonstrate the prediction of the ROM architectures, Fig. 7 shows the prediction of the surface pressure distribution of two test airfoils with a shock wave occurring on the upper surface. The prediction results show that the ROM architectures can predict the pressure distribution of the airfoil and the shock location and shock strength to a high degree of accuracy. As expected, the supervised ROM trained with a weighting of 0.99 generally captures the pressure distribution better than a ROM trained with a weighting of 0.25.

5 Conclusion

This work proposes a supervised ROM architecture that uses a supervised embedding procedure to embed high-dimensional aerodynamic flow information into a latent space that is informed using supervised labels. This introduces a multi-task formulation into the ROM architecture that will improve the accuracy of the model. A parametric map is created between the parameter space and the latent space using an MTGP model to enable predictions for new design points in the parameter space.

The proposed model is demonstrated on a transonic airfoil modeling problem. The supervised ROM architecture shows significant improvements in the prediction of the lift and drag coefficients of the airfoil and outperforms a standard neural network architecture designed to predict only the coefficients. Upon evaluating the prediction of the pressure distributions, it was found that the supervised ROM architecture can match the performance of the unsupervised ROM architecture. In general, a higher weighting must be given to the quantity that is more difficult to predict. In this case, the pressure distribution is more difficult to predict and a higher weighting must, therefore, be given to the pressure distribution to ensure sufficient accuracy. The prediction of the aerodynamic coefficients is less sensitive to the value of the weighting.

Future work will be dedicated to fine-tuning the parameters of the architecture and exploring different formulations to further improve the architecture. Efforts will also be dedicated to creating optimization loops that incorporate the supervised ROM architecture to perform constrained aerodynamic shape optimization. This will demonstrate the use of such supervised ROM in important aerodynamic design tasks.

Acknowledgments. The work done in this paper was supported in part by the U.S. National Science Foundation (NSF) award number 2223732 and the Icelandic Centre for Research (RANNIS) award number 239858.

References

1. Balandat, M., et al.: BOTORCH: a framework for efficient Monte-Carlo bayesian optimization. In: Proceedings of the 34th International Conference on Neural Information Processing Systems, Vancouver, Canada, 6–12 December 2020, pp. 21524–21538 (2020)
2. Bonilla, E.V., Chai, K., Williams, C.: Multi-task gaussian process prediction. In: Advances in Neural Information Processing Systems, Vancouver, Canada, 3–6 December 2007, vol. 20 (2007)
3. Chang, C.M., et al.: Proper-orthogonal-decomposition–based emulation of spatiotemporal evolution of turbulent wakes. AIAA J., 1–13 (2025). https://doi.org/10.2514/1.J064779
4. Decker, K., Iyengar, N., Rajaram, D., Perron, C., Mavris, D.: Manifold alignment-based nonintrusive and nonlinear multifidelity reduced-order modeling. AIAA J. **61**, 454–474 (1 2023). https://doi.org/10.2514/1.J061720
5. Dikshit, A., Leifsson, L.: Efficient design of transonic airfoils using non-intrusive reduced order models and composite bayesian optimization. In: AIAA Aviation Forum and ASCEND 2024, Las Vegas, NV, 30 July–02 August 2024. https://doi.org/10.2514/6.2024-3666
6. Elfwing, S., Uchibe, E., Doya, K.: Sigmoid-weighted linear units for neural network function approximation in reinforcement learning. Neural Netw. **107**, 3–11 (2018). https://doi.org/10.1016/j.neunet.2017.12.012
7. Franz, T., Zimmermann, R., Görtz, S., Karcher, N.: Interpolation-based reduced-order modelling for steady transonic flows via manifold learning. Int. J. Comput. Fluid Dyn. **28**, 106–121 (2014). https://doi.org/10.1080/10618562.2014.918695
8. Gardner, J.R., Pleiss, G., Bindel, D., Weinberger, K.Q., Wilson, A.G.: GPyTorch: blackbox matrix-matrix gaussian process inference with GPU acceleration. In: Proceedings of the 32nd International Conference on Neural Information Processing Systems, Montreal, Canada, 3–8 December 2018, pp. 7587–7597 (2018)
9. Hajdik, H.M., et al.: pyGeo: a geometry package for multidisciplinary design optimization. J. Open Source Softw. **8**(87), 5319 (2023). https://doi.org/10.21105/joss.05319
10. Halder, R., Fidkowski, K.J., Maki, K.J.: Non-intrusive reduced-order modeling using convolutional autoencoders. Int. J. Numer. Methods Eng. **123**, 5369–5390 (2022). https://doi.org/10.1002/nme.7072
11. Hinton, G.E., Salakhutdinov, R.R.: Reducing the dimensionality of data with neural networks. Science **313**(5786), 504–507 (2006). https://doi.org/10.1126/science.1127647
12. Iyengar, N., Rajaram, D., Mavris, D.: Uncertainty quantification in flows with discontinuities: a probabilistic approach on nonlinear manifolds. In: AIAA Aviation and Aeronautics Forum and Exposition, 2023. San Diego, CA, 12–16 June 2023 (2023). https://doi.org/10.2514/6.2023-4092
13. Iyengar, N., Rajaram, D., Mavris, D.: Domain decomposition strategy for combining nonlinear and linear reduced-order models. AIAA J. **62**, 1375–1389 (2024). https://doi.org/10.2514/1.J063361

14. Jang, B., Lee, W., Lee, J.J., Jin, H.: Artificial neural network-based temperature prediction of a lunar orbiter in thermal vacuum test: data-driven reduced-order models. Aerosp. Sci. Technol. **145** (2024). https://doi.org/10.1016/j.ast.2023.108867
15. Jin, R., Chen, W., Sudjianto, A.: An efficient algorithm for constructing optimal design of computer experiments. J. Stat. Plan. Inference. **134**, 268–287 (2005). https://doi.org/10.1016/j.jspi.2004.02.014
16. Kingma, D.P., Ba, J.: Adam: a method for stochastic optimization. In: International Conference on Learning Representations (ICLR), San Diego, CA, 12 May 7 - 9 2015 (2015)
17. Kulfan, B.M.: Universal parametric geometry representation method. J. Aircr. **45**, 142–158 (2008). https://doi.org/10.2514/1.29958
18. Le, L., Patterson, A., White, M.: Supervised autoencoders: improving generalization performance with unsupervised regularizers. In: Advances in Neural Information Processing Systems, Montreal, Canada, 2–8 December 2018, vol. 31 (2018)
19. Li, J., Du, X., Martins, J.R.: Machine learning in aerodynamic shape optimization. Progr. Aerosp. Sci. **134** (2022). https://doi.org/10.1016/j.paerosci.2022.100849
20. Mader, C.A., Kenway, G., Yildirim, A., Martins, J.: ADflow: an open-source computational fluid dynamics solver for aerodynamic and multidisciplinary optimization. J. Aerosp. Inf. Syst. **17**(9), 508–527 (2020). https://doi.org/10.2514/1.I010796
21. Paszke, A., et al.: PyTorch: an imperative style, high-performance deep learning library. In: 33rd Conference on Neural Information Processing Systems (NeurIPS 2019), Vancouver, Canada, 8–14 December 2019 (2019)
22. Rasmussen, C.E., Williams, C.: Gaussian Processes for Machine Learning. MIT Press, Cambridge (2005)
23. Rumelhart, D.E., Hinton, G.E., Williams, R.J.: Learning representations by back-propagating errors. Nature **323**, 533–536 (1986). https://doi.org/10.1038/323533a0
24. Sabater, C., Stürmer, P., Bekemeyer, P.: Fast predictions of aircraft aerodynamics using deep-learning techniques. AIAA J. **60**, 5249–5261 (2022). https://doi.org/10.2514/1.J061234
25. Secco, N., Kenway, G., He, P., Mader, C.A., Martins, J.: Efficient mesh generation and deformation for aerodynamic shape optimization. AIAA J. **59**, 1151–1168 (2021). https://doi.org/10.2514/1.J059491
26. Spalart, P.R., Allmaras, S.R.: One-equation turbulence model for aerodynamic flows. In: 30th Aerospace Sciences Meeting and Exhibit, pp. 5–21. Reno, NV, 6–9 January 1992 (1992). https://doi.org/10.2514/6.1992-439
27. Wang, J., He, C., Li, R., Chen, H., Zhai, C., Zhang, M.: Flow field prediction of supercritical airfoils via variational autoencoder based deep learning framework. Phys. Fluids **33** (2021). https://doi.org/10.1063/5.0053979
28. Xu, Z., Saleh, J.H., Yang, V.: Optimization of supercritical airfoil design with buffet effect. AIAA J. **57**, 4343–4353 (2019). https://doi.org/10.2514/1.J057573
29. Yildirim, A., Kenway, G.K., Mader, C.A., Martins, J.R.: A Jacobian-free approximate Newton–Krylov startup strategy for rans simulations. J. Comput. Phys. **397**, 108741 (2019). https://doi.org/10.1016/j.jcp.2019.06.018
30. Zhang, Y., Yang, Q.: A survey on multi-task learning. IEEE Trans. Knowl. Data Eng. **34**(12), 5586–5609 (2022). https://doi.org/10.1109/TKDE.2021.3070203

Exact and Approximate Methods for Solving the Edge-Strength Problem

Eduardo Rodriguez-Tello[1](✉), Eric Monfroy[2], and Claudia Vasconcellos-Gaete[2]

[1] Cinvestav Unidad Tamaulipas, Km. 5.5 Carretera Victoria-Soto La Marina, 87130 Victoria Tamps., Mexico
ertello@cinvestav.mx
[2] LERIA, Université d'Angers, 2 Bd de Lavoisier, 49000 Angers, France
{eric.monfroy,claudia.vasconcellos}@univ-angers.fr

Abstract. The *Edge-Strength* (*ES*) problem is a graph labeling problem where the goal is to assign integer labels to the edges of a finite undirected graph in such a way that the maximum sum of labels between any two adjacent edges, known as *edge-strength*, is minimized. This work introduces the first methods to solve the *ES* problem exactly and approximately, including two constraint satisfaction problem (CSP) models and a simulated annealing (SA<small>ES</small>) metaheuristic. The first CSP model is based on constrained optimization using the AllDifferent global constraint, while the second employs extensional constraints. Computational experiments on 40 standard topology graph instances demonstrate the effectiveness and robustness of these approaches. The CSP models provide exact solutions for smaller instances, while the SA<small>ES</small> algorithm efficiently approximates solutions for larger and complex graphs. These contributions advance the state-of-the-art in solving the *ES* problem and pave the way for further research.

Keywords: Edge-strength problem · Exact solution methods · Constraint programming · Simulated Annealing · Metaheuristics

1 Introduction

There has been a growing interest in studying graph labeling problems (GLP) in recent years. Theoretically, these problems are essential since they generally belong to the $\mathcal{NP}$-hard class. They are also relevant in the industry because they can be used as abstract models that allow engineers to solve diverse practical application problems.

The *Edge-Strength* (*ES*) is a GLP, which was first stated in [6]. It can be formally stated as follows. Let $G(V, E)$ be a finite undirected graph of order $n = |V|$ and size $m = |E|$. Given an injection $\varphi : E \to \{1, 2, \ldots, m\}$, which represents a labeling of the edges of the graph, the *edge-strength* (cost) of G for

φ is defined as:

$$estr(G, \varphi) = \max\{\varphi(e_1) + \varphi(e_2) : e_1, e_2 \text{ are adjacent edges of } G\}, \quad (1)$$

where $\varphi(e_i)$ denotes the label associated to the edge $e_i \in E$. Thus, the *ES* problem consist in finding a labeling φ^*, such that $estr(G, \varphi^*)$ is minimized, i.e.,

$$\varphi^* = \arg\min_{\varphi \in \Phi}\{estr(G, \varphi)\},$$

where Φ is the set of all the possible labelings. The labeling φ^* satisfying this condition is known as optimum.

For example, consider the graph $G(V, E)$ of order $n = 7$ and size $m = 6$ depicted in Fig. 1(a) with the labeling φ given by the numbers shown over each edge. The distance between each pair of adjacent edges is calculated using the expression $\varphi(e_1) + \varphi(e_2)$ and presented in column 2 of Table 1. For this particular labeling φ, the edge-strength (cost) of G is $estr(G, \varphi) = 11$. For the labeling φ' of G, presented in Fig. 1(b), the edge-strength is $estr(G, \varphi') = 9$ and represents the optimal solution for this particular graph (see column 3 of Table 1).

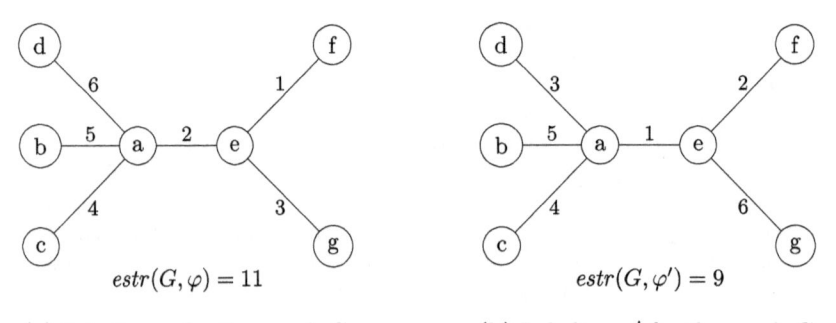

(a) Labeling φ for the graph G.   **(b)** Labeling φ' for the graph G.

Fig. 1. Example of an *Edge-Strength* (*ES*) problem instance.

Given its recent introduction, the *ES* problem has received less attention than other well-known graph labeling problems. Up to now, most of the research on this problem has been concentrated on the theoretical study of its properties to find exact solutions for certain specific families of graphs: paths, cycles, stars, complete graphs, bipartite complete graphs [6,7]. A list of formulas for computing the optimal edge-strength for those specific graph families is presented in Table 2. In the case of general graphs, a lower bound can be computed using the following formula [5]:

$$estr(G) \geq m + 2(\delta(G) - 1),$$

Table 1. Computing the edge-strength for two different graph labelings shown in Fig. 1

Adjacent edges	$\varphi(e_1) + \varphi(e_2)$	$\varphi'(e_1) + \varphi'(e_2)$
ab, ac	$5 + 4 = 9$	$4 + 5 = 9$
ab, ad	$5 + 6 = 11$	$4 + 3 = 7$
ab, ae	$5 + 2 = 7$	$4 + 1 = 5$
ac, ad	$4 + 6 = 10$	$5 + 3 = 8$
ac, ae	$4 + 2 = 6$	$5 + 1 = 6$
ad, ae	$6 + 2 = 8$	$3 + 1 = 4$
ae, ef	$2 + 1 = 3$	$1 + 2 = 3$
ae, eg	$2 + 3 = 5$	$1 + 6 = 7$
ef, eg	$1 + 3 = 4$	$2 + 6 = 8$

where m is the size of G and $\delta(G)$ its minimum degree. Despite this fundamental advancement, many different graph topologies exist for which optimal solutions remain unknown.

Table 2. Formulas reported in the literature for computing the optimal edge-strength for the following graph families: Path (P_n), Cycle (C_n), Star (S_n), Double star (S_{n_1,n_2}) and Complete bipartite (K_{n_1,n_2}).

Graph type	$|V|$	$estr(G)$	Ref.
P_n $(n \geq 3)$	n	n	[6]
C_n $(n \geq 3)$	n	$n + 2$	[6]
S_n $(n \geq 2)$	n	$2n - 3$	[6]
S_{n_1,n_2}	$n_1 + n_2$	$2(n_1 + n_2) - 5$	[6]
K_{n_1,n_2} $(n_1 > n_2 \geq 2)$	$n_1 + n_2$	$2n_1 n_2 - 2n_2 + 1$	[7]

As far as we know, no exact or approximate solution algorithms have been reported in the literature for tackling the edge-strength problem. This article addresses this gap by introducing two new constraint satisfaction problem (CSP) models designed to find exact solutions for small instances of the problem. Moreover, we propose an approximation approach based on the simulated annealing algorithm for dealing with medium-scale benchmark instances. These contributions provide a comprehensive framework for tackling the edge-strength problem across different scales, advancing the state-of-the-art and opening new avenues for further research.

The remainder of this work is organized as follows. Section 2 introduces two CSP models – one using arithmetic constraints and the other using extensional constraints – designed for the exact resolution of the *ES* problem. Section 3 outlines the key implementation details of the proposed simulated annealing algorithm, called SAES. In Sect. 4, we describe the experimental setup used to assess the practical performance of both the exact and heuristic methods. This is followed by an analysis of the computational results, demonstrating the efficiency and robustness of our approaches across the selected benchmark problems. Finally, Sect. 5 summarizes the key contributions of this work and discusses potential directions for future research.

2 Modeling the Problem as a COP or CSP

2.1 A COP - Arithmetic Model

We propose a first model based on constrained optimization (COP) that relies on the efficiency of the global constraint *AllDifferent* to treat inequalities [1,14].

In the *ES* problem, we have m variables $l_{e_1}, \ldots, l_{e_m}$; each l_{e_i} has a domain $[1..m]$, and represents the label of edge e_i. Like in any other labeling problems, we must first state that each label is unique (2). Then, the COP model proposed minimizes the maximum sum of two labels of adjacent edges (3), where $\mathcal{L}^*(G)$ represents the edge-strength of G.

$$AllDifferent(\{l_{e_1}, \ldots, l_{e_m}\}), \quad (2)$$

$$\mathcal{L}^*(G) = minimize\left(\max\{l_{(u,v)} + l_{(v,w)} \mid \forall((u,v),(v,w)) \in E(G) \times E(G)\}\right). \quad (3)$$

However, our preliminary experiments showed that this COP model could be inefficient. Indeed, all the constraints (except the *AllDifferent* constraint) are in the objective function, and thus, constraint solvers cannot prune the search space properly. We thus propose a CSP model based on extensional constraints, i.e., table constraints [11].

2.2 A CSP Model with Extensional Constraints

The main idea of this CSP model is to convert the problem into a constraint satisfaction problem, looking for a labeling $\mathcal{L}(G) \leq k$ for a given positive integer k. Thus, given k, we can deduce which pairs of labels are permitted for adjacent edges. Let's call $\mathcal{T}(m,k)$ this set of pairs of permitted labels:

$$\mathcal{T}(m,k) = \{(l,l') \in [1..m] \times [1..m] \mid l + l' \leq k\}$$

We need the same variables as in the COP model. Thus, we consider m finite domain variables $l_{e_1}, \ldots, l_{e_m}$ with domains $[1..m]$, representing the labels of the edges of G. We still need to state that each label is unique by using constraint (2).

Then, we have to enforce that the sum of labels of each pair of adjacent edges is smaller than or equal to k:

$$\forall((u,v),(v,w)) \in E(G) \times E(G), \ (l_{(u,v)}, l_{(v,w)}) \in \mathcal{T}(m,k) \quad (4)$$

The CSP model is thus given by:

$$\mathcal{M}_{CSP,m,k} = Constraint~(2) \land Constraint~(4)$$

Note that the model $\mathcal{M}_{CSP,m,k}$ is parameterized by the number of edges m of the graph and the integer k.

By instantiating $\mathcal{M}_{CSP,m,k}$ for a graph G with $|E(G)| = m$, we obtain a finite domain CSP instance $\mathcal{I}_{CSP,G,m,k}$. Solving $\mathcal{I}_{CSP,G,m,k}$ with a standard finite domain solver, we are thus able to decide whether G has an edge-strength less than or equal to a given integer k.

2.3 Minimizing k with the CSP Model

A standard incremental optimization can be used: starting with k equal to the lower bound of the instance, we try to find a labeling with an edge-strength value smaller than k. If satisfiable, we have the minimum edge-strength; otherwise, we increment k and continue the same iterative process. However, this optimization is not efficient at all.

We thus propose an optimization based on a property of the *ES* problem.

Property 1. If there is a labeling with edge-strength value k, there is also a labeling with edge-strength value $k+1$. If no labeling results in a cost $k+1$, there is also no solution for value k.

Proof (Sketch of the proof).
$\implies$ consider two adjacent edges (u,v) and (v,w) such that $l_{(u,v)} + l_{(v,w)} = \max\{l_{(u,v)} + l_{(v,w)} \mid \forall ((u,v),(v,w)) \in E(G) \times E(G)\} = k$. Consider $l_{(u,v)} < l_{(v,w)}$ and thus, $l_{(u,v)} < m$. Then, we swap the label $l_{(u,v)}$ with the label $l_{(u,v)} + 1$ which was on edge (x,y). Thus, the sum of each pair of adjacent edges with x or y in common is decreased by one and is thus strictly less than k. The sum of each pair of adjacent edges with u or v in common is increased by 1, and a fortiori, $l_{(u,v)} + l_{(v,w)} = k+1$. We thus obtain a labeling of cost $k+1$.
$\impliedby$ we have shown that "labeling of cost k" $\rightarrow$ "labeling of cost $k+1$", which is the same as "no labeling of cost k" $\lor$ "labeling of cost $k+1$", which is also the same as "no labeling of cost $k+1$" $\rightarrow$ "no labeling of cost k".

Using the previous property, we propose a dichotomic algorithm (see Algorithm 1) for optimization of k between the standard lower and upper bounds, using the model $\mathcal{M}_{CSP,m,k}$:

3 Simulated Annealing Algorithm

Simulated Annealing (SA) is a versatile probabilistic optimization method proposed independently in [10,16]. Since its introduction, it has demonstrated its effectiveness in approximating globally optimal solutions for numerous NP-hard problems [4,8,9]. The core principle of this optimization approach involves occasionally accepting a neighboring solution that worsens the current one, with

Algorithm 1: Optimize Function

Data: G, lb, ub
1 $best\_k \leftarrow ub$;
2 **while** $lb < ub$ **do**
3 $\quad k \leftarrow (ub + lb) \div 2$;
4 $\quad$ **if** solve($\mathcal{I}_{CSP,G,|E(G)|,k}$) is SAT **then**
5 $\quad\quad ub \leftarrow k$;
6 $\quad\quad best\_k \leftarrow ub$;
7 $\quad$ **else**
8 $\quad\quad lb \leftarrow k + 1$;

9 **return** $best\_k$

this acceptance governed by a carefully managed probability. As the algorithm progresses, the likelihood of accepting such non-improving moves gradually decreases [2].

We illustrate in Algorithm 2 the pseudo-code of our SA implementation for solving the *ES* problem. Next, we describe its main components.

Search Space. Given a graph $G = (V, E)$ of size $|E| = m$, the search space Φ for the *ES* problem is composed of all possible labelings (solutions) of G, $\varphi : E \rightarrow \{1, 2, \ldots, m\}$. Therefore, $m!/2$ possible solutions exist for such a graph, given that every one of the $m!$ possible labelings can be inverted to obtain an equivalent *edge-strength* (cost).

Solutions Representation. In our SAES algorithm a labeling (solution) φ is represented as a vector l of integers with length m, which is indexed by the edges and whose i-th value $l[i]$ denotes the label assigned to the edge e_i.

Evaluation Function. The quality $estr(G, \varphi)$ of the labeling φ is evaluated by using the evaluation function (1). Every edge in the graph G must be analyzed to compute it. As a result, $\mathcal{O}(|E|)$ instructions must be executed by this *complete evaluation scheme*. Nevertheless, the proposed SAES algorithm employs an *incremental evaluation* of neighboring solutions. To this end, the edge-strength of each edge in the graph is stored using an appropriate data structure. Indeed, suppose that the labels of two different non-adjacent edges (e_u, e_v) are exchanged in a labeling φ to produce a neighboring solution φ'. We should only recompute the edge costs that change to obtain the new edge-strength of φ'. It takes only $\mathcal{O}(|\mathcal{A}(e_u)| + |\mathcal{A}(e_v)|)$ operations, where $|\mathcal{A}(e_i)|$ represents the number of adjacent edges to e_i. Therefore, our SAES algorithm can analyze thousands of neighboring solutions, employing only a small fraction of the time required by the complete evaluation scheme.

Algorithm 2: Simulated annealing algorithm (SA)

Data: $\mathcal{N}, f, l, \alpha$

1 $t_i, t_f \leftarrow$ ComputeTemperatures();
2 $\varphi \leftarrow$ GenerateInitialSolution();
3 $\varphi^* \leftarrow \varphi$;
4 $t_k \leftarrow t_i$;
5 $k \leftarrow 0$;
6 **while** ¬StopCriterion() **do**
7   Select randomly $\varphi' \in \mathcal{N}(\varphi)$;
8    $\Delta f \leftarrow f(\varphi') - f(\varphi)$;
9   Generate a random $u \in [0,1]$;
10   **if** $(\Delta f <= 0)$ **or** $(e^{-\Delta f / t_k} > u)$ **then**
11     $\varphi \leftarrow \varphi'$;
12   **if** $f(\varphi') < f(\varphi^*)$ **then**
13       $\varphi^* \leftarrow \varphi'$;
14   **if** TemperatureLength() **then**
15     $t_k \leftarrow \alpha t_{k-1}$;
16   $k \leftarrow k+1$;
17 **return** φ^*

Initial Solution. The initial solution is the starting labeling used for the algorithm to begin the search for better configurations in the search space Φ. In our implementation, the starting solution is randomly generated.

Initial and Final Temperatures. Our algorithm's initial and final temperatures are computed using the following formulas proposed in [3].

$$t_i = \Delta f_{min} + \frac{\Delta f_{max} - \Delta f_{min}}{10}, \tag{5}$$

$$t_f = \Delta f_{min}, \tag{6}$$

where Δf_{min} and Δf_{max} are the minimum and maximum cost difference between consecutive visited labelings during a random walk, of length $\frac{m(m-1)}{4}$, through the search space.

Neighborhood Function. The main objective of the neighborhood function in a local search algorithm is to identify the set of potential solutions that can be reached from the current solution [13]. In our SAES algorithm, we implemented the neighborhood function $\mathcal{N}_1(\varphi)$ that is formally defined as follows:

$$\mathcal{N}(\varphi) = \{\varphi' = swap(\varphi, e_u, e_v) : e_u, e_v \in E \text{ and } e_v \notin \mathcal{A}(e_u)\}, \tag{7}$$

where $swap(\varphi, e_u, e_v)$ is a function allowing the exchange of the labels of a pair of non-adjacent edges e_u and e_v to produce a new labeling φ'. The operation that replaces the incumbent solution with a new neighboring labeling is called a move.

Acceptance Criterion. Our SAes algorithm employs the Metropolis condition [12]. It systematically accepts moves to improving or equal quality neighboring labelings and could execute worsening moves with a probability that depends on the increase of the evaluation function value and the temperature t_k (see line 10 in Algorithm 2).

Temperature Length. The maximum number of neighboring solutions visited at each temperature value ($maxConfigurations$) depends directly on the number of edges (m) of the graph because more moves are required for denser graphs: $maxConfigurations = \beta \times m$.

Cooling Schedule. A geometry cooling schedule is used in our simulated annealing implementation: $T_k = \alpha \times T_{k-1}$ [10].

Termination Condition. The SAes algorithm stops when the current temperature reaches the computed final temperature T_f value.

4 Computational Experiments

In this section, computational experiments for assessing the practical usefulness of the COP/CSP models and the SAes algorithm introduced above are presented. We employed a benchmark set composed of 40 instances. It includes graphs with standard topologies (path, cycle, star, double star, complete bipartite, cycle power, hypercube, complete, Möbius ladder, triangulated grid, r-level t-ary tree, wheel, Petersen, Cartesian products). These graphs' order ($|V| = n$) ranges from 5 to 20, while their size ($|E| = m$) spans from 4 to 32 edges.

These experiments were run on a computer equipped with an Intel® Xeon® E5-2630 v4 processor at 2.20 GHz, 64 GB of RAM, and a Linux operating system. The COP/CSP models were coded in Python using the PyCSP³ v2.4 library [15]. The allotted execution time for completing the `optimization()` function over each benchmark instance was 120 h, which can lead to several calls to the CSP solver (same instance, with different k values).

Our SA algorithm was implemented in C++ and compiled using g++ v14.2 with the -$O3$ optimization flag. Given its stochastic nature, we conducted 10 independent sequential runs for each selected benchmark instance. The algorithm requires only two input parameters: β and α. The parameter β determines the maximum number of neighboring solutions evaluated at each temperature level ($maxConfigurations$), while α controls the temperature reduction after processing $maxConfigurations$ solutions. The values for α and β were chosen experimentally and fixed to 0.99 and 5, respectively.

The detailed results from our experiments can be found in Table 3. The first three columns in this table display the name of the graph, its order (n), and its size (m). Column 4 lists the optimal edge-strength cost (Opt) found by the proposed COP model. The lower (lb) and upper (ub) bounds established by the CSP model, as well as the best cost ($B^\star$) attained by it, are presented in columns 6 to 8. Columns 10 to 13 register the average cost (Avg), standard deviation (Std), average CPU time in seconds, and the best cost ($B^\star$) reached

Table 3. Experimental results reached with the introduced COP/CSP models and the SAES algorithm over 40 graphs with diverse standard topologies.

Instance	n	m	COP		CSP				SAES			
			Opt	t	lb	ub	B^*	t	Avg	Std	B^*	t
path10	10	9	10	0.8	10	10	10	0.9	10.0	0.0	10	0.1
path15	15	14	15	1.3	15	27	15	0.9	15.2	0.4	15	0.1
path20	20	19	20	78.5	20	20	20	2.6	20.5	0.5	20	0.1
cycle10	10	10	12	1.7	11	12	12	0.9	12.0	0.0	12	0.1
cycle15	15	15	17	93.0	16	29	17	1.1	17.2	0.4	17	0.1
cycle20	20	20	–	–	21	30	22	1.6	22.5	0.5	22	0.1
star5	5	4	7	2.2	5	7	7	0.9	7.0	0.0	7	0.1
star7	7	6	11	2.4	7	11	11	0.9	11.0	0.0	11	0.1
star10	10	9	17	5.5	10	17	17	0.9	17.0	0.0	17	0.1
star15	15	14	–	–	15	27	27	0.6	27.0	0.0	27	0.2
dStar6-3	9	8	13	1.6	9	15	13	0.9	13.0	0.0	13	0.1
dStar5-5	10	9	15	4.3	10	17	15	0.9	15.0	0.0	15	0.1
dStar6-6	12	11	19	87.2	12	21	19	0.9	19.0	0.0	19	0.1
dStar8-4	12	11	19	71.6	12	21	19	0.9	19.0	0.0	19	0.1
bipartite3x3	6	9	14	2.5	10	14	14	0.9	14.0	0.0	14	0.1
bipartite4x3	7	12	19	79.6	13	23	19	0.9	19.0	0.0	19	0.1
bipartite5x3	8	15	–	–	16	29	25	0.7	25.0	0.0	25	0.1
bipartite4x4	8	16	–	–	17	24	26	1.0	26.0	0.0	26	0.1
cyclePow15-2	10	30	–	–	31	59	47	1.2	47.0	0.0	47	0.2
hypercube3	8	12	–	–	13	23	18	1.5	18.0	0.0	18	0.1
k5	5	10	17	8.1	11	17	17	0.9	17.0	0.0	17	0.2
k6	6	15	–	–	16	29	26	0.6	26.0	0.0	26	0.2
mobLadder6	6	9	14	2.2	10	17	14	0.9	14.0	0.0	14	0.1
mobLadder8	8	12	18	53.0	13	23	18	0.9	18.0	0.0	18	0.1
mobLadder10	10	15	22	15139.2	16	29	22	0.9	22.0	0.0	22	0.1
mobLadder12	12	18	–	–	19	35	26	0.6	26.0	0.0	26	0.1
triangle6	6	9	14	3.1	10	17	14	0.9	14.0	0.0	14	0.1
triangle8	8	13	20	295.8	14	25	20	1.0	20.0	0.0	20	0.1
tree2x2	7	6	9	2.3	7	9	9	0.9	9.0	0.0	9	0.1
tree2x3	13	12	19	316.0	13	23	19	0.9	19.0	0.0	19	0.1
tree3x2	15	14	19	724.9	15	19	19	0.9	19.0	0.0	19	0.2
wheel5	5	8	13	2.6	9	13	13	0.9	13.0	0.0	13	0.1
wheel7	7	12	–	–	13	18	19	1.0	19.0	0.0	19	0.2
petersen	10	15	22	14425.1	16	29	22	1.0	22.0	0.0	22	0.1
c4xc4	16	32	–	–	33	63	50	2.3	50.0	0.0	50	0.4
p2xp3	6	7	–	–	8	11	10	0.6	10.0	0.0	10	0.1
p3xp3	9	12	17	18.5	13	17	17	0.9	17.0	0.0	17	0.1
p2xc3	6	9	–	–	10	14	14	0.6	14.0	0.0	14	0.1
p3xc3	9	15	–	–	16	23	23	0.7	23.0	0.0	23	0.1
st3xst3	9	12	17	15.3	13	23	17	1.0	17.0	0.0	17	0.1

by our SAes algorithm in 10 executions with each instance. The computational time (t) in seconds expended by the COP/CSP models are also presented in columns 5 and 9.

We observe several key findings based on data provided in Table 3. To begin, we will delve into the analysis of the COP model, which uses basic arithmetic constraints (see Sect. 2.1). It successfully found the optimal solutions for simpler graph structures such as path, cycle, and star graphs with a small number of edges (m). However, as the complexity of the graph increased, the expended computational time rose significantly. For instance, the path graph with 19 edges took 78.5 s, and the cycle graph with 15 edges took 93.0 s, indicating a substantial computational burden. Despite these challenges, the COP model demonstrated its capability of reaching the optimal solutions when given sufficient time, making it suitable for small to medium-sized graphs. Nonetheless, it is important to note that it did not resolve 13 out of 40 instances within the maximum allotted 120 h (rows marked with symbol "–"), highlighting its limitations when dealing with larger and more complex graphs.

Turning to the CSP model (see Sects. 2.2 and 2.3), which employs extensional constraints and a dichotomic search algorithm, the results indicated a more efficient handling of larger graph instances than the COP model. The CSP model found the optimal or near-optimal solutions with relatively lower computational times for all the 40 tested instances. For example, it efficiently solved instances like the complete bipartite and double star graphs. However, for highly complex graphs like cycle power and hypercube, the CSP model's bounds (lb and ub) still left some gaps, indicating that while it is faster, it may not always reach the exact optimal solution quickly. The CSP model was, in fact, able to handle graphs with up to 32 edges in a much shorter time frame.

Analyzing the results of our Simulated Annealing algorithm implementation (SAes), it is evident that this heuristic approach is highly efficient for larger instances, providing near-optimal solutions within very short computational times. For instance, it consistently found optimal or close to optimal solutions for all graph types, including highly complex graphs such as the Möbius ladder and Cartesian products for which their solutions were still unknown. The average CPU time for SAes was the lowest among the three tested solution approaches, demonstrating its effectiveness in quickly approximating solutions. The standard deviation values were also minimal, indicating the robustness and reliability of our metaheuristic in providing consistent results across multiple executions. Notably, even for graphs with up to 32 edges, SAes showed remarkable performance.

5 Conclusions and Future Work

This work addressed the edge-strength (ES) problem in graphs, presenting the first exact and approximate methods to solve this problem. The key contributions are the development of two constraint satisfaction problem (CSP) models and a metaheuristic based on simulated annealing (SAes). The CSP models include an arithmetic constraints-based model and a model using extensional constraints, while the SAes metaheuristic is designed to handle medium-sized instances efficiently.

The experimental results demonstrate the effectiveness of these methods. The CSP models can find exact solutions for smaller instances. In contrast, the SAes algorithm performs well for larger and more complex graphs, providing near-optimal solutions in significantly less computational time. These findings indicate that our proposed approaches offer a comprehensive framework for tackling the *ES* problem across different scales.

For future research, several avenues can be explored to advance the study of the *ES* problem further:

- Developing other metaheuristic algorithms, such as genetic algorithms or particle swarm optimization, to compare their performance against our simulated annealing implementation.
- Investigating the fitness landscape of the es problem to understand its structure and optimization challenges, potentially leading to more efficient solution methods.
- Studying graph instances with more edges and other complex topologies to evaluate the proposed models' and algorithms' scalability and applicability in various industrial contexts.

Acknowledgments. The first author thankfully acknowledge a sabbatical leave granted by Cinvestav (01/09/2024–31/08/2025), as well as the courtesies and facilities of the LERIA, Université d'Angers France.

References

1. Beldiceanu, N., Carlsson, M., Rampon, J.X.: Global constraint catalog. Technical report. Research Report SICS T2005-08, Swedish Institute of Computer Science, Kista (2005). https://hal.science/hal-00485396
2. Blum, C., Roli, A.: Metaheuristics in combinatorial optimization: overview and conceptual comparison. ACM Comput. Surv. **35**(3), 268–308 (2003). https://doi.org/10.1145/937503.937505
3. Connolly, D.T.: An improved annealing scheme for the QAP. Eur. J. Oper. Res. **46**(1), 93–100 (1990). https://doi.org/10.1016/0377-2217(90)90301-Q
4. Franzin, A., Stützle, T.: Revisiting simulated annealing: a component-based analysis. Comput. Oper. Res. **104**, 191–206 (2019). https://doi.org/10.1016/j.cor.2018.12.015
5. Ichishima, R., Oshima, A., Takahashi, Y.: Bounds for the edge-strength of graphs. Mem. Kokushikan Univ. Inf. Sci. **41**, 9–15 (2020)
6. Ichishima, R., Oshima, A., Takahashi, Y.: The edge-strength of graphs. Discret. Math. Lett. **3**, 44–49 (2020)
7. Ichishima, R., Oshima, A., Takahashi, Y.: Some new results on the edge-strength and strength of graphs. Discret. Math. Lett. **12**, 22–25 (2023)

8. Johnson, D.S., Aragon, C.R., McGeoch, L.A., Schevon, C.: Optimization by simulated annealing: an experimental evaluation; Part I, graph partitioning. Oper. Res. **37**(6), 865–892 (1989). https://doi.org/10.1287/opre.37.6.865
9. Johnson, D.S., Aragon, C.R., McGeoch, L.A., Schevon, C.: Optimization by simulated annealing: an experimental evaluation; Part II, graph coloring and number partitioning. Oper. Res. **39**(3), 378–406 (1991). https://doi.org/10.1287/opre.39.3.378
10. Kirkpatrick, S., Gelatt, C.D., Vecchi, M.P.: Optimization by simulated annealing. Science **220**(4598), 671–680 (1983)
11. Lecoutre, C.: Optimization of simple tabular reduction for table constraints. In: Stuckey, P.J. (ed.) Principles and Practice of Constraint Programming, CP 2008. LNCS, vol. 5202, pp. 128–143. Springer (2008)
12. Metropolis, N., Rosenbluth, A.W., Rosenbluth, M.N., Teller, A.H.: Equation of state calculations by fast computing machines. J. Chem. Phys. **21**(6), 1087–1092 (1953)
13. Talbi, E.: Metaheuristics: From Design to Implementation. Wiley (2009)
14. van Hoeve, W.: The all different constraint: a survey. In: 6th Annual Workshop of the ERCIM Working Group on Constraints, pp. 1–12 (2001). https://doi.org/10.48550/arXiv.cs/0105015
15. XCSP3 Team: PyCSP^3 v2.4 (2024). https://www.pycsp.org/
16. Černý, V.: A thermodynamic approach to the traveling salesman problem: an efficient simulation algorithm. J. Optim. Theory Appl. **45**(1), 41–51 (1985)

Author Index

A
Achtelik, Wojciech   301
Amblard, Frédéric   208
Asao, Shinichi   3

B
Bączkiewicz, Aleksandra   176
Bagirov, Adil M.   392
Bartman-Szwarc, Piotr   392
Bayer, David   67
Beresnev, Artem   141

C
Chen, Ye   24
Cherifi, Hocine   269
Choudhary, Harshita   239

D
Dellen, Babette   54
Dignum, Eric   239
Dikshit, Abhijnan   406

E
Evtyukhov, Dmitrii   128

G
Garibay, Victoria M.   224
Giabbanelli, Philippe J.   208
Gijsbers, Sacha   83
Golubev, Alexander   141
Gusarova, Natalia   141
Guseva, Elizabetty   149

H
Halawa, Krzysztof   11
Halliday, Ian   98
Harjule, Priyanka   39
Huang, Dongdong   24
Husni, Zyad   54

I
Ikeda, Takahiro   3
Iwai, Misaki   3

J
Jaekel, Uwe   54
Japkowicz, Nathalie   208
Jaros, Jiri   67

K
Kobayashi, Yusei   3
Kopanitsa, Georgy   128
Koratikere, Pavankumar   347
Korzin, Andrey   164
Kosela, Piotr M.   255
Koshkareva, Maria   149
Kovalchuk, Sergey   113, 128
Kozera, Ryszard   316
Koziel, Slawomir   347, 362, 406
Krzhizhanovskaya, Valeria   83
Kurant, Łukasz   193

L
Lees, Mike   239
Leifsson, Leifur   347, 362, 406
Leonenko, Vasiliy   149, 164
Li, Chao   113
Lim, Chung   98
Łoś, Marcin   332

M
Maczuga, Paweł   332
Malawski, Maciej   98
Mengesha, Isaak   284
Metsker, Oleg   128
Mogilevskii, Aleksandr   128
Monfroy, Eric   421

© The Editor(s) (if applicable) and The Author(s), under exclusive license to Springer Nature Switzerland AG 2025
M. Paszynski et al. (Eds.): ICCS 2025 Workshops, LNCS 15908, pp. 433–434, 2025.
https://doi.org/10.1007/978-3-031-97557-8

N
Narracott, Andrew 98
Nishiya, Yusuke 3
Noakes, Lyle 316

O
Ochal, Anna 392
Olsak, Ondrej 67
Otta, Magdalena 98

P
Paszyński, Maciej 332
Pietrenko-Dabrowska, Anna 347, 362, 406

R
Ren, Fei 113
Rodriguez-Tello, Eduardo 421
Roy, Debraj 284
Rudawska, Iga 176
Rusiecki, Andrzej 11

S
Sharma, Rinki 39
Sharova, Alyona 149
Shen, Yuqiang 24
Sheng, Lin 24
Sheraton, Vivek M. 83
Singh, Anurag 269
Służalec, Tomasz 332
Smołka, Maciej 301
Szeliga, Danuta 332
Sztangret, Łukasz 332

T
Takeuchi, Seiichi 3
Takii, Ayato 3
Trebeschi, Stefano 83
Tsui, Janice 98
Tyagi, Dhruv 269

U
Ulyanov, Pavel 141

V
Vamosi, Ralf 377
Vasconcellos-Gaete, Claudia 421
Vatian, Aleksandra 141

W
Wang, Kai 24
Wątróbski, Jarosław 176
Wilkołazka, Magdalena 316
Wubineh, Betelhem Zewdu 11

X
Xu, Haoyun 24

Y
Yakovlev, Alexey 128
Yamakawa, Masashi 3

Z
Zając, Karol 98
Zhang, Dutao 113
Zhong, Stephen 208
Zhou, Qingli 24
Zubanenko, Alexey 141

GPSR Compliance

The European Union's (EU) General Product Safety Regulation (GPSR) is a set of rules that requires consumer products to be safe and our obligations to ensure this.

If you have any concerns about our products, you can contact us on

ProductSafety@springernature.com

In case Publisher is established outside the EU, the EU authorized representative is:

Springer Nature Customer Service Center GmbH
Europaplatz 3
69115 Heidelberg, Germany